AF528593

Advanced Structured Materials

Volume 79

More information about this series at http://www.springer.com/series/8611

Mushahid Husain · Zishan Husain Khan
Editors

Advances in Nanomaterials

Springer

Editors
Mushahid Husain
M.J.P. Rohilkhand University
Bareilly, Uttar Pradesh
India

Zishan Husain Khan
Department of Applied Sciences
& Humanities
Jamia Millia Islamia
New Delhi
India

ISSN 1869-8433 ISSN 1869-8441 (electronic)
Advanced Structured Materials
ISBN 978-81-322-2666-6 ISBN 978-81-322-2668-0 (eBook)
DOI 10.1007/978-81-322-2668-0

Library of Congress Control Number: 2016930552

Printed on acid-free paper

This Springer imprint is published by SpringerNature
The registered company is Springer (India) Pvt. Ltd.

Foreword

There have been remarkable new developments related to amazing properties of materials when their particle sizes reduce to nanometer scale. There are reports of applications of several of these in widely different disciplines of science ranging from physical sciences to engineering, life sciences, agriculture, textiles, cosmetics, medical and biomedical fields as well as in environmental protection. Several technological developments, particularly those related to nanoelectronics with feature sizes entering below ten nanometers have raised expectations to unprecedented levels. The potential of wide ranging applications is posing challenges in the fundamental sciences regarding clear understanding of relationship between properties and basic physical and chemical characteristics including composition, trace impurities, crystallographic structure, and defects. There are numerous reputed R&D groups in India and worldwide, which are focusing their R&D activities on different aspects of nanoscience and technology. The number of scientists engaged in this field has increased substantially and is increasing at a rapid pace. As a consequence, the scientific literature being produced is also increasing day by day.

In view of the enormous increase in activities in nanomaterials, several top international organizations have started getting involved to promote its growth. The International Standards Organization (ISO) has a standard on nanomaterials, according to which a nanomaterial is:

"Material with any external dimension in the nanoscale or having internal structure in the nanoscale. Nanoscale is, in turn, defined as: size range from approximately 1 to 100 nm." A group constituted by International Council for Science (ICSU) has been engaged in deliberations on nomenclature. There are also serious concerns about health hazards associated with production of nanomaterials and their applications. Their impact on quality of environment is also attracting attention. Nanosize particles can penetrate to any organ of human body including brain, and therefore, our natural defenses against undesirable external matter of low dimensions are ineffective.

A book covering different aspects of nanoscience and nanotechnology is very welcome, keeping in view the continuous increase in the scientific literature and interest in this field. I congratulate Professor Mushahid Husain and Dr. Zishan Husain Khan for their initiative in bringing out this document. Besides a chapter on the fundamentals of nanomaterials, this book has covered several important topics in this field. The latest developments in carbon-based nanomaterials such as graphene, carbon nanotubes, and their applications have been included. Metal matrix nanocomposites for control of corrosion and metal oxides are important topics that have been discussed. A chapter deals with silicon nanowire arrays for solar cell applications. Grinding is a major industrial area, which has wide ranging applications from most advanced mechanical engineering to building constructions. Applications of nanodiamond grinding from industrial perspective have been included. Nanolayers of materials such as gallium nitride are widely used for fabrication of solid-state light-emitting diodes. Epitaxial growth of GaN layers is the subject of one of the chapters. Several important applications of nanomaterials in different biomedical areas have been discussed. These include application of gelatin nanoparticles and exploring graphene for drug delivery and study of antibacterial properties of nanomaterials. One of the chapters deals with optical coherence tomography as glucose sensor in blood.

In the end, I thank all the authors who have contributed articles reviewing fundamentals and applications in different aspects of nanomaterials. Several of them are well-recognized experts. I highly appreciate the efforts of editors Prof. Mushahid Husain and Dr. Zishan Husain Khan for contributing important knowledge and in bring out this book. I hope this book will be widely used as a reference by experts as well as beginners in the field.

Krishan Lal
President, The Association of Academies and Societies of Sciences in Asia [AASSA]
Visiting Professor, University of Delhi, Delhi
Immediate Past President, Indian National Science Academy New Delhi
Former Director, National Physical Laboratory, New Delhi
Foreign Member, Russian Academy of Science;
Past President, ICSU CODATA

Preface

Evolution is an ongoing process that replaces what is existing gradually with something better. With the rapid development in the field of nanomaterials, it would not be wrong if we call this time of rapid evolution, the Nano Era. Nanomaterials are one of the most significant research areas to emerge in the past decade or so. It is an interdisciplinary field that draws on knowledge and expertise and covers a vast and diverse array of devices derived from engineering, physics, chemistry, and biology. As the research on nanomaterials matures, an increasing number of applications become commercially viable. Numerous approaches have been utilized in successfully developing different types of nanomaterials, and it is expected that with advancement of technology, new approaches may also emerge. The approaches employed thus far have generally been dictated by the technology available and the background experience of the researchers involved. It is a truly multidisciplinary field involving chemistry, physics, biology, engineering, electronics, and social sciences, which need to be integrated together in order to generate the next level of development in nanomaterials research. The "top-down" approach involves fabrication of nanomaterials via monolithic processing on the nanoscale and has been used with spectacular success in the semiconductor devices used in consumer electronics. The "bottom-up" approach involves the fabrication of nanomaterials via systematic assembly of atoms, molecules, or other basic units of matter. This is the approach nature uses to repair cells, tissues, and organ systems in living things and indeed for life processes such as protein synthesis. Tools are evolving which will give scientists more control over the synthesis and characterization of novel nanostructures yielding a range of new products in the near future. There are many applications of nanomaterials which are still in the realm of scientific fiction and will be made possible in the near future. Every day, the research on nanomaterials is enriched with new innovations/discoveries and scientists are putting serious efforts in bringing out more advancement to this research field.

This book includes some of the latest advancements and applications in the field of nanomaterials. It provides an overview of the present status of this rapidly developing field. The book includes twelve chapters authored by the experts in the field of nanomaterials and applications. Chapter 1 is an introductory chapter and presents an introduction of nanomaterials. It presents an overview of nanomaterials, their classification, different methods of synthesis of nanomaterials, and potential applications. Chapter 2 shows the significant progress made in the field of carbon nanomaterials specifically carbon nanotubes (CNTs). This chapter summarizes various fabrication techniques, characterization, and potential applications of carbon nanomaterials. It includes various methods to fabricate CNT fibers or yarns via spinning from CNT solutions, spinning from vertically aligned CNT arrays on substrates, direct spinning from CNT aerogels synthesized in chemical vapor deposition (CVD) chambers, spinning from cotton-like precursors, spinning with dielectrophoresis and rolling from CNT films/sheets. The chapter also provides the difference among thin films composed of free standing CNTs of different thicknesses known as membranes, sheets, buckypapers or papers. It also presents a brief discussion about 3D nanomaterials based on CNTs that include bulks, foams, and gels.

Chapter 3 summarizes the significant development in theoretical and experimental study of doped graphenes. The chapter discusses various doping methods, doping levels, heteroatom sources, chemical bond structures between heteroatom and graphene, their synthesis by several techniques such as thermal CVD, arc discharge approach, graphite oxide post-treatment, and plasma treatment synthesis. It also includes important properties such as electrical stability, quality of doped material, and technological applications.

Chapter 4 discusses the effect of varying size from chalcogenides to nanoscale chalcogenides, i.e., nanochalcogenides on physical properties of chalcogenides. It presents a brief discussion of methods for preparation of chalcogenide thin films via physical vapor condensation, sputtering, pulsed laser deposition, and chemical vapor deposition. Various models such as CFO model and Davis–Mott model for describing the electrical properties of nanochalcogenides have been discussed briefly in this chapter. It also includes the optical and thermal properties of nanochalcogenides. The applications of chalcogenides in memories based on phase change and electrical switching has also been discussed in detail in this chapter.

Chapter 5 presents a review on metal oxide nanostructures, their growth, and applications. The chapter includes the introduction of metal oxide nanostructures with chemical growth process CVD technique. The growth of indium oxide nanostructures with effect of ambient conditions such as tunable growth of nanowires, nanotubes, and octahedrons, effect of time, pressure, gas flow dynamics has been discussed in this chapter. This chapter also includes the growth of 3-D indium zinc oxide and gallium oxide nanostructures. Finally, the applications of metal oxide nanostructures such as environmental sensors and photodetector have been discussed in detail at the end of this chapter.

Chapter 6 gives an overview on metal matrix nanocomposites and their applications in corrosion control. The chapter includes the introduction of nanocomposites, its various types such as CMNCs, MMNCs, and PMNSCs. Solid- and liquid-state methods for synthesizing routes for fabrication of nanocomposites have been described in this chapter. This chapter also discusses major application of nanocomposites mainly corrosion and its various forms, mechanism, calculation, and control of corrosion.

Chapter 7 describes the process of diamond nanogrinding. The chapter includes the principle of the process of nanogrinding using coated piezoelectric materials. The chapter also discusses about the bonds in porous tools engineered to minimize abrasive grain loss and the ways to process the vitrified bonding bridges using a laser to form extremely sharp nanoscale cutting wedges.

Chapter 8 presents the epitaxial growth of GaN layer by using laser molecular beam epitaxy technique. It includes the structural and optical properties of the epitaxial GaN layers by using HRXRD, AFM, FTRAMAN, SIMS, and Pl spectroscopy techniques.

Chapter 9 presents a review on aperiodic SiNWs array fabrication by silver-assisted wet chemical etching method. The chapter includes the light trapping properties of aperiodic SiNWs array and PV applications with emphasis on SiNWs array-based solar cells. This chapter also discusses the challenges in use of SiNWs arrays in PV devices and its future perspective.

Chapter 10 presents the recent trends in gelatin nanoparticles (GNPs) and its biomedical applications. It includes chemical structure, methods used to synthesize GNPs and the characterization of these GNPs by SEM, AFM, and HRTEM. The use of GNPs for target delivery of drug and gene for a range of diseases such as cancer, malaria, and infectious diseases has also been included in this chapter. It also discusses ocular, pulmonary drugs delivery as well as nutraceutical, proteins, peptides delivery and their application in tissue engineering.

Chapter 11 presents the studies on graphene and its application in drug delivery. The chapter gives a brief introduction of method of synthesis and functionalization of graphene and deals with the applications of graphene in medicine and biomedical.

Chapter 12 describes the optical coherence tomography (OCT) as glucose sensor in blood. The chapter includes basic principle of OCTs and application of OCT for glucose monitoring. This chapter describes the use of OCT technique for measuring glucose in liquid phantoms, whole blood (in vitro *and* in vivo) based on temporal dynamics of light scattering.

Mushahid Husain
Zishan Husain Khan

Acknowledgments

We would like to praise Almighty God for granting us the wisdom, health, and strength to undertake this challenging work and enabling us to accomplish it.

We sincerely offer our gratitude to the authors, who have contributed to this book for their comprehensive coverage of current research in the exciting area of nanotechnology that fulfill a wide variety of application in nanomedicine, nano-electronics, and renewable energy.

We are extremely grateful to the Vice-Chancellor, Jamia Millia Islamia (Central University), New Delhi (India), for encouraging academics in the university and inculcating the habit of research among the young faculty members. His support throughout the process of editing this book is worth mentioning. His pearls of wisdom, coupled with valuable suggestions, have contributed largely to the completion of this work.

We take this opportunity to express our deep sense of gratitude to Prof. Vikram Kumar, Former Director, National Physical Laboratory, New Delhi (India), for his constant support and encouragement for establishing nanotechnology research in Jamia Millia Islamia.

Our special thanks are also due to our families whose unconditional support and encouragement have been a source of inspiration for us. The technical support provided by research and graduate students especially Dr. (Mrs.) Samina Husain, Dr. Avshish Kumar, Dr. Javid Ali of Department of Physics and Center for Nanoscience and Nanotechnology, and Mr. Mohd. Bilal Khan of Nanotechnology and Renewable Energy Research Laboratory at Department of Applied Sciences & Humanities, Jamia Millia Islamia (Central University), New Delhi, during the compilation of this book is also acknowledged. One of the editors, Prof. Mushahid Husain, is grateful to Department of Electronics and Information Technology, Ministry of Information Technology, Government of India, for sponsoring a research project on synthesis of single-wall carbon nanotubes for semiconducting applications and strengthening nanotechnology research at Center for Nanoscience and Nanotechnology. The establishment of this center of excellence of higher research in nanotechnology at Jamia Millia Islamia (Central University) with the support of University Grants Commission, Government of India, is a landmark and inspired us to excel the higher impacted research in nanotechnology. The world-class facilities established in this center with the untiring efforts of one of the editors, Prof. Mushahid Husain, are praiseworthy and provided a huge platform for nanotechnology researchers at the university to utilize them for benefit of the institution as well as society.

One of editors is also thankful to the senior faculty members of M.J.P. Rohilkhand University, Bareilly (UP), for their constant support.

Mushahid Husain
Zishan Husain Khan

Contents

Editors and Contributors

About the Editors

Prof. Mushahid Husain is the vice-chancellor of M.J.P. Rohilkhand University, Bareilly. Formerly, he was a senior professor and director in "Centre for Nanoscience and Nanotechnology" at Jamia Millia Islamia (Central University), New Delhi. He has been instrumental in starting M.Tech. Nanotechnology course in 2007 in Jamia Millia Islamia. Professor Husain has completed several major research projects on amorphous semiconductors funded by various government agencies such as DST and DRDO. He has been a vice chairman of Semiconductor Society of India for two consecutive terms (1999–2003). He is also holding various positions in different academic societies. In addition, he also held the office of the vice president of Indian Physical Society during the session 1990–1992. At present, he is the president of Society for NanoScience and Technology and secretary of one of the prestigious "Society for Semiconductor Devices." He has made about 27 foreign visits including University of Cambridge—UK; University of Southampton—UK; University of Princeton, New Jersey; and ICTP, Italy; UNAM, Mexico; SIRIM, Malaysia; National University, Singapore; National Tsing Hua University, Hsinchu, Taiwan; US Naval Research Laboratory, Washington, DC, and University of Arkansas, Fayetteville, where he had delivered lectures. He has delivered 111 invited talks at national and international forums. A number of popular talks on All India Radio and the National TV Channel (Doordarshan) have also been presented by him. Professor Husain has about 182 research papers in reputed international journals to his credit. He has produced 36 Ph.Ds. He has also edited one book on "Advances in Physics of Materials."

Zishan Husain Khan is working as an associate professor (applied physics) in the Department of Applied Sciences & Humanities, Faculty of Engineering and Technology, Jamia Millia Islamia, New Delhi-110025 (India). He specializes in nanotechnology with special emphasis of carbon nanotubes, semiconducting nanostructures, and nanochalcogenides. His work on the fabrication of a FET based on an individual carbon nanotube using e-beam lithography and its I–V characteristics was significant. He also studied the electrical transport properties and field emission properties of bulk carbon nanotubes. Dr. Khan has also contributed significantly in the field of oxide semiconductors and synthesized ZnO nanostructures for various applications. His contribution in the field of carbon nanotubes-based sensors is notable. His recent work on OLEDs based on the organic semiconductors and nanomemory devices based on the nanochalcogenides is expected to have significant impact on emerging field of nanotechnology. Dr. Khan has earned his Ph.D. on amorphous semiconductors from Jamia Millia Islmia, New Delhi-110025 (India) in 1996. He joined as a postdoctoral researcher at the Center of Nanoscience and Nanotechnology, National Tsing Hua University (NTHU), Hsinchu, Taiwan in December 2001 and continued there up to February 2005. While pursuing his postdoc, Dr. Khan worked extensively on different types of nanostructures with special focus of carbon nanotubes. He has published around 84 research papers in outstanding and high impact factor journals and made over 40 presentations in conferences and symposia, which includes invited talks in the conferences. Dr. Khan has been one of founder members of Center of Nanotechnology, King Abdulaziz University, Jeddah, Saudi Arabia). He has been the guest editor of International Journal of Nanoparticles (UK), International Journal of Nano-Biomaterials (UK), International Journal of Nanomanufacturing (UK), Journal of Nanomaterials (USA), and Advanced Science Letters (USA). He is also a regular reviewer of many reputed international journals.

Contributors

Faisal A. Agel Department of Physics, University of Hail, Hail, Kingdom of Saudi Arabia

Ejaz Ahmad Laser and Optronics Laboratory, Department of Physics, Bahauddin Zakariya University, Multan, Pakistan

Waqar Ahmed School of Medicine and Dentistry, University of Central Lancashire, Preston, UK

Yu-Cheng Chang Department of Chemical Engineering, National Taiwan University of Science and Technology, Taipei, Taiwan ROC

Wei-Hung Chiang Department of Chemical Engineering, National Taiwan University of Science and Technology, Taipei, Taiwan ROC

Pallav Gupta Department of Mechanical and Automation Engineering, A.S.E.T, Amity University, Noida, U.P., India

Yizhong Huang School of Materials Science and Engineering, Nanyang Technological University, Singapore, Singapore

M. Husain Center for Nanoscience and Nanotechnology, Department of Physics, Jamia Millia Islamia, New Delhi, India; MJP Rohilkhand University, Bareilly, UP, India

Samina Husain Center for Nanoscience and Nanotechnology, Jamia Millia Islamia, New Delhi, India

Fayyaz Hussain Material Simulation Research Laboratory (MSRL), Department of Physics, Bahauddin Zakariya University, Multan, Pakistan

Mark J. Jackson Micro Machinists Corporation, Cambridge, MA, USA

Shamshad A. Khan Department of Physics, St. Andrews PG College, Gorakhpur, India

Zishan H. Khan Department of Applied Sciences and Humanities, Jamia Millia Islamia, New Delhi, India

Ling Bing Kong School of Materials Science and Engineering, Nanyang Technological University, Singapore, Singapore

Avshish Kumar Center for Nanoscience and Nanotechnology, Jamia Millia Islamia, New Delhi, India

Devendra Kumar Department of Ceramic Engineering, Indian Institute of Technology (Banaras Hindu University), Varanasi, UP, India

Mukesh Kumar Functional and Renewable Energy Materials Laboratory, Department of Physics, Indian Institute of Technology Ropar, Rupnagar, Punjab, India

Vikram Kumar Department of Physics, Indian Institute of Technology Delhi, New Delhi, India

Sunil S. Kushvaha Physics of Energy Harvesting Division, CSIR-National Physical Laboratory, New Delhi, India

Sean Li School of Materials Science and Engineering, The University of New South Wales, Sydney, NSW, Australia

Jonathan S. Morrell Technology Development, Oak Ridge, TN, USA

Om Parkash Department of Ceramic Engineering, Indian Institute of Technology (Banaras Hindu University), Varanasi, UP, India

Wenxiu Que Electronic Materials Research Laboratory, School of Electronic and Information Engineering, Xi'an Jiaotong University, Xi'an, Shaanxi, People's Republic of China

M.A. Quraishi Department of Chemistry, Indian Institute of Technology (Banaras Hindu University), Varanasi, UP, India

C.M.S. Rauthan CSIR-National Physical Laboratory, New Delhi, India

Grant M. Robinson Micro Machinists Corporation, Cambridge, MA, USA

Numan A. Salah Center of Nanotechnology, King Abdulaziz University, Jeddah, Kingdom of Saudi Arabia

M. Senthil Kumar Physics of Energy Harvesting Division, CSIR-National Physical Laboratory, New Delhi, India

Anshu Sharma Special Centre for Nano Sciences, Jawaharlal Nehru University, New Delhi, India; School of Physical Sciences, Jawaharlal Nehru University, New Delhi, India

Prem Prakash Sharma Special Centre for Nano Sciences, Jawaharlal Nehru University, New Delhi, India

P.K. Singh CSIR-National Physical Laboratory, New Delhi, India

Pratima R. Solanki Special Centre for Nano Sciences, Jawaharlal Nehru University, New Delhi, India

O.N. Srivastava Nanoscience and Nanotechnology Unit, Physics and Bioscience Division, Banaras Hindu University, Varanasi, India

Sanjay K. Srivastava CSIR-National Physical Laboratory, New Delhi, India

Mahe Talat Nanoscience and Nanotechnology Unit, Physics and Bioscience Division, Banaras Hindu University, Varanasi, India

Hafeez Ullah Laser and Optronics Laboratory, Department of Physics, Bahauddin Zakariya University, Multan, Pakistan

Michael D. Whitfield Micro Machinists Corporation, Cambridge, MA, USA

Weili Yan School of Materials Science and Engineering, Nanyang Technological University, Singapore, Singapore

Tianshu Zhang Anhui Target Advanced Ceramics Technology Co. Ltd., Hefei, Anhui, People's Republic of China

Chapter 1
Introduction to Nanomaterials

Zishan H. Khan, Avshish Kumar, Samina Husain and M. Husain

Abstract The applications of nanomaterials have been enormous, which not only encompasses a single discipline but it stretches across the whole spectrum of science right from agricultural science to space technology. New approaches to synthesize nanomaterials in order to design new devices and processes are being developed and the techniques of fabrication of nanomaterials involve analyzing and controlling the matter at atomic scales. This fascinating research field has started a new era of integration of basic research and advanced technology at the atomic scale which has a potential to bring the technological innovations at highest level. The rudimentary capabilities of nanomaterials today are envisioned to evolve in our overlapping generations of nanotechnology products: passive nanostructures, active nanostructures, systems of nanosystems, and molecular nanosystems. This chapter presents the basic introduction to nanomaterials and their popular applications.

Keywords Nanomaterials · Fullerene · Carbon nanotubes · Graphene · Nanodiamond · ZnO nanostructures

Z.H. Khan (✉)
Department of Applied Sciences and Humanities, Jamia Millia Islamia,
New Delhi 110025, India
e-mail: zishanhk@jmi.ac.in

A. Kumar · S. Husain · M. Husain
Center for Nanoscience and Nanotechnology, Jamia Millia Islamia,
New Delhi 110025, India
e-mail: mush_phys@rediffmail.com

M. Husain
MJP Rohilkhand University, Bareilly, UP, India

M. Husain and Z.H. Khan (eds.), *Advances in Nanomaterials*,
Advanced Structured Materials 79, DOI 10.1007/978-81-322-2668-0_1

1.1 Introduction

Different materials and structures having low dimensions show outstanding properties. This facilitates these materials and structures to play a decisive role in the rapid progress in the fields of science and technology. With these remarkable properties, nanomaterials has become the back bone of research in the field of science and technology. The continuous interest in the research on nanomaterials is due to the infinite possibilities that they can offer when they are manipulated on atomic scale. The interest of these materials arose due to their unique mechanical and chemical properties, which are very different in comparison with material at micrometer scale with the same composition. All the branches of science and engineering have been employed to explain many phenomenon based on size effect, that these materials exhibit. Material science basically utilized those elements (iron, silicon, etc.) that are available in nature for development new compounds in past few years. As a result of this the researcher have learned to make devices using synthetic structures in which they deposited atoms layer by layer and after they are structured creatively following redesigned architectures and manipulated molecules individually. This allows making system with novel and different properties.

Nanomaterials have attracted a lot of attention among researchers because of its new physical properties and new technologies that enables the development of new generation of scientific and technological approaches, research and devices [1]. The essence of the technologies is the fabrication and utilization materials and devices at the level of atoms, molecules and supra-molecular structures and the exploitation of the unique properties and phenomena of matter at the nanoscale (1–100 nm) with the fact that, at this scale, materials behave very differently from when they are in bulk form [1–4]. In early 1980's, this technology was popularized with the statement "building machines on the scale of molecules" by Eric Drexler when was talking about, a few nanometers wide. This technology has very rapidly captured the public imagination, but they are desperately hard to pin down. In the early decades of twenty-first century, concentrated efforts have brought together nanotechnology and the new technologies based in cognitive science [5, 6]. Although, nanomaterials are known to be used for centuries, but the realization of these materials could have possible with the advent of nanotechnology. Since the ancient times, nanomaterials are being used due to their extraordinary properties without knowing the scientific facts behind them. One of the examples is Lycurgus cup, which is currently located in British Museum. It is about 1600 years old and looks jade green in natural light. This cup is the only complete sample of an extremely uncommon kind of glass, known as dichroic, which switches shading when held up to the light. The hazy green cup turns to a gleaming translucent red when light is shone through it. The glass contains minor measures of colloidal gold and silver, which provide these unordinary optical properties. The glass changes its color due to presence of nanoparticles contained in the glass. Therefore, nature has already been the leader in this technology [7–9]. Lizards hang upside down on the roof due to nanoscale "hairs" on their toes. Every "hair" holds with a miniscule power, however when there are a huge number of these “hairs” on every toe, it has the capacity backing its own particular weight when it is upside down [8–10].

In the medieval period of history, there were instances where nanomaterials were used. Many items like the Persian Khanjar and the Damascus steel were made by unconsciously applying nanomaterials. While examining these ancient products, we find the existence of carbon nanotubes in them. India, too, produced many products by unconsciously deploying nanomaterials. The attractive colouring on ancient Czech glasses is found to contain nanoparticles. This shows that nanomaterials were used for spraying and making a product look attractive and beautiful.

Researchers are continuously involved in the development of new product which use the exciting properties of nanomaterials. Many chemicals and chemical processes have nanoscale features. Different types of polymers and macromolecules made up of nano-sized components have been efficiently synthesized to be used in nanoscale systems. Computer chips have become smaller with higher memory contain per unit area in last thirty years [11–14]. A beetle which lives in Namibian deserts fulfill its need of water using nanostructures. Its back has hydrophobic surface which repels water but its back consists of nanostructured bumps. The moisture on the atmosphere condenses on these bumps, takes the shape of water droplets and reaches direct to its mouth running down on its water repelling back. This type of water repellent nanomaterials are being used in textile industries to produce waterproof cloth material. Gortex are using nanomaterials for wind and water proofing since a long time. The exciting colors on butterfly wings and pearl shells are due to nano-photonic structures. These structures reflects some particular wavelengths of light, which are responsible for their attractive colors [15–17].

It is very difficult to estimate the timescale in which the products based only on nanomaterial become reality. But there is a certain possibility that there will not be a single aspect of life which will remain unaffected by nanomaterials. Some experts predicts that the nanomaterials will act as the boon for mankind while the others forecast that the excessive use of nanomaterials may become a curse to the humanity. We are entering into a new era in which the known materials will be used more efficiently due to their extraordinary properties at the nanoscale. More complex systems based on nanomaterials may be developed in future which can work as efficiently as the natural systems. This technology has the tendency to bring revolutionary changes in the world that the human eyes have not seen so far.

1.2 Nanomaterials: A Revolution in 21st Century

Recently, nanomaterials has become one of the focused research area and has the potential to provide one of the key technologies of the new world. Nanotechnology involves the investigation and design of materials or devices close the atomic and molecular levels. One nanometer, a measure equal to one billionth of a meter, spans approximately 10 atoms and one may be able to rearrange matter with atomic precision to an intermediate size. Adopting one of the preceding definitions may however unduly restrictive. In a broad survey, it is profitable to include anything and everything that has been described as nanotechnology. One may then evolve a framework for classifying and understanding the different types of nanomaterials.

The extraordinary properties of nanomaterials are based on the size effect. Reduction of size induces various extraordinary properties and phenomena, which cannot be seen at the bulk level i.e., micron scale and up. The origin of size effect is the quantum confinement which changes the electronic properties of materials such as density of states, discrete energy bands, changes in the edges of conduction and valence bands. The change in these electronic properties of the materials lead to the drastic changes in the materials properties to what they show at the bulk scale. For example, opaque substances become transparent (copper); inert materials become catalysts (platinum); stable materials turn combustible (aluminum); solids turn into liquids at room temperature (gold); insulators become conductors (silicon). The important examples of the change in material properties are as follows:

1.2.1 Melting Point of the Material

The materials at nanoscale melt at lower temperatures than bulk materials; a phenomenon termed as "Melting Point Depression". This effect of melting point depression at nanoscale was studied since the 1950's, when it was first noticed that materials having small grain size shows a lower melting point than their bulk counterparts [12–16]. Generally, the melting temperature of a bulk material is not dependent on its size. But the melting temperature of the materials at nanoscale decreases with the decrease in grain size. The decrease in melting temperature can be on the order of tens to hundreds of degrees for metals with nanometer dimensions. The change in melting point can be attributed to the large surface to volume ratio than bulk materials that affects their thermodynamic properties. Melting point depression can be seen in nanowires, nanotubes and nanoparticle, which all melt at lower temperatures than their bulk counterparts.

1.2.2 Optical Studies

Optical properties of nanostructures are known to be sensitive to their size. For example, bulk lead sulfide (PbS) is a semiconductor with an optical band gap of 0.41 eV with continuous optical absorption at shorter wavelengths. However, increment of the band gap of PbS is observed from 0.41 to 5.4 eV as the crystallite size is reduced from 20 to 2 nm. The changes in electronic and optical properties due to size effect drastically alter the macroscopic properties such as optical band gap, conductivity type, carrier concentrations, electrical resistivity, and device characteristics [1–6].

1.2.3 The Giant Magneto-Resistance (GMR) Effect

Giant Magneto resistance (GMR) is a quantum mechanical effect. It is observed in thin film structures composed of alternating ferromagnetic and nonmagnetic metal layers. The GMR effect depends on the thickness and number of the alternating layers. GMR can be considered one of the first real applications of the promising field of nanotechnology. Primarily, induction coils were used in read-out heads due to the fact that a changing magnetic field induces a current through an electric coil. A read out head based on GMR effect can convert very small changes into differences in electrical resistance which leads to the changes in the current emitted by the read out head.

Inspite of rapid advancements, this technology has not been able to meet the demands of shrinking hard disks, we still use induction coils for data storage. A prerequisite for the discovery of the GMR-effect was provided by the new possibilities of producing fine layers of metals on the nanometre scale which started to develop in the 1970s. Nanotechnology is the study of few layers consisting of only individual atoms. The behavior of the matter will be different at this scale and the nano-sized structures show extraordinary properties, which will be different from their bulk counterparts. The size effect not only alters the magnetic and electrical properties but it also changes the strength of materials, chemical and optical properties. Therefore, at nanometer scale, GRM is one of the interesting applications of nanotechnology and may be applied to the future data storages devices. The size of hard disk is continuously reducing rapidly day by day [9–15].

The use of GMR materials has successfully enhanced the storage capacity of hard disks from one to 20 gigabits. In 1997 IBM launched read heads based on GMR, worth around one billion dollar into the market. The field of spintronics is relatively new but it is expected to have the entire requisite potential to be evolved as an extremely successful technology. There are several new materials and technologies such as magnetic semiconductors and exotic oxides, which are in development phase and are expected to show many interesting phenomenon such as colossal magneto-resistance [8].

1.3 Classification of Nanomaterials

The classification of the nanomaterials is based on the number of dimensions, which are not confined to the nanoscale range (<100 nm). Nanostructures demonstrate an essential quality of quantum mechanics known as quantum confinement [18–21]. Quantum confinement means that electrons are trapped in a small area. It occurs when electrons and holes are confined by a potential well in 1D (quantum well), 2D (quantum wire) or 3D (quantum dot). In other words, the size of the nanocrystal is made very small so that it approaches the size of an exciton (bound state of electron-hole pair) called the Bohr exciton radius. For effective confinement, the size should be less than 30 nm.

Quantum dot are 0-D structures where the electrons are confined in all the three dimensions, thus having 3-D confinement. In other words, here all the dimensions are near the Bohr exciton radius making a small sphere. Only atoms in nature have 3-D confinement. Thus, quantum dot can be described as an 'artificial atom'. Atoms, since they are small, are difficult to be isolated, but quantum dots can be easily manipulated.

2-D confinement is exhibited by nanowires. It is a structure, where the length is long while height and breadth are small. It is an electrically conducting wire in which quantum effects affects transport properties. Here, the conduction electrons are confined in the transverse direction of the wire and the energy is quantized into a series of discrete values. The other 1-D structures include: nanorods, nanotubes and quantum wires.

1-D confinement is restricted only in one dimension, resulting in quantum well or plane. The length and breadth of the quantum well can be large in comparison to the height, which is about the Bohr exciton radius. Due to quantum confinement effect, the carriers (generally electrons and holes) possess discrete energy values, when the thickness of the quantum well becomes comparable at the de-Broglie wavelength of the carriers the electronic states get quantized. Owing to their sharper density of states because of their quasi-two dimensional structure, quantum wells are being widely used in LASERs.

1.4 Types of Nanomaterials

Nanomaterials show very different properties compared to the bulk structures [22–31]. There are large numbers of nanomaterials but we will discuss only important ones.

1.4.1 Carbon Nanomaterials

The synthesis of carbon based nanoscale-materials have influenced many researchers and scientists worldwide to understand its potential applications. Fullerenes, carbon nanotubes (CNTs), carbon onions, carbon nanocones, nanohorns and nanowires are all members of this family. Amongst these members, CNTs are an example of true nanotechnology: only a nanometer in diameter, but molecules that can be manipulated chemically and physically. They are termed as the true materials of the 21st century.

1.4.1.1 Fullerenes

It is known that the condensed phase of carbon has a hexagonal ground state, which is represented by graphite with sp^2 bonding. It is highly anisotropic and two dimensional semimetal. Diamond is a three dimensional material, which lies close to graphite and is also isotropic. The discovery of fullerene, a zero dimensional form of carbon, stimulated great interest in carbon materials. The identification of C60 as a molecule having the shape of a regular truncated icosahedron by Kroto et al. [32] was the most influential discovery on carbon materials. With these findings, a lot of theoretical as well as experimental work has been reported. The chemistry of fullerenes was developed in the mid 1980's [32]. Fullerene is a cage like hollow, spherical, or ellipsoid super-molecule made up of sp^2-hybridized carbon atoms (each carbon atom is bonded to three carbon atoms) composed of 12 pentagonal and 20 hexagonal [40]. This structure was similar to geodesic dome designed by architect Richard Buckminster Fuller and it was named after his name as Fullerene, which is often referred to as "buckyballs" [24]. Fullerenes were the seventh allotropic form of carbon [together with the two forms of diamond, the two forms of graphite, chaoit, and carbon (IV)]. Curl, Kroto, and Smalley received the Nobel Prize for Chemistry in 1996 for the discovery of fullerenes. Figure 1.1 presents a graphical representation of the 60-carbon atom containing C-60 fullerene. The C60 molecule has two bond lengths in which the 6:6 ring bonds can be considered "double bonds" and are shorter than the 6:5 bonds. C60 is not "superaromatic" as it has tendency to avoid double bonds in the pentagonal rings, which results in poor electron delocalisation. As a result, C60 behaves like an electron deficient alkene, and reacts readily with electron rich species. The synthesis of large quantities of fullerenes and measurement of its properties [33] is also reported, which opens the new field for the researchers. It is interesting to note that this material is also used for producing alkali metal (M) doped M3C60 compounds (highest observed at early stage Tc = 33 K) and a relatively high Tc

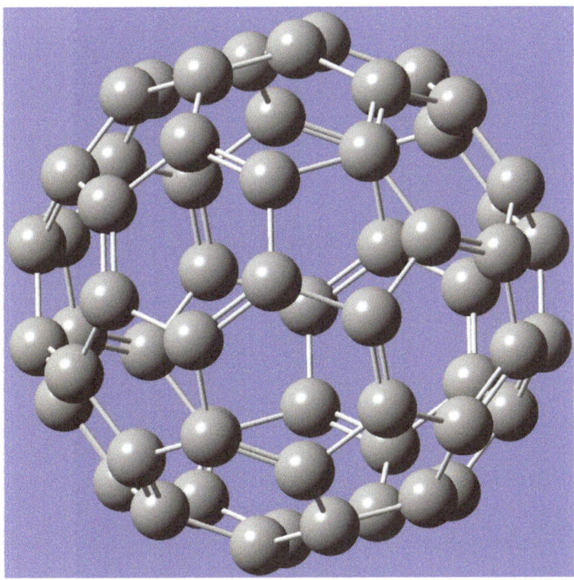

Fig. 1.1 Image representation of C-60 fullerene [35]

superconductivity was reported, which is an important breakthrough [34]. With these interesting results, a lot of research on doped fullerenes is under progress (Fig. 1.1).

1.4.1.2 Carbon Nanotubes

Carbon nanotubes (CNTs) are tubular carbon structures of nanoscale diameter having length up to several micron to centimeters. CNTs were discovered by SuniyoIijima in 1991. Since then, it remains a hot topic of research among the researchers working on nanomaterials. Owing to their unique structure, CNTs have a wide range of applications. CNTs are 100 times stronger and five times lighter than steel. They show that electrical conductivity is six order higher than copper and thermal conductivity is five times higher than copper. Due to these unique properties, researchers from all over the world are being involved in the research related to CNTs and the devices based on CNTs [36–40].

Carbon nanotubes are considered as nearly one-dimensional structures according to their high length to diameter ratio. There are three types of CNTs (Fig. 1.2):

1. Single wall carbon nanotubes (SWCNTs)
2. Double wall carbon Nanotubes (DWCNTs)
3. Multi wall carbon nanotubes (MWCNTs).

Multiwalled CNTs were accidently discovered by Iigima in 1991 when he was trying to make fullerenes by arc discharge method. MWNTs are made of several

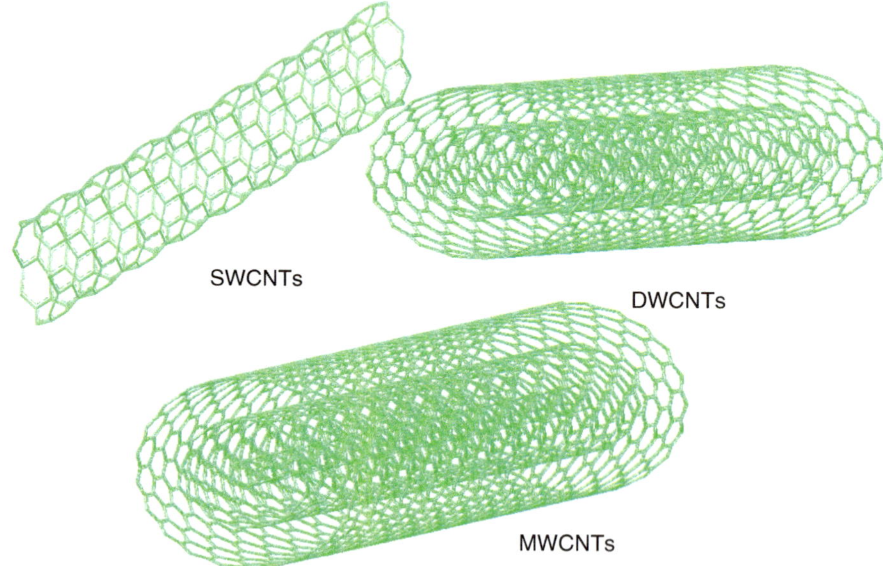

Fig. 1.2 Different types of carbon nanotubes [35]

concentric cylinders with the inter wall spacing nearly equal to the interplanner spacing of graphite (0.34 nm). Their inner diameter varies from 0.4 nm up to a few nm and their outer diameter ranges typically from 2 up to 20–30 nm depending on the number of layers. MWNTs usually have closed tips by insertion of pentagonal defects into the graphite network. The lengths of MWNTs are ranging from several micron to centimeters. Single walled CNTs has the seamless cylindrical structure made up of rolled graphene sheet. It was first reported in 1993. Their diameters range from 0.4 to 2–3 nm, and their length is usually of the micrometer order. SWNTs usually exist in form of bundles. In a bundle they are hexagonally arranged to form crystalline structure.

There are many methods to synthesize CNTs, which includes arc discharge, LASER ablation and chemical vapour deposition method. In arc discharged method, electrical arc is created between two carbon electrodes. Carbons vapours are created because of the high temperature of arc. These vapours self assemble to form Carbon Nanotubes. The LASER ablation method uses a high power laser, which is illuminated on a volume of carbon containing feedstock gas. This method produces a small quantity of clean nanotubes while arc discharge method produces large quantity of impure nanotubes. Out of these methods, the chemical vapour deposition method is most popular and widely used for the synthesis of CNTs. In this method, carbon containing gases are decomposed in the presence of metal catalyst on a high temperature which is maintained using resistive heating. Carbon atoms are diffused to the substrate and produces a film on the substrate. If growth parameters such as growth temperature, gas flow rate etc. are properly maintained, CNTs are formed on the substrate.

Kumar et al. [41, 42] synthesized SWCNTs and MWCNTs by plasma enhanced chemical vapour deposition (PECVD) system and low-pressure chemical vapour deposition (LPCVD) system. They used thermal evaporation to deposit different catalysts on silicon substrate. A mixture of gases C_2H_2/H_2 with a flow rate of 50/50 sccm were used as reactive gases. Throughout the experiment, the chamber pressure was kept at 10 Torr and the temperature was set at 800 °C. The growth time was varied from 5 min to 1 h. Electron Cyclotron Resonance Chemical Vapour Deposition (ECR-CVD) method is an another method for the growth of CNTs at low temperature and low pressure. Khan et al. [43, 44] synthesized CNTs and studied the electrical properties of CNTs. They [45, 46] have also employed CNTs for gas sensing applications.

Owing to their extra ordinary mechanical, chemical, electrical properties, CNTs find a wide range of applications in different fields. Its excellent adsorption properties makes it useful materials for chemical sensors. CNTs will find vast applications in different fields such as micro-electronics/semiconductors, controlled drug delivery/release, catalysts support for fuel cells and sensors. Other potential application of CNTs includes its use in solar cell, nanoporous fibers, catalyst supports and coatings. CNTs offer big promise for medicine, providing better contrast agents for MRI and localized heaters that can induce a target tumor cell dead. They also offer a new approach to gene therapy.

1.4.1.3 Nanodiamond

Carbon is known as a charcoal or soap material for the latest several hundred years. It has been used of reduction of metal oxide since long. Graphene was discovered in 1979 by sheele and is found to contain the pure form of carbon. It was Lavoisier to realise that carbon is also one of the chemical element and identify Diamond as one of the allotropic carbon. The old form of carbon either exist in nature or prepared artificially using different thermodynamic condition. The hexagonal graphite is found to be one of the most allotropes of carbon act ambient pressure. Since the density of diamond is higher than of graphite, the conversion of graphite to diamond take place at high pressure [47]. Initially the diamond is formed in metastable state to avoid its decomposition to graphite, it is needed to lower the temperature before releasing the pressure. Graphite (sp^2 bonded) and diamond (cubic, sp^2 bonded) are the two purist crystalline form of carbon known for the first scientist to produce diamond like carbon in 1969, it is also form of the carbon as identify by these scientist. Following these work on the use of energetic species for the deposition, they study the epitaxial deposition of silicon by sputtering of silicon in Ar plasma. The same apparatus was employed to sputter carbon electrodes to produce carbon ion for the deposition of transparent, hard, insulating carbon films. They found that these as deposited carbon films show property singular to natural diamond but there structure was amorphous. Due to this amorphous structure these films are not suppose to be the pure diamond but a new name diamond like carbon was given to these films. After this pioneer work ashenburg and chebot many methods have been used to deposit diamond like carbon (DLC) films [45, 46]. The film deposited using hydrocarbon plasma at lower deposition rate are hard, transparent, and contain 20–50 % hydrogen. They are different from hydrogen free DLC but are demoted as DLC without monitoring there hydrogen content more proper and carefully names are a-C:H (amorphous carbon hydrogen), DLHC (diamond like hydrocarbon). During the past several years DLC films involving hydrogen have been study more incomperision to the DSC involving hydrogen free films. Due to the diversity of different deposition system used [45–50] and with the help of powerful characterization technique, a nominee cloture such as (a-DLC, a-C.a-CH, ta-C, a-D, i-C) are carbon etc. was given the most important term used in nominee cloture in amorphous diamond or amorphic diamond (a-D), which is logically same as amorphous crystal. Due to the variety of synthesis method and characterization technique used for these films, the expression data available on films structure, the optimal energy for the production of sp^3 rich films, the effect of the substrate temperature and the property of as deposited films are sometimes conflict each other and become controversial data. Last several years has seen many report of the deposition of DLC films [51–54] in these reports the deposition parameter were optimised to get the perfect DLC films. These studies provide a better understand of deposition process, growth mode [54–56] and computer simulation of film growth. The research work on the hydrogen free DLC gain momentum during last few years there for in the present scenario DLC is the mostly refers to hydrogen free amorphous carbon with some sp^3 content in it. A pre dominantly sp^3 films ($sp^3 \geq 70$ %) with a tetrahedral local carbon configuration (Fig. 1.3) will be demoted as ta-a-carbon, amorphous or nanodiamond.

The potential applications of diamond and related materials are enormous which includes hard coatings, optical windows, surface acoustic wave (SAW) devices, electrochemical electrodes for micro-mechanical systems and electron emitting surfaces for flat panel displays. The applications of diamond realized until now are based on the physical strength and hardness of diamond such as cutting tools, protective coatings and composite additives, many other applications of these materials in development stage [57, 58]. In-spite of many techniques available for the growth of diamond films, but the growing smooth and defect free film is still difficult. Since most of applications such as protective optical coating require smooth and defect free films, we need to overcome this problem. This problem can be over by using nanocrystalline or amorphous diamond, tetrahedral amorphous carbon (ta-C) and diamond-like carbon films, which are much smoother, hard as conventional diamond and can be grown at lower temperature [59, 60].

Due to these unique properties, nanometric sized crystallites diamond films have drawn a lot of attention. The nano-diamond films are useful for protective coatings on mechanical tools, the antireflection coatings for IR optics which require low optical absorption in the IR region. The application of nanodiamond varies from abrasives for semiconductors to lubricating material for hard disks. These films also find application in semiconductor devices related to the microelectronic industry.

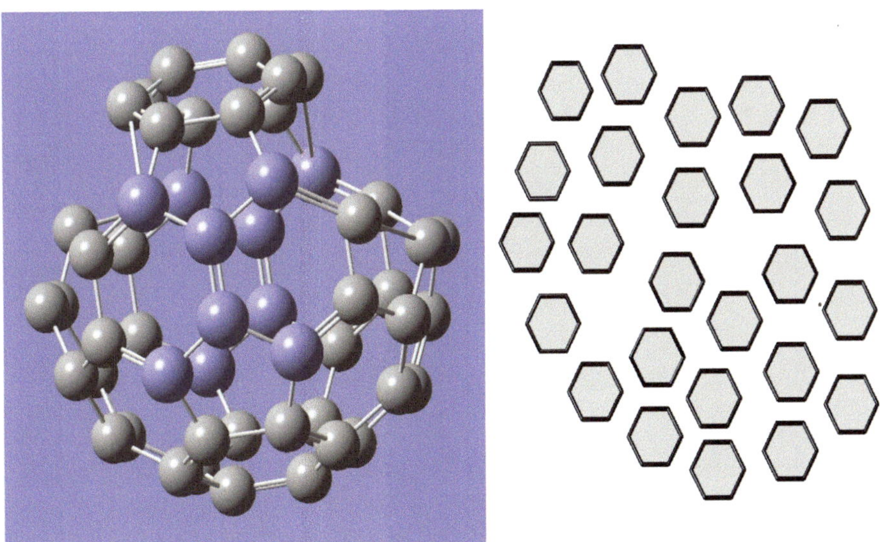

Fig. 1.3 Image representation of nanodiamond [35]

1.4.1.4 Graphene

Graphene is one of the most rigorously researched material in materials science and condensed-matter physics. It is a quasi-two dimensional material made up of sp^2 hybridized carbon atoms with high quality crystal structure (Fig. 1.4). Its unusual electronic properties draw a new light on quantum electrodynamics which enlightens various quantum relativistic phenomena, which are unobservable in high-energy physics. The phenomena can now be mimicked and tested in easy experiments. From the point of view of its electronic properties, graphene is a two-dimensional zero-gap semiconductor and its low-energy quasiparticles are mass-less and described by the Dirac-like Hamiltonian rather than usual Schrodinger's Hamiltonian. Electron waves in graphene travel within a mono-atomic layer, which allows them to interact with various scanning probes [61]. Electronic waves in graphene are also responsive to the presence of various materials such as high-k materials, superconductors etc. Its electrons can propagate significant distances without scattering. Due to massless carriers and little scattering, quantum mechanical aspects of Graphene can be realized at normal temperatures. These extra ordinary properties led to the interpretation of the half-integer QHE and the prognosis of several phenomena such as Klein tunneling, zitterbewegung, the Schwinger production [62], supercritical atomic collapse [63], and Casimir-like interactions [64].

Its charge carriers show zero effective mass, have large mobility and get less scattered while travelling. The current densities through graphene can attain the values approximately six times higher than that of copper. It also shows good

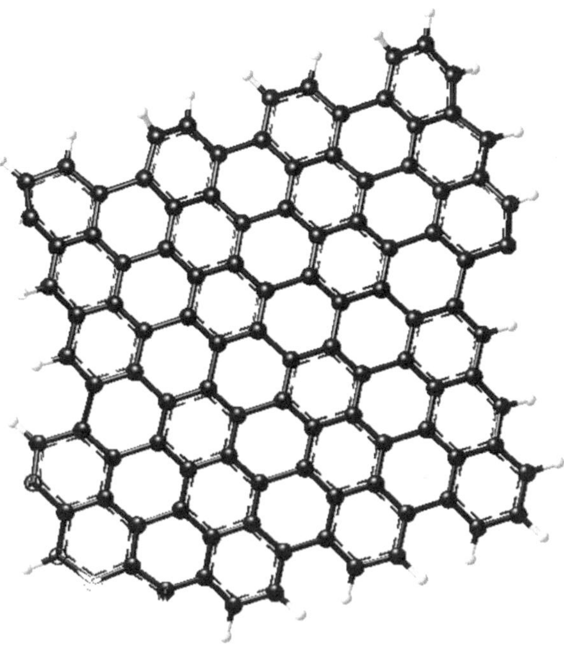

Fig. 1.4 Schematic representation a graphene [35]

thermal conductivity, stiffness, resistant to gases and reconciles such conflicting qualities as brittleness and ductility. Graphene is an exciting material which has a large theoretical specific surface area (2630 m^2 g^{-1}), high intrinsic mobility (200,000 cm^2 v^{-1} s^{-1}) conductivity (5000 Wm^{-1} K^{-1}), and its optical transmittance (97.7 %) and good electrical conductivity lead to its applications as transparent conductive electrodes, among many other potential applications [64, 65].

Initially, graphene was synthesized by micromechanical exfoliation from graphite flakes. Since then, various methods are being used to prepare Graphene. Single layer graphene can be synthesized by micromechanical cleavage of HOPG, CVD on metal surfaces, Epitaxial growth on SiC and Dispersion of graphite in water, while few layer Graphene can be produced by the chemical reduction of exfoliated graphene oxide (2–6 layers), thermal exfoliation of graphite oxide (2–7 layers), intercalation of graphite Aerosol pyrolysis (2–40 layers) and arc discharge in H_2 atmosphere [65].

As now, graphene is not being used in commercial application, but it is a hot topic of research worldwide due to its extraordinary properties. Various applications of graphene in electronics, solar energy production, energy harvesting, hydrogen storage etc. has been proposed or are under development Graphene in powder form can be dispersed in polymer matrix to form advanced composites, paints, coatings, 3-D printing materials etc. [66].

1.4.2 ZnO Nanostructures

The research interest in ZnO, (one of wide band gap semiconductors,) has increased in recent years. The first enthusiasts who started studies of the lattice parameter were M.L. Fuller in 1929 [67] and C.W. Bunn' in 1935 [68]. During the last decade, the ZnO related research has received an increased impetus. The number of articles published on ZnO has steadily increased every year and ZnO became the second most popular semiconductor material (after Si) during 2007–2008. This popularity, to a large extent, is due to the improvements in growth related techniques of single crystalline ZnO (in both epitaxial layers and bulk form). Another reason is emergence of novel electrical, mechanical, chemical and optical properties with reduction in size. They are largely believed to be the result of surface phenomena and quantum confinement effects.

Zinc oxide is II-VI compound semiconductor. It can be available in a variety of crystal structures including wurtzite, zincblende and rocksalt, but the stable structure at ambient temperature and pressure is the hexagonal, wurtzite structure [69]. Wurtzite zinc oxide has a hexagonal structure (space group C6mc) with two lattice parameters a = 0.3296 and c = 0.52065 nm. The distance between Zn^{2+} ion and O^{2-} ion along the c-axis is 0.1992 nm, and the distance between these two ions along other three axes is 0.1973 nm. ZnO naturally forms a wide variety of nanostructures with excellent crystal quality (superior to thin films and even bulk crystals in some cases) by a self organization process. Recently, significant research has been dedicated to explore the different aspects of ZnO material as part of the general trend towards the study of nanotechnology and self-organized processes based on this material. This wide range of high crystal quality nano-morphologies open the possibilities for more potential applications [67–74]. A great deal of attention has been focused on synthesis and analysis of ZnO nanostructures due to their interesting material properties. The combination of the material properties and self-organized nanostructure synthesis open up the new possibilities for the applications of these materials in optoelectronics, nanoelectronics, nanomechanics, nanoelectrochemical systems, and sensors [78]. ZnO forms a variety of nanostructures (Fig. 1.5) using a range of growth techniques, including, nanorods [71], nanowires [72], nanobelts [73–75] nanorod/nanowall [76], nanotubes [77], tetrapods [78] and nanoribbons [79], which are ideal systems for investigating the dependence of optical properties, electrical transport and mechanical properties on dimensionality and size. They are likely to play an important role as both interconnects and functional components in the invention of nanoscale electronic and optoelectronic devices.

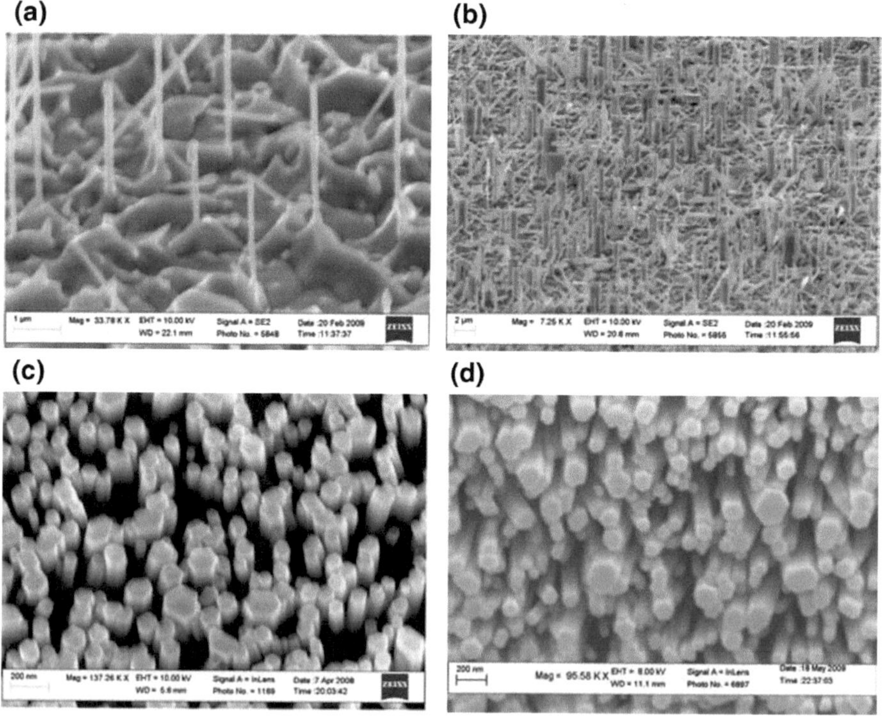

Fig. 1.5 SEM images of aligned ZnO nanowires grown on different substrates. **a**, **b** ZnO nanowires grown on SiC and silicon substrates by the high temperature vapor liquid solid (*VLS*) method. **c**, **d** ZnO nanowires grown on SiC and Si substrates by the low temperature approach aqueous chemical growth (*ACG*) method [80]

A variety of methods has been used for synthesis of ZnO nanostructures. These methods have been categorized as either physical vapor deposition (PVD), chemical vapor deposition (CVD), or solution based chemistry (SBC), PVD and CVD fall into the category of gas-phase approach where high vacuum and/or elevated temperature are normally required. ZnO nanowires, dendritic side-branched/comblike structures, and nanosheets can be synthesized under the identical conditions only with increased oxygen content [81]. The three most common and important PVD technologies for ZnO deposition are thermal evaporation, pulsed laser deposition (PLD), and sputtering.

ZnO is often used in paint, paper, rubber, food and drug industries. It is bio-safe and biocompatible, and it can be directly used for biomedical applications without coating [82, 83]. It is also a potential material in nano-electronics and nano-robotic technology. ZnO can be useful for electronic and photonic devices, as well as high-frequency applications due to its wide band gap, high exciton binding energy and high breakdown strength. By controlling electronic properties, it is easy to produce various optoelectronic devices based on ZnO. Other applications of ZnO includes the ZnO based sensors for detecting various gases such as H_2, NO_2, NH_3 etc., LEDs, cantilevers, solar cells field emission displays [84–87]. A ZnO nanorod based H_2 sensor has already been developed by Wang et al. [88]. Also Lee et al. [89] have already proposed the use of well-aligned single-crystalline nanowires to be used as sharp atomic force microscopy (AFM) tips. ZnO is suitable for an UV photo-detector because of its direct wide band gap and large photoconductivity. ZnO epitaxial film-based photoconductive and schottky type UV photodetectors have been demonstrated by Liang et al. and Liu et al. [90, 91]. Lee et al. [92] were the first group to study the field emission properties of ZnO nanowires in 2002. Field effect transistor (FET) based on a single ZnO nanobelt has also been reported [93]. As a wide band gap semiconductor, ZnO has also been investigated as a hydrogen storage material [94, 95]. Under a pressure of 5 MPa, the maximum hydrogen storage capacity of about 2.94 wt% is achieved for Al-doped ZnO nanobelts [96]. Optically pumped ZnO nanowire laser arrays as well as single ZnO nanowire lasers have been demonstrated [97–99].

A single nanowire light-emitting diode was fabricated by Bao et al. [100]. They used a focus ion beam (FIB) system to define pattern. They measured the current-voltage characteristics, photoluminescence, and electroluminescence of a single nanowire by depositing metallic contact onto the top surface of a single nanowire. ZnO has been employed as an electrode material in dye-sensitized solar cells recently [101–104]. Studies on nanoporous dye-sensitized ZnO films have suggested that the ultrafast electron injects from the dye into the conduction band of the ZnO particles [105–108], which is comparable to the timescale of electron injection into TiO_2 layers [109]. The main requirement for the efficient dye-sensitization and light harvesting is that a materials should have a wide band gap and high charge carrier mobility. High surface to volume ratio is also one of the interesting properties which can be attained by producing nanoscale materials. Therefore, the nanostructures of ZnO is one of the promising materials for solar cell applications.

1.5 Potential Applications of Nanomaterials

Nanomaterials have potential applications in various fields such as electronics, optical communications and biological systems. The factors which support these applications are particularly their physical and chemical properties, high surface to volume ratio and small size which provide best possibilities for manipulation and room for accommodating multiple functionalities. Nanomaterial provides a basic platform towards the understanding and production of nanodevices. Nanomaterials are basically an intermediate between molecules and isolated atoms, which have dimensions in the range of largest molecules. It is also possible to put together them and act together keenly with in a device. Nanomaterials are formed of cluster of atoms or cluster of molecules.

Nanomaterials can also be used in UV photo detector because of their large photoconductivity and direct wide band gap. ZnO nanowires shows reversible switching shows promising behaviour for optoelectronic switches. A broad application of UV laser source includes the bio-agent detection, UV photonics, etc. Nanowire UV laser source could be used as light source for quantum computing, optical interconnections and environmental applications such as the integration of "lab-on-chip".

Non-polluting nature and easy synthesis property of hydrogen makes it efficient and clean fuel for future prospects. In Nanostructure, hydrogen storage can be used for the development of transportation technology such as H_2 fuel cell vehicles. Several attempts have been made to improve the hydrogen storage properties such as doping with elements and modify their surface properties. An increasing demand of hydrogen storage material, ZnO nanostructure pure and doped with metals such as Sb and Al can be used for energy storage material.

Nanomaterials have high specific area as well as its electronic processes are strongly influenced by the surface processes, which shows high sensitivity to chemical environment. Various nanomaterials can be widely used for sensing applications. CNTs have shown sensitivities toward gases such as NH_3, NO_2, H_2, C_2H_4, CO, SO_2, H_2S, and O_2.

AFM consist of cantilever with a sharp tip at the end which is used to scan the specimen surface. Sharp atomic force microscopy (AFM) tips can be fabricated using well-aligned single-crystalline nanowires [110]. Under typical operating conditions, ZnO nanowires are structurally compatible with AFM cantilevers, and are expected to provide high-aspect-ratio probes for AFM. The flexural mode of the nanowire in situ in a transmission electron microscope (TEM) was studied by Huang et al. [111]. They suggested that single ZnO nanowire can be used as a nano-resonator and a nanoscale cantilever.

Highly efficient next-generation LED lighting systems could be possible with the help of nanotechnology. Single nanowires light emitting diodes have been constructed by various different groups. They are dispersing nanowires of made of different nanomaterials on a substrate (say Si) and then thin film of Poly (methyl methacrylate) PMMA, is spin coated on the substrate. Imagine the nanowire on a focus ion beam (FIB) system and a pattern for e-beam exposure of PMMA is defined. Remove the unexposed PMMA and then a metallic contact is deposited onto the top surface of nanowire. In the similar way, various groups [100, 111] are able to measure I-V, PL and electroluminescence of single nanowire.

Medical applications of nanotechnology covers the areas such as nanoparticle drug delivery and nanovaccinology. Nanoscale systems possess the potential to detect diseases and deliver treatment to the body. Nanostructures can efficiently sense and repair the damage parts of our body like naturally occurring biological nanostructures such as the white blood cells. Nano-biotechnology may be helpful in curing cancer by attacking directly the tumor, maintenance and improvement of human organs and biocompatible implants.

Now a day's pharmaceutical science are using nanoparticles to reduce toxicity and side effects of drugs and pharmacologist did not realize that carrier systems themselves may impose risks to the patient. Human body shows a significant variation in rendering medical treatments which is needed to be coordinated with the biological patterns. Research on drug delivery should be more specific to target an infection, reduce toxicity while maintaining therapeutic effects and biocompatible.

Nanotechnology can be used as a medicine to kill cancer cells and heal patients without any harm. Scientist are using CNTs as they can absorb near-infrared light waves, which are slightly longer than visible rays and pass harmlessly through the cells. Electrons produce in the nanotube become excited and release excess amount of energy in the form of heat. When nanotubes were placed inside cells and radiated by the laser beam, the cells were quickly destroyed by the heat. CdSe Nanoparticles glows when exposed to UV and while injecting in human body they directly seep into tumours. Surgeon can see the glowing tumour, and use it as a guide for more accurate tumour removal.

Researchers are using nanotechnology for building nanorobots and utilizing them for health related problems. Most exciting use of nanorobots is to detect damages and repairing our body at cellular level. Primary element used to build these nanorobots is carbon and other forms of carbon (diamond/fullerene composites) due to its inherent strength.

1.6 Toxicity of Nanomaterials

No doubt the nanotechnology is an emerging technology and it has vast applications in all walks of life. It has remarkable impact in medical sciences, space program, and military technology. The possible danger of nanotechnology lies in how these tiny particles interact with the environment, and more importantly, with the functioning of human body. Some experts say that the elements encountered at the nano level behave differently than their larger size particles. For example properties of graphite are well known and they hold a specific position in toxicology guidelines. Physicist Richard Smalley of Rice University discovered fullerenes (nanoparticles of carbon) are legally categorized as graphite, yet they behave in ways unlike graphite and make its classification a potentially dangerous.

Environmental toxicologist Eva Oberdörster (March 2004) conducted a test with Southern Methodist University in Texas found extensive brain damage to fish exposed to fullerenes for 48 h. Fish exhibit changes in their gene markers indicate towards its affected physiology. Fullerenes also killed water fleas which is an important link in the marine food chain. CBEN (Center for Biological and Environmental Nanotechnology) in 2002 indicated that nanoparticles accumulate in the body of laboratory animals and travel freely through soil and could easily be absorbed by earthworms. Lab. animals, soil and earthworm link up to form the food chain to human and hazardous to health which is biggest dangers of nanotechnology.

Size and shape of nanoparticles complicates the dangers of nanotechnology as their ability to interact with other living systems increases because they can easily cross the skin, lung, and crosses the blood/brain barriers. Nano compact devices could seriously damage to society that includes several varieties of remote assassination weapons that would be difficult to detect or avoid. Small-integrated computers, tiny weapons could be aimed at targets remote in time and space from the attacker will not only impair defence, but also reduce post-attack detection and accountability. Also the molecular manufacturing raises the possibility of horrifically effective weapons. For example the smallest insect (200 microns) creates a plausible size estimate for a nanotech-built antipersonnel weapon capable of seeking and injecting toxin to protect humans.

Thus the question rises here is what can be done? The answer lies in the fact that there must be some defined ethical guidelines for utilizing nanotechnology products. There should be close monitoring and coordination of all nanotechnology research for potential hazards and also an extensive testing for hazards of all new nanotechnology products should be undertaken.

1.7 Concluding Remarks

Last few years have seen rapid developments in nanomaterials research and some of the potential applications of nanomaterials have been realized. It is expected that the nanomaterials has potential to provide one of the key technologies in near future. A variety of nanomaterials developed so far, are the special type of materials having unique properties. The coverage of this field is massive, ranging from the use of nanoparticles in cosmetics to additives for solid rocket propulsion applications. The field has developed so rapidly that it is almost impossible to find even a section of any technological domain where the implications of nanomaterials have not been explored so far. Therefore, the applications of nanomaterials are enormous and with these applications, we may imagine significant changes in all spheres of our life in coming few years. Therefore, the nanotechnology is expected to bring a revolution to our society and coming few years will witness the revolution of nanotechnology.

With the advancement of nanotechnology, there are some serious issues of toxicity of nanomaterials, which can not be ignored. These areas require continuous attention of nanotechnologists. These issues have to be addressed satisfactorily for complete realization of applications of nanotechnology.

References

1. A.S. Edelstein, R.C. Cammarata, *Nanomaterials: Synthesis, Properties and Applications* (Institute of Physics Publishing, Bristol, 1998)
2. K.E. Geckeler, E. Rosenberg (eds.), *Functional Nanomaterials* (American Scientific Publishers, Valencia, 2006), p. 488
3. B. Bhushan, *Handbook of Nanotechnology* (Springer, Berlin, 2004)
4. M. Wilson, K. Kannangara, G. Smith, M. Simmons, B. Raguse, *Nanotechnology: Basic Science and Emerging Technologies* (CRC Press, Boca Raton, 2002)
5. R. Valiev, Materials science: nanomaterial advantage. Nature **419**(6910), 887–889 (2002)
6. W.G. Kreyling, M. Semmler-Behnke, Q. Chaudhry, A complementary definition of nanomaterial. Nano Today **5**(3), 165–168 (2010)
7. N.C. Seeman, DNA in a material world. Nature **421**, 427 (2003)
8. G. Taubes, Double helix does chemistry at a distance—but how? Science **275**, 1420 (1997)
9. A. Okamoto, K. Tanaka, I. Saito, Rational design of a DNA wire possessing an extremely high hole transport ability. J. Am. Chem. Soc. **125**, 5066 (2003)
10. J.R. Peralta-Videa, L. Zhao, M.L. Lopez-Moreno, G. de la Rosa, J. Hong, J.L. Gardea-Torresdey, Nanomaterials and the environment: a review for the biennium 2008–2010. J. Hazard. Mater. **186**(1), 1–15 (2011)
11. B.A. Magnuson, T.S. Jonaitis, J.W. Card, A brief review of the occurrence use, and safety of food related nanomaterials. J. Food Sci. **76**(6), R126–R133 (2011)
12. R.P. Adams, *Nanotechnology: understanding small system* (CRC Press, Taylor and Francis Group, Boca Raton, 2007)
13. M.S. Rajan, *Nano: The Next Revolution* (National Book Trust, New Delhi, 2005)
14. M.J. O'Connell, *Carbon Nanotubes: Properties and Applications* (CRS Taylor and Francis, Boca Raton, 2006)
15. M.A. Ratner, D. Ratner, *Nanotechnology: A Gentle Introduction to the Next Big Idea, Technology and Engineering* (Prentice Hall, Upper Saddle River, 2003)
16. Samori, Bruno. Plenty of Room for Biology at the Bottom. An Introduction to Bionanotechnology. By Ehud Gazit. 236–237 (2008)
17. J.L. de la Fuente, G. Mosquera, R. París, High performance HTPB-based energetic nanomaterial with CuO nanoparticles. J. Nanosci. Nanotechnol. **9**(12), 1–7 (2009). 685
18. B.K. Teo, X.H. Sun, Classification and representations of low-dimensional nanomaterials: terms and symbols. J. Cluster Sci. **18**(2), 346–357 (2007)
19. A.N. Guz, Y.Y. Rushchitskii, Nanomaterials: on the mechanics of nanomaterials. Int. Appl. Mech. **39**(11), 1271–1293 (2003)
20. T. Tervonen, I. Linkov, J.R. Figueira, J. Steevens, M. Chappell, M. Merad, Risk-based classification system of nanomaterials. J. Nanopart. Res. **11**(4), 757–766 (2009)
21. V.V. Pokropivny, V.V. Skorokhod, Classification of nanostructures by dimensionality and concept of surface forms engineering in nanomaterial science. Mater. Sci. Eng. C **27**(5), 990–993 (2007)
22. L.M. Liz-Marzan, P. Mulvaney, The assembly of coated nanocrystals. J. Phys. Chem. B **107**, 7312 (2003)
23. X.H. Sun, N.B. Wong, C.P. Li, S.T. Lee, T.K. Sham, Chainlike silicon nanowires: morphology, electronic structure and luminescence studies. J. Appl. Phys. **96**, 3447 (2004)

24. X.H. Sun, C.P. Li, N.B. Wong, C.S. Lee, S.T. Lee, B.K. Teo, Templating effect of hydrogen-passivated silicon nanowires in the production of hydrocarbon nanotubes and nanoonions via sonochemical reactions with common organic solvents under ambient conditions. J. Am. Chem. Soc. **124**, 14856 (2002)
25. A.I. Hochbaum, R. Fan, R.R. He, P.D. Yang, Controlled growth of Si nanowire arrays for device integration. Nano Lett. **5**, 457 (2005)
26. Z. Zhong, F.X. Chen, A.S. Subramanian, J.Y. Lin, J. Highfield, A. Gedanken, Assembly of Au colloids into linear and spherical aggregates and effect of ultrasound irradiation on structure. J. Mater. Chem. **6**, 489 (2006)
27. V. Svrcek, C. Pham-Huu, M.J. Ledoux, F. Le Norman, O. Ersen, S. Joulie, Filling of single silicon nanocrystals within multi-walled carbon nanotubes. Appl. Phys. Lett. **88**, 033112 (2006)
29. W.Z. Li, S.S. Xie, L.X. Qian, B.H. Chang, B.S. Zou, W.Y. Zhou, R.A. Zhao, G. Wang, Large-scale synthesis of aligned carbon nanotubes. Science **274**, 1701 (1996)
30. B.K. Teo, H.X. Sun, Silicon-based low-dimensional nanomaterials and nanodevices. Chem. Rev. **107**, 1454 (2007)
31. A.N. Khlobystov, K. Porfyrakis, M. Kanai, D.A. Britz, A. Ardavan, H. Shinohara, T.J.S. Dennis, G.A.D. Briggs, Molecular motion of endohedral fullerenes in single-walled carbon nanotubes. Angew. Chem. Int. Ed. **43**, 1386 (2004)
32. H.W. Kroto, J.R. Heath, S.C. O'Brien, R.F. Curl, R.E. Smalley, Reactivity of large carbon clusters: spheroidal carbon shells and their possible relevance to the formation and morphology of soot. Nature **318**, 162 (1985)
33. W. Kratschmer, L.D. Lamb, K. Fostiropoulos, D.R. Huffman, C60: a new form of carbon. Nature **347**, 354 (1990)
34. R.C. Haddon, A.F. Hebard, M.J. Rosseinsky, D.W. Murphy, S.J. Duclos, K.B. Lyons, B. Miller, J.M. Rosamilia, R.M. Fleming, A.R. Kortan, S.H. Glarum, A.V. Makhija, A. J. Muller, R.H. Eick, S.M. Zahurak, R. Tycko, G. Dabbagh, F.A. Thiel, Conducting films of C60 and C70 by alkali-metal doping. Nature **350**, 320 (1991)
35. M. Shahid Khan, *Figures Simulated* (Department of Physics, Jamia Millia Islamia, New Delhi, India) (2015)
36. S. Iijima, Helical microtubules of graphitic carbon. Nature **354**, 56 (1991)
37. S. Iijima, T. Ichihashi, Single-shell carbon nanotubes of 1-nm diameter. Nature **363**, 603 (1993)
38. D.S. Bethune, C.H. Kiang, M.S. de Vries, G. Gorman, R. Savoy, J. Vazquez, R. Beyers, Cobalt-catalysed growth of carbon nanotubes with single-atomic-layer walls. Nature **363**, 605 (1993)
39. S. Frank, P. Poncharal, Z.L. Wang, W.A. deHeer, Carbon nanotube quantum resistors. Science **280**, 1744 (1998)
40. M.K. Rai, S. Sarkar, Carbon nanotube as VLSI interconnect, in *Electronic Properties of Carbon Nanotubes*, ed. by J.M. Marulanda (Intech, Rijeka, Croatia, 2011)
41. A. Kumar, S. Parveen, S. Husain, J. Ali, M. Zulfequar, Harsh, M. Husain, Effect of oxygen plasma on field emission characteristics of single-wall carbon nanotubes grown by plasma enhanced chemical vapour deposition system. J. Appl. Phys. **115**, 084308 (2014)
42. A. Kumar, S. Husain, J. Ali, M. Husain, Harsh, M. Husain, Field emission study of carbon nanotubes forest and array grown on Si using Fe as catalyst deposited by electro-chemical method. J. Nanosci. Nanotech. **12**(3), 2829 (2012)
43. Z.H. Khan, M. Husain, Carbon nanotube and its possible applications. Indian J. Mat. Sci. Eng. **12**, 529–551 (CSIR, New Delhi)
44. Z.H. Khan, S. Khan, M. Husain, Variable range hopping in carbon nanotubes. Curr. Nanosci. **6**, 1–16 (2010)
45. Z.H. Khan, N. Salah, S.S. Habib, A. Azam, M.S. Al-Shahawi, Multi-walled carbon nanotubes film sensor for carbon mono-oxide gas. Curr. Nanosci. **8**, 274 (2012)

46. Z.H. Khan, N. Salah, S.S. Habib, M.S. Ansari, M.S. Al-Shahawi, Cobalt catalyzed multi-walled carbon nanotubes film sensor for carbon mono-oxide gas. Dig. J. Nanomater. Biostruct. **6**(4), 1947 (2011)
47. A.F. Hollemann, E. Wiberg, *Lehrbuch der Anorganischen Chemei* (Walter de Gruyter, Berlin, 1985), p. 701
48. Y. Lifshitz, DLC-present status. Diamond Relat. Mater. **3–5**, 388 (1996)
49. D.R. Mckenzie, Tetrahedral bonding in amorphous carbon. Rep. Prog. Phys. **59**, 1611 (1996)
50. H. Tsai, D.B. Bogi, Characterisation of diamond-like carbon films and their application as overcoats on thin-film media for magnetic recording. J. Vac. Sci. Technol. A **5**(6), 3287 (1987)
51. J.P. Hirvonen, J. Koskinen, R. Lappalainen, A. Anttila, Preparation and properties of high density hydrogen free hard carbon films with direction beam or arc discharge deposition mater. Sci. Forum **52–53**, 197 (1990)
52. Y. Liftshitz, S.R. Kasi, J.W. Rabalais, Carbon (sp3) film growth from mass selected ion beams: parametric investigations and subplantation model. Mater. Sci. Forum **52–53**, 237 (1990)
53. Y. Lifshitz, G.D. Lempert, E. Grossman, I. Avigal, C. UzanSaguy, R. Kalish, J. Kulik, D. Marton, J.W. Rabalais, The influence of substrate temperature during ion beam deposition on the diamond-like or graphitic nature of carbon films. Diam. Relat. Mater. **4**, 287 (1995)
54. C.A. Davis, A simple model for the formation of compressive stress in thin films by ion bombardment. Thin Solid Films **226**, 30 (1993)
55. J. Robertson, Deposition mechanisms for promoting sp3 bonding in diamond-like carbon. Diam. Relat. Mater. **2**, 984 (1993)
56. Y. Lifshitz, S.R. Kasi, J.W. Rabalais, Subplantation model for film growth from hyperthermal species: application to diamond. Phys. Rev. Lett. **62**, 1290 (1990)
57. B. Bhushan, Chemical, mechanical and tribological characterization of ultra-thin and hard amorphous carbon coatings as thin as 3.5 nm: recent developments. Diam. Relat. Mater. **8**, 1985 (1999)
58. A.K. Sikder, T. Sharda, D.S. Misra, P. Selvam, Chemical vapour deposition of diamond on stainless steel: the effect of Ni-diamond composite coated buffer layer. Diam. Relat. Mater. **7**, 1010 (1998)
59. M. Chhowalla, Y. Yin, G.A.J. Amaratunga, D.R. McKenzie, Th Fraurnheim, Highly tetrahedral amorphous carbon films with low stress. Appl. Phys. Lett. **69**, 2344 (1996)
60. C.B. Collins, F. Davanloo, T.J. Lee, D.R. Jander, J.H. You, H. Park, J.C. Pivin, The bonding of protective films of amorphic diamond to titanium. J. Appl. Phys. **71**, 3260 (1992)
61. A.K. Geim, Graphene: status and prospects. Science **324**(5934), 1530 (2009)
62. M.I. Katsnelson, K.S. Novoselov, A.K. Geim, Chiral tunnelling and the Klein paradox in graphene. Nat. Phys. **2**, 620 (2006)
63. J.C. Slonczewski, P.R. Weiss, Band structure of graphite. Phys. Rev. **109**, 272 (1958)
64. G.W. Semenoff, Condensed-matter simulation of a three-dimensional anomaly. Phys. Rev. Lett. **53**, 244 (1984)
65. C.N.R. Rao, U. Maitra, H.S.S. Ramakrishna Matte, Synthesis, characterization, and selected properties of graphene, in *Graphene: Synthesis, Properties, and Phenomena*, 1st edn, ed. by C.N.R. Rao, A.K. Sood (wiley, Chichester, 2013)
66. M.A. Rafiee, J. Rafiee, Z. Wang, H. Song, Z.Z. Yu, N. Koratkar, Enhanced mechanical properties of nanocomposites at low graphene content. ACA Nano **3**(12), 3884–3890 (2009)
67. M. Fuller, The axial ratio and lattice constants of zinc oxide. Science **70**, 196 (1929)
68. C. Bunn, The lattice-dimensions of zinc oxide. Proc. Phys. Soc. **47**, 835 (1935)
69. U. Ozgur, Ya.I. Alivov, C. Liu, A. Teke, M.A. Reshchikov, S. Dogan, V. Avrutin, S.-J. Cho, H. Morkoc, A comprehensive review of ZnO materials and devices. J. Appl. Phys. **98**, 41301 (2005)
70. J. Grabowska, K.K. Nanda, E. McGlynn, J.-P. Mosnier, M.O. Henry, A. Beaucamp, A. Meaney, Synthesis and photoluminescence of ZnO nanowires/nanorods. J. Mater. Sci. Mater. Electron. **16**, 397 (2005)

71. J. Grabowska, K.K. Nanda, E. McGlynn, J.-P. Mosnier, M.O. Henry, Control of ZnO nanorod array density by Zn supersaturation variation and effects on field emission. Surf. Coat. Technol. **200**, 1093 (2005)
72. C. Ronning, P.X. Gao, Y. Ding, Z.L. Wang, Manganese-doped ZnO nanobelts for spintronics. Appl. Phys. Lett. **84**, 782 (2004)
73. W.Z. Wang, B.Q. Zeng, J. Yang, B. Poudel, J.Y. Huang, M.J. Naughton, Z.F. Ren, Aligned ultralong ZnO nanobelts and their enhanced field emission. Adv. Mater. **18**, 3275 (2006)
74. Z.W. Pan, Z.R. Dai, Z.L. Wang, Nanobelts of semiconducting oxides. Science **291**, 1947–1949 (2001)
75. J. Grabowska, A. Meaney, K.K. Nanda, J.-P. Mosnier, M.O. Henry, J.R. Duclere, E. McGlynn, Surface excitonic emission and quenching effects in ZnO nanowire/nanowall systems: limiting effects on device potential. Phys. Rev. B **71**, 115439 (2005)
76. Y.J. Xing, Z.H. Xi, Z.Q. Xue, X.D. Zhang, J.H. Song, Optical properties of the ZnO nanotubes synthesized via vapor phase growth. Appl. Phys. Lett. **83**, 1689 (2003)
77. Y. Qiu, S. Yang, ZnO Nanotetrapods: controlled vapor-phase synthesis and application for humidity sensing. Adv. Func. Mater. **17**, 1345 (2007)
78. W.I. Park, D.H. Kim, S.-W. Jung, G.C. Yi, Fabrication and electrical characteristics of high-performance ZnO nanorod field-effect transistors. Appl. Phys. Lett. **80**, 4232 (2002)
79. X. Fan, M.L. Zhang, I. Shafiq, W.J. Zhang, C.S. Lee, S.T. Lee, ZnS/ZnO heterojunction nanoribbons. Adv. Mater. **21**, 2393 (2009)
80. M. Riaz, J. Song, O. Nur, Z.L. Wang, M. Willander, Study of the piezoelectric power generation of ZnO nanowire arrays grown by different methods. Adv. Funct. Mater. **XX**, 1–6 (2010). doi:10.1002/adfm.201001203
81. J.J. Wu, S.C. Liu, Catalyst-free growth and characterization of ZnO nanorods. Adv. Mater. **14**, 215 (2002)
82. X. Wang, J. Song, J. Liu, Z.L. Wang, Direct-current nanogenerator driven by ultrasonic waves. Science **316**, 102 (2007)
83. Z.L. Wang, J. Song, Piezoelectric nanogenerators based on zinc oxide nanowire array. Science **312**, 242 (2006)
84. C.J. Lee, T.J. Lee, S.C. Lyu, Y. Zhang, H. Ruth, H.J. Lee, Field emission from well-aligned zinc oxide nanowires grown at low temperature. Appl. Phys. Lett. **81**, 3648 (2002)
85. E. Comini, G. Faglia, G. Sberveglieri, Z.W. Pan, Z.L. Wang, Stable and highly sensitive gas sensors based on semiconducting oxide nanobelts. Appl. Phys. Lett. **81**, 1869 (2002)
86. M. Zhao, Z.L. Wang, S.X. Mao, Piezoelectric characterization on individual zinc oxide nanobelt under piezoresponse force microscope. Nano Lett. **4**, 587 (2004)
87. W. Hughes, Z.L. Wang, Nanobelts as nanocantilevers. Appl. Phys. Lett. **82**, 2886 (2003)
88. H.T. Wang, Hydrogen-selective sensing at room temperature with ZnO nanorods. Appl. Phys. Lett. **86**, 243503 (2005)
89. W. Lee, Catalyst-free growth of ZnO nanowires by metal-organic chemical vapour deposition (MOCVD) and thermal evaporation. Acta Mater. **52**, 3949 (2004)
90. S. Liang, H. Sheng, Y. Liu, Z. Huo, Y. Lu, H. Shen, ZnO Schottky ultraviolet photodetectors. J. Cryst. Growth **225**, 110 (2001)
91. Y.P. Liu, Y. Guo, J.Q. Li, M. Trunk, A.Y. Kuznetsov, J.B. Xu, Z.X. Mei, X.L. Du, Temperature dependence of surface plasmon mediated near band-edge emission from Ag/ZnO nanorods. J. Opt. **13**, 075003 (2011)
92. C.J. Lee, T.J. Lee, S.C. Lyu, Y. Zhang, H. Ruth, H.J. Lee, Field emission from well-aligned zinc oxide nanowires grown at low temperature. Appl. Phys. Lett. **81**, 3648 (2002)
93. M.S. Arnold, P. Avouris, Z.W. Pan, Z.L. Wang, Field-effect transistors based on single semiconducting oxide nanobelts. J. Phys. Chem. **107**, 659 (2003)
94. P. Hui, L. Jizhong, S. Han, F. Yuanping, P. Cheekok, L. Jianyi, Hydrogen storage of ZnO and Mg doped ZnO nanowires. Nanotechnology **17**, 2963 (2006)
95. Q. Wan, C.L. Lin, X.B. Yu, T.H. Wang, Room-temperature hydrogen storage characteristics of ZnO nanowires. Appl. Phys. Lett. **84**, 124 (2004)

96. M. Ahmad, J. Zhu, ZnO based advanced functional nanostructures: synthesis, properties and applications. J. Mater. Chem. **21**, 599 (2010)
97. H. Yan, R. He, J. Johnson, M. Law, R.J. Saykally, P. Yang, Dendritic nanowire ultraviolet laser array. J. Am. Chem. Soc. **25**, 4728 (2003)
98. J.C. Johnson, H. Yan, P. Yang, R.J. Saykalley, Optical cavity effects in ZnO nanowire lasers and waveguides. J. Phys. Chem. B **105**, 8816 (2001)
99. M.H. Huang, S. Mao, H. Feick, H. Yan, Y. Wu, H. Kind, E. Weber, R. Russo, P. Yang, Room-temperature ultraviolet nanowire nanolasers. Science **292**, 1897 (2001)
100. J. Bao, M.A. Zimmler, F. Capasso, X. Wang, Z.F. Ren, Broadband ZnO single-nanowire light-emitting diode. Nano Lett. **6**(8), 1719–1722 (2006)
101. I. Bedja, P.V. Kamat, X. Hua, A.G. Lappin, S. Hotchandani, Photosensitization of nanocrystalline ZnO films by Bis(2,2′-bipyridine)(2,2′-bipyridine-4,4′-dicarboxylic acid) ruthenium(II). Langmuir **13**, 2398 (1997)
102. K. Keis, C. Bauer, G. Boschloo, J. Photochem et al., Nanostructured ZnO electrodes for dye-sensitized solar cell applications. Photobiol. A **148**, 57 (2002)
103. K. Keis, E. Magnusson, H. Lindström, S.-E. Lindquist, A. Hagfeldt, A 5% efficient photoelectrochemical solar cell based on nano structured ZnO electrodes. Sol. Energy Mater. Sol. Cells **73**, 51 (2002)
104. K. Keis, Photoelectrochemical properties of nano-to microstructured ZnO electrodes. J. Electrochem. Soc. **148**, 149 (2001)
105. R. Katoh, A. Furube, Y. Tamaki, T. Yoshihara, M. Murai, K. Hara, S. Murata, H. Arakawa, M. Tachiya, Microscopic imaging of the efficiency of electron injection from excited sensitizer dye into nanocrystalline ZnO film. J. Photochem. Photobiol. A **166**, 69 (2004)
106. R. Katoh, A. Furube, T. Yoshihara, K. Hara, G. Fujihashi, S. Takano, S. Murata, H. Arakawa, M. Tachiya, Efficiencies of electron injection from excited N3 dye into nanocrystalline semiconductor (ZrO_2, TiO_2, ZnO, Nb_2O_5, SnO_2, In_2O_3) films. J. Phys. Chem. B **108**, 4818 (2004)
107. A. Furube, R. Katoh, K. Hara, S. Murata, H. Arakawa, M. Tachiya, Ultrafast stepwise electron injection from photoexcited Ru-complex into nanocrystalline ZnO film via intermediates at the surface. J. Phys. Chem. B **107**, 4162 (2003)
108. H. Horiuchi, R. Katoh, K. Hara, M. Yanagida, S. Murata, H. Arakawa, M. Tachiya, Electron injection efficiency from excited N3 into nanocrystalline ZnO films: effect of ($N3–Zn^{2+}$) aggregate formation. J. Phys. Chem. B **107**, 2570 (2003)
109. B. O'Regan, M.A. Grätzel, Low-cost, high-efficiency solar cell based on dye-sensitized colloidal TiO_2 films. Nature **353**, 737 (1991)
110. W. Lee, M.-C. Jeong, J.-M. Myoung, Fabrication and application potential of ZnO nanowires grown on GaAs (002) substrates by metal–organic chemical vapour deposition. Nanotechnology **15**, 254 (2004)
111. Y. Huang, X. Bai, Y. Zhang, In situ mechanical properties of individual ZnO nanowires and the mass measurement of nanoparticles. J. Phys. Condens. Mat. **18**, 179–184 (2006)

Chapter 2
Carbon Nanomaterials Based on Carbon Nanotubes (CNTs)

Ling Bing Kong, Weili Yan, Yizhong Huang, Wenxiu Que, Tianshu Zhang and Sean Li

Abstract A new group of nanomaterials, in free-standing form, such as fiber/yarns, paper/sheet and bulk, made of carbon nanotubes (CNTs), has emerged in recent years. These materials have shown special and unique mechanical, thermal and electrical properties, with potential applications in various aspects. This chapter is aimed to summarize the significant progress that has been made and the importance of potential applications of these carbon nanomaterials. CNTs, of single-walled (SW), double-walled (DW) and multiwalled (MW), processing strategies (spinning, filtration, deposition and SPS), morphologies, properties (mechanical, electrical and thermal) and potential applications, as well as their inter-relationships, will be presented and discussed in detail.

Keywords Carbon nanotubes · Carbon fibres · Carbon yarns · Spinning · Filtration · Twisting/rolling · Fabrication technologies

---

L.B. Kong (✉) · W. Yan · Y. Huang
School of Materials Science and Engineering, Nanyang Technological University, 50 Nanyang Avenue, Singapore 639798, Singapore
e-mail: elbkong@ntu.edu.sg

W. Que
Electronic Materials Research Laboratory, School of Electronic and Information Engineering, Xi'an Jiaotong University, Xi'an 710049, Shaanxi, People's Republic of China

T. Zhang
Anhui Target Advanced Ceramics Technology Co. Ltd., Hefei, Anhui, People's Republic of China

S. Li
School of Materials Science and Engineering, The University of New South Wales, Sydney, NSW 2052, Australia

M. Husain and Z.H. Khan (eds.), *Advances in Nanomaterials*, Advanced Structured Materials 79, DOI 10.1007/978-81-322-2668-0_2

2.1 Introduction

Since the work on carbon nanotubes (CNTs) reported by Iijima [1], there have been tremendous progresses in preparation and characterization of single-walled (SW), double-walled (DW) and multiwalled (MW) CNTs, which have various potential applications, due to their extraordinary electrical, thermal and mechanical properties [2–6]. For example, elastic modulus and strength of individual CNT are in the order of as high as 1.0 TPa and 50 GPa, respectively [7]. CNTs are also good electrical and thermal conductors. SWCNTs have electrical conductivities of 10^6 S cm^{-1}, while MWCNTs' conductivities are up to 10^4 S cm^{-1} [8, 9]. Room-temperature thermal conductivities of SWCNTs and MWCNTs are about 3500 and 3000 W m^{-1} K^{-1}, which are much higher than that of bulk-graphite conductivity that is about 2000 W m^{-1} K^{-1} [10]. To make use of these mechanical and thermal properties, various nanomaterials based on CNTs, including fibers/yarns, papers/sheets and bulks, have been fabricated.

Review articles on carbon nanomaterials started to appear in the open literature. For example, Behabtu et al. [11] summarized fabrication, characterization and applications of CNT fibers. A more detailed description on CNT nanomaterials, including fibers, films, thin sheets and arrays, was reported by Liu et al. [12], with an emphasis on mechanical properties, as well as summaries on certain applications. However, much less information is available on CNT buckypapers while buck CNT nanomaterials, such as those made by using spark plasma sintering (SPS) was not mentioned until now. A similar review on films, sheets and yarns was reported by Jiang et al. [13] more recently, but mainly focusing on their own group's works. In addition, bulk materials are not systematically described till now. Noting the significant progress that has been made and the importance of their potential applications, it is timely to provide a thorough overview to summarize the progress in this special group of nanomaterials. The nanomaterials presented in this review include those that are made by using CNTs as precursors and some of the as-grown CNTs materials that can be potentially characterized or used in free-standing forms.

2.2 Fibers and Yarns (1D)

2.2.1 Brief Introduction

Various methods have been developed to fabricate CNT fibers or yarns, which can be classified into six main groups [13–15]: (i) spinning from CNT solutions [16–19], (ii) spinning from vertically aligned CNT arrays on substrates [20–23], (iii) direct spinning from CNT aerogels synthesized in chemical vapor deposition (CVD) chambers [24–28], (iv) spinning from cotton-like precursors [29–31], (v) spinning with dielectrophoresis [32, 33] and (vi) rolling from CNT films/sheets

[34, 35]. Although there is difference between fiber and yarn, they will not be differentiated in this discussion. Details and characteristics of these technologies are discussed as follows.

2.3 Fabrication Technologies

2.3.1 Spinning from Solutions

Spinning of CNT fibers was initially adopted from coagulation spinning, in which polymer solutions containing high contents of CNTs are extruded into a liquid that can dissolve the solvent while remaining the polymers [36, 37]. CNTs are not soluble in in organic or aqueous solvents. They are prone to agglomerate due to the strong van der Walls forces of their side walls. To have solutions that are suitable for the spinning of CNT fibers, it is necessary to use surfactants [38] or super-acids [18, 39–41].

Figure 2.1 shows SEM images of representative fibers spun out from solution of arc-grown SWCNTs, which were dispersed in water using 1 % sodium dodecyl sulfate (SDS) as surfactant [38]. During the wet spinning, the CNT dispersion was injected into the center of a cylindrical flow of the coagulating mixture where it solidified into a fiber. In this study, the coagulating flow was a mixture of ethanol/glycerol or ethanol/glycol, with volume ratio of 1:1 and 1:3 respectively, so that the density of the coagulating flow matched that of the nanotube dispersion. It was found that continuous fibers could be obtained by using the alcohol mixtures, while aqueous aluminum nitrate solution as coagulation agent led only rod-shaped segments. The wet fibers were swollen and flexible, whereas the dried ones were brittle and rigid, with diameters of 20–30 μm [38]. The dried CNT fibers did not swell when they immersed into water. As shown in Fig. 2.1, the CNT fibers exhibited bundle structures. Electrical characterization indicated that the fibers possessed semiconductor behaviors, whose resistivity was sensitive to adsorbed molecules and atmospheres.

Another example was to disperse MWCNTs in ethylene glycol to form liquid crystalline dispersions, which were extruded into a diethyl ether bath to form CNT fibers, as shown in Fig. 2.2 [42]. The dispersions contained 1–3 wt% CNTs with pure nematic phase, as shown by the cross-polar optical microscopy in Fig. 2.2b, in which the domains of the ordered nanotubes are clearly demonstrated. The extrusion of the dispersions into fibers could be at rates of 0.03–0.3 mL min^{-1} through a needle into diethyl ether. When the dispersions were passed through the needle, the domains of the nematic dispersions would have shear orientation. As the wet fibers were injected into the ether bath, the ethylene glycol rapidly diffused out of CNT fibers into the ether, while the ether back-diffused into the fibers. As the ether-swollen fibers were collected from the ether bath, the ether evaporated rapidly, so that continuous fibers could be readily obtained, as shown in Fig. 2.2c.

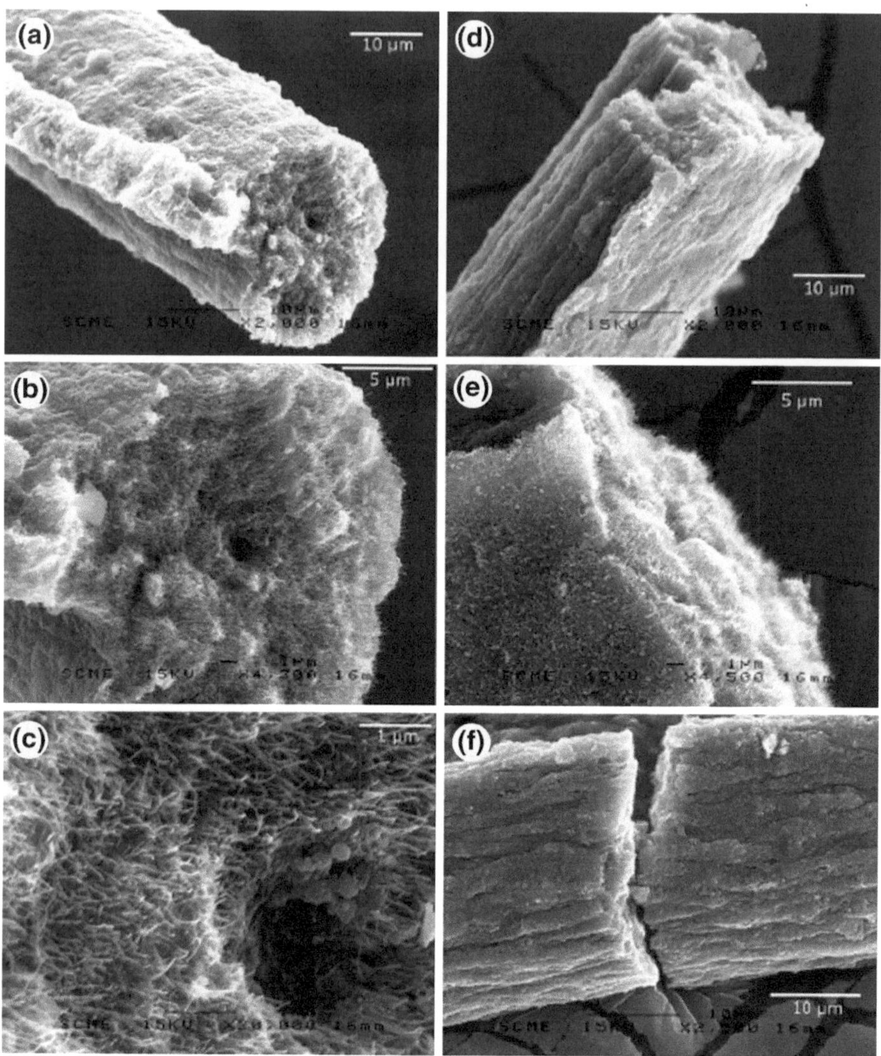

Fig. 2.1 SEM images of the CNT fibers: **a–c** side views, **d** view after drying on aflat substrate, **e** detailed side view showing individual CNTs and **f** view of a broken fiber. Reproduced with permission from [38], Copyright © 2005, Elsevier

The residual ethylene glycol in the fibers could be removed after annealing at 280 °C, due to its decomposition at 220—260 °C [42].

The MWCNT fibers were highly birefringent with polarization along the axis of the fibers (Fig. 2.3a). These images indicated that both the MWCNTs and N-MWCNTs were highly aligned along the fiber axis in almost one single domain and had no obvious defects or disclinations. The fibers had diameters of 10–80 mm, which could be controlled through the processing parameters (Fig. 2.3b). They were

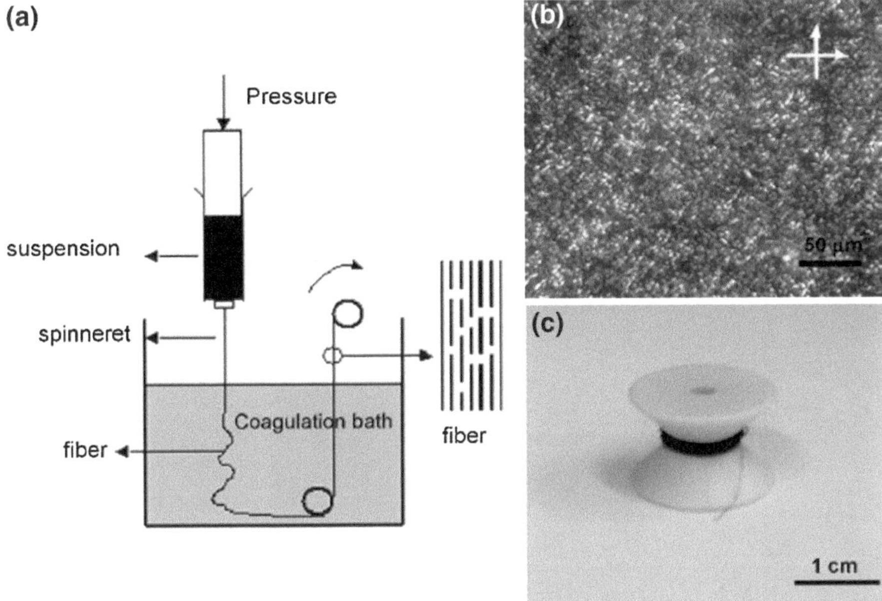

Fig. 2.2 **a** Schematic of the experimental set-up used to the spin CNT fibers. **b** Optical image of the 3 wt% MWCNT suspension before extruding under crossed polars. **c** A 3-m-long MWCNT fiber collected on a small winder. Reproduced with permission from [42], Copyright © 2008, John Wiley & Sons

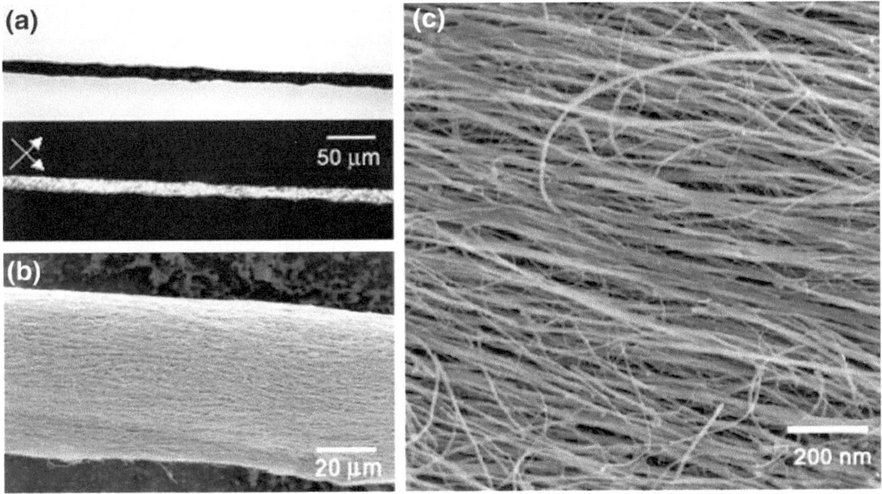

Fig. 2.3 **a** Optical images of a MWCNT fiber with and without crossed polars, where the fiber showed strong nematic-like birefringent under crossed polars. **b** SEM image of the MWCNT fiber spun from a 3 wt% suspension in glycol and coagulated in ether. **c** High-magnification surface SEM image of the fiber individual nanotubes well aligned along the fiber long axis. Reproduced with permission from [42], Copyright © 2008, John Wiley & Sons

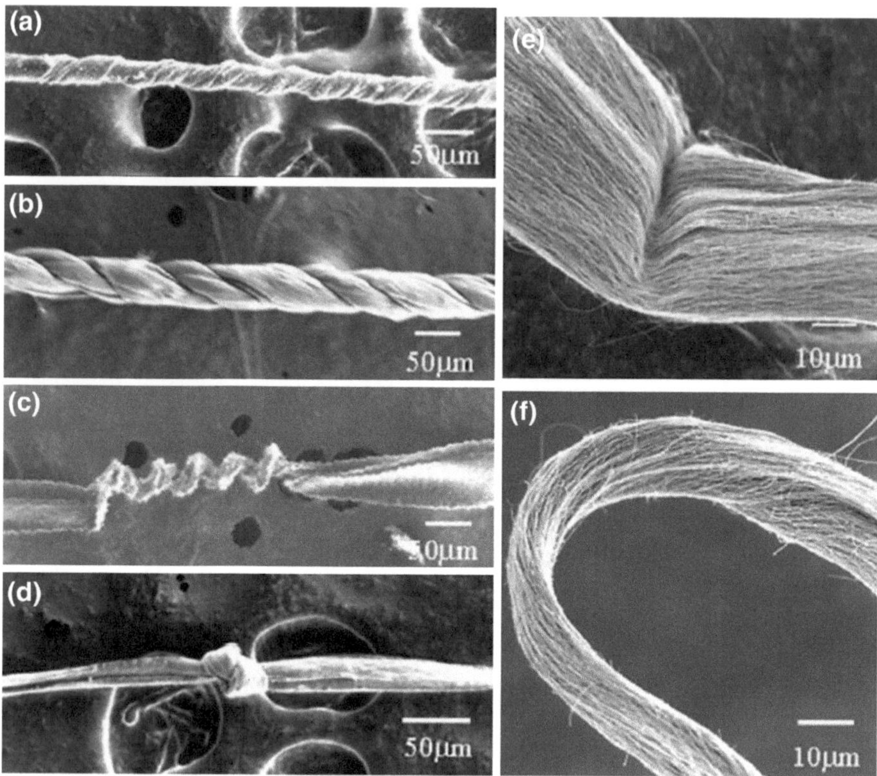

Fig. 2.4 SEM images of the MWCNT fibers: **a** single twist, **b** double twist, **c** spiral deformation due to overtwist, **d** tying knot, **e** deformation due to cutting and **f** bending. Reproduced with permission from [42], Copyright © 2008, John Wiley & Sons

free of contamination with degree of alignment (Fig. 2.3c). The presence of collapsed ribbon structure with internal voids was the microstructural characteristics of the MWCNT fibers, which were closely related to the processing.

The spun fibers had very high flexible and thus could be manipulated feasibly, as shown in Fig. 2.3 [42]. The obtained fibers were flexible and could be readily manipulated. They could be easily twisted and knotted, which makes them suitable for weaving and knitting into various technical textile fibers, as shown in Fig. 2.4a–d. The cut fibers showed flatten cross section, while no permanent change was observed after bending, as illustrated in Fig. 2.4e, f. The MWCNT fibers had a Young's modulus of 69 ± 41 GPa and a yield strain of 0.3 %. The high stiffness of the fibers was ascribed to the high degree of orientation of the CNTs. Both local deformation and pullout of the nanotubes were observed at the fractured surfaces of the fibers. Room temperature conductivity along length axis of the MWCNT fibers was as high as 8.0×10^3 S m^{-1}, but the conductivity was anisotropic, due to the high alignment of the CNT.

Although CNTs can be dispersed with the aid of surfactants, the concentration is usually about 1 vol%, which is still undesirable, because large volume of surfactants should be used during the processing. In contrast, by using superacids, such as sulfuric acid with excessive SO_3, chlorosulfonic acid and triflic acid, the dispersion concentration of CNTs can be up to 10 wt%, which is attributed to the protonation of the CNTs [18, 19, 40, 41]. In superacids, the CNTs are stabilized and thus prevented from aggregation due to the formation of the electrostatic double layers of positive protons and negative counterions [43].

A systematic study on SWCNTs dispersed in sulphuric and chlorosulphonic acid at weight concentrations of up to 0.5 wt%, for fiber spinning, was reported by Davis et al. [40]. Figure 2.5 shows phase diagram of the SWCNTs in the superacids, together with cross-polarized light micrographs, with phases including isotropic (I), liquid-crystalline (LC), crystal solvate (CS) and solid (S). Experimental results (black symbols) were in good agreement with theoretical predictions (red symbols). Circles designate Similar agreement was observed between the experiment (filled circles) and model (open circles) isotropic concentrations. Initial system concentration before phase separation was also studied (black and red diamonds). Experimental phase boundaries, such as the LC/LC + S vertical boundary, LC + S/CS + S horizontal boundary and I + LC/I + CS diagonal boundary, could be identified by connecting the experimental data points (black dotted lines). There was also an LC + CS regime in the phase diagram (shaded region). The concentration ϕ_i, of the isotropic phase in equilibrium with the aligned phase increased with increasing fractional charge (or solvent quality), for both experimental data (filled circles) and theoretical predictions (open circles). The critical concentration ϕ_n (black squares) was derived from the results of light microscopy, dynamic rheometry, steady shear rheometry and differential scanning calorimetry [40]. If the solvent had high acidities, ϕ_n showed a weak dependence on acidity, which was different from ϕ_i. Morphology of the liquid crystals was controlled by the acidity level. The higher the acidity, the larger the domains and the fewer the defects would be present and thus the higher the ordering of the macroscopic materials.

The phase diagram also provided insight into the solvent removal process, i.e., coagulation, which determines micro- and macrostructure of the fibers [40]. As shown in Fig. 2.6a, the fibers spun from the dispersion of SWCNTs in 102–123 % H_2SO_4 and coagulated with water had a hierarchical microstructure, which consisted of large bundles with micrometer thickness that contained smaller ones of tens to hundreds of nanometers of nanotubes. Smooth and dense film with coalesced bundles could be obtained if the dope was sandwiched between two glass slides after the acid was evaporated, as shown in Fig. 2.6b. This wet-spinning technique is quite productive. Tens of meters of continuous fibers could be produced in several minutes.

It can be used to spin CNT fibers with special shapes, e.g., coiled structure shown in Fig. 2.6c. It was found that, when spinning fibers from 8.5 vol% SWCNT in pure chlorosulphonic acid, if the fibers were spun into a stagnant viscous coagulation bath of 96 % sulphuric acid, they would be uniformly composed of fine fibrils, with fewer small-scale defects than the fibers spun from sulphuric acid,

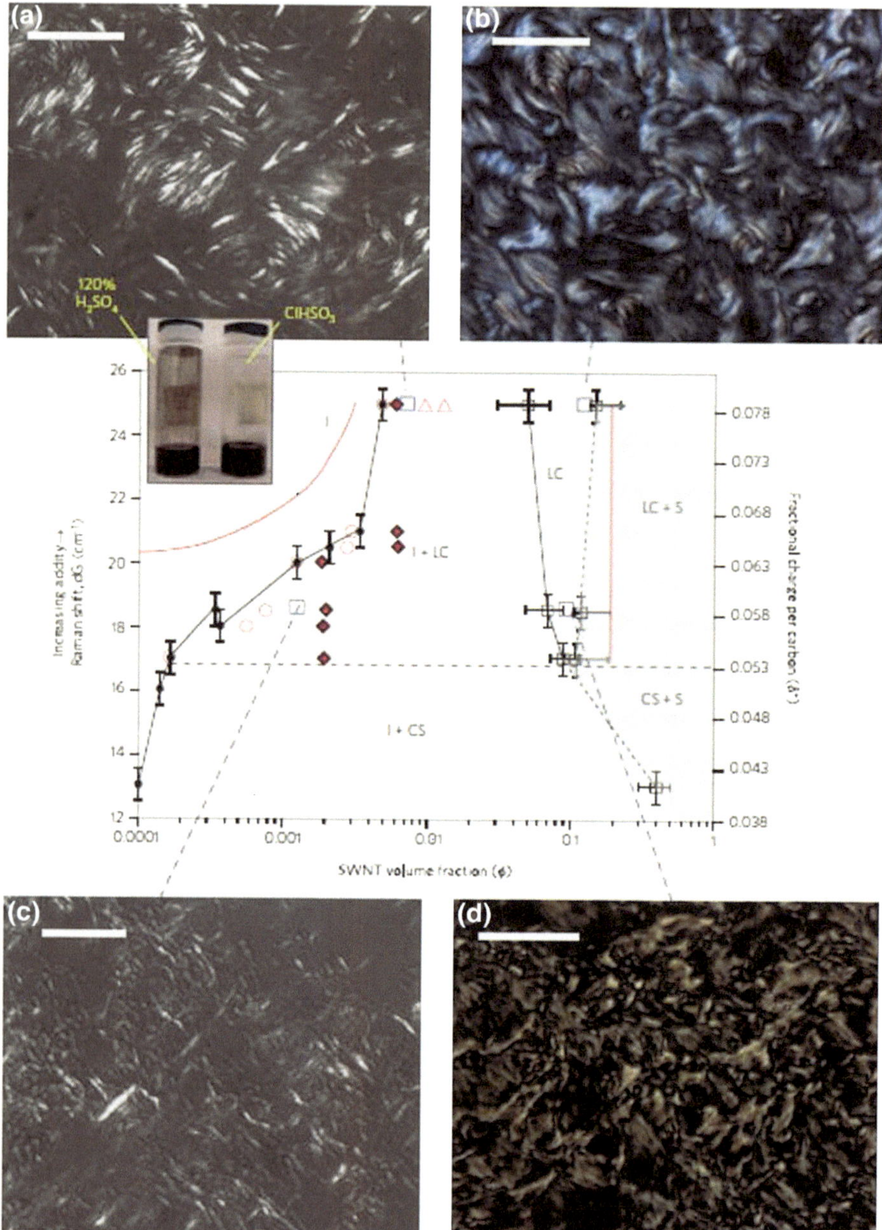

Fig. 2.5 Phase diagram of the SWCNTs in superacids and cross-polarized light micrographs: morphology just inside the biphasic region at 1.21 vol% in $ClSO_3H$ (**a**) and 0.132 vol% in 120 % H_2SO_4 (**c**), and in the liquid-crystalline phase at 12.1 vol% in $ClSO_3H$ (**b**) and 10.6 vol% in 120 % H_2SO_4 (**d**). Scale bars are 50 μm for (**a**) and 20 μm for (**b**–**d**). Reproduced with permission from [40], Copyright © 2009, Nature Publishing Group

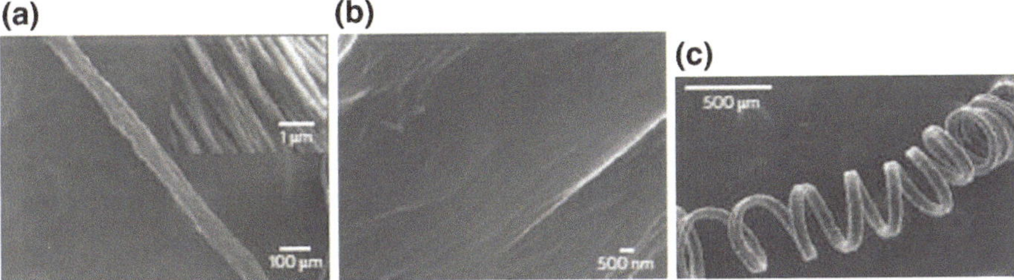

Fig. 2.6 Dispersions of the SWCNTs (10.8 vol%) in sulphuric acid processed using different coagulation conditions: **a** straight fiber coagulated in water, showing hierarchical microstructure of bundles (detail shown in inset), **b** film slowly evaporated in anhydrous air, showing a smooth microstructure and **c** coiled and pleated fiber coagulated in ether. Reproduced with permission from [40], Copyright © 2009, Nature Publishing Group

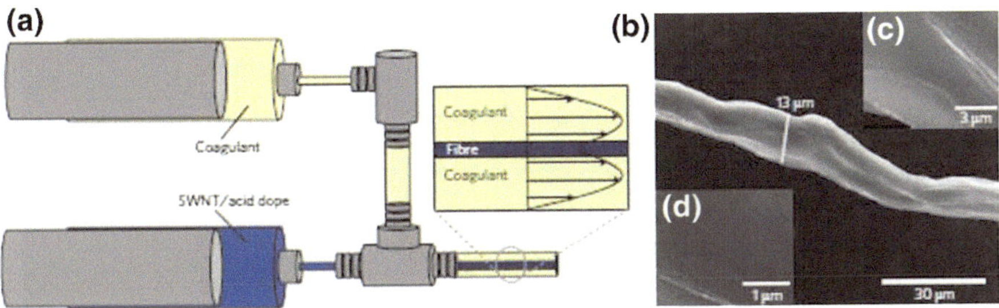

Fig. 2.7 Fibers spun from 8.5 vol% SWCNTs in chlorosulphonic acid and coagulated in 96 % aqueous sulphuric acid. The SWCNT/acid dope was extruded into a coagulant flowing faster than the dope (**a**), which provided tension and draw the coagulating fiber, yielding thin fibers with a smooth surface (**b**). The large liquid-crystalline domains in the fluid dope led to uniform microstructure of the aligned fibrils (**c**, **d**). Reproduced with permission from [40], Copyright © 2009, Nature Publishing Group

because of the more regular microstructure of the liquid-crystalline phase. Viscous forces at the coagulant–proto-fiber interface caused surface buckling-like irregularities. This could be avoided by using a low-viscosity coagulant, e.g., chloroform or dichloromethane or by co-flowing the coagulant with the coagulating proto-fiber, so that fibers with uniform microstructure and aligned macrostructure could be achieved, as shown in Fig. 2.7 [40].

2.3.2 *Spinning from Arrays*

Spinning from CNT arrays, first reported by Jiang et al. [22] in 2002, has been widely and extensively used to produce CNT fibers [21, 44–53]. The CNT arrays for spinning of fibers are previously grown on substrates, such as silicon wafers,

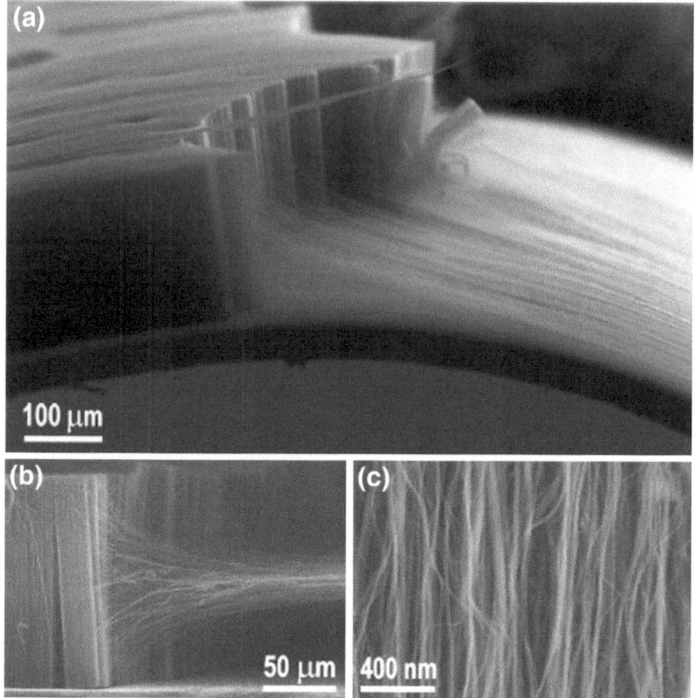

Fig. 2.8 SEM images of the drawable CNT forests grown on stainless steel sheets: **a** low magnification SEM image of a CNT ribbon being drawn from a forest, **b** cross-sectional view of the CNT forest's edge showing the CNT self-assembly to form a sheet and **c** high magnification image showing the alignment of the CNTs in the forest side-wall. Reproduced with permission from [44], Copyright © 2010, Elsevier

ceramics and stainless steel sheets, by using various chemical vapor deposition (CVD) methods, including traditional CVD, plasma enhanced CVD (PECVD), floating catalyst CVD techniques, and so on. CVD methods usually produce MWCNTs. Feasibility of this technology is demonstrated by the following examples.

Figure 2.8 shows representative SEM images of MWCNT arrays grown on stainless steel sheet that can be spun into CNT fibers [44]. The stainless steel sheets are commercial products. Before the growth of the CNT arrays, the substrates were cleaned and deposited with buffer layers, either Si or SiO_x. CNT arrays were grown by using PECVD, with C_2H_4 as the carbon source. Single-ply fibers spun from the MWCNT arrays had electrical conductivity and tensile strength of 350 S cm^{-1} and 303 MPa, respectively. It was found that surface roughness of the stainless steel sheet had no significant effect on spinnability of the MWCNT arrays, while the catalyst layer could be reused.

Another example is the use of silicon substrate, which was coated with a thin buffer layer of Al_2O_3 and then 1 nm thick Fe catalytic layer [45]. The MWCNT

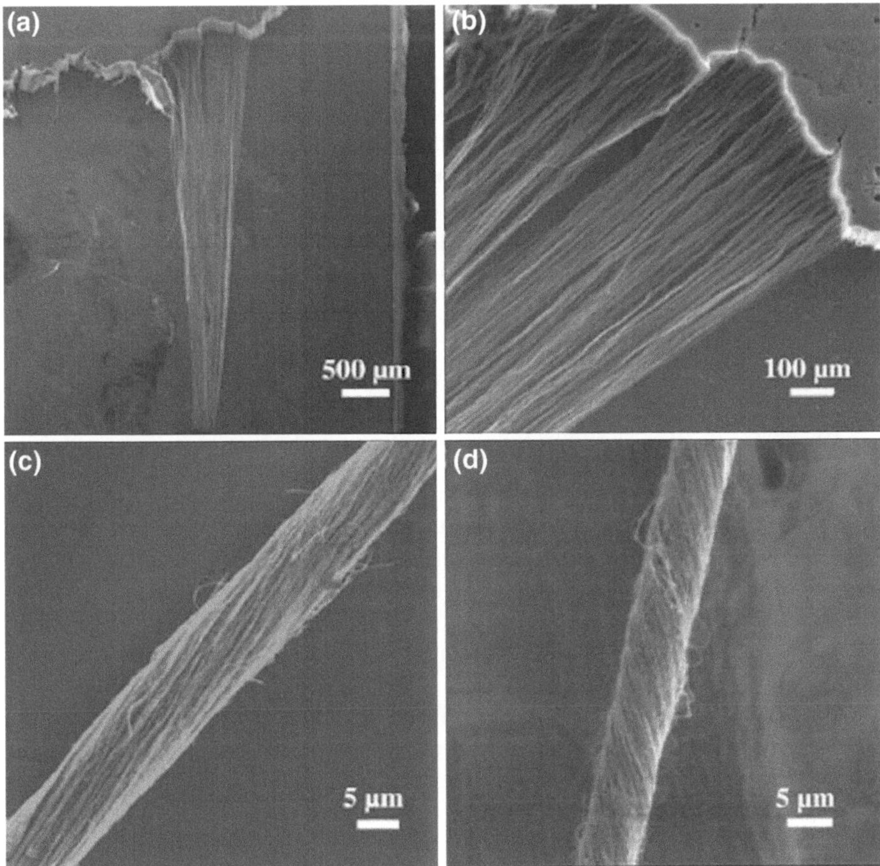

Fig. 2.9 SEM images of the CNT ribbons and fibers: **a**, **b** CNT ribbons initiated from the CNT array for spinning, **c** as-spun CNT fiber and **d** the fiber after post-spin twisting. Reproduced with permission from [45], Copyright © 2007, John Wiley and Sons

arrays grown by using a CVD method had a thickness of as large as 0.65 mm, with average diameter of 10 nm. As shown in Fig. 2.9a, b, CNT ribbons with pretty good continuity could be readily pulled out from the arrays, showing their high spinnability. Figure 2.9c, d shows SEM images of an as spun and a twisted CNT fibers. The as spun fiber had a loose microstructure, with relatively small twist angle, defined as the angle between the longitudinal direction of individual CNTs and the axis of the CNT fiber. When the as spun fiber was twisted, the twister angle was increased from about 10°–21°, while thickness was decreased from 10 to 7 μm. The twisted CNT fiber exhibited greatly improved mechanical properties (strength of 1.91 GPa and Young's modulus of 330 GPa) and electrical conductivity (from 1.7 to 4.1×10^4 S m^{-1}).

In summary, spinning from well aligned CNT arrays is a versatile technology to fabricate CNT fibers. The performances of the CNT fibers are influenced by both the properties of the CNTs and post treatment techniques, including CNT structures (such as tube length, tube diameter and wall thickness and tube waviness), diameter of the CNT fibers, degree of twisting, liquid densification treatment and polymer infiltration, as detailed in Ref. [14]. Noting the fact that significant progress has been made in fabrication of CNT fibers and the CNTs synthesized with CVD methods have a wide range of properties and characteristics, it is necessary to establish theoretical understanding, which can be used to govern and guide the further development and improvement of this technology. In this light, several theoretical models have been developed to describe the assembly of CNTs into fibers, two of which are discussed as representative examples [54, 55].

Figure 2.10 shows two-dimensional schematic drawing of the model proposed by Kuznetsov et al. [54]. In this model, the original forest was considered to consist of vertically oriented forest trees, known as large bundles, which were interconnected by smaller bundles or individual nanotubes, called connects. If the direction of the CNTs in the forest is defined as the "z" axis, the draw direction is "x" axis and "y" is the axis perpendicular to both "x" and "z". Before the first bundle is entirely detached from the forest, it pulls out its adjacent one that is in the same z-x plane. Once the first bundle is peeled off from the bottom of the CNT forest, the process will occur in the opposite direction. Therefore, the first bundle (forest tree) is peeled off (unzipped) from the bottom of the forest, while the next bundle will be pulled off from the top, which is repeated so that continuous fibers could be obtained. If there are more interconnections between adjacent bundles, all these interconnections unzip in the same way and will be concentrated at the bottom or on the top alternatively. When the number of interconnection reaches a critical value, they will be strong enough to pull out the next bundle, as shown in Fig. 2.11 [54].

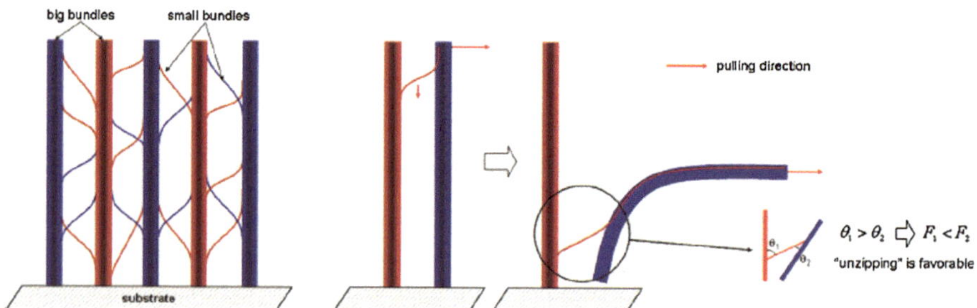

Fig. 2.10 *Left* Schematic draw of a nanotube forest consisting of vertically oriented large bundles (forest trees) and small interconnecting bundles or individual nanotubes (connects). *Right* The unzipping zipping process causes the connects to move along the lengths of the forest tree, so as to concentrate these connects at ends of the forest tree. In the detail, the angles of peeling at adjacent bundles, with the respective forces F1 and F2 indicated. *Red arrow* represents the external force used to draw the bundles. Reproduced with permission from [54], Copyright © 2011, American Chemical Society

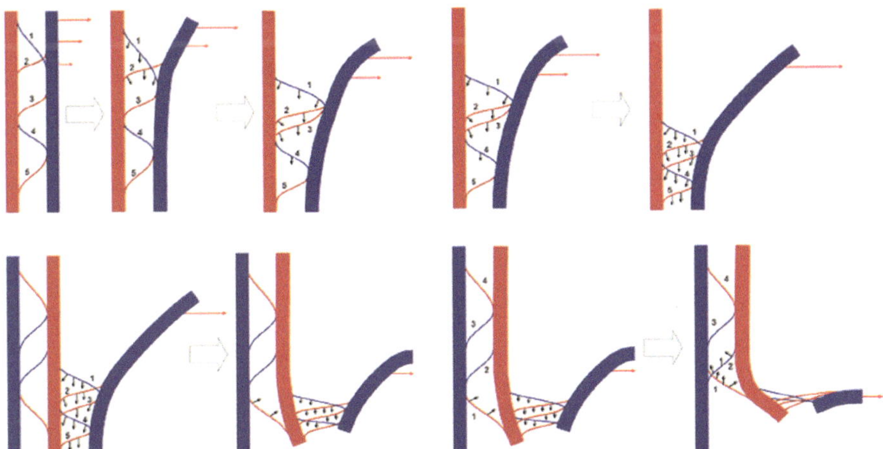

Fig. 2.11 Scheme of the process of pulling out a forest tree from the forest. *Blue* and *red bars* represent forest trees or bundles of the CNTs, whereas the separation distance between them is increased for clarity. *Red arrows* show the external force applied to the first bundle. *Small black arrows* show the direction of the net movement of the interconnections. Reproduced with permission from [54], Copyright © 2011, American Chemical Society

In this model, the number of interconnections between the forest trees is the critical parameter. For both the top and bottom of the bundles, there are limit interconnection densities, corresponding to limiting minimal and maximal heights of the CNT arrays. If the interconnection density is too low, as shown in Fig. 2.12a, the external pulling force divided by the number of the interconnections will be too large, so that every interconnection is entirely off the next bundle and thus pulling (drawing) process will be interrupted. However, if the interconnection density is high, as shown in Fig. 2.12b, the adjacent forest tree will be pulled out from somewhere near the middle, instead of bottom or top, so that the pulling process will also be interrupted. Because the CNT bundles in the forest trees are very stiff, it is very difficult to stretch them. Therefore, the second forest tree cannot be continuously bent in the x direction, it will keep its vertical parts at the same distance from the third forest tree without conserving its total length. As a result, the second forest tree will completely detach from the forest or if the interconnections between the second and the third forest tree are very strong, the third forest tree will be entirely pulled out. In these two cases, fibers cannot be formed. In real evaluation, three-dimensional structure of the interconnections was considered, by including the z-y plane also [54].

With this model, the minimum and maximum numbers of interconnections have been estimated, which can be used to explain (i) the presence of the lateral interconnections, (ii) unzipping process and (iii) critical number of interconnections. The conclusions have been supported by experimental observations, as shown in Fig. 2.13 [54].

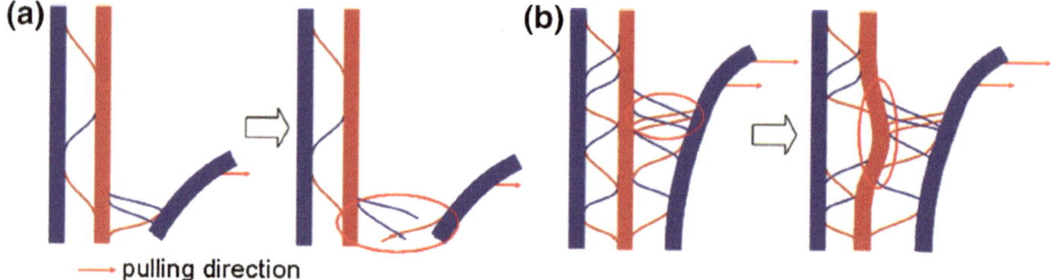

Fig. 2.12 Predicted effects of low (**a**) and high (**b**) interconnection density on the pulling-out process. Reproduced with permission from [54], Copyright © 2011, American Chemical Society

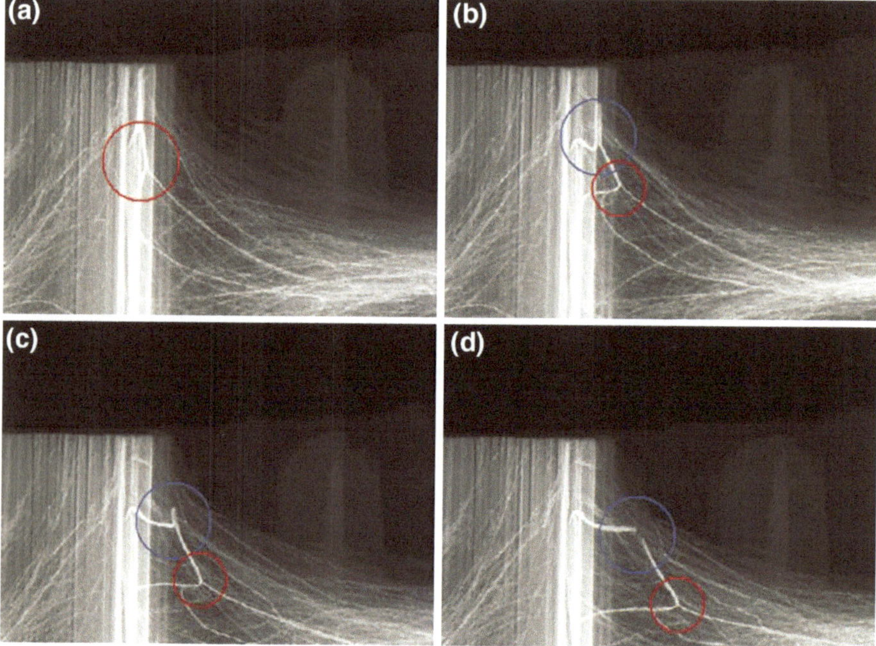

Fig. 2.13 Consecutive SEM images of the side view of the CNT forest during the pulling-out process. *Red circles* show the sequences of points where a pulled-out bundle started pulling the next bundle. *Blue circles* show the rupture of two adjacent bundles. Reproduced with permission from [54], Copyright © 2011, American Chemical Society

In a separate study, Zhu et al. [55] proposed a self-entanglement model to explain the spinning of CNT fibers from CNT arrays, according an in situ electron microscopy observations. It was found that the formation of entangled structures at the ends of the CNT bundles during the pulling was critical to ensure a continuous fiber drawing process. The entangled structures were formed when the drawing

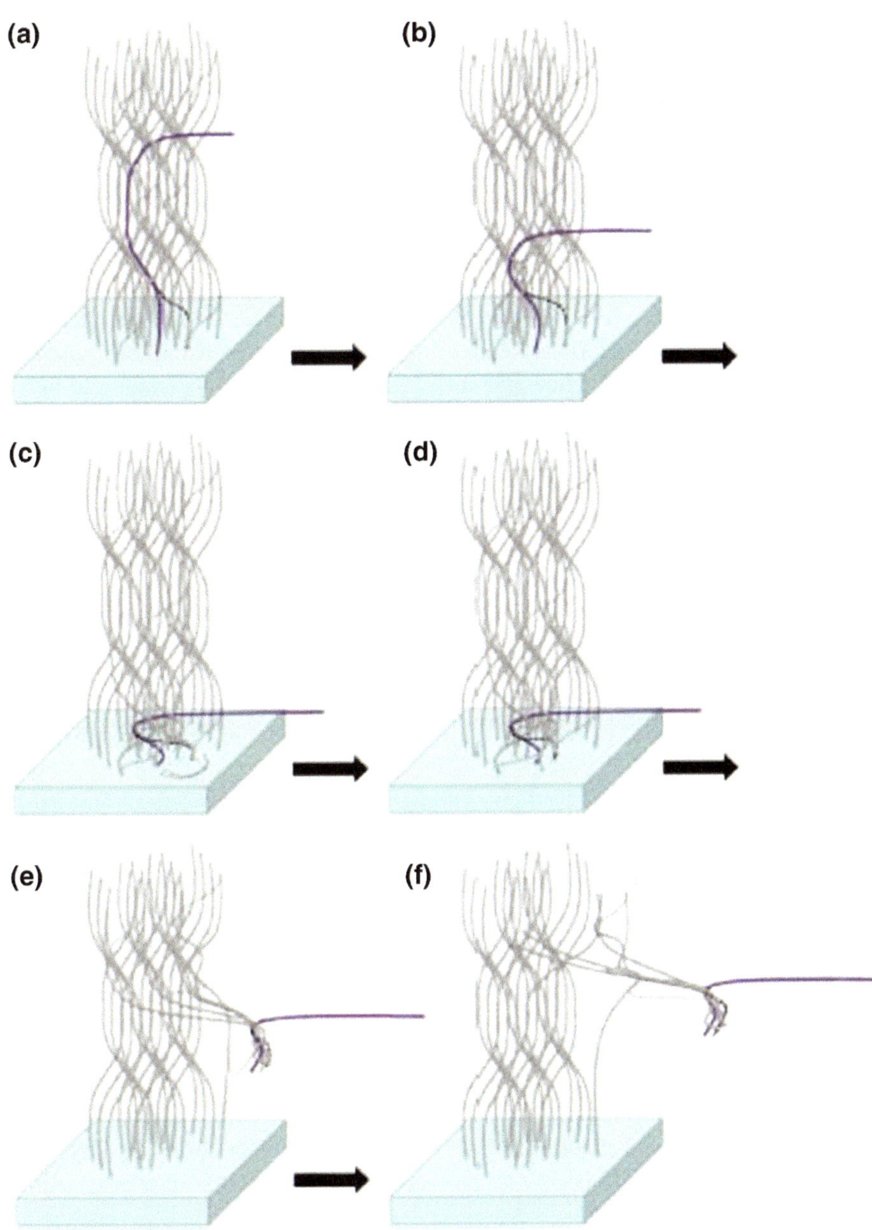

process approached the bottom or top ends of the CNTs due to the self-entanglement effect, as shown in Fig. 2.14 [55].

When a bundle standing at the front of the array is pulled out of from a CNT array, it detaches from other bundles, as shown in Fig. 2.14a. Because the CNT

◀ **Fig. 2.14** Self-entanglement mechanism for continuous pulling of CNT yarns. The CNT array consists of crossing over (*solid gray lines*) and branched (*dashed gray lines*) CNT bundles: **a** pulling out of the bundle (*purple*) and then detaching from other bundles, **b** cleaving of the front branched bundles and shifting of the crossing over structure, **c** occurrence of self-entanglement, **d** entwining of the twisted bundles nearby together with the branched bundles to form an entangled structure, **e** pulling out of more bundles due to the entangled structures and the connection between the extracted bundles and becoming firmer with further pulling and **f** a similar entangled structure formed on the top surface region. Reproduced with permission from [55], Copyright © 2011, Elsevier

bundles tend to cross over whenever they are in contact with, this bundle is likely coming from somewhere among other bundles. Pulling out this bundle could cause squeezing of the front branched and crossing over bundles, thus leading to twist of these bundles, which will be cleaved after further pulling that pushes crossing over structure towards the end regions, either top or bottom surface regions of the array, as shown in Fig. 2.14b. Once the pulling process approaches the ends of the bundles, self-entanglement occurs (Fig. 2.14c). Due to the weak bonding between CNTs and the substrate, entangled structure can be formed at the end region, as shown in Fig. 2.14d. The formation of the entangled structures ensures the pulling out of more bundles with firmer connection between the extracted CNT bundles (Fig. 2.14e). This step is repeated during the continuous spinning of the CNT fibers, as seen in Fig. 2.14f. The formation of the entangled structures well explains the fact that CNT yarns usually become slightly wider during pulling, because more and more branched are CNTs involve. The model was confirmed by experiments [55].

2.3.3 *Spinning from Aerog*

An alternative way to fabricate CNT fibers is the direct formation from CNTs synthesized in CVD chambers. In 2002, Zhu et al. [26] used a floating catalyst CVD method to directly synthesize long strands of ordered SWCNTs by using a vertical furnace presented in Ref. [56], as shown in Fig. 2.15. With hydrogen as carrier gas, n-hexane solution in ferrocene and thiophene was used as carbon source. The obtained SWCNT strands were up to 20 cm in length with a diameter of about 0.3 mm, which showed metallic behavior with resistivity of $5\text{–}7 \times 10^{-6}\ \Omega$ m over the temperature range of 90–300 K. However, this is not a continuous spinning of CNT fibers.

A instantaneous continuous spinning method to spin CNT fibers from aerogel of CNTs synthesized in CVD chamber was reported by Li et al. [24]. In this direct spinning process, the precursor material was mixed with hydrogen and injected into the hot zone of the CVD chamber, where aerogels of CNTs were produced. The precursor material is typically a liquid carbon source, such as ethanol, together with the presence of ferrocene as the source of catalysts, because iron nanoparticles are

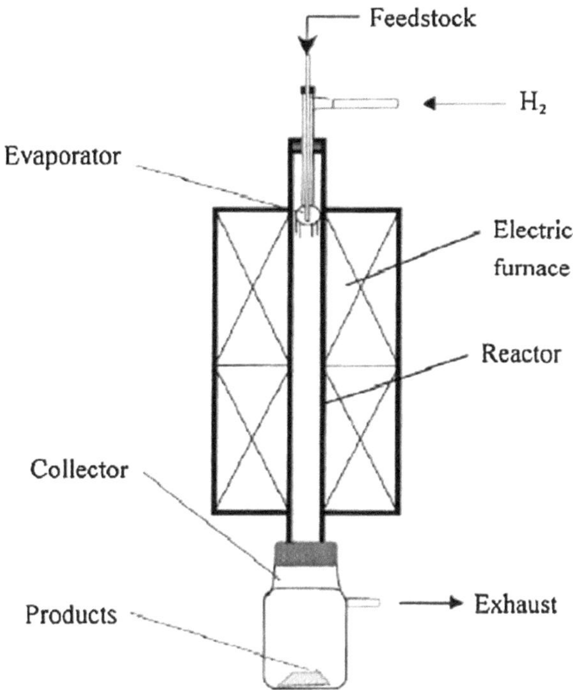

Fig. 2.15 Schematic of the floating catalyst CVD method to grow CNTs. Reproduced with permission from [56], Copyright © 2000, Elsevier

formed that act as nucleation sites for the growth of nanotubes. The CNT aerogel can be captured and spun continuously out of the hot zone as fibers or films. CNT fibers, which were spun directly and continuously from the aerogels, have shown high strength and high stiffness [25, 28, 57, 58].

By using the mixture of acetone and ethanol as the carbon source, Zhong et al. [59] produced continuous CNT fibers with a multilayered structure, by using a facility shown in Fig. 2.15a. In this case, the CNTs self-assembled into a multi-layered CNT sock in the gas flow, which could be observed as shown in Fig. 2.15b. The assembly of the CNTs was associated with the interaction of the gas molecules in the chamber, which drove the CNTs towards the outer circumference of the gas flow. The formation of the multilayer structure was attributed to the fact that higher concentration of CNTs was obtained in the gas flow, with the CNT yield (240 mg h^{-1}) from the mixed carbon source to be double that (110 mg h^{-1}) from the single source of ethanol.

The layered CNT sock was densified due to the presence of water once it was drawn out of the reaction chamber, when a water tank was placed at the end of the CVD chamber, as shown in Fig. 2.16a [59]. The water level in the tank was sufficiently high to seal the reaction chamber, so that the CNT assembly could be drawn out continuously from the chamber in a safe and controlled manner. Once the CNT sock touched the surface of the water, it shrunk into fiber, as shown in Fig. 2.16c. The fiber was directed to be around a rotator in the water tank, which

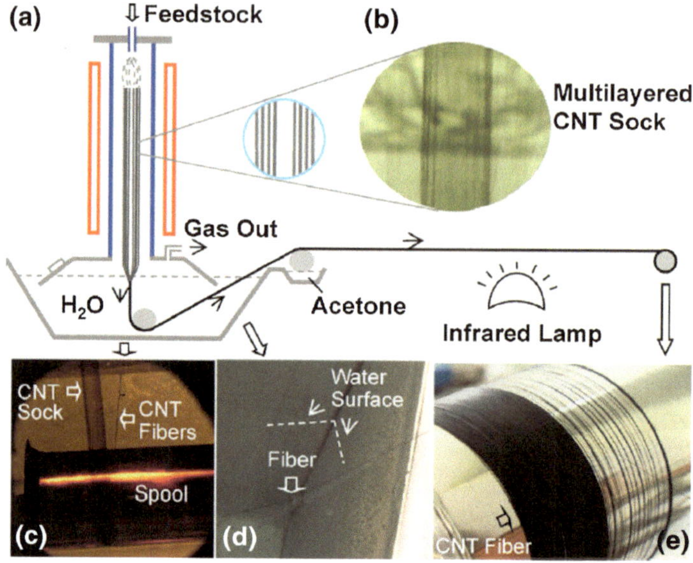

Fig. 2.16 CVD synthesis and spinning set-up for the fabrication of continuous CNT yarns. **a** Schematic diagram of the synthesis and spinning set-up. **b** Photograph of a layered CNT sock formed in the gas flow, with at least eight layers to be observed. Photographs showing the transformation of the layered sock into a fiber, during the water densification step (**c**), drawing from the water at the other side of the water tank (**d**) and final spinning of the fiber (**e**). Reproduced with permission from [59], Copyright © 2010, John Wiley & Sons

was then pulled out into air from the other side continuously, as shown in Fig. 2.16d. The fiber was then gone through acetone over the second rotator, in order for washing and further densification. Finally, the fiber was dried by using an infrared lamp, at about 100 °C (Fig. 2.16e). The productivity, morphology and dimension of the CNT fiber can be readily controlled by controlling the spinning speed (5–20 m min^{-1}).

Figure 2.17 shows microstructures of the CNT fibers [59]. The hollow and multilayered structure of the CNT fibers after densification through acetone is revealed in Fig. 2.17a–d. The CNTs were densely packed and aligned along the fiber axis on the yarn surface (Fig. 2.17b). It was found that without acetone densification the CNT layers showed spontaneous detachment, as shown in Fig. 2.17e, due to the hydrophobic characteristics of the CNTs. The CNT films exhibited homogeneous microstructures with a uniform thickness of about 50 nm (Fig. 2.17f).

The multilayered structure of the fibers was further confirmed by the TEM image of Fig. 2.17g. The CNTs were DWCNTs with a large diameter of 8–10 nm (Fig. 2.17h, left), which became flattened and stack in 50-nm-thick bundles. Figure 2.17h (right) shows a bundle packed with 6 nm DWCNTs which were

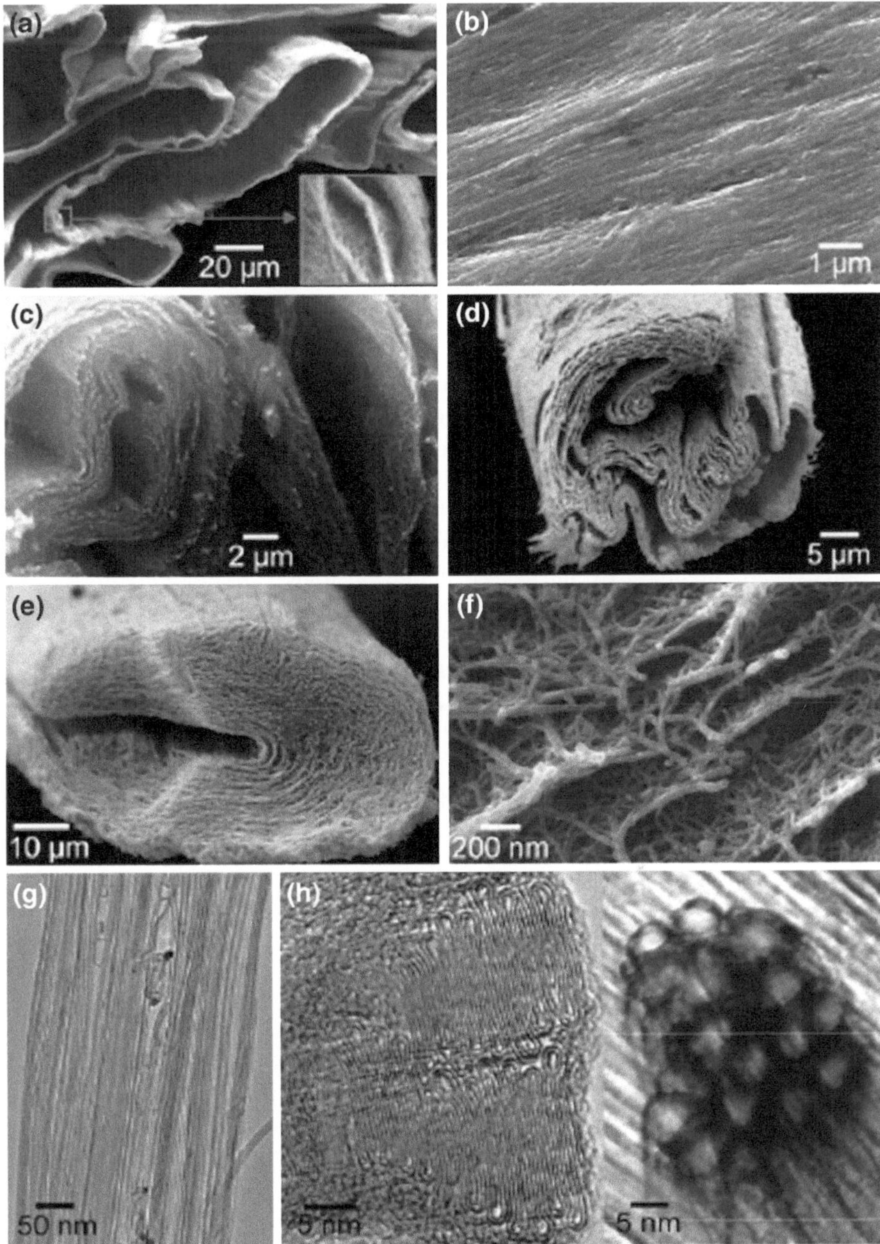

obtained at a lower concentration of thiophene of 0.4 wt%. It means that the structure of the CNTs could be further modified by varying the synthesis conditions. These multilayered fibers have shown unique mechanical, structural, surface morphological and electrical properties [59].

◂ **Fig. 2.17** SEM (**a**–**f**) and TEM (**g**, **h**) images of the multilayered CNT fiber: **a** cross-sections of the as-spun yarns, showing a hollow and multilayered microstructure (inset), **b** surface of the fiber, showing densely packed CNT bundles aligned along the fiber axis, **c**, **d** cross-sections of the multilayered yarn after mechanical rolling, **e** detailed view of the yarn, with more than forty layers packed along the fiber axis, **f** enlarged view of the fiber in **e**, showing an individual layer, with a single layer of crossed CNT bundles, **g** TEM image of the CNT bundles and **h** end view of a CNT bundle, showing double-walled characteristic with a collapsed structure packed in the bundle (*left*) and a round structure (*right*). Reproduced with permission from [59], Copyright © 2010, John Wiley & Sons

2.3.4 Spinning from Cotton-Like Precursors

Spinning cotton or cotton-like CNT precursors to fabricate CNT fibers has been found to be a simple and cost-effective method [29–31]. It is to mimic the ancient cotton spinning process, thus very practical and promising for large-scale production. Two examples are discussed in this section.

The first example is the spinning of cotton like DWCNTs, which were synthesized by using a CVD method, with xylene and ferrocene as carbon and catalyst sources [29]. Figure 2.18 shows schematic illustration of the spinning process, together with optical photographs of the as-spun DWCNT fibers. Heat treatment in argon gas was conducted for 1 h at 150 °C to completely remove the water contained in the fibers, which also made the fibers to be more compacted.

Figure 2.19 shows microscopic characteristics of the DWCNT fibers. Low-magnification SEM images indicated that the as-spun DWCNT fibers had a uniform thickness along their axis, with diameters of 10–100 μm. High-magnification SEM image showed that most of the nanotube bundles inside

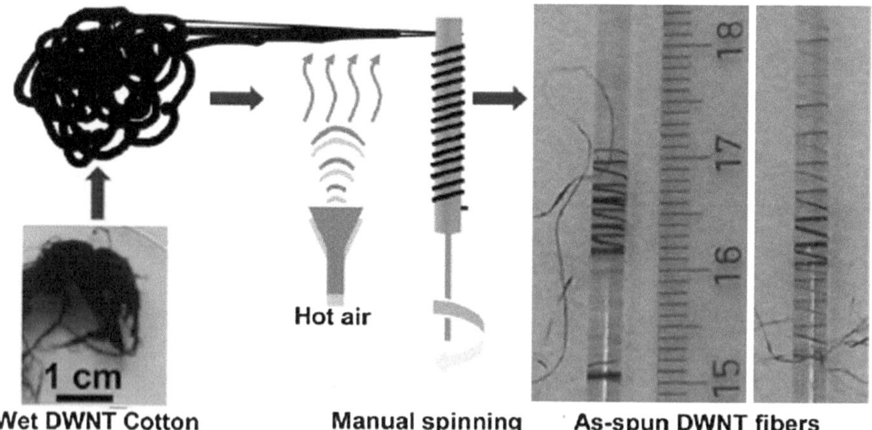

Fig. 2.18 Schematic illustration of the drawing–drying spinning process to fabricate the DWCNT fibers and photographs of the as-spun DWCNT fibers. Reproduced with permission from [29], Copyright © 2007, John Wiley & Sons

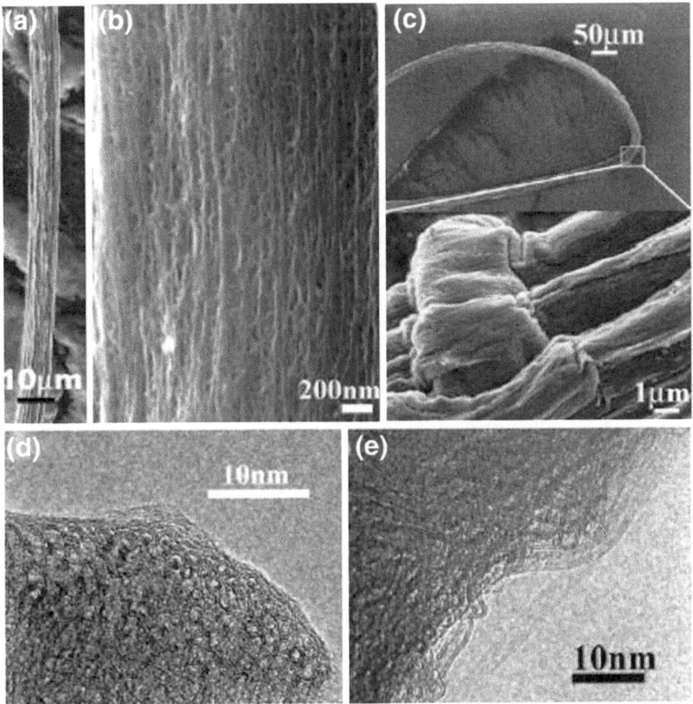

Fig. 2.19 **a** Low-magnification SEM image of the DWCNT fiber. **b** High magnification SEM image, showing that most of the CNT bundles are aligned along axis of the fiber. **c** SEM image of the bent DWCNT fiber together with its local buckling. **d**, **e** Cross-sectional high-resolution TEM images of the DWCNTs of the fibers. Reproduced with permission from [29], Copyright © 2007, John Wiley & Sons

the fibers were aligned along the axis of the fibers and some of them were entangled among the aligned bundles. The DWCNT fibers showed a local buckling behavior when they were bent or knotted, as shown in Fig. 2.19c. Several buckling locations were observed at the inner edge of the bending site, due to the protrusion of the nanotube bundles at surface. Due to the similarity to the actuation of muscles, this flexible buckling behavior implied that the DWCNT fibers could be used to develop artificial muscles. As confirmed by the HRTEM images in Fig. 2.19d, e, the fibers consisted of DWCNTs.

The DWCNT fibers exhibited a tensile strength of 299 MPa, Young's modulus of 8.3 GPa and elongation-to-break values of up to 5 % [29]. It was found that slipping out of the nanotubes were responsible for the failure of the fibers. Mechanical properties could be further improved by using various strategies: such as twisting to increase the frictional force between nanotube bundles and annealing at high temperatures (>2000 °C) to structurally coalesce the nanotubes inside the fibers to increase the bulk strength. The fibers also showed promising electron field emission effect.

The second example was reported by Zheng et al. [30], who used another type of CNT cotton to spin fibers. The CNT cotton was prepared by using a CVM method. Figure 2.20a shows photograph of the fluffy and gray CNT cotton on a quartz support with a dimension of 15 mm × 35 mm. The CNT cotton had very low density with large space in between individual CNTs. In this respect, it is similar to the conventional cotton. The gray color of the CNT cotton was caused by the light-scattering effect, because of its low spatial density. From the side view of Fig. 2.20b, the collection of the CNT cotton was 2 mm thick, which composed of millimeter long individual CNTs. The CNT cotton was hydrophobic, as shown in Fig. 2.20c, in which a water droplet with a diameter of 2.5 mm was sitting on the surface of the CNT cotton. It sank into the CNT cotton because of the gravity effect, but it maintained a quasi-spherical shape. After the water-droplet was completely evaporated, the CNTs in the area covered by the water droplet were rearranged into a web-like mesh, as shown in Fig. 2.20d, whereas the CNTs beneath the web still retained the form of cotton, as demonstrated in Fig. 2.20e, confirming the hydrophobic behavior of the CNT cotton.

The average space between individual CNTs in the CNT cotton was about 10 μm, giving a 2D density of 10^4 mm^{-2}, which was much lower than that of vertically aligned CNT arrays, usually about 10^9 mm^{-2}. TEM analysis indicated that the CNTs were multi-walled, with diameters of 100–380 nm and an average diameter of 250 nm. The formation of the CNT cotton was attributed to the uneven distribution of the catalysts. All the CNTs in the cotton were initially grown from the catalyst area, while those areas without the catalyst, there were no CNTs.

CNT fibers were spun from the CNT cotton by using a spinning set-up built by the authors [30]. SEM images of the CNT fibers during and after the spinning are shown in Fig. 2.21a, b. The individual CNTs in the cotton were easily collected during the fiber spinning, which were well aligned along the pulling direction. Figure 2.21c shows a plot of stress versus strain of a representative CNT fiber. It exhibited very high mechanical strength. Figure 2.21d shows SEM image of the end of the fractured fiber. The individual CNTs were slided one another, rather than structural breaking, so that the fiber became thinner and thinner until the final failure. The loose attachment among the CNTs and the long length of individual CNTs were responsible for the breaking behavior of the CNT fiber. This was because the longer the individual CNTs, the longer the distance that they could slide. However, the CNTs themselves did not show any obvious fracture during tensile tests, which implied that the strength of the CNT fibers could be further increased.

2.3.5 *Spinning with Dielectrophoresis*

Although rarely used in the open literature, dielectrophoresis method is an interesting technology to produce CNT fibers. Figure 2.22 shows a schematic diagram illustrating a dielectrophoresis process reported by Tang et al. [33]. A W tip

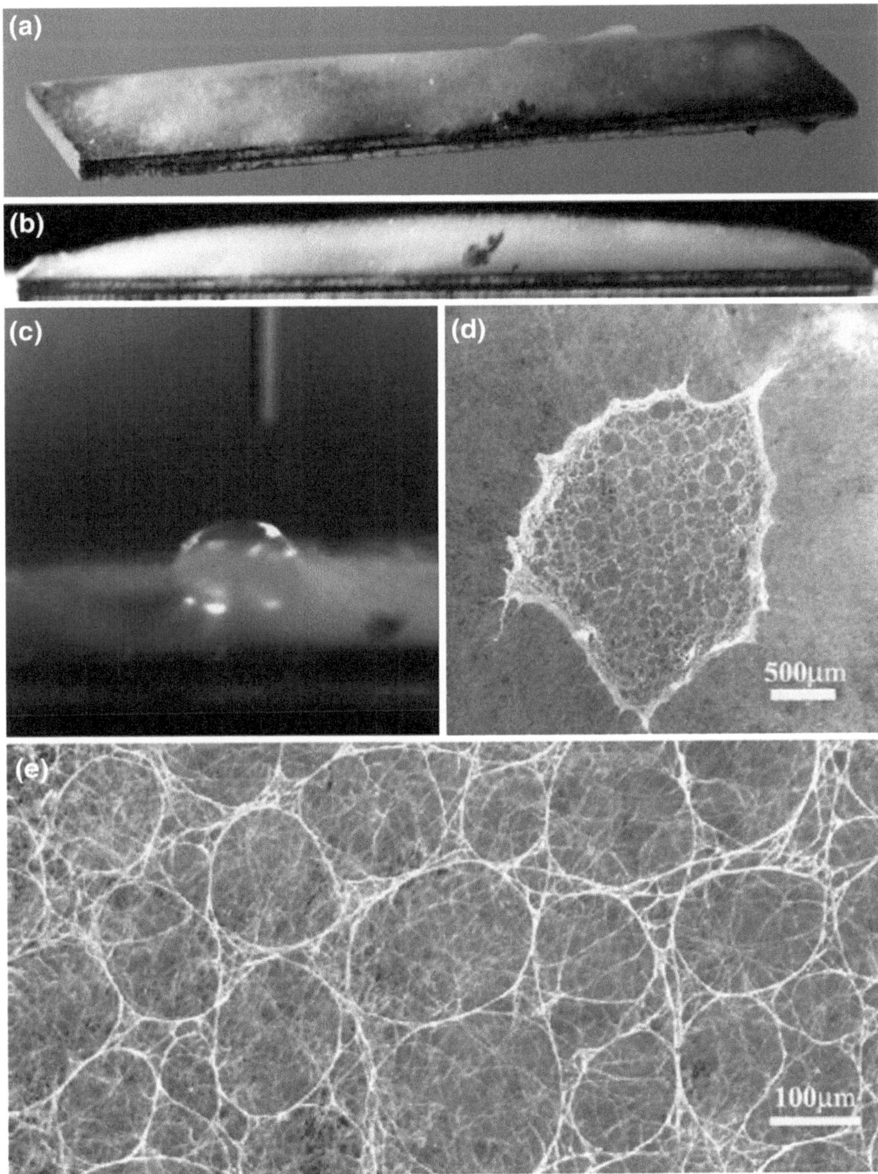

Fig. 2.20 **a** Top view. **b** Side view of the CNT cotton on a 35 mm long quartz support. The CNT cotton showed hydrophobic nature. **c** A water droplet suspended on CNT cotton. **d** Web-like mesh formed after water is evaporated. **d** Cotton form retained beneath the web. Reproduced with permission from [30], Copyright © 2007, John Wiley & Sons

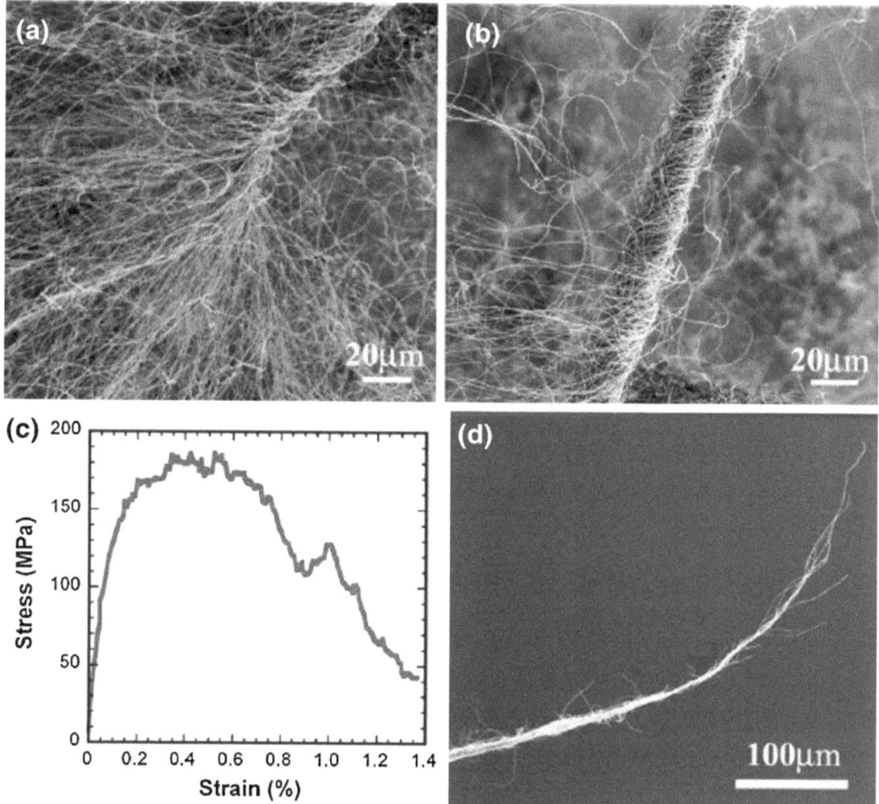

Fig. 2.21 **a** SEM images illustrate that CNT cotton can be easily spun into fibers. **b** A segment of a CNT fiber spun from the CNT cotton. **c** A plot of stress versus strain for a typical spun CNT fiber. **d** The failure mode of the CNT fiber: gradual thinning due to inter-nanotube sliding. Reproduced with permission from [30], Copyright © 2007, John Wiley & Sons

prepared by using chemical etching was used as the working electrode, while a small metal ring acted as the counter electrode, both of which were mounted on separate translation stages and thus placed under an optical microscope. Purified SWCNTs were dispersed in de-ionized waster. A droplet of the CNT suspension was placed inside the metal ring. An AC electric field of 10 V at 2 MHz was applied between the two electrodes. The W electrode was translated horizontally to contact the CNT-water suspension and then was withdrawn under the electrical field until a fiber of desired length was formed.

CNT fiber was formed due to the interaction between the polarizable CNTs and the applied AC electric field. Figure 2.23 shows the mechanism proposed to explain the formation of the CNT fibers. As an AC electric field was applied, the polarized CNTs in water were first aligned along the electric field direction due to the torque action of the induced dipole, which is step (1). With the presence of an asymmetric

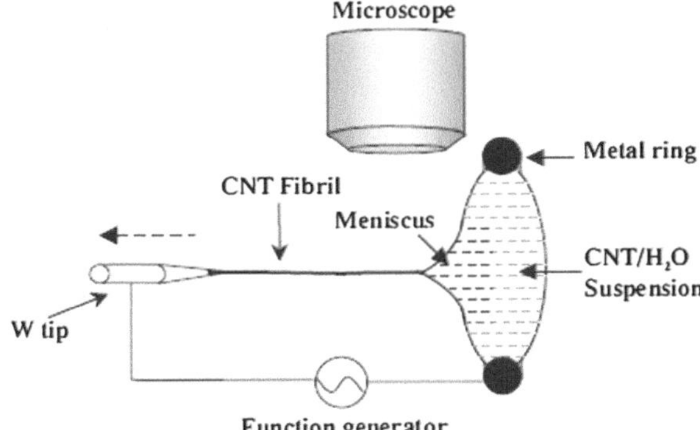

Fig. 2.22 Schematic diagram of the fiber formation through the dielectrophoresis process. A function generator was used to supply an AC field of 10 V at h MHz between the W tip and the metal ring that contained the CNT-water suspension. The W tip is translated to contact the surface of the suspension and was thengradually withdrawn at the AC field to draw out the CNT fiber. The fiber was anchored on the apex of the W tip. Reproduced with permission from [33], Copyright © 2003, John Wiley & Sons

inhomogeneous electric field, the AC frequency and the dielectric constant of the medium could be adjusted, such that the aligned CNTs were driven by the field gradient towards the high field region near the W tip, then it came to step (2). The first group of CNTs was adhered to apex of the W tip, forming new outermost surface of the electrode, to which new CNTs would be precipitated. Continuous migration and precipitation of CNTs facilitated the growth of the fiber along the electric field direction, i.e., step (3). By gradually pulling the W electrode from the liquid in such a way that only the tip of the CNT fiber was in contact with the liquid surface, precipitation and growth could be confined within the liquid meniscus. As a result, fiber with desired dimension could be formed. This is sept (4).

Figure 2.24 shows SEM images of the SWCNT fibers drawn from the CNT suspension [33]. The fiber was about 100 μm long and about 0.2 μm thick, as shown in Fig. 2.24a. The SWCNT bundles were adhered on surface of the W tip, with very few CNT bundles in the radial direction, as shown in Fig. 2.24b. All the CNT bundles were aligned along the longitudinal axes direction of the fiber, which was also the direction of electric field (Fig. 2.24c). The end of the fiber consisted of only a single CNT bundle, as shown in Fig. 2.24d. The fiber was uniform and smooth, whose diameter could be controlled by controlling experimental parameters, including pulling rate, electric field, concentration of the suspension and so on. The technique has been demonstrated to be feasible in other aspects. However, it is obvious that it is not very suitable for large-scale production.

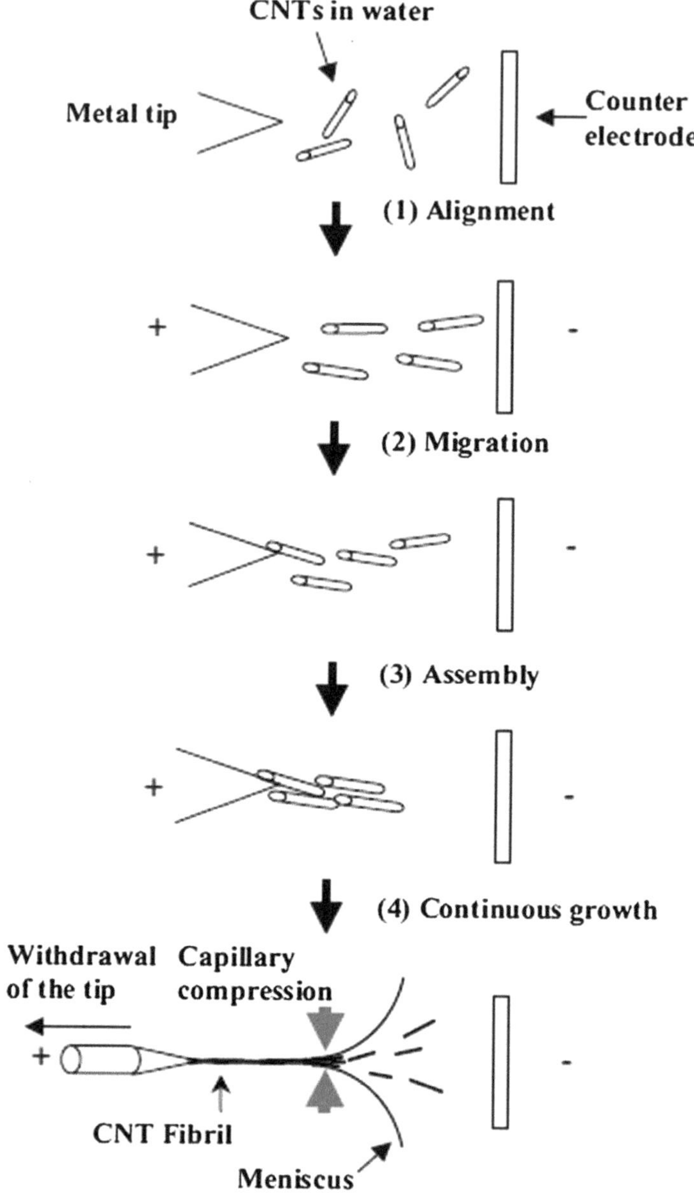

Fig. 2.23 Mechanism of the dielectrophoresis deposition of the CNT fiber at an asymmetric electric field. Step (1): the CNTs dispersed in water are first aligned along the field direction due to their anisotropic polarizability. Step (2): the CNTs migrate towards the high field region, at appropriate frequencies in propermedium. Step (3): the fiber nucleates at the apex of the metal tip and grows towards the counter electrode direction. Step (4): the W electrode is gradually withdrawn from the liquid so that only the tip is in contact with the suspension. The capillary pressure from the surface of the meniscus the alignment and the adhesion of the fiber. Reproduced with permission from [33], Copyright © 2003, John Wiley & Sons

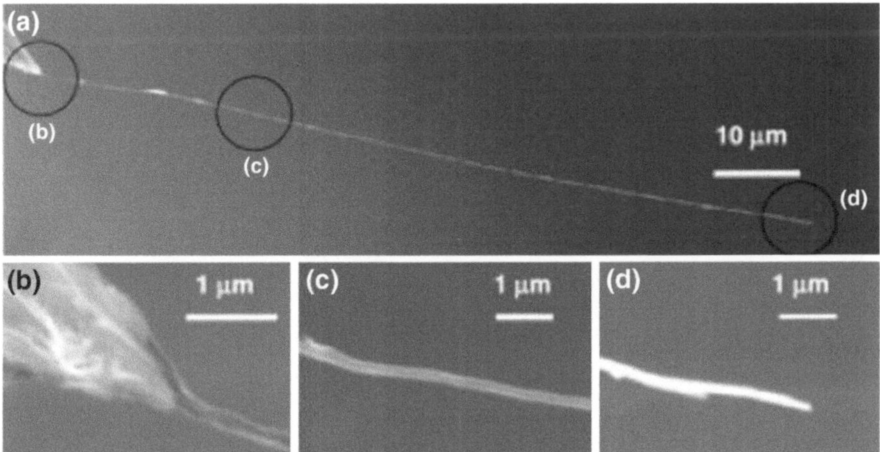

Fig. 2.24 SEM images of the SECNT fiber drawn from the SWCNT-water suspension by using the dielectrophoresis technique: **a** overall morphology, **b** interface between the fiber and the W tip, **c** mid-segment of the fiber and **d** end of the fiber. Reproduced with permission from [33], Copyright © 2003, John Wiley & Sons

2.3.6 *Twisting/Rolling of Films*

CNT fibers have also been fabricated by twisting or rolling CNT films [34, 35, 60, 61]. The fabrication of CNT fibers by using this method was first reported by Ma et al. [34] with SWCNT films. The SWCNT films synthesized by using a floating catalyst CVD method had a reticulate architecture. A CNT strip was sliced from the thin film and then twisted into CNT fibers, as shown in Fig. 2.25. The length and diameter of the fabricated fibers were determined by the width and length of the strip of the CNT films. The CNT fibers prepared in this way had typical diameters of 30–35 µm and lengths of 4–8 cm, showing Young's modulus of 9–15 GPa and tensile strengths of 500–850 MPa. The mechanical properties of the SWCNT fibers were closely related to the intrinsic waviness and inter-bundle slippage of the CNTs.

Figure 2.26 shows schematic diagram of the twisting process [34]. During the twisting of the film into a fiber, the part near the center of the fiber experienced strong twists, so that it collapsed into a cylindrical core, while the portion on surface of the fiber was extended by a factor of 1/cosθ, due to the requirement of matching the length of the center part. For fibers with diameters of 35–40 µm, if the twist angle was 20°, the pre-extension of the film on the fiber's surface was about 6.5 %, corresponding to Raman downshift of 4 cm^{-1}. Due to the pre-elongation of the meshes, CNT bundles in the fibers were pre-tightened and helically aligned along the axis of the fibers. Therefore, the strain of the fibers could be transferred to the axial elongation of the bundles more effectively, so that the moduli of the fibers were increased as a consequence of the twisting process [34]. Another method to

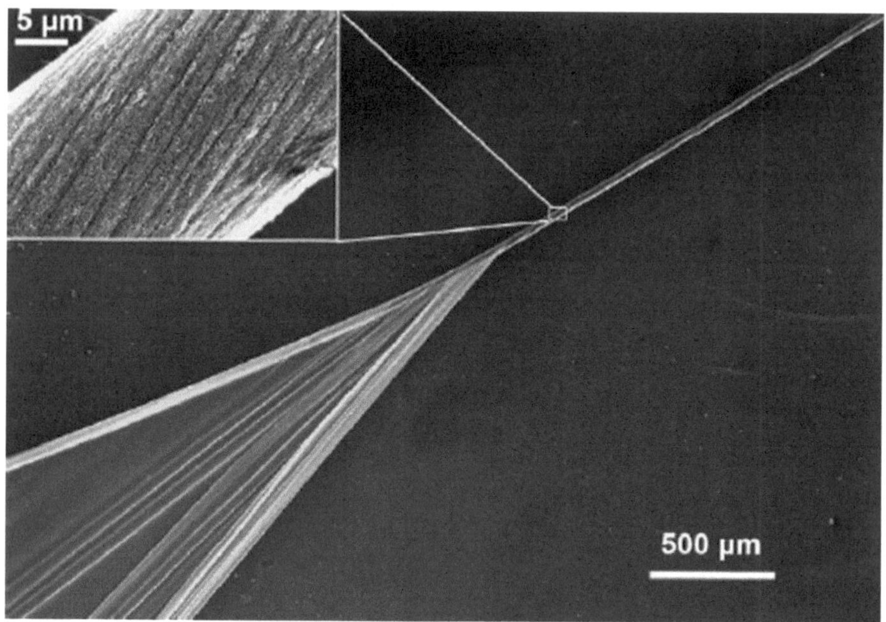

Fig. 2.25 SEM image of a piece of the as-synthesized SWCNT film with a width of 2 mm and a thickness of 200 nm being twisted into a fiber. Reproduced with permission from [34], Copyright © 2009, John Wiley & Sons

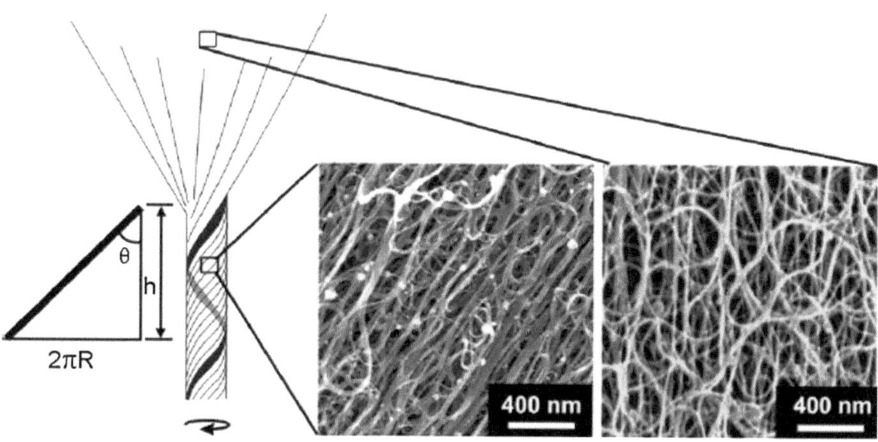

Fig. 2.26 Schematic diagram (*left*) of the twisting process, showing the geometrical elongation of the film near the surface of the fiber, with the two SEM images (*right*) of the unstrained and strained part of the film. Reproduced with permission from [34], Copyright © 2009, John Wiley & Sons

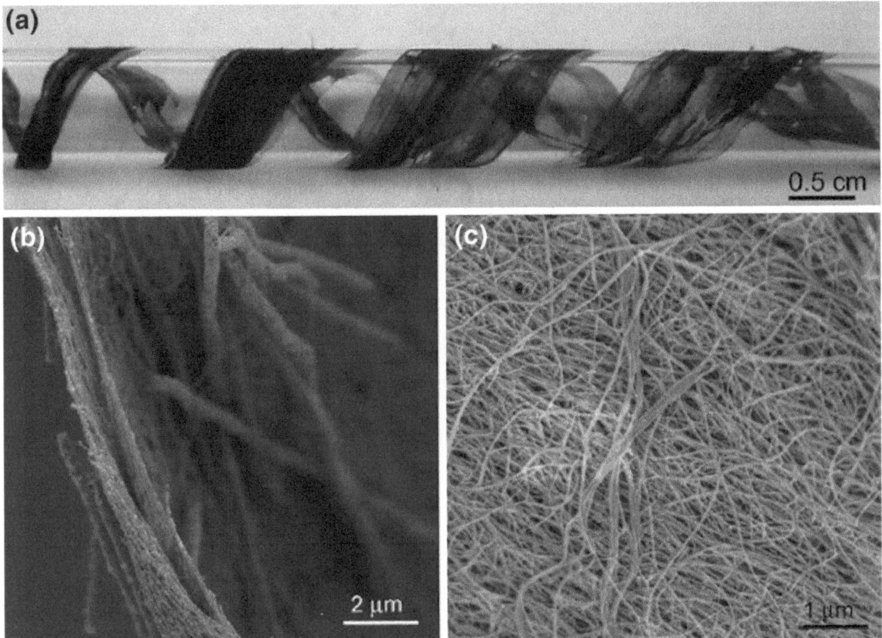

Fig. 2.27 **a** Photograph of the DWCNT films wrapped on a glass rod. **b** Crosssectional SEM image of the film. **c** In-plane view SEM image of the film. Reproduced with permission from [35], Copyright © 2010, Elsevier

fabricate DWCNT fibers by rolling was reported by Feng et al. [35]. The DWCNT thin films used to roll into fibers were synthesized by using one step catalytic CVD gas-flow reaction process, with acetone as the carbon source in argon flow. Figure 2.27a shows a photograph of the free-standing CNT films, with width of 6 mm and length of 20 cm, derived from the spindle and wound onto a glass bar. Figure 2.27b shows cross-sectional SEM image of the films, demonstrating a double layered structure with an equal thickness of about 100 nm. The double layered structure of the CNT films was attributed to the two walls of the CNT sock, which was folded on the spindle during the spinning process. The CNT films consisted of uniformly distributed nanofibers, with diameters of about 30 nm and length of 200 μm. The CNT nanofibers were aligned along the film length direction, as shown in Fig. 2.27c.

Figure 2.28 shows cross-sectional SEM of a fiber derived from the DWCNT films. The fibers exhibited a multilayered structure. They were highly flexible and mechanically strong, with a linear density of 1.78 g km^{-1}. Their bulk density was about 1.11 g cm^{-3}, estimated according to the volume and weight of a film. Specific strength of the fibers was 70 MPa (g $cm^{-3})^{-1}$, corresponding to strength of 112 MPa from the linear density (1.67 mg m^{-1}) and the cross-sectional area of the

Fig. 2.28 SEM image of the DWCNT fiber prepared by rolling the DWCNT film, showing multilayered structure. Reproduced with permission from [35], Copyright © 2010, Elsevier

film (1.0×10^{-9} m^2). The fibers also showed promising electrical conductivity, 8.0×10^4 S m^{-1} [35].

More recently, Song et al. [60] developed a motorized pulling/twisting stage to spin CNT fibers from SWCNT films. The SWCNTs films were synthesized by using a floating catalyst CVD method. Figure 2.29a shows representative photographs of the as-synthesized SWCNT films. The films were uniform and continuous, with areas of up to several tens of square centimeters. The SWCNTs had diameters of 1–1.5 nm. Before spinning, the SWCNT films were immersed into hydrochloric acid (3 wt%) for two days and then washed with deionized water. When the films were put into isopropanol, they were spread and floated on surface of the solvent. A sharp tip was used to contact edge of the film and draw it, so that fiber was spun out. Continuous spinning was realized by using a motor, as shown in Fig. 2.29b, c.

Figure 2.29d shows SEM image of the SWCNT fiber with diameter of about 70 μm [60]. The diameter of the fibers could be controlled by controlling the area initially covered by the film, the spinning/pulling speed of the motors and the ratio of the speed of spinning and pulling. During the spinning/pulling process, the deionized water/isopropanol mixture that were trapped inside the films was squeezed out, so that droplets were formed on surface of the fibers, as shown in the inserted image of Fig. 2.29c. The fibers were dried by removing the remaining moisture by blowing nitrogen gas.

2.3.7 Summary

CNT fibers are made of axially aligned and highly packed CNTs, so that they are expected to have much higher specific modulus and specific strength than the commercial carbon and polymeric fibers. They are also more flexible and exhibit

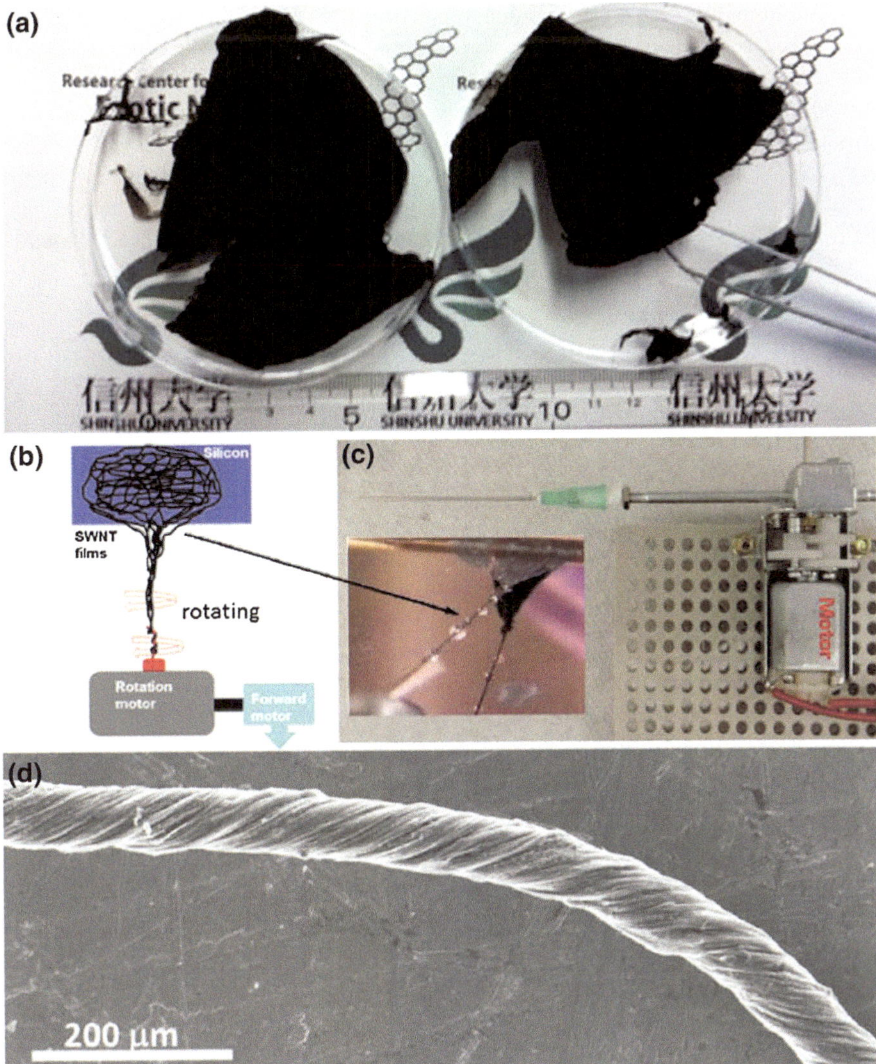

Fig. 2.29 **a** Photographs of the as synthesized SWCNT films with floating catalyst CVD. **b** Schematic of the spinning process. **c** Photograph of the spinning stage with rotation motor, with the inset showing an image of the fiber spinning process. **d** SEM image of a SWCNT fiber with diameter of about 70 μm. Reproduced with permission from [60], Copyright © 2012, Elsevier

higher energy to break than some commercial fibers. More importantly, CNT fibers possess higher electrical and thermal conductivities than the conventional fibers. Due to these novel properties, CNT fibers could find applications in various aspects, such as reinforced composites, mechanical and biosensors, transmission lines and microelectrodes [14].

Although significant achievements have been made in fabrication of CNT fibers and they have shown promising performances for potential applications, CNT fibers are challenged by the conventional carbon fibers. For example, there is still no processing route to synthesize CNTs with high and consistent quality and controllable microstructures, due to the diversity of CNTs, including, SWCNTs, DWCNTs, MWCNTs, metallic CNTs, semiconducting CNTs, diameters, lengths and aspect ratios. Also, large-scale production of high quality spinnable CNTs is still a challenge. Therefore, much effort is necessary to address these problems.

2.4 Films, Membranes, Sheets and Papers (2D)

2.4.1 Brief Introduction

In general, film means a thin layer of materials that are deposited on certain substrates, which are included in this review. Instead, the CNT films discussed here are those that are free-standing, with relatively larger thickness than those with substrates [12, 35, 62–67]. With different thicknesses, the CNT films are also called membranes, sheets, Buckypapers or papers, which will be not intentionally differentiated in this discussion unless otherwise specifically mentioned. Due to their excellent properties, such as mechanical stability, flexibility, chemical stability, electrical and thermal conductivities, CNT films have shown various potential applications [66–68]. Free-standing CNT films have been fabricated by using either suspension-based deposition (filtration or casting) or directly growing, which will be described as follows.

2.4.2 Fabrication Technologies

2.4.2.1 Wet Suspension Methods

Suspension processing of CNT films includes mainly vacuum filtration [69–83], as well as suspension casting, droplet-drying and electrophoretic deposition (EPD) [63, 84–86]. Various nanoporous membranes, such as anodized aluminum oxide, polycarbon, nylon and PTFE, have been used for vacuum filtration. The pore size and density of the filtration membranes can be different, so as to be used for CNTs with different thicknesses and lengths. CNT films can be peeled off after drying for further studies and applications. Therefore, this is a simple and effective technique to fabricate CNT films.

Since the first free standing macroscopic SWCNT film, prepared by using vacuum-filtration of SWCNT suspension [69], the method has been widely used to fabricate various CNT films. In this method, SWCNTs were first dispersed in a solution with the aid of surfactants and ultrasonication. The SWCNT suspensions were filtrated to form films, whose thickness was controlled through the concentration of the nanotube and the volume of the suspension. This is a simple and versatile method suitable for nanoscale fibrous materials. By using this method, ultrathin, transparent, electrically conducting SWCNT films, which can be transferred onto various substrates [70]. Electric field-activated optical modulator made of the SWCNT films exhibited an optical analog to the nanotube-based field effect transistor. Another free-standing SWCNT film with a thickness of less than 200 nm over an area of several square centimeter was reported by Hennrich et al. [71]. Figure 2.30 shows that the SWCNT film prepare by using this method was smooth surface and homogeneous in microstructure.

MWCNT papers have also been prepared by using the membrane filtration technology [76–78, 83]. For example, Kukovecz et al. [76] used CNTs with different lengths to prepare MWCNT Buckypapers with apparent mean pore diameter of 24–39 nm. Figure 2.31 shows SEM images of an unused nylon filter membrane and the filters after being used to filtrate the longer and shorter MWCNTs. It was found that after filtration the longer (>2 μm) MWCNTs had almost no effect on the pore structure of the membrane, while the shorter ($\sim$230 nm) ones blocked the pores. Surface SEM images of the Buckypapers made with the longer and shorter MWCNTs are shown in Fig. 2.32. The shorter nanotubes were packed more densely with smaller pores. This work provides a guidance in fabrication of Buckypapers with desired pore characteristics for given applications. Also, controlling the length of CNTs by using ball milling is a simple and effective way.

Besides filtration method, CNT films have been derived from suspensions by using various other strategies. For instance, a drop-drying method to prepare SWCNT film was reported by Duggal et al. [63]. Figure 2.33 shows a schematic of the drying process of SWCNT suspension. The solvent to disperse the SWCNT was F68 Pluronic which was dissolved in water at a concentration of 2 wt% or 2.4 mM. The surfactant had two PEO chains that were connected by a PPO chain. The hydrophobic segment of the surfactant (PPO) sited at surface of the individual CNTs, while the two hydrophilic PEO segments expanded in water. The polymer provided a steric barrier to the bundling induced by van der Waals forces between CNTs. The suspension was dropped on glass slides to form SWCNT films after drying the loss of the solvent due to evaporation, transport of the SWCNT–F68 driven by the advection and preferential adsorption of the surfactant at the air/liquid interface caused an increase in the local concentration of the SWCNT–F68 complex at the free surface, which resulted in the formation of the crust. The presence of the inhomogeneity due to SWCNTs in the thin crust served as nucleation sites of fractures. As the crust ruptured at isolated points, volcanic landscape features were formed, as shown in Fig. 2.34a (panel 2) and Fig. 2.34b. The evaporation front moved across the drop from the rupture sites, leading to the formation of aligned liquid-crystalline domains, which could be observed under polarization, as shown in Fig. 2.34a (panels 3–6) and Fig. 2.34c. When two evaporation fronts met, defect planes would be created, which was responsible for the formation of grain boundaries and defects in the crystals, as shown in Fig. 2.3a (panel 4). If water was

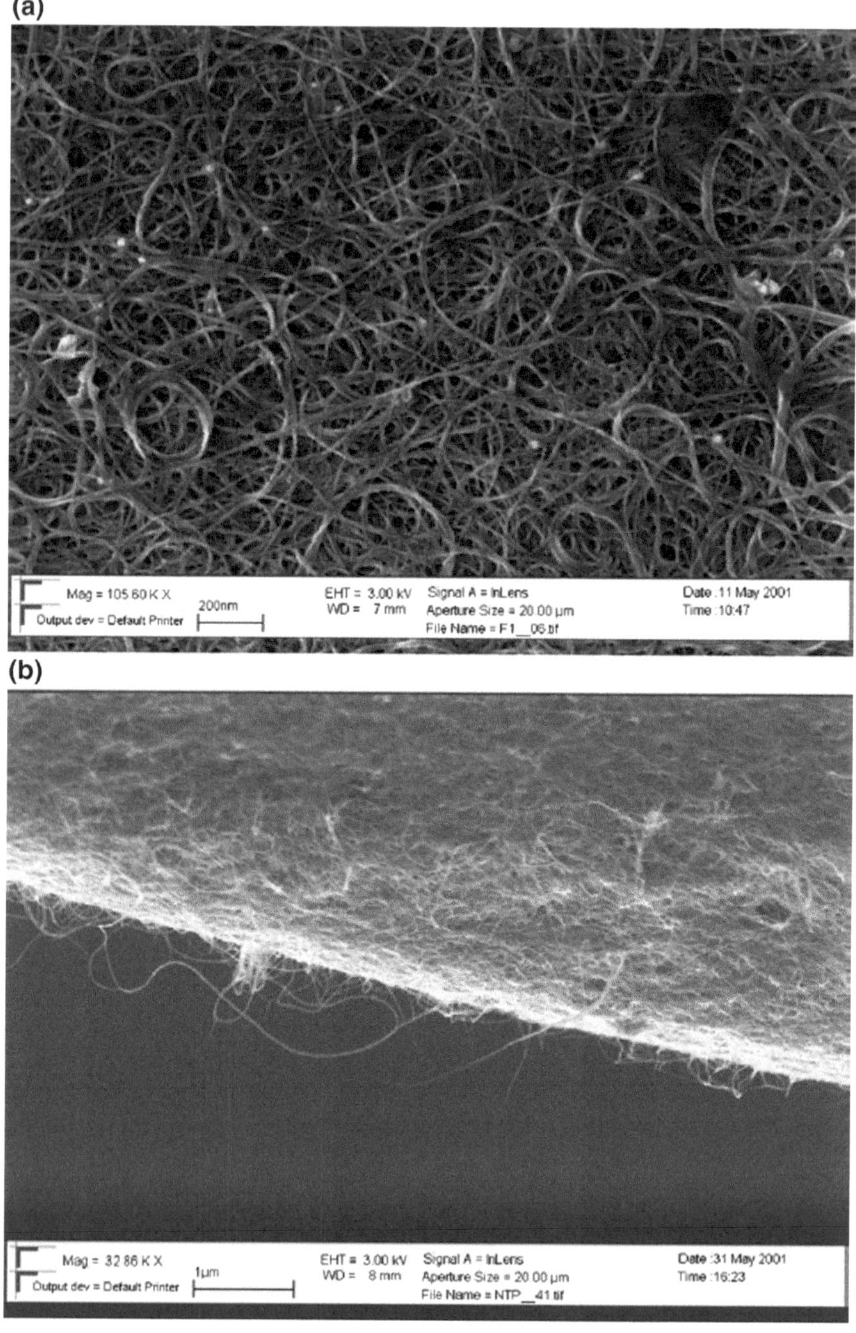

Fig. 2.30 SEM images of the SECNT free-standing film: **a** front surface at the suspension side and **b** back surface in contact with the filter membrane. Reproduced with permission from [71], Copyright © 2002, Royal Society of Chemistry

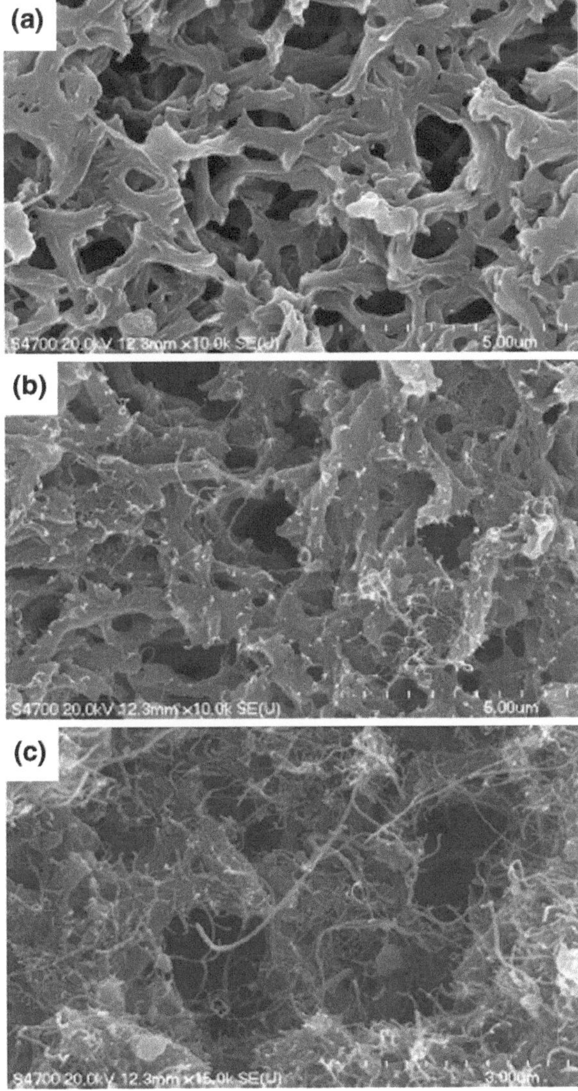

Fig. 2.31 SEM images of the fresh nylon filter with 450 nm nominal pore size (**a**), the filter after being used to make Buckypapers from the long (>2 μm) MWCNTs (**b**) and the filter after being used to fabricate Buckypapers from the short (~230 nm) MWCNTs (**c**). Reproduced with permission from [76], Copyright © 2007, Elsevier

further evaporated while the horizontal dimensions were fixed, the crust would become thinner and thinner. In this case, the crust experienced gradually increased tension, because it tended to shrink, which would be then fractured as a result, i.e., forming cracks, as shown in Fig. 2.34a (panel 6). The direction of the cracks at the surface was dependent on the packing of the SWCNT–F68 micelles.

Alternative suspension CNT film formation method, frit compression, was reported by Whitby et al. [84]. In this method, polypropylene syringe and polypropylene frit with pore size of 50 μm were used. Purified MWCNTs were

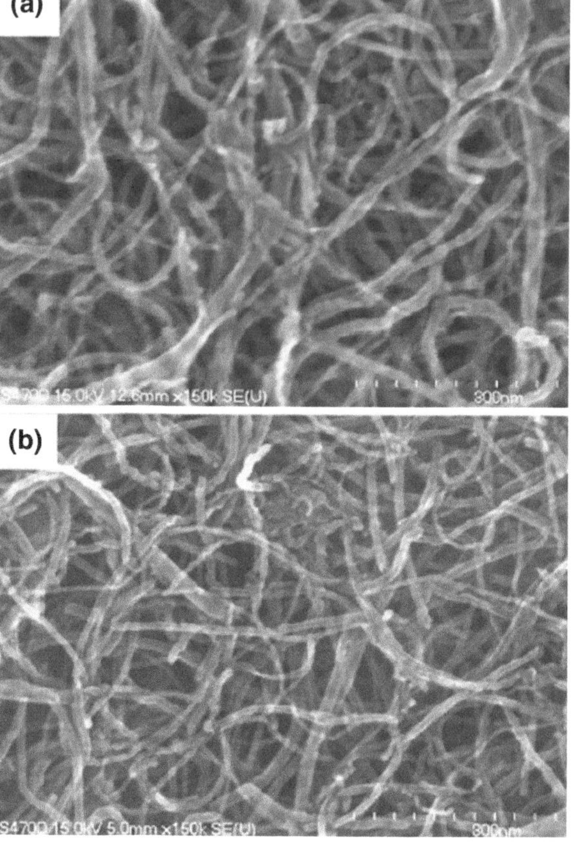

Fig. 2.32 Surface SEM images of the Buckypapers made of pristine (>2 μm) MWCNTs (**a**) and shortened (~230 nm) MWCNTs (**b**). Reproduced with permission from [76], Copyright © 2007, Elsevier

ultrasonicated in a solvent to form suspension. After a polypropylene frit was inserted inside a syringe, the MWCNT suspension with an appropriate quantity was put into the syringe, than another frit was placed on top of the suspension, which was then pressed by pushing the plungers. The solvent was squeezed out through the porous frits. After drying, MWCNT Buckpapers were obtained. Different solvents were used in that study.

Figure 2.35a–c shows photographs of representative Buckpapers prepared by using the frit compression method [84]. Appearances of the Buckypaper made in this were related to the type and properties of the solvent used, concentration and quantity of the suspensions. Unbalanced forces on two surfaces caused by evaporation of the solvent led to domed-shaped Buckypapers, as shown in Fig. 2.35b, c. Thickness of the Buckpapers could be readily controlled by the quantity and concentration of the CNT suspensions. Obviously, this method can be easily scaled up.

Another example of non-filtration suspension method to prepare CNT films is electrophoretical deposition (EPD) [86]. Figure 2.36 shows schematic diagram of

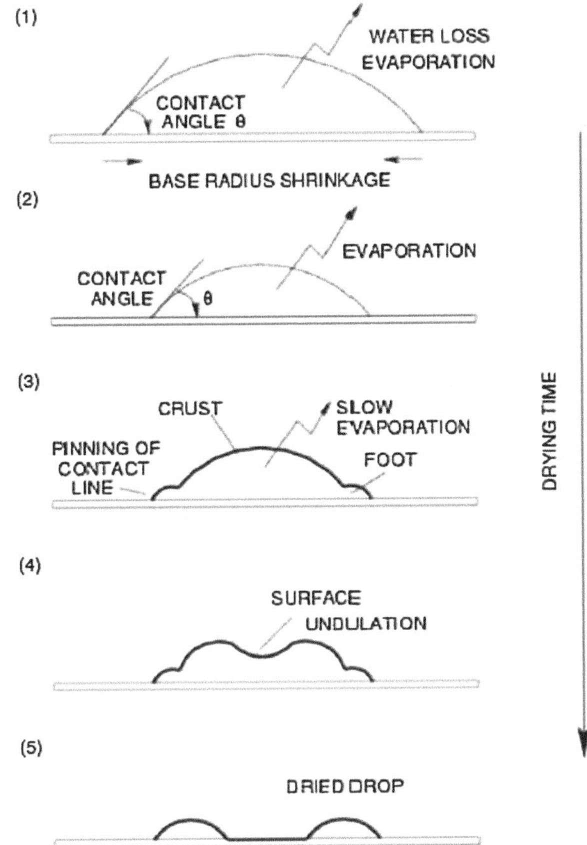

Fig. 2.33 Schematic of the drying process of SWCNT suspension drops. (*1*, *2*) The drop maintains a fixed contact angle as the base radius shrinks. (*3*) The contact line is pinned and a gelled foot appears. A crust is formed at the free surface, reducing the evaporation rate. (*4*) The volume that the crust envelops decreases as evaporation proceeds and a surface undulation appears. (*5*) The crust sits on the surfactant-SWCNT deposit on the substrate as the drying reaches completion. Reproduced with permission from [63], Copyright © 2006, John Wiley & Sons

the EPD set-up used in that study. It consisted of vertically aligned electrodes in a parallel plate configuration with a separation of about 1 cm. Electrode pairs were inserted into the aqueous MWCNT suspension for 10 min, at an applied DC voltage of 2.8 V. After the EPD process, the electrodes were extracted from the suspension and were dried, while the applied voltage of 2.8 V was maintained for an additional duration of 5 min, in order to enable further densification of the CNT deposit, which would improve the homogeneity of the dried films as compared with those dried only through the evaporation processes. 316 stainless steel sheets, with a thickness of 1 mm and dimensions of 5 cm × 2.5 cm, were used as electrodes.

Multiple-deposition was used to control thickness of the MWCNT Buckypapers. For each deposition, all the experimental parameters were kept to be identical. For example, to make a two-time deposited film, two beakers containing identical suspensions of MWCNTs would be used. In this case, the concentration of the suspension was constant for each single deposition. The films after air-drying were peeled off from the stainless steel electrode by merging the anode in DI water for several hours. The mechanical cleavage was realized by using a razor, which was

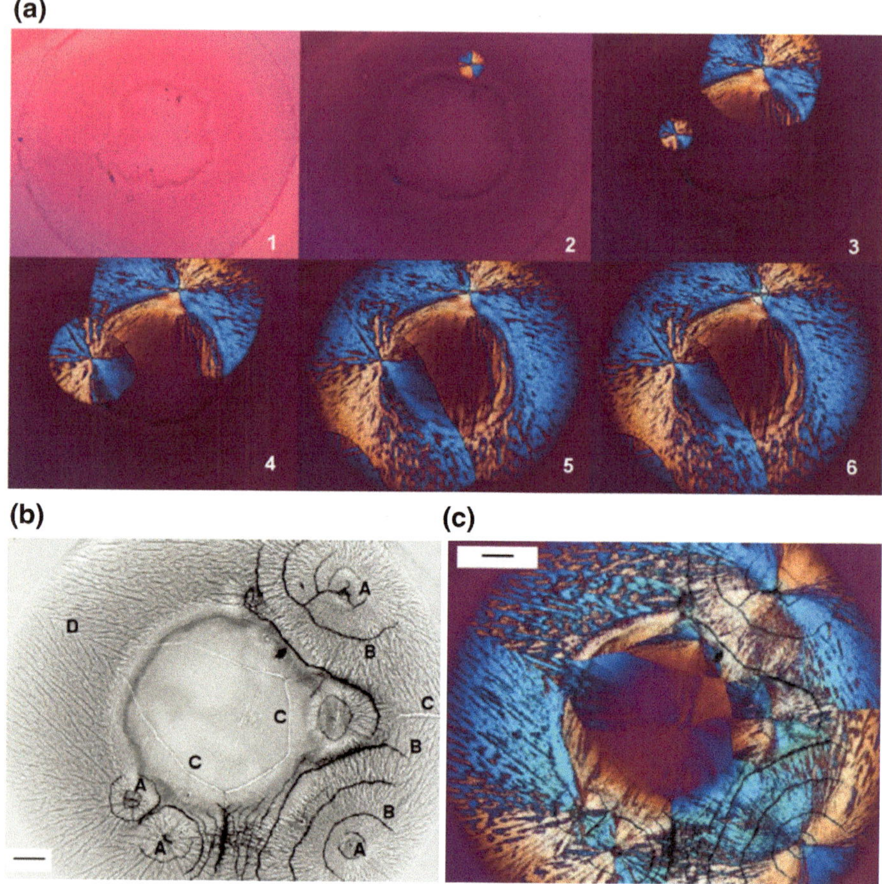

Fig. 2.34 Dried drops of the SWCNT–F68 on glass. **a** Progression of drying under crossed polarizers. (*1*) Contact line is pinned and the drop forms a gelled foot. Undulation of the free surface in the middle can be observed. (*2*) Crust ruptures at a point and birefringent patterns start appearing. (*3*) Crust ruptures at another point and the drying fronts move radially. (*4*) The drying fronts meet to form a grain boundary. (*5*) The drying fronts move across the whole drop. (*6*) A circular crack appears around the rupture site (at the top). **b** Bright-field image showing the foot of the drop, volcanic landscape and cracks: (*A*) rupture sites, (*B*) cracks, (*C*) grain boundaries and (*D*) foot. Scale bar is 100 μm. **c** Image under crossed polarizers. The micelles exhibit birefringent fan-like arrangements and characteristic of hexagonal liquid-crystalline domains. Scale bar is 100 μm. Reproduced with permission from [63], Copyright © 2006, John Wiley & Sons

inserted between the film and stainless steel substrate, so that minimal fraying at the films edges was caused. After peeling off from the stainless steel substrates, the films were placed on Teflon plates and then were dried into Buckypapers. Thickness of the MWCNT Buckypapers could be controlled through controlling the thickness of the single deposition and the number of the multiple deposition.

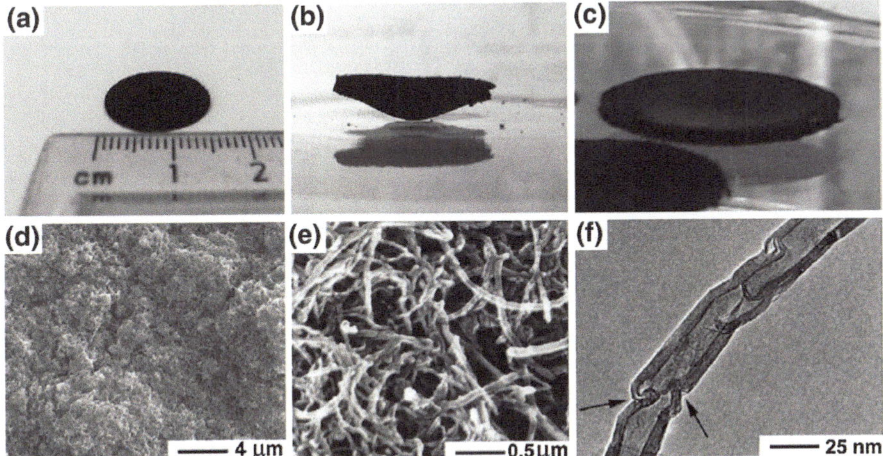

Fig. 2.35 Photographs and microstructures of the MWCNT Buckypapers made by using frit compression. Depending on the surface tension of the solvent, the MWCNTs Buckypapers had different appearances: **a** planar, **b** convex lower surface and **c** concave upper surface of dome-shaped Buckydiscs (thick Buckypapers), **d** SEM image showing the undulated surface of the Buckypapers, **e** cross-sectional SEM image of the Buckypaper, with MWCNTs randomly orientated, **f** TEM image showing compartmentalized nature of the MWCNT and tube geometry defects indicated by the *arrows*. Reproduced with permission from [84], Copyright © 2008, Elsevier

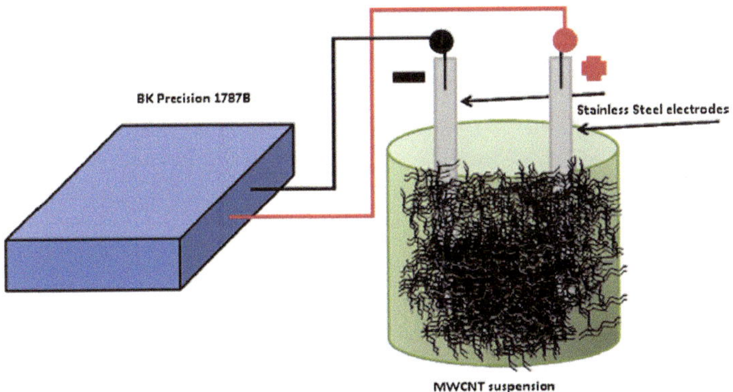

Fig. 2.36 Schematic of the EPD process to fabricate MWCNT films fromsuspension. The MWCNTs were negatively charged due to the proprietary surfactant and deposited onto the positive electrode. Electrode spacing between electrodes was 1 cm. The voltage applied across the electrodes was 2.8 V. Reproduced with permission from [86], Copyright © 2010, Elsevier

Preliminarily, the Buckypapers fabricated by using this method showed average tensile strength and Young's modulus of 14.5 MPa and 3.3 GPa, respectively [86].

2.4.2.2 Dry Methods

Dry methods to fabricate CNT films include continuous drawing of CNT sheets from aligned CNT arrays [87–93] and unidirectional Buckypaper from aligned CNT arrays [94].

The direct drawing method to produce CNT sheets from super-aligned MWCNT arrays is similar to the direct spinning to fabricate CNT fibers [87]. The only difference is that the sheets are drawn without applying any twisting force. The CNT sheets were drawn from a sidewall of MWCNT forests synthesized by using catalytic CVD method. The MWCNTs were about 10 nm in diameter, with forest heights of 70–300 mm. Draw was initiated with an adhesive strip, to contact the MWCNTs that were teased from the forest sidewall. MWCNT sheets with length of several meters and width of up to 5 cm could be drawn manually out at about 1 m min^{-1}. They had a density of only about 2.7 mg cm^{-2}, but strong enough to self-supportive when the sheet was as large as 500 cm^2. A forest of 1 cm long and 245 μm high yielded a 3 m long free-standing MWCNT sheet. The sheet production was robust and consistent. The drawing processing, e.g., maximum drawing rate, was related to the properties or qualities of the CNT forests. Thickness of the as-drawn MWCNT sheets was determined by the height of the forests. The MWCNT sheets could be densified by immerging in liquids, like ethanol, for further characterization.

Figure 2.37 shows SEM images of the MWCNT sheets during the direct drawing process and after densification [93]. It is clearly demonstrated that the free-standing sheet was drawn laterally from the side of the CNT forest, Fig. 2.37a. Either glasses or flexible plastics could be used as substrates to place the as-drawn sheet, because the sheet was weakly bound to these substrates due to the van der Waals forces.

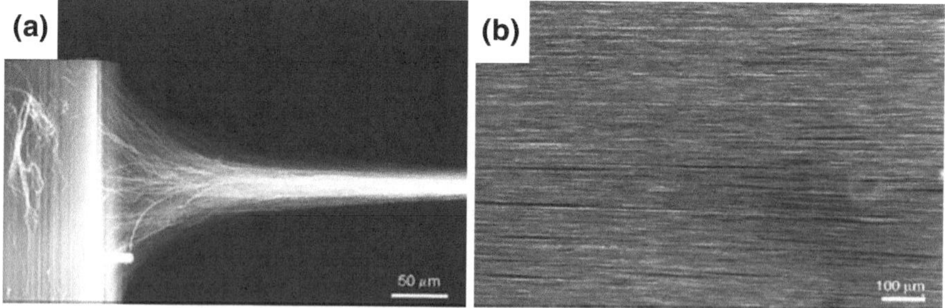

Fig. 2.37 SEM images of **a** drawing of MWCNT sheet from a nanotube forest and **b** a densified MWCNT sheet on a glass substrate (top view). Reproduced with permission from [93], Copyright © 2010, Elsevier

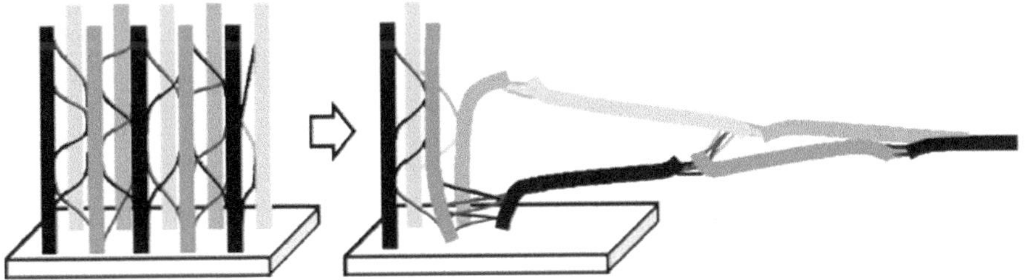

Fig. 2.38 Schematic diagram of MWCNT sheets dry spinning process. Reproduced with permission from [93], Copyright © 2010, Elsevier

Methanol was used to densify the CNT sheet, when the solvent was rapidly evaporated. The densified sheet consisted of well-aligned MWCNT bundles, with tens of 5–10 nm diameter MWCNTs in each bundle, as shown in Fig. 2.37b. The sheet exhibited transmission of 70 % for polarization parallel to nanotube orientation and 90 % for perpendicular polarization at 450 nm.

Figure 2.38 shows a schematic of the dry drawing process to fabricate MWCNT sheets from the CNT arrays on substrates [93]. As the MWCNT sheet was drawn out from the forest, the vertically oriented bundles in the forest became to be horizontally oriented in the sheet, as shown in Fig. 2.38a. During the drawing, smaller bundles that were present in initial forest redistributed, so as to interconnect bigger bundles. As a result, continuous drawing process was maintained [54]. The MWCNT sheets could be drawn from the forests with different densities and heights of 100–500 μm. Generally, the number of free ends of bundles was inversely proportional to the height of the forest, meaning that the shorter the forest the more free ends of bundles per unit area would be in the drawn MWCNT sheets. The MWCNT sheets showed promising electric field emission.

Another dry method is called domino pushing, which can be used to produce aligned Buckypapers directly from aligned MWCNT arrays [94]. Figure 2.39 shows schematic illustration of the domino pushing method, consisting of three steps. Firstly, the CNT array was covered with a piece of microporous membrane, so that all the CNTs of the CNT array were forced down to one direction by pushing a cylinder which was placed upon the CNT array with constant pressure. In this case, all the CNTs in the array would be put together, due to the strong van der Waals forces. Therefore, a Buckypaper with high degree of alignment was obtained. Secondly, the aligned Buckypaper was peeled off from the silicon substrate together with the membrane. Finally, ethanol was spread on the microporous membrane, which then permeated through the membrane to the Buckypaper. The ethanol saturated Buckypaper could be readily peeled off from the membrane. Representative photographs of the aligned MWCNT Buckypapers are shown in Fig. 2.40. The Buckypapers fabricated in this way had very smooth surface, with a density of about 0.032 g cm^{-3}.

Figure 2.41 shows microstructural characteristics of the MWCNT arrays and the Buckypapers [94]. The CNT array shown in Fig. 2.41a had a height of about

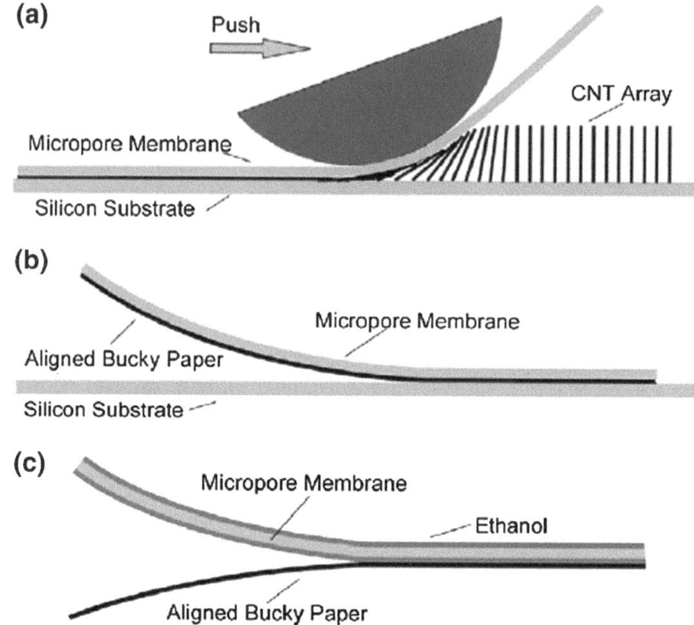

Fig. 2.39 Schematics of the domino pushing method: **a** Formation of aligned Buckypaper, **b** peeling the Buckypaper off from the silicon substrate and **c** peeling the Buckypaper off from the microporous membrane. Reproduced with permission from [94], Copyright © 2008, IOP Publishing

500 μm, whereas the individual MWCNT was of a diameter of about 15 nm, as shown as an inset in Fig. 2.41a. Figure 2.4b indicates that Buckypaper exhibited a highly aligned surface microstructure, confirming that all the CNTs were well aligned in the Buckypapers, so that their density was much higher than that of the CNT array. This was because all the CNT tips were pushed down one by one. In addition, the Buckypapers had a surface roughness of less than 100 nm. These MWCNT Buckypapers possessed promising thermal transport properties. Their axial electrical conductivity was 200 S m^{-1} at room temperature, which was higher than the value of conventional Buckypaper (~ 150 S m^{-1}).

2.4.2.3 Direct Growth with CVD

Besides the use of CVD to synthesize CNT powders and arrays, it has also been found that CNT films can be directly grown by using CVD [95–97]. Song et al. [95] reported a floating catalyst CVD approach to synthesize SWCNT films. Figure 2.42 shows schematic of the CVD set-up, which consisted of a two-stage furnace system with a special quartz tube structure and a tube with smaller diameter. Ferrocene and sulfur powder with a molar ratio of 20:1 were mixed and used catalyst. Methane

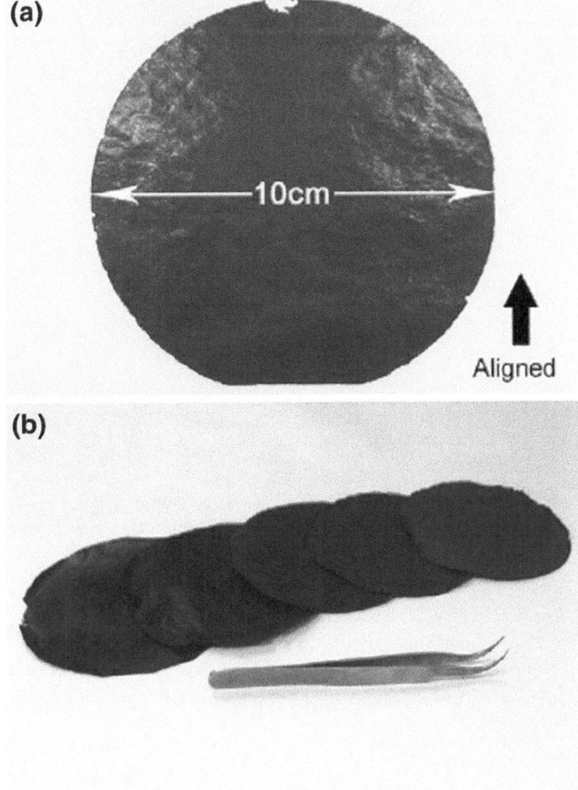

Fig. 2.40 **a** Photograph of a round aligned Buckypaper with a diameter of 10 cm. **b** A pile of Buckypapers. Reproduced with permission from [94], Copyright © 2008, IOP Publishing

was used as carbon source, which was carried by using argon gas, to react at 1100 °C for 6–9 h. SWCNT films were peeled off from the inside walls of the reactor tube. Fe catalyst particles in the films were oxidized and then removed by using washing with HCl solution. The films were thoroughly cleaned with water. Photograph and SEM image of the as-grown SWCNT films are shown in Fig. 2.43a, b, respectively. The films consisted of highly entangled CNT bundles, decorated with many nanoparticles, which were catalysts encapsulated by graphite layers. The purified SWCNT films exhibited promising mechanical properties.

The same group also reported a floating-catalyst CVD that could be used to synthesize free-standing SWCNT films with interesting optical, mechanical and electrical properties [96]. In this method, thickness of the SWCNT films could be well controlled by precisely controlling the sublimation rates of the catalysts and gas flows. Figure 2.44 shows photographs and SEM images of the as-grown SWCNT films. The SWCNTs were homogeneously distributed and entangled in the films, and the CNT bundles were preferentially aligned along the flow direction, forming Y-shaped junctions. Electrical measurements indicated that the conductivity of the SWNT films could reach up to 2026 S cm^{-1}, which is over 60 times

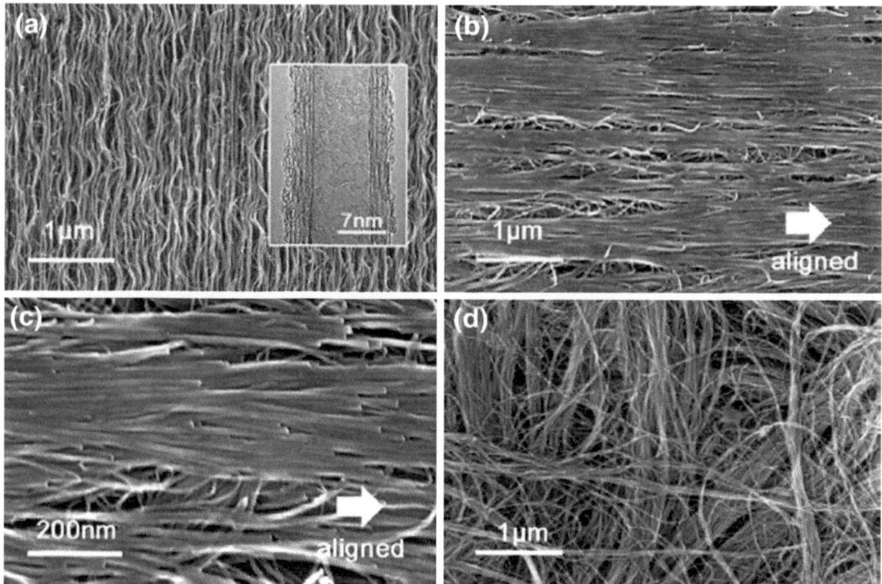

Fig. 2.41 SEM images of the CNT array and Buckypaper: **a** side view of an aligned MWCNT array, with the inset being a HRTEM image showing an individual CNT, **b** micrograph of an aligned Buckypaper surface, **c** higher magnification image of (**b**) and **d** micrograph of a random Buckypaper surface. Reproduced with permission from [94], Copyright © 2008, IOP Publishing

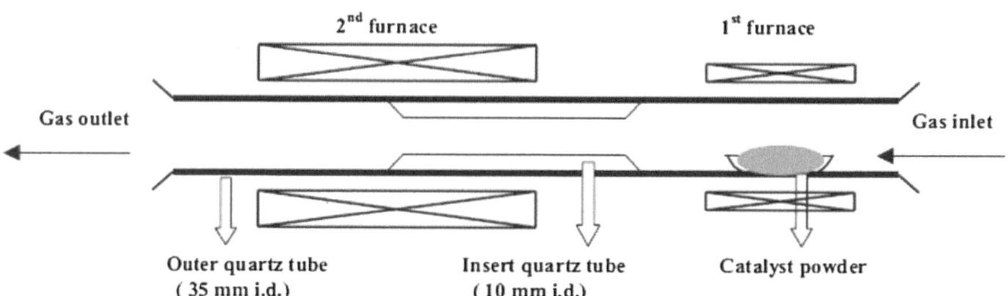

Fig. 2.42 Schematic set-up of floating catalyst CVD to synthesize SWCNT films. Reproduced with permission from [95], Copyright © 2004, John Wiley & Sons

larger than that of unpurified HiPCO (high pressure carbon monoxide) transparent films and nearly 3 times that of purified arc-discharge SWNT films. The superior electric performance was attributed to the lower inter-tube contact resistance, high purity and the anisotropic characteristics of the SWCTN films. Mechanical tensile characterization results indicated that the failure strength of the film was 360 MPa and the Young's modulus was about 5 GPa.

Liu et al. [97] further modified the CVD method so as to synthesize macroscopic SWCNT sheets with multilayer structures, which were derived from synthesized

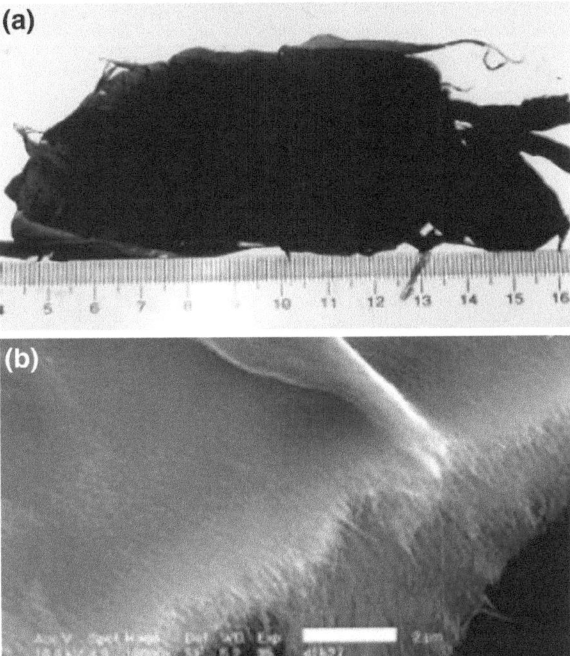

Fig. 2.43 Photograph (**a**) and SEM image (**b**) of the as-grown SWCNT films by using the floating catalyst CVD. Reproduced with permission from [95], Copyright © 2004, John Wiley & Sons

Buckybooks. The synthesis and assembly of SWCNTs were performed in a quartz tube reactor inside an electrical furnace. In order to synthesize the Buckybooks, specifically designed porous membranes, made from various porous materials, such as carbon fabric felt, gauze, and so on, with the same diameter as that of the inner diameter of the quartz tube reactor, were used as substrates. The substrates were placed vertically at the outlet of the quartz reaction tube. When the SWCNTs were continuously grown and formed in the reaction zone, they were immediately transported to the outlet end and in situ assembled into a Buckybook on the specifically designed porous membranes.

The as-fabricated Buckybook was about 40 mm in diameter and up to 3 mm in thickness derived from the sample with a reaction time of 30 min, as shown in Fig. 2.45a, b [97]. Freestanding monosheets could be peeled off from the Buckybooks. Figure 2.45c shows that the homogeneous monosheets were optically transparent. One of the advantages of this method was that the Buckybooks could be easily engineered, i.e., the thickness of the final Buckybooks was controlled directly by adjusting the reaction duration time, while the size of the Buckybooks could be scaled up simply by using larger reaction quartz tubes. At the same time, the diameter of individual SWCNTs in the Buckybooks and packing density of the SWCNT sheets, could be designed through the synthesis parameters.

Figure 2.46 shows SEM characterization results of the Buckybooks. They were comprised of hundreds of homogeneous thin monosheets, as shown in Fig. 2.46a–c. Figure 2.46d indicates that the monosheet peeled from the Buckybooks had a very

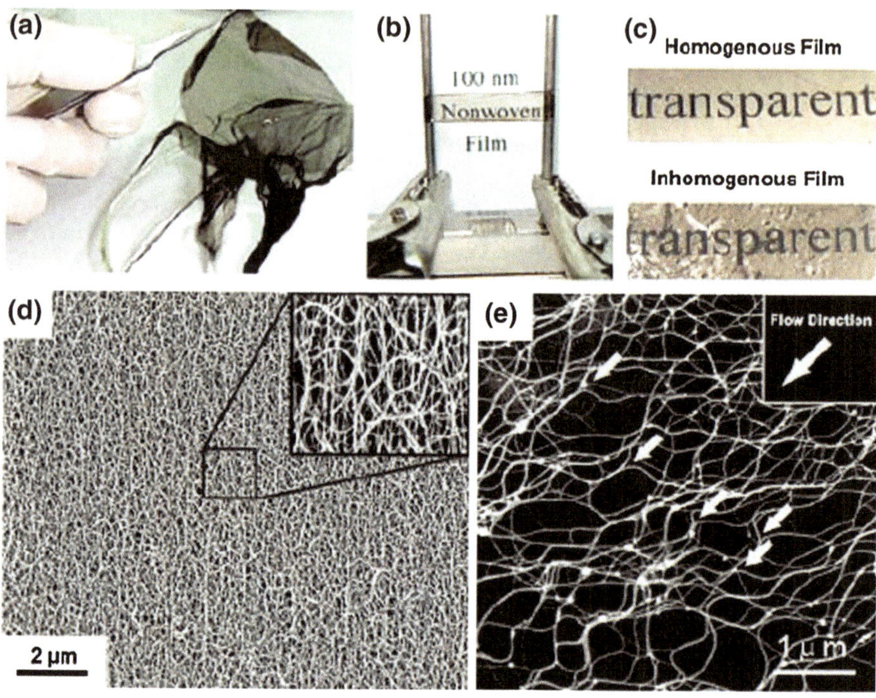

Fig. 2.44 **a** Photograph of an as-grown 250 nm thick SWCNT film. **b** A freestanding transparent 100 nm thick film held between two metallic pillars. **c** A 150 nm thick homogeneous (*upper panel*) film and an inhomogeneous film (*lower panel*). **d** SEM image of a 250 nm thick film, with the inset image taken at higher magnification. **e** SEM image of SWCNT network in a single layer, with white arrows in the image pointing out the Y-type junctions and the flow direction. Reproduced with permission from [96], Copyright © 2007, American Chemical Society

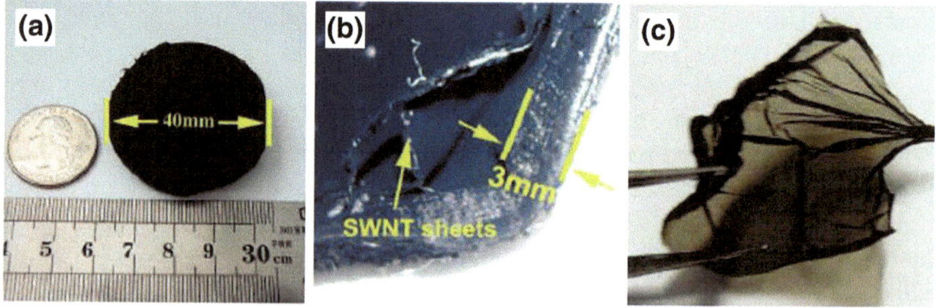

Fig. 2.45 Photographs of a Buckybook of ~40 mm in diameter (**a**) and ~3 mm in thickness (**b**), and a freestanding SWCNT monosheet peeled off from a Buckybook with tweezers (**c**). Reproduced with permission from [97], Copyright © 2009, American Chemical Society

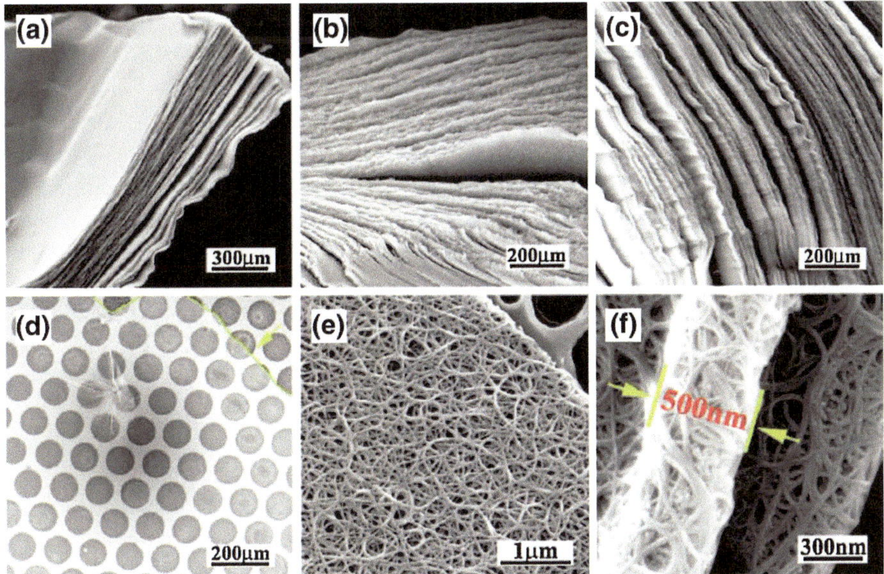

Fig. 2.46 SEM images of **a** the top, **b** side and **c** cross-section of a SWCNT Buckybook, **d** low-magnification SEM image of a peeled SWCNT monosheet placed on a copper grid and **e** high-magnification SEM image of the same monosheet with a monolayer of **f** about 500 nm in thickness, exhibiting abundant entangled SWCNT bundles with clean surfaces. Reproduced with permission from [97], Copyright © 2009, American Chemical Society

clean, transparent and smooth surface. The sheet consisted of numerous compactly inter-entangled SWCNT bundles (Fig. 2.46e), with a thickness of about 500 nm (Fig. 2.46f). The SWCNT bundles had strong interaction so that the sheets possessed high mechanical flexibility and could withstand ultrasonication for long time. The thickness of a monosheet in the Buckybooks was determined by the dispersion concentration of the SWNTs in the carrier gas during the reaction, which was determined by growth rate of the SWCNTs and flow rate of the carrier gas. It was found that the higher carbon feeding rates, higher sublimation temperatures of ferrocene and lower flow rates of carrier gas, the thicker the sheets would be in Buckybooks [97]. The SWCNT sheets showed potential applications as high-efficiency molecular separation filters and high-capacitance electrodes.

Although these SWCNT sheets synthesized by using floating catalyst CVD have shown to possess high electrical conductivity and mechanical flexibility, the method has a significant problem of low productivity. Also, there are too main factors that have influence on morphology, microstructure, quality and properties of the resulting CNT films, such as types of carbon source, property of catalyst, flow rate of carrier gas, growth rate, substrate, external field, reaction temperature and time duration. Therefore, further studies are necessary to understand the interrelations between properties of the CNT sheets and the experimental parameters.

2.5 Foams, Gels and Bulks (3D)

2.5.1 *Brief Introduction*

3D nanomaterials based on CNTs that will be discussed in the section include bulks [98–112], foams [113–121] and gels [122–125]. Bulk CNT-based nanomaterials are mainly prepared by using could-pressing or hot-pressing (HP) and spark plasma sintering (SPS) technique, while foams are made directly by using CVD method and gels are assemblies of CNTs through wet-chemical processing. Representative examples for each group will be briefly discussed in the following sub-Sections.

2.5.2 *3D Bulks*

Hou et al. [98] reported could-pressing densification of MWCNTs for hydrogen storage applications. The MWCNTs were synthesized by using the floating catalyst CVD method. Morphology and dimension of the MWCNTs before and after purification are shown in Fig. 2.47a, b, respectively. Figure 2.47c shows a HRTEM image of the MWCNTs. The tubes had discontinuous layers instead of ideal structure, which was responsible for the high capability of hydrogen storage, because the open structures provided the access for hydrogen to be stored between the graphite layers. Free-standing CNT compact were formed at 60 MPa, with a density of 0.47 g cm^{-3}, as shown in Fig. 2.47d. Hydrogen storage of the CNT compact was evaluated after it was annealed at 1000 °C in Ar.

Fabrication of 3D bulks by using hot-pressing technique is also not very popular. One example was reported by Ma et al. [99]. The MWCNTs used in this study, with diameters of 30–40 nm, as shown in Fig. 2.48a, were synthesized by using the conventional catalytic method, with Ni particles as catalyst. Purified CNT powder was hot-pressed with a graphite die at 25 MPa and 2000 °C for 1 h in Ar. As shown in Fig. 2.48b, the consolidated CNT assembly consisted of randomly entangled and cross-linked CNT. Samples with diameter of 50 mm and thickness of 5 mm had a density of about 1.1 g cm^{-3} and resistivity of 2–3 × 10^{-4} Ω cm. The MWCNT bulk assembly showed promising electric field emission performance.

Spark plasma sintering is also known as electric current activated/assisted sintering (ECAS) [126, 127]. During experiments, loose powders or cold formed compacts to be consolidated are loaded into a container, which is heated to a targeted temperature and then held at the temperature for a given period of time, while a pressure is applied and maintained at the same time. Thermal energy is supplied by applying an electrical current that flows through the powders and/or their container, thus facilitating the consequent Joule heating effect. Because the sintering is conducted in an encapsulated chamber, oxidization can be prevented, so the CNT powders can be sintered by using this method.

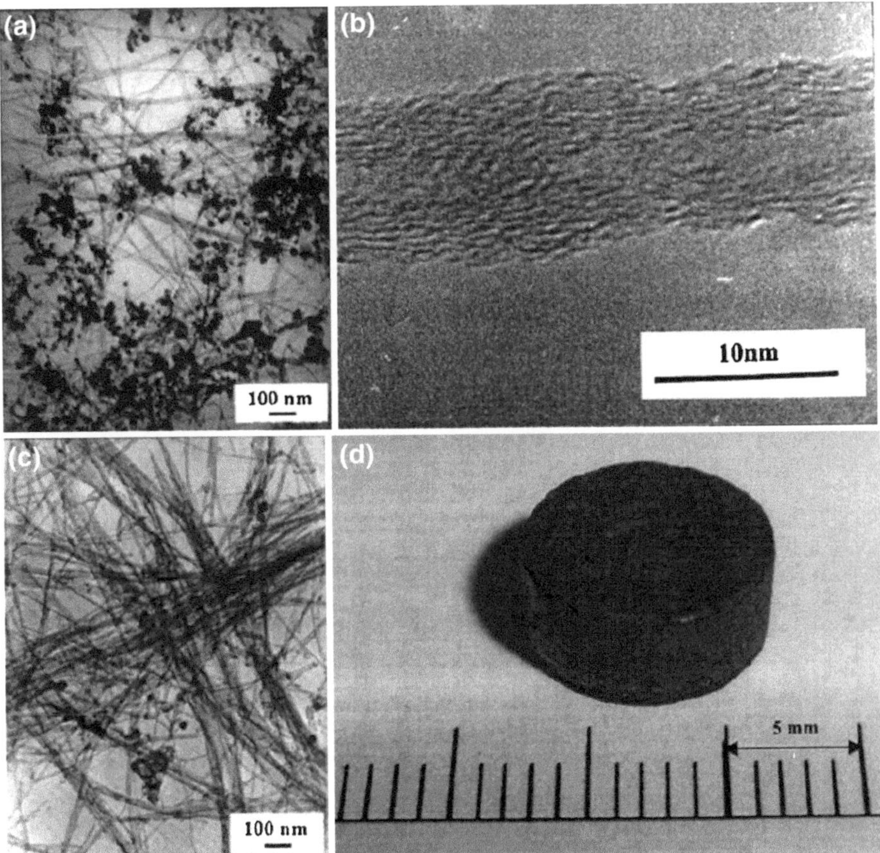

Fig. 2.47 TEM images of the as-synthesized (**a**) and purified (**b**) MWCNTs. **c** HRTEM image of the purified CNT with imperfect outer graphite layers. **d** Photograph of the bulk WMCNT compact prepared at 60 MPa without using binder. Reproduced with permission from [98], Copyright © 2002, American Chemical Society

Li et al. [100] used SPS to consolidate MWCNTs, which had promising mechanical properties. The MWCNTs used in this study were prepared by using CVD method, with Fe particles as catalyst. The as-synthesized CNT powders were purified with acid solution. SPS was conducted in a vacuum by using a Dr. Sinter 1050 SPS apparatus, one of the most popular SPS facilities in the open market. The sintering temperature was 1600 °C, at a heating rate of 100 °C min^{-1} at a pressure of 60 MPa for 1 min. The MWCNTs had diameters of 10–30 nm and lengths of several microns. The CNTs appeared as relatively straight tubes, as shown in Fig. 2.49a. After sintering, the CNTs became entangled structure, as shown in Fig. 2.49b, which was due to the pressure and the high sintering temperature. Figure 2.49a shows polished surface SEM image of the as-sintered bulk CNT assembly. Only some pores with a size of 2–5 μm could be observed, while the

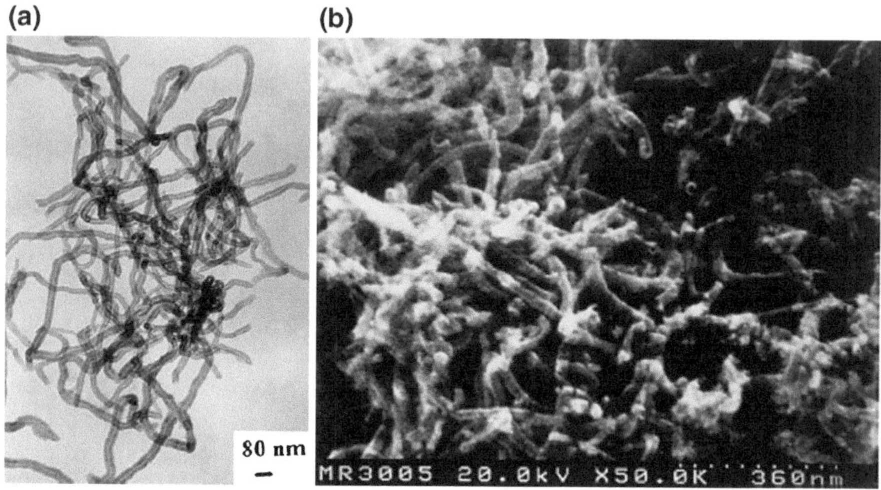

Fig. 2.48 TEM image of the CNTs used (**a**) and SEM image of the hot-pressed CNT assemble (**b**). Reproduced with permission from [99], Copyright © 1999, Elsevier

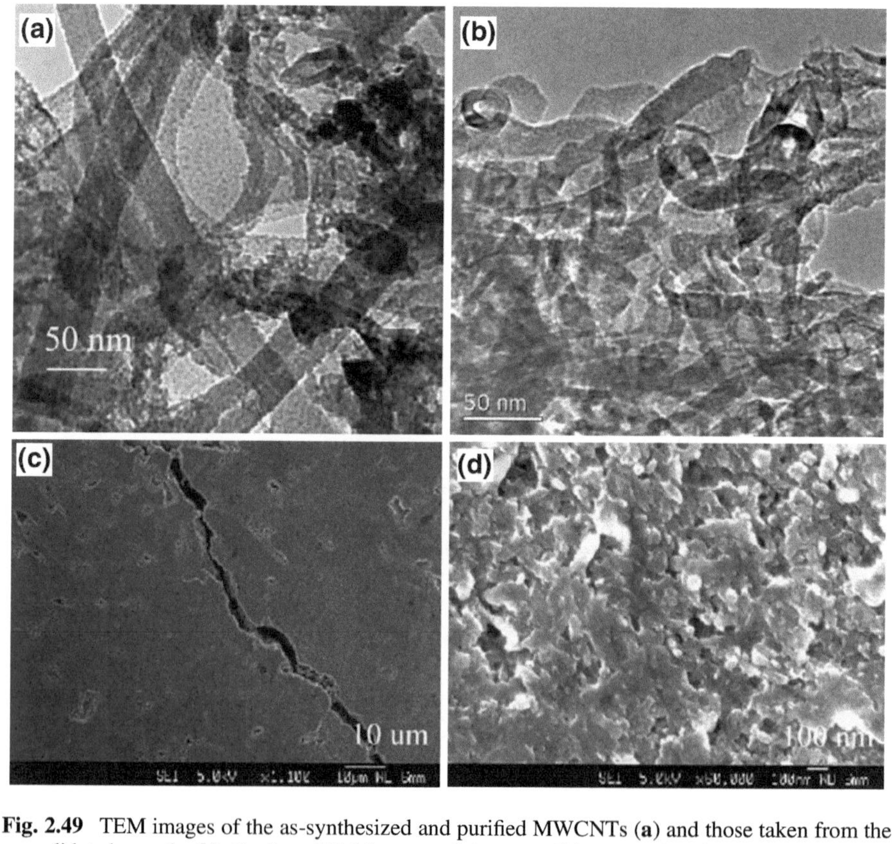

Fig. 2.49 TEM images of the as-synthesized and purified MWCNTs (**a**) and those taken from the consolidated sample (**b**). Surface SEM images of the consolidate sample with low (**c**) and high (**d**) magnifications. Reproduced with permission from [100], Copyright © 2005, Elsevier

crack observe in the figure was caused by indentation measurement. High magnification SEM image indicated that the CNT assembly had a rough surface at the nanometer scale, as shown in Fig. 2.49d. The presence of the nanopores and nanoholes attributed to the pullout of the CNTs during the polish process. However, the CNT samples possessed relatively low mechanical strength due to the weak bounding of the CNTs. Mechanical, thermal and electrical properties of the CNT assemblies consolidated by using SPS have been widely reported in the open literature.

2.5.3 Foams

CNT foams, either in the form of arrays [113, 128–131] or entangled assemblies, have been mainly directly synthesized by using CVD method [114–118, 132–134]. In the case of arrays, which are similar to those of sheets as discussed before, the length of the CNTs is sufficiently long, so that the arrays can be easily separated from the substrate. In contrast, the entangled assemblies are disordered and usually grown directly on the walls of the CVD reaction chambers or sometimes on substrates that are only used as supports. In addition, the entangled assemblies are also known as sponges. More recently, there have been also reports on combination of ordered arrays and entangled assemblies, to form hierarchical structures [119–121, 135]. Representative examples of each type of the CNT foams will be briefly discussed as follows.

Hata et al. [128] demonstrated an efficient CVD method to synthesize SWCNTs arrays with ultra-length and super-density, with the aid of water. Water-stimulated enhanced catalytic activity was responsible for the massive growth of the vertically aligned SWCNT forests with heights up to 2.5 mm, which could be readily separated from the catalysts, leading to CNTs with carbon purity of as high as 99.98 %. The water-assisted synthesis method addressed the critical problems that currently limit the synthesis of high quality CNTs by using CVD methods.

This significant achievement was attributed to the fact that water had completely reactivated each and every dead catalyst, so that the catalysts would have high catalyst activity and long lifetime and thus provided high growth rate [136]. During the reaction process of conventional hydrocarbon CVD, there are two competing reactions: production of SWCNTs, i.e., pathway one or synthesis, and creation of carbon that coats and deactivates the catalysts, i.e., pathway two or deactivation. This is the reason why the hydrocarbon CVD synthesis always has low efficiencies. The introduction of water would clean the catalyst particles by removing the carbon coating through its oxidization. Therefore, the catalysts were cleaned and thus reactivated, i.e., there was pathway three or reactivation. Because of this, the synthesis could be constantly maintained. In addition, it was found that oxidized Fe catalysts showed improved growth efficiency. As a result, the presence of water could oxidize the Fe catalysts, thus offering an additional positive effect on the growth of the CNTs.

This method was also used to synthesize ultra-thick DWCNT arrays [129]. By tailoring the catalyst nanoparticles, the selectivity of the growth of DWCNT could be maximized. The significance of the selective growth of DWCNTs is related to their electron emission behavior as compared with that of SWCNTs. SWCNTs have an unsatisfactory emission lifetime for real applications, although they have the advantage of low threshold voltage. MWCNT emitters have high durability which is desired for real applications, but their threshold voltages are relatively high. Therefore, DWCNTs are the best candidate for real FED applications because they exhibit the advantages of SWCNT and MWCNT emitters, i.e., low threshold voltage and high durability.

Figure 2.50a shows photograph of the DWCNT forest, with 2.2 mm in height on a 20 × 20 mm Si substrate. It was found that the ability to grow the CNT forests was dependent on thickness of the catalyst layers [129]. There was a threshold in the catalyst thickness, i.e., 0.8 nm, below which forests were not available. However, above the critical thickness and within the DWCNT growth region, the height of the forest showed very weak dependence on the thickness of Fe catalytic layers. Figure 2.50b shows edge of the DWCNT forest, indicating that the DWCNTs were densely packed and vertically aligned. High resolution TEM images revealed that only clean nanotubes were observed, which are mainly DWCNTs in the form of small bundles, as shown in Fig. 2.50c. The high quality characteristics of the DWCNTs were also confirmed by medium-scale wide-view TEM results. Figure 2.50d shows low-resolution TEM image of the as-grown forest, further demonstrating the absence of metallic particles and supporting materials. The TEM images also indicated that the difference between the radii of the inner tubes and outer tubes was close to the spacing of the layers in graphite. Such DWCNT forests/arrays could be used to make highly organized structures by using lithographically patterned catalyst islands.

When CNTs synthesized with CVD are randomly arranged as porous networks, CNT sponges are formed [114–118]. Figure 2.51a shows a schematic diagram of the CVD set-up to directly synthesize CNT sponge reported by Gui et al. [114]. Mixtures of dichlorobenzene as carbon source and ferrocene as catalyst were injected into the CVD chamber at given rates. As the source injection rate was varied from 0.10 to 0.25 mL min^{-1}, the densities of the sponges was increased from 5.8 to 25.5 mg cm^{-3}. At the same time, the average outer diameter of the CNTs was increased from 32 to 44 nm, whereas the inner diameter was almost unchanged, which was about 16 nm. The increase in the density of the sponge was attributed to the increased formation rate and diameter of the CNTs, with increasing feeding rate. Although a 5-fold increase in density was observed, the porosity of sponges was only slightly decreased, dropping from 99.8 to 98.9 %. Figure 2.51b shows a piece of as-grown CNT sponge, which was so soft and flexible that it could be tightly scrolled without breaking. Figure 2.51c shows two sponge bricks with densities of 5.8 and 11.6 mg cm^{-3}, which were placed in series to support a 200 g standard weight. The brick of lower density showed much larger deformation than the denser one. They also showed different degrees of volume recovery after removing the weight. The sponges with lower-densities showed very high compressibility of up

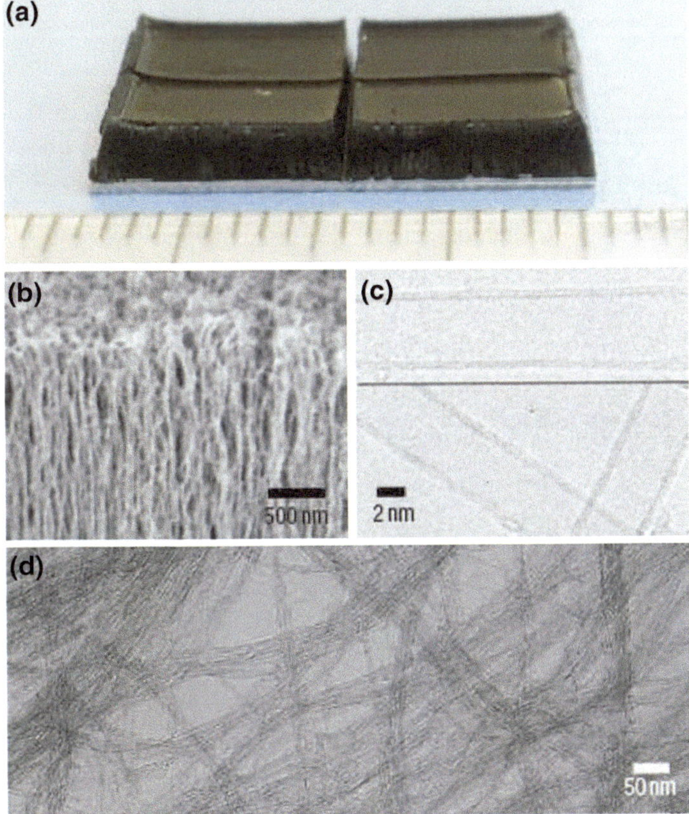

Fig. 2.50 Photograph of the DWCNT arrays and SEM/TEM images of the DWCNTs: **a** 20 × 20 mm vertically standing DWCNT forest with a height of 2.2 mm grown using a 1.69 nm Fe catalyst film, where the scale along the bottom is marked in mm. **b** SEM image of the edge of the DWCNT forest shown in (**a**). **c** High-resolution TEM image of the DWCNTs without the presence of amorphous carbon and metal particles. **d** Low-resolution TEM image of the DWCNTs showing the absence of metallic particles and support materials. Reproduced with permission from [129], Copyright © 2006, Nature Publishing Group

to 90 % volume reduction, while those with higher-density could recover to 93 % of the original volume after compression. Electrical resistivity of the CNT sponges changed linearly and reversibly during large-strain compression cycles. The sponges could be used to fabricate conductive composites with resistivity comparable with that of pure CNT scaffolds [114].

Figure 2.52a shows another piece of similar CNT sponges reported by the same authors [115]. The sponges consisted of CNTs that were randomly self-assembled and interconnected to form highly porous three-dimensional frameworks, as shown in Fig. 2.52b. The morphology and distribution of the CNTs were almost the same at top surface and side-walls. The CNTs were multi-walled, with diameters

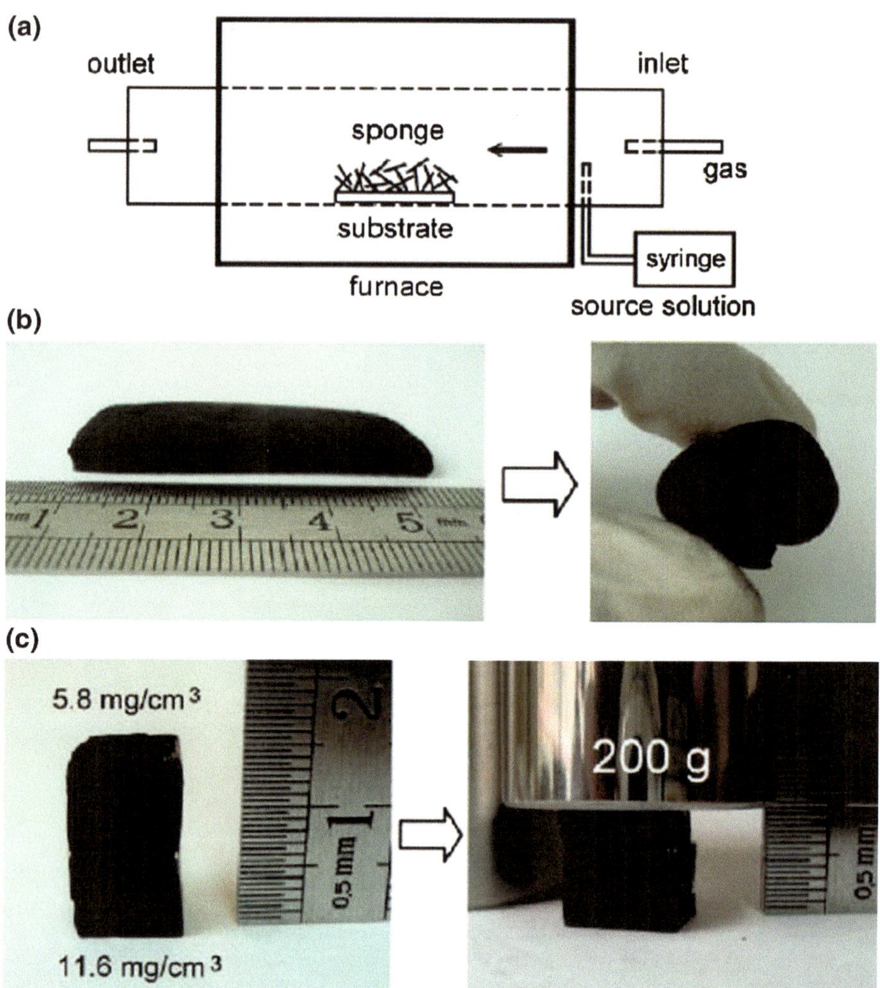

Fig. 2.51 **a** Schematic of the CVD set-up to produce the soft CNT sponges with controlled density and porosity. The source solution consisting of ferrocene and dichlorobenzene was constantly injected into the furnace chamber by using a syringe pump. Carbon nanotubes were grown on the quartz substrate and overlapped into porous sponges. **b** Photographs of a soft carpet-like sponge which could be rolled up tightly. **c** Photographs of two stacked sponges supporting a 200 g weight, in which the top sponge was compressed to a larger degree due to its lower density (5.8 mg cm^{-3}). Reproduced with permission from [114], Copyright © 2010, American Chemical Society

30–50 nm and lengths of tens to hundreds of micrometers, as shown in Fig. 2.52c. Many catalyst particles were encapsulated inside the frame of the CNT sponges. Growth rate in the thickness direction was about 2–3 mm per hour, which was almost constant during the entire process, which indicated that the sponge could be produced in massive production. The key to the formation of the random

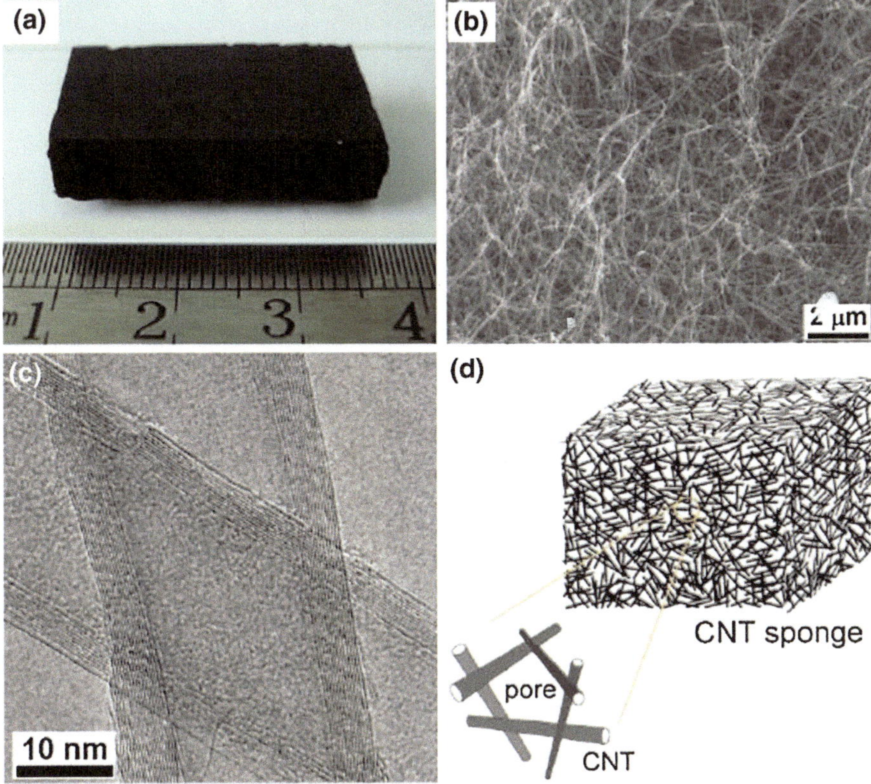

Fig. 2.52 **a** Photograph of a monolithic light, porous and flexible CNT sponges. Sponge with a size of 4 cm × 3 cm × 0.8 cm and a bulk density of 7.5 mg cm^{-3}. **b** Cross-sectional SEM image of the sponge showing a porous morphology and overlapped CNTs. **c** TEM image of large-cavity thin-walled CNTs. **d** Illustration of the sponge consisting of CNTs, with piles (*black lines*) as the skeleton and open pores (void space). Reproduced with permission from [115], Copyright © 2010, John Wiley & Sons

entanglement of the CNTs in this method was the use of dichlorobenzene as the carbon source, which disturbed the CNTs to grow all in one direction. Figure 2.52d shows a schematic diagram, demonstrating the formation of the CNT sponges.

Density of the as-grown sponges was about 5–10 mg cm^{-3}, which was comparable with those of the un-densified vertically aligned arrays and solution-processed aerogels slightly higher than that of the aerogel sheets drawn from CNT forests. The CNT sponges had surface area of 300–400 m^2 g^{-1} and average pore size of about 80 nm, which was close to the average inter-tube spacing. The pristine sponges were hydrophobic, with a contact angle of about 156° for water droplets, so that they could be floated on water due to their low densities, while the conventional cleaning sponge was sank when being in contact with water, because it was made of hydrophilic micro polyurethane fibers, as shown in Fig. 2.53a [115]. The CNT

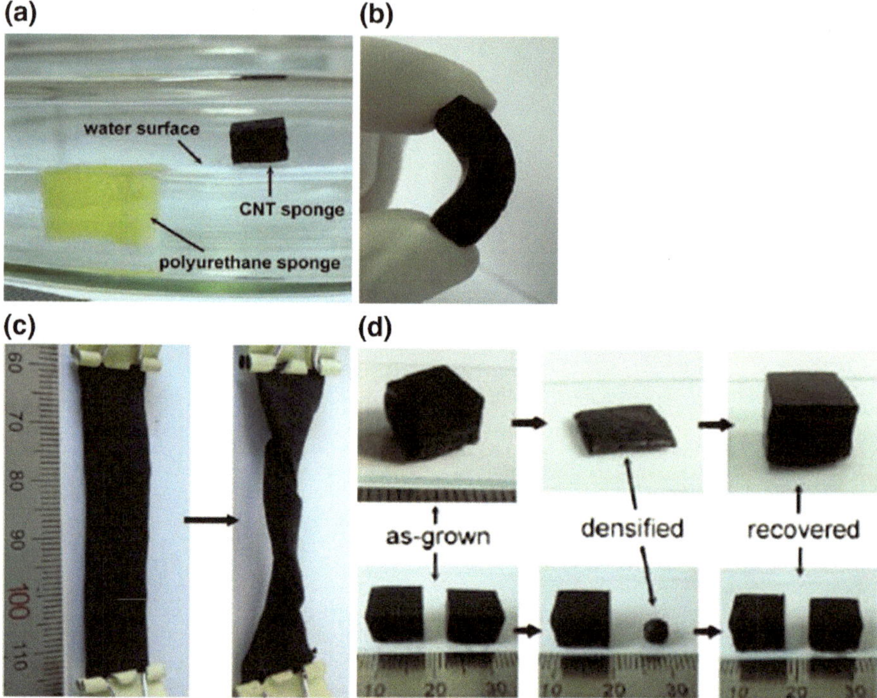

Fig. 2.53 **a** Photograph of a CNT sponge and a polymeric sponge placed in a water bath. The CNT sponge was floating on the top while the polyurethane sponge absorbed water and sank to below the surface level. **b** A CNT sponge bent to arch-shape at a large-angle by finger tips. **c** A 5.5 cm × 1 cm × 0.18 cm sponge twisted by three turns at the ends without breaking. **d** Densification of two cubic-shaped sponges into small pellets, a flat carpet and a spherical particle, respectively, and full recovery to original structure upon absorbing ethanol. Reproduced with permission from [115], Copyright © 2010, John Wiley & Sons

sponges could be easily manually compressed to >50 % volume reduction. They were mechanically flexible and strong, so that they could be bent or twisted without breaking and nearly entire recovery in shape, as shown in Fig. 2.53b, c. The mechanical robustness of the CNT sponges was attributed to three-dimensionally isotropic configuration formed by the highly interconnected CNTs, which prevented the sliding or splitting between CNTs in any direction.

The CNT sponges could be densified by using ethanol [115]. Figure 2.53d shows that the volume of the sponges could be reduced by 10–20 times. The densified pellets would instantaneously swell back to the original volume by adding ethanol. The densification and swelling could be repeated and high reversible. This was because the randomly distributed individual CNTs prevented the formation of van der Waals interactions, so that the original configuration could be easily recovered through liquid absorption.

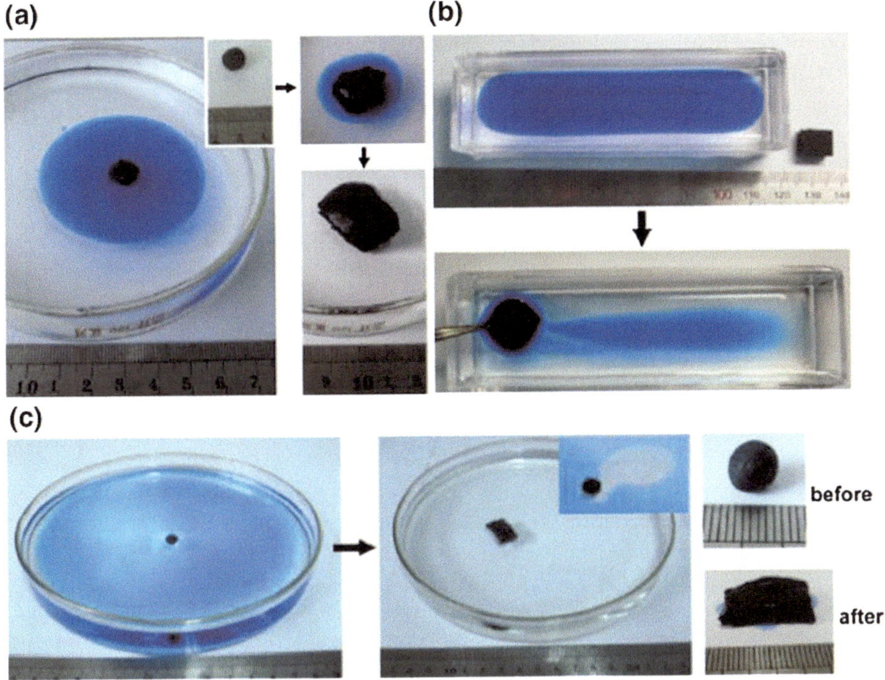

Fig. 2.54 **a** Snapshots showing the absorption of a 28 cm^2 and mm-thick vegetable oil film (dyed with *Oil Blue*) distributed on water by a small spherical sponge. **b** Active absorption of a continuous 20 cm^2 oil strip distributed in a rectangular water channel by a small non-densified sponge (<1 cm^3) placed near the left edge and in contact with the oil. **c** Large-area oil cleanup. *Left panel* a diesel oil film (2 g in total weight) with an area of 227 cm^2 spreading on water and a densified sponge pellet placed at the center. The oil area was about 800 times that of the projected sponge area. *Middle panel* clean water surface after complete oil absorption by the sponge, which grew to a larger size and changed to a rectangular shape. Inset shows the floating pellet during the absorption process. The *white-color* region along the path of the drifting pellet indicates the removal of oil film and exposure of fresh water. *Right panel* photographs of the CNT sponge before and after the oil absorption. The sponge was swollen from a 6-mm-diameter spherical particle to a rectangular monolith of 2 cm × 1.4 cm × 0.6 cm after collecting all the oil from water. Reproduced with permission from [115], Copyright © 2010, John Wiley & Sons

The CNT sponges demonstrated very high absorption capacities for a wide range of solvents and oils, which were 80–180 times their own weight [115]. The mechanism was physical absorption of organic molecules that could be stored in the pores of the sponges. The sponges exhibited much better absorption performance than various porous materials. For instance, the authors used various natural fibrous materials, like cotton towel and loofah, as comparison. More importantly, the absorption behavior was highly reversible. Oil absorption performances of the pristine and densified CNT sponges are demonstrated in Fig. 2.54 [115]. Figure 2.54a shows that the oil film kept shrinking toward the center while the size of the sponge increased gradually due to the oil absorption. After the oil was completely absorbed, the sponge still floated on water and recovered to its original dimension. The pristine sponge could continuously attract and suck an oil film when it was just in contact with the edge of the film, as shown in Fig. 2.54b. A small piece of densified CNT sponge was able to remove a spreading diesel oil film with a very large area in minutes, as shown in Fig. 2.54c. The area of the oil that could be completely cleaned was nearly 800 times larger than the size of the densified sponge, showing huge oil absorption capability and efficiency of the CNT sponges. Furthermore, the CNT sponges exhibited low thermal conductivity which was similar to that of most porous materials, while they had much higher electrical conductivity.

Another interesting feature of CNT sponges is the creep recovery over a wide range of temperature, from as low as −190 °C to as high as 970 °C, as demonstrated by Xu et al. [116, 117]. The CNT sponges were synthesized by using a water-assisted CVD at 750 °C, with ethylene as carbon source, combined with reactive ion etching (RIE) of Al_2O_3/Fe catalyst layer [116]. They had a density of 0.009 g cm^{-3}, with ~68 % DWCNTs. The as-synthesized sponges of 4.5 mm thick could be compressed to ~1.1 mm, corresponding to a density of 0.036 g cm^{-3}. The sponges were extremely mechanically flexible, as shown in Fig. 2.55a, which was attributed to entangled network microstructure, as shown in Fig. 2.55b. Three point bending testing results were conducted from −190 to 970 °C. The high temperatures were created by using butane torch, while liquid nitrogen was used to cool the sponges during the testing. The CNT sponges could be bent repeatedly with large

Fig. 2.55 The CNT sponge showing elasticity and compliance over a wide temperature range. **a** Photograph of the compliant CNT sponge bent at a large degree without fracture. **b** SEM image showing the non-aligned entangled structure. Photographs of the elasticity and compliance of the CNT sponge with three point bending testing at **c** 25 °C. **d** 970 °C (produced by using butane torch) and **e** −190 °C (surrounded by using liquid nitrogen). **f** Optical real-time images captured from a movie to show the creep recovery at 970 °C. Reproduced with permission from [117], Copyright © 2011, John Wiley & Sons

deformation without fracture, as shown in Fig. 2.55c–f. After stress was released, the CNT sponges all recovered to nearly its original state progressively, i.e., they had strong creep recovery at the extreme temperatures.

The CNT sponges also exhibited high thermal stability [117]. In contrast, most conventional viscoelastic materials would experience structural distortion even at stresses below their failure stress, if they are stressed at high temperatures for long time durations. For example, the CNT sponges could withstand the stress at 20 kPa for 4 h at temperatures of up to 450 °C. However, silicone rubber was broke down in 5 min at 350 °C. Therefore, these CNT sponges are promising candidates for applications at extreme conditions, such as space crafts, rockets and engines.

More recently, CNT bulk or bulk-like materials, comprising of ordered arrays and disordered assemblies, with interesting mechanical properties, have been reported [119, 121]. Composite structures, with various combinations of the two types of CNTs, including sponge-on-array double layers, sponge-array-sponge and array-sponge-array triple layers, could be directly synthesized by using CVD, by controlling the synthesis conditions [121]. For example, the aligned arrays were grown by using source solution of ferrocene dissolved in xylene with concentration of 0.02 g mL^{-1} and injection rate of 0.4 mL min^{-1}. To grow sponges, 0.06 g mL^{-1} ferrocene dissolved in dichlorobenzene was used, with injection rate of 0.3 mL min^{-1}.

The steps to grow the composite structures are schematically shown in Fig. 2.56a, while Fig. 2.56b summarizes the individual single-layers and the composite structures, two-layer and three-layer, synthesized by alternatively injecting the two carbon sources during the CVD process [121]. For instance, after an array layer was first grown by feeding xylene solution, a sponge layer would be grown on surface of the array by switching to dichlorobenzene solution, while another sponge layer on the other side of the array was grown by flipping the array, leading to a sponge–array–sponge sandwich composite structure. Because CNT sponges are grown in vapor phase, they can be deposited on an array that acts as substrate. In contrast, it is more difficult to grow an array on sponge. Nevertheless, the authors successfully fabricated various composite structures, as shown in Fig. 2.56c–f. It was reported that the growth rate could reach 25 μm min^{-1} and composites millimeter thickness could be readily detached from the substrate as freestanding bulk materials. Therefore, it is a versatile yet simple fabrication process.

Figure 2.57 shows representative SEM and TEM characterization results of the CNT composite structures [121]. Figure 2.57b shows cross-sectional SEM image of a double-layer structure, consisting of a sponge layer sitting on an array layer. SEM images of two typical triple-layer composites are shown in Fig. 2.57a, c, with clearly observable sponge-array interfaces, at which the sponge and the array exhibited different appearances. The sponge consisted of random CNT network, as shown in Fig. 2.57d. The CNTs from the sponge occasionally penetrated into the empty space of the array, as shown in Fig. 2.57f, g. Parallel and continuous CNTs were formed in the array (Fig. 2.57h). TEM results indicated that both the sponge and the array consisted of MWCNTs. The CNTs in the sponge had thinner walls

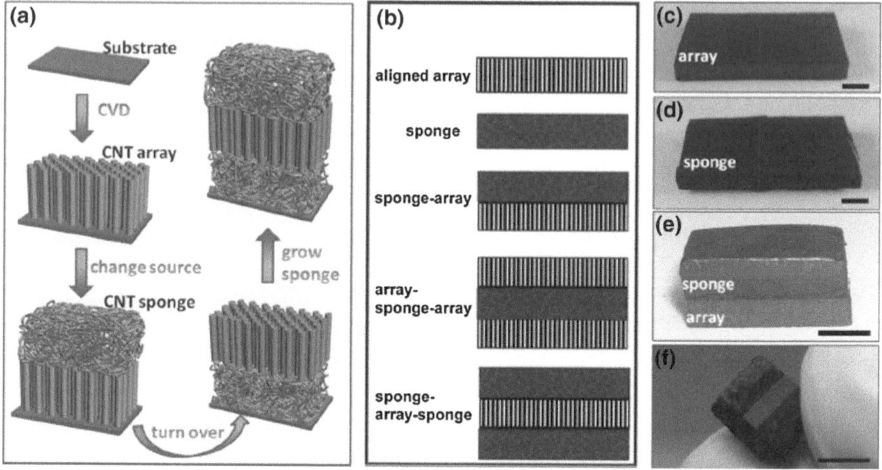

Fig. 2.56 Schematic of the CNT sponge-array composite structures and steps of the growth process: **a** CVD process to synthesize a sponge–array–sponge triple layer composite, involving the growth of an aligned array (first layer) on a quartz substrate, the growth of a sponge layer on top of the array by switching the carbon source and the growth of another sponge layer on the other side of the array. **b** Schematic of single-layers and composite structures, including individual array and sponge, double-layer and triple-layer structures. Photographs of the as-synthesized aligned CNT array (**c**), CNT sponge (**d**), sponge-on-array double-layer (**e**) and sponge-array-sponge triple-layer structures (**f**). Scale bars in (**c**–**f**): 5 mm. Reproduced with permission from [121], Copyright © 2013, John Wiley & Sons

and larger cavity with diameters of 25–40 nm Fig. 2.57e, whereas those in the array had larger diameters (50–80 nm) and thicker walls or narrower cavity Fig. 2.57i. The difference in dimension between the CNTs in the two layers was attributed to the difference in the two carbon sources.

The sponge with the randomly oriented thinner CNTs of large cavity was soft, while the array with the thicker CNTs of small cavity was much stronger. During compression, the soft sponge layer deformed first at low stresses, whereas the rigid array deformed later at higher stresses. The multilayered structure and sequential deformation mechanism were beneficial to mechanical energy absorption, where a large strain was generated by the sponge and a high stress was available due to the aligned array. As a result, the composite structures exhibited a wider stress range with relatively low cushioning coefficients, when compared with individual arrays or sponges. This is an effect method to fabricate CNT composite structures that could find interesting applications.

The above authors further extended the method to fabricate 3D architectures, in which the CNT arrays and sponges were arranged into different morphologies and orientations with clear interfaces [119]. Various complex and hierarchical systems have been synthesized by combining the two types of CNT assemblies. Representative structures, including series (3 array blocks connected by 2 sponge layers), parallel (sponge filled into space between vertical arrays), package (an array

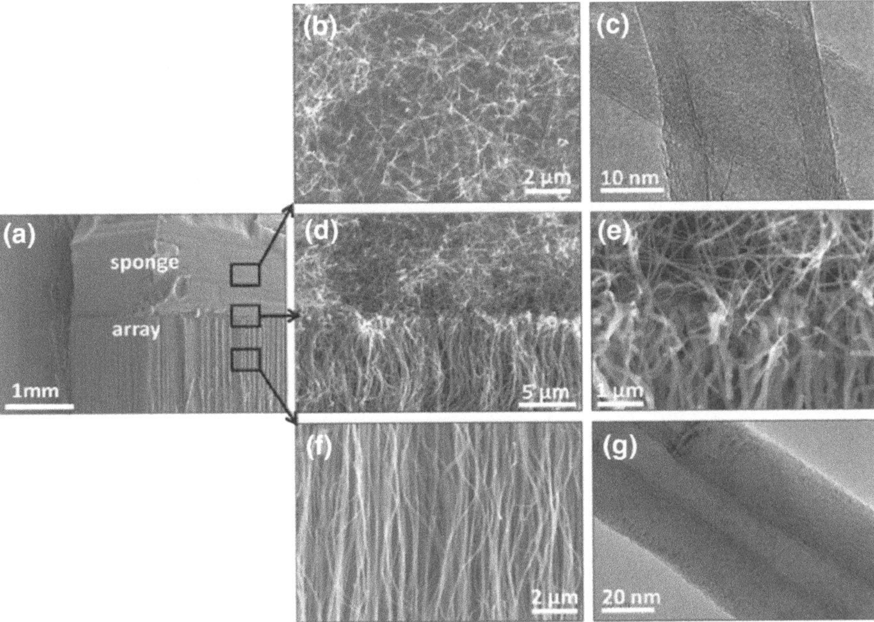

Fig. 2.57 Structural characterization of the CNT composites and the CNTS in the arrays and sponges: **a** SEM image of a sponge-array-sponge triple-layer composite. **b** SEM image of a sponge-on-array double-layer composite structure. **c** SEM image of a array-sponge-array triple-layer composite. **d** Close view of the upper sponge in (**b**). **e** TEM image of the MWCNTs in the sponge. **f** SEM image of the interface in (**b**). **g** SEM image of the interface at high magnification. **h** SEM image of the bottom array. **i** TEM image of the MWCNTs in the array. Reproduced with permission from [121], Copyright © 2013, John Wiley & Sons

with exposed surfaces covered with a conformable sponge coating) and sandwich (sponge filled among array blocks with different orientations), are shown schematically in Fig. 2.58a–d [119]. The arrays and sponges could be combined to form monoliths with pretty good adherence and interface due to the direct growth process. The sharp interfaces between the aligned and the random CNTs in each type of structure were observed, as shown in Fig. 2.58e–g. The sponge partly contacted the tips of the CNTs in the array, i.e., in series shown in Fig. 2.58e, or the side of the array, i.e., in parallel shown in Fig. 2.58f. When an array block was wrapped by a layer of sponge, a core-shell structure was formed, in which the core was hard while the shell was soft, as shown in Fig. 2.58g. Figure 2.58d shows a complex integrated structure, with arrays having different orientations and sponges to fill the empty spaces. It is believed that the versatility of the method is only limited by our imagination.

The method has also been developed to fabricate 3D heterostructures by patterning the substrate, so that selective growths of arrays or sponges could be realized. One example is shown in Fig. 2.59a [119]. Thin Au film on quartz substrate was patterned with square blocks by using copper mesh as hard mask.

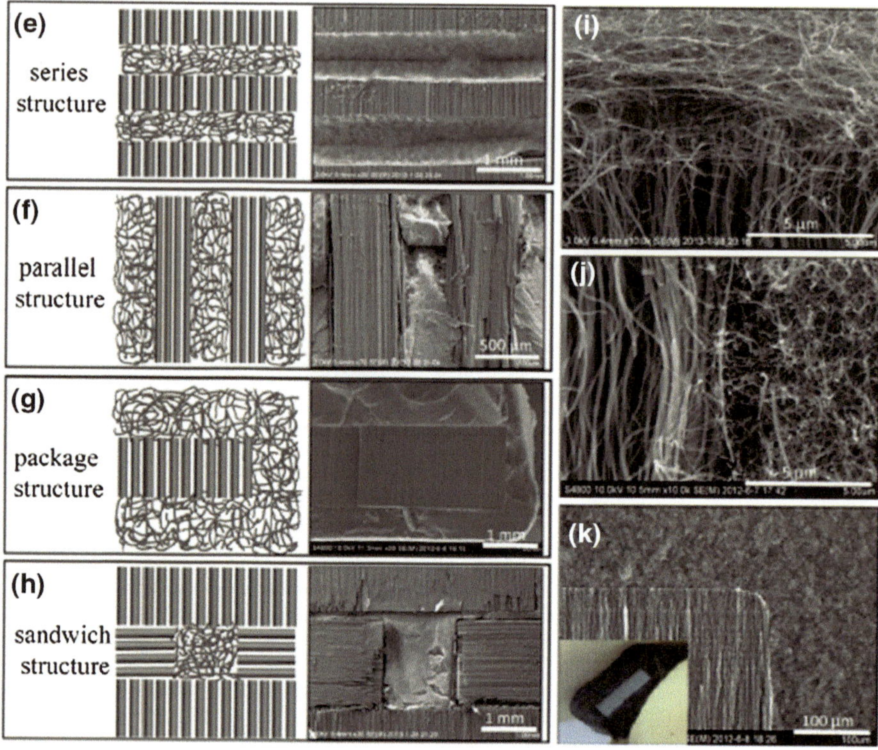

Fig. 2.58 Morphologies of the 3D CNT sponge-array architectures. **a–d** Schematic diagrams and SEM images of the series, parallel, package and sandwich structures. **e–g** SEM images of the array-sponge interfacial regions in a series, parallel and package structures, respectively. Inset of (**g**) is photograph of a package structure. Reproduced with permission from [119], Copyright © 2014, John Wiley & Sons

Because of the decatalytic behavior of Au nanoparticles to CNT arrays, only the uncovered regions were covered by CNT arrays. As a result, after the first CVD growth, a cross-wall array network with built-in square holes was obtained, as shown in Fig. 2.59b, c. However, CNT sponges could be grown in on the metal film areas. Therefore, this combination led to a new structure, in which sponge units were embedded in the patterned arrays. Figure 2.59d shows that the patterned CNT arrays had walls of 80 μm in width and 200 μm in height. The empty areas among the array network could be completely filled with the sponge of similar height to the array, as shown in Fig. 2.59e, f. Sharp interface was observed between the array and the sponge, as shown in Fig. 2.59g [119].

Mechanical properties and energy absorption capabilities of the CNT structures could be designed by tailoring the configurations of the structures. For example, the parallel structure could absorb more energy by maintaining a stable stress plateau, but density of the material could not be selected according to the desired maximum

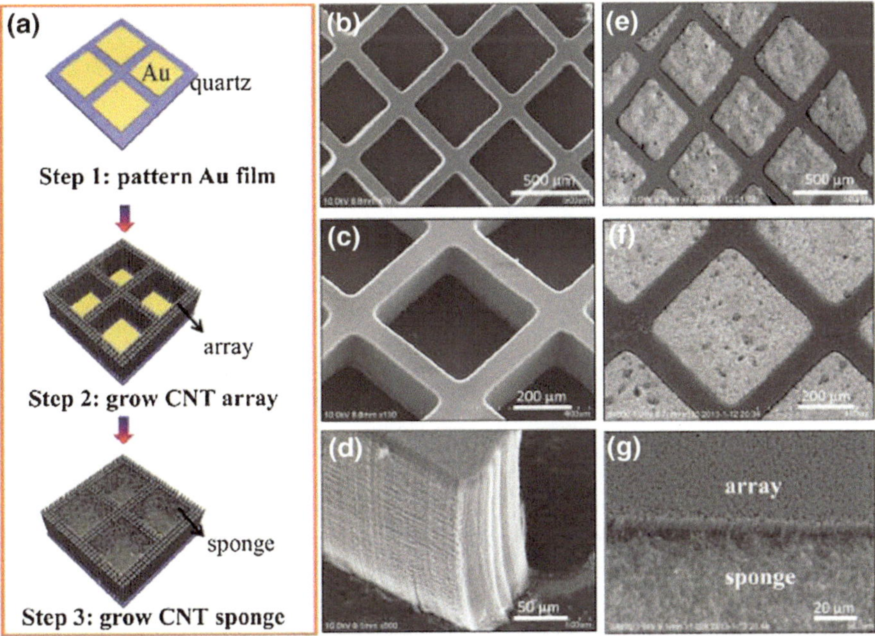

Fig. 2.59 **a** Fabrication steps of patterned parallel structures, including patterning of Au film on quartz substrates and growth of CNT arrays on the quartz area and CNT sponges on the Au film. **b–d** Low and high magnification SEM images of the CNT arrays grown as grid patterns by using a copper mesh mask. **e**, **f** SEM images of the parallel structure where the empty areas among the array were completely filled with sponges. **g** SEM image of the sponge-array interface. Reproduced with permission from [119], Copyright © 2014, John Wiley & Sons

strain or stress and the amount of energy to be absorbed. Energy absorption materials need to dissipate kinetic energy while keeping the maximum impact force below certain limits. In energy–strain curves, the parallel structures had a typical shoulder. For a given energy absorption, at different strain ranges, the materials should have different densities. If the density of the material is too high, the force cannot exceed the critical value before the energy is completely absorbed, so that the material stays within the low compressive strain range. In contrast, if the density is too low, the materials will be densified (collapsed). To achieve efficient energy absorption, the density of the parallel structure should be optimized, which can be achieved by using this synthesis method [119]. For instance, parallel structures with a wide range of densities from 74.1 to 163.3 mg cm^{-3}), having high energy loss coefficients of 0.74–0.80, have been constructed by the authors [119]. Such structures can find potential applications for energy dissipation.

2.5.4 Gels

Gels or aerogels based on CNTs formed a new group of 3D carbon nanomaterials [122–125]. Figure 2.60 shows a schematic illustrate of the mechanism to explain the formation of CNT aerogels by using freeze drying [122]. Firstly, ice nuclei initiated freezing in the supercooled solution or suspension. After that, the ice crystals gradually grew up from bottom to top of the sample. During the freezing process, the carboxymethyle cellulose (CMC) sodium salt surfactant, which was attached to the CNTs, was pushed into the inter-dendritic spaces by the ice dendrites. As a result, the CMC sodium salt with the embedded CNTs was assembled onto surface of the ice crystals. Once the bulk matrix was entirely solidified, the sample was lyophilized to remove the ice crystals, so that macroporous solid foam was developed. This mechanism can be used to explain the formation of similar CNT aerogels.

A representative MWCNT aerogel, with elastic feature, is shown in Fig. 2.61 [122]. It exhibited sponge-like characteristic due to its elasticity. Such CNT gels had very high porosity of 97 % and low bulk density of 0.05–0.06 g cm^{-3}. The MWCNTs used were commercial product, with average outer diameter of 13-16 nm and length of up to tens of micron. CMC sodium salt was used as dispersant. Aqueous dispersions of the MWCNTs with different concentrations were freeze dried at −40 °C and then lyophilized at −20 °C. The MWCNT gels were formed due to the presence of the CMC sodium salt surfactant, which facilitated the incorporation of the MWCNTs in the walls of the foams. Therefore, the mechanical strength of the CNT gels was dependent on both the concentration of the surfactant and the content of the MWCNTs. In addition, other processing parameters, especially cooling rate, also had strong influence on properties of the CNT gels. For example, a fast cooling rate resulted in a high degree of homogeneity and a small macropore network. The MWCNT gels exhibited gas sensing capability, with rapid response and high sensitivity.

Kim et al. [124] reported a method to prepare SWCNT aerogels. The SWCNTs had diameters of 0.8 ± 0.1 nm and lengths of 0.45–1 μm. The CNTs were dispersed in deionized water solution of 0.1 wt% sodium dodecylbenzene sulfonate (SDBS) as surfactant. Gels were formed from concentrated suspension due the van der Waals interactions between the CNTs. The surfactant SDBS was removed by washing with 1 M nitric acid at 50 °C for 20 min. The water contained in the

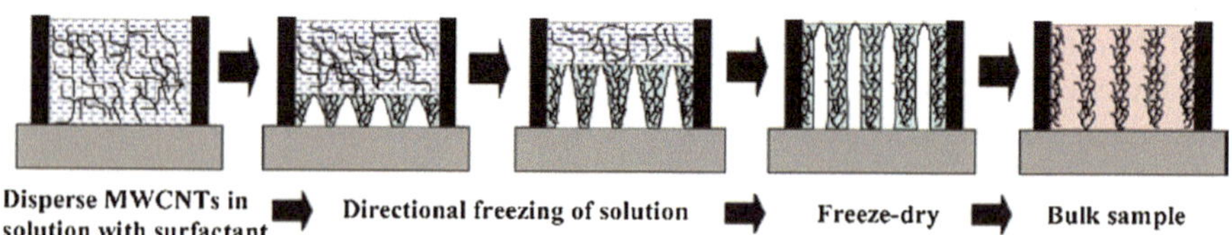

Fig. 2.60 Sketch of a mechanism for the formation of CNT aerogels (solid foams). Reproduced with permission from [122], Copyright © 2008, Elsevier

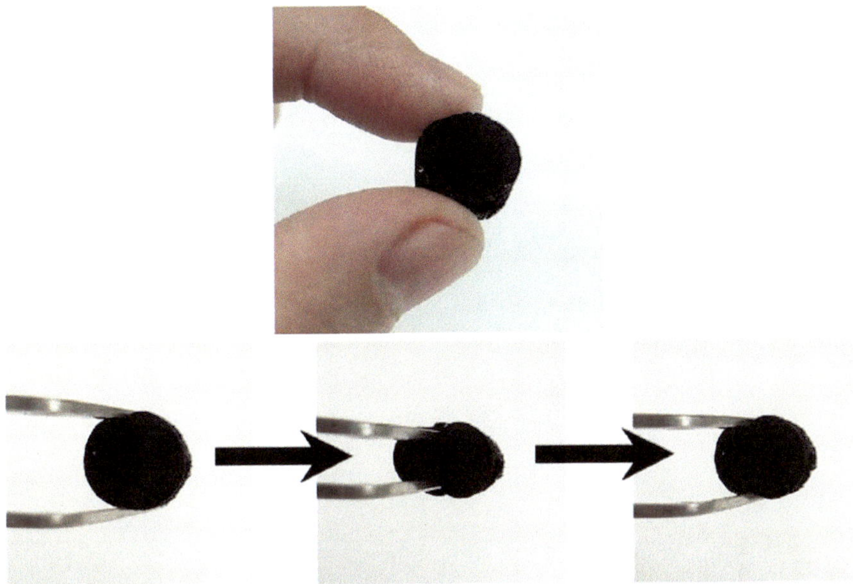

Fig. 2.61 Elastic feature of the obtained MWCNT gels. Reproduced with permission from [122], Copyright © 2008, Elsevier

hydrogels was removed by exchanging with ethanol, which was then removed by using crystal point drying to form aerogels.

Figure 2.62a shows photographs of the free-standing SWCNT aerogels, with various shapes and sizes, which were obtained by using different molds [124]. The aerogels were formed as porous isotropic CNT network with an open-cell structure, as shown in Fig. 2.62b. TEM images of the CNTs are illustrated in Fig. 2.62c, d. The SWCNT aerogels had high electrical conductivities, which were dependent on their density. For instance, as the density of the aerogels was increased from 7.3 to 14.1 mg ml^{-1}, electrical conductivity was linearly increased from 23 to 71 S m^{-1}. Due to their high porosity, they could be compressed by larger than 90 %. Because of the filamentous cell walls, the aerogels could be compacted along the compression direction without expansion normal to the compression direction. Additionally, the aerogels showed enhanced heat transfer when compared with air in forced convective looling method.

Sun et al. [125] developed a method, sol-cryo, to fabricate CNT aerogels with ultra-flyweight by incorporating with giant graphene oxide (GGO) nanosheet. GGO aqueous dispersion and CNTs aqueous dispersion were mixed, which was then freeze-dried for two days to form GGO/CNTs foam. The as-obtained GGO/CNTs forms were reduced in hydrazine vapor at 90 °C for 24 h and then vacuum dried at 160 °C for 24 h. The aerogels prepared in this way had densities of 0.16–22.4 mg cm^{-3}, depending on the concentrations of the GGO and CNTs in the aqueous suspensions. Because the assembly process in the mothed was simple and

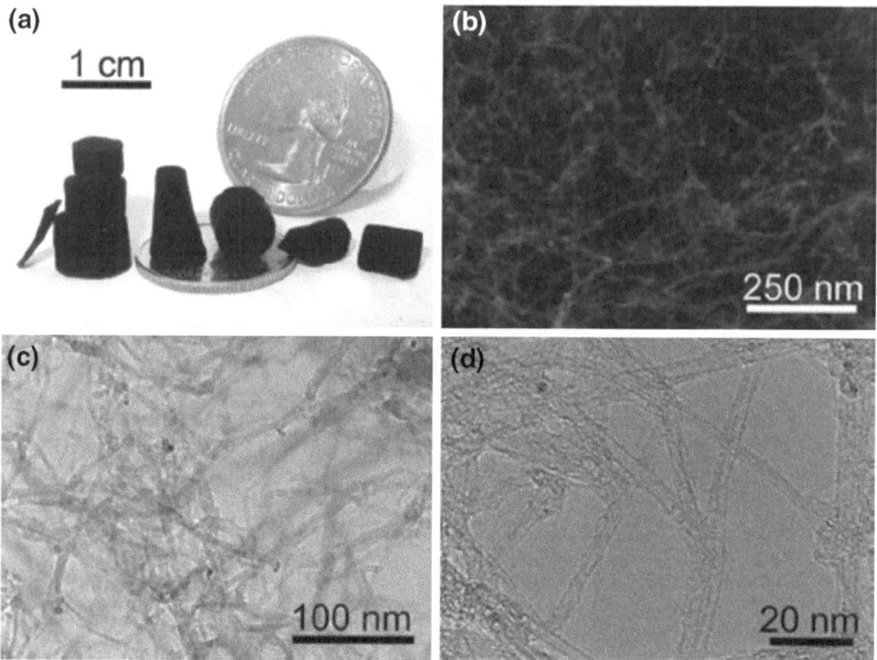

Fig. 2.62 Images of SWCNT aerogels. **a** Our fabrication method allowed production of aerogels in various sizes and shapes, such as *cylinders*, *rectangles*, *cubes*, *truncated cylinders*, and *truncated cones*. **b** An SEM image of an aerogel cross-section showing networks of nanotubes. **c** Low and **d** high resolution TEM images show that the network composed of mostly isolated and entangled nanotubes. Reproduced with permission from [124], Copyright © 2013, John Wiley and Sons

template-free, it could be easily scaled up and was able to readily control the density and shape of the aerogels, as shown in Fig. 2.63a–c [125].

The aerogels possessed an interconnected, randomly oriented, porous 3D framework structure, crinkly sheets with continuous macropores with sizes ranging from hundreds of nanometers to tens of micrometers, as shown in Fig. 2.63d. Figure 2.63e shows that the graphene cell walls were covered by entangled spaghetti-like CNT networks. Multiform interconnections of sheets were observed in the microstructure: overlapping, twisting and enwrapping, as shown in Fig. 2.63f–h, respectively. Their schematic illustrations, demonstrating the CNT-coated graphene building block and the interconnections, are shown in Fig. 2.63i–l. The CNTs with lengths of hundreds of nanometres to tens of micrometers were tightly adhered to the graphene substrates (Fig. 2.63m). The single layer graphene had a thickness of about 0.3 nm, with curled edges, as shown in the inset of Fig. 2.63m. A set of hexagonal diffraction spots in selected area electron diffractions (SAED) was observed, as shown as the inset of Fig. 2.63n, confirming the presence of monolayer graphene.

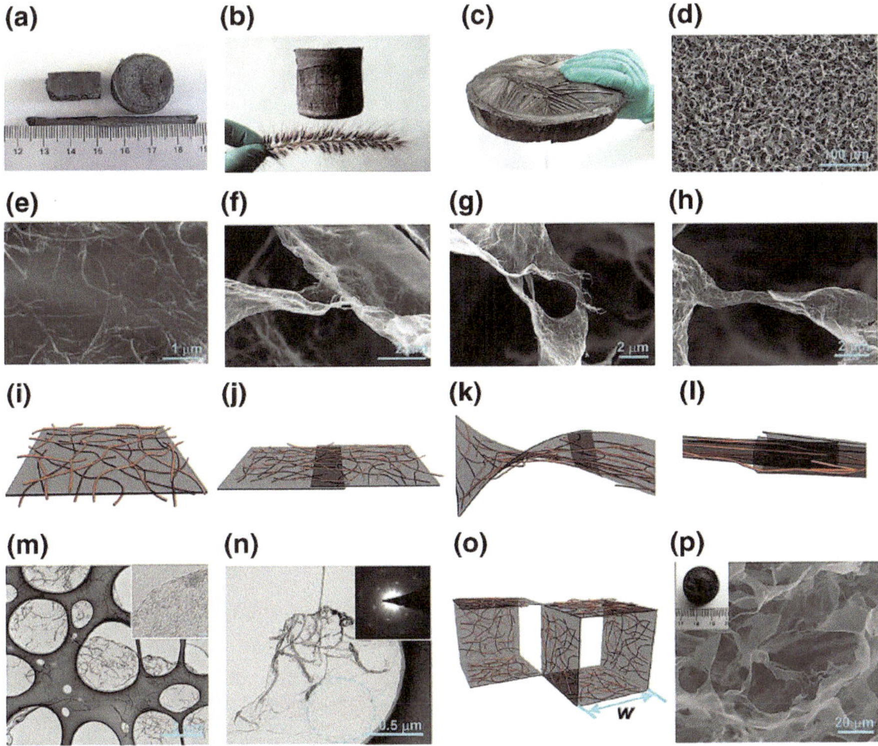

Fig. 2.63 Macroscopic and microscopic structures of the GGO/CNTs aerogels. **a** Photograph of the aerogels with various shapes. **b** A 100 cm^3 aerogle cylinder standing on a flower like dog's tail. **c** A ~ 1000 cm^3 aerogel cylinder (21 cm in diameter and 3 cm in thickness). **d–e** Microscopically porous architectures of the aerogel at different magnifications, showing the CNT-coated graphene cell walls. **f–h** SEM images of different interconnections of the CNTs-coated graphene (graphene@CNTs) cell walls: overlapping (**f**), twisting (**g**) and wrapping (**h**). **i** Schematic illustration of a flattened CNT-coated graphene cell wall. **j–l** Schematic illustrations of three types of interconnection corresponding to (**f–g**) respectively. The gray lamellar, orange wires and brown wires represent the graphene sheet, CNTs coatings on top and the CNTs coatings on back of the graphene nanosheets. **m**, **n** TEM images of the CNT-coated graphene cell walls, with HRTEM image of a cell wall edge (inset in (**m**), scale bar of 2 nm) and the SAED patterns. The inset in (**n**)) was taken at the labeled area. **o** Schematic illustration of idealized building cells of the aerogels through the synergistic assembly of the graphene nsnosheets and CNTs. **p** SEM image and photograph (inset in (**p**)) of the lightest neat graphene aerogel (ρ = 0.16 mg cm^{-3}). Reproduced with permission from [125], Copyright © 2013, John Wiley & Sons

The cyclic patterns were due to the CNTs that covered the graphene nanosheets [125].

Figure 2.63o shows a structural model of the aerogels, which could be used to calculate density of the aerogels [125]. The aerogels could nearly completely recover after 50–82 % compression, demonstrating their super-elasticity. The elastic properties, including storage modulus, loss modulus and damping ratio, were almost constant over a wide range of temperature from −190 to 300 °C. They also exhibited a typical plastic deformation at room temperature under tensile strength. In the elastic region, the aerogels had a typical Young's modulus of 329 kPa at small deformation (<0.5 %). In addition, they had high toughness with fracture elongation of as high as 16.5 %, with fracture strength to be 10.9 kPa, making them suitable for practical applications. The 3D structural integrity of the aerogels under large deformations was ascribed to the isotropic assembly of the graphene nanosheets entangled with the CNTs, which prevented the graphene nanosheets from sliding and splitting along a certain direction. The synergistic effect of the graphene cell was and the CNT ribs were responsible for the elasticity of the aerogels. Together with high thermal stability, excellent absorption capacity of organic liquids and phase change materials, as well as high electrical conductivity, the GGO/CNT hybrid aerogels have formed a new class of multifunctional materials.

2.6 Potential Applications

The ultimate goals of materials research and development are practical applications, so are the CNT-based carbon nanomaterials. Various applications, including multifunctional composites, sensing/biosensing devices, transmission lines and electrochemical devices, and so on, by using CNT-base carbon nanomaterials, have been reported. Detailed descriptions on such applications can be found in Refs. [12, 14] and references therein. A brief description of each type of material will be presented as follows.

The most straightforward application of CNT fibers is the use as reinforcement agent to fabricate polymer based composites. The nanoscale diameters and microscale lengths of individual CNTs have made it difficult to control their orientation and distribution in polymer matrix. The specific strength and stiffness of CNT fibers are higher than those of commercial fibers. Also, CNT fibers are much more flexible and possess higher strain to failure. Mora et al. were among the those first to use CNT fibers to reinforce epoxy-based composites [137]. The CNT fiber reinforced composites had high compressive strength, making them suitable for loadbearing applications. They also exhibited high toughness, so that they could find applications in anti-meteorite/anti-ballistic shields for satellites, anti-ballistic vests, explosion-proof blankets for aircraft cargo bays, and safety belts [138]. Gao et al. [139] found that the stiffening and strengthening effects of CNT fibers inside polymeric matrices were determined by the characteristics of the CNTs. The molecular coupling between CNTs and polymer chains effectively impeded the initiation of buckling of the CNTs at lower compressive strains. The CNT fibers had high flexibility without experiencing any permanent deformation or failure under compressive stresses.

Due to their piezo-resistive effect, CNT yarns have been used as sensors with excellent repeatability and stability [140]. Electrical resistance of the CNT yarns had response to the change in elastic strain. The resistance–strain behavior was repeatable with strain up to 3.3 %. The resistance of the yarn was very stable during 100 min of measurements at different temperatures. Because CNT yarns have low density, they can be permanently integrated in composites with minimal invasiveness and weight penalty. The embedded CNT yarn sensors can be used to real time monitor the deformation and crack propagation of the composites.

Since metallic CNTs have high electrical conductivity, metallic CNT fibers have the potential to be next-generation conducting wires and power transmission lines. Wei et al. used SWCNT and DWCNTs filaments (fibers) as conducting wires for household light bulbs [141]. The CNT-enabled bulbs have lower threshold voltage for light emission and higher brightness at high voltages. Electrons flow ballistically along the CNTs, without scattering. This characteristic of the CNT fibers is superior to that of metallic wires, such as Cu and Al. Scattering increases the resistance of metals, which causes the power lines to heat up, expand and then sag. Therefore, for such purposes, CNT fibers should be made of pure metallic CNTs. Sundaram et al. [142] found that metallic properties and chiral angle of SWCNTs could be controlled through controlling the sulfur precursor, when spinning CNT fibers directly from the CVD synthesis reaction zone. Electrical conductivity of CNTs could be increased by using doping [143].

Their high specific surface area, excellent mechanical properties and high electrical conductivity have made CNT fibers promising candidates for applications in electrochemical devices such as microelectrodes, supercapacitors and actuators. One example of CNT fiber electrochemical biosensor was demonstrate by Zhu et al. [144]. The CNT fiber (~28 μm) consisted of bundles (~50 nm) of DWCNTs (8–10 nm) entangled to form concentrically compacted multiple layers of nano-yarns along the CNT fiber axis. The electrode end tip of the CNT fiber was freeze-fractured to obtain a unique brush-like nanostructure. Glucose oxidase (GOx) enzyme was immobilized at the brush-like end of the CNT fiber, while the enzyme layer was encapsulated by the epoxy-polyurethane semi-permeable membrane. The CNT fiber microelectrodes exhibited much better sensitivity, detection range and linearity for detecting glucose than Pt–Ir coil electrodes. Significant progress in using CNT fibers for renewable energy harvesting and storages has been made by the Peng's group [145–151].

An early example to demonstrate the application of CNT fibers as actuator was reported by Baughman et al. [152]. The study indicated that CNT fibers could expand and contract upon charge injection and ejection, based on which electromechanical actuators was facilitated. Foroughi et al. used CNT fibers for torsion actuation [153]. They found that an electrolyte-filled twist-spun CNT yarn, much thinner than a human hair, could function as a torsional artificial muscle in a simple three-electrode electrochemical system, with a reversible 15,000° rotation and 590 RPM (revolutions or rotations per minute). The hydrostatic actuation was attributed to the simultaneous occurrence of lengthwise contraction and torsional rotation during the increase in volume of the yarn, which was caused by electrochemical double-layer charge injection. The use of a torsional yarn muscle as a mixer for a fluidic chip was demonstrated. Lima et al. [154] fabricated guest-filled, twist-spun carbon nanotube yarns as electrolyte-free muscles, which exhibited fast, high-force, large-stroke torsional and tensile actuation. More than a million torsional and tensile actuation cycles were demonstrated, in which a muscle could spin a rotor at an average 11,500 RPM or delivered 3 % tensile contraction at 1200 cycles min^{-1}. The hybrid CNT yarns responded to electrical, chemical or photonic excitations.

Free-standing CNT thin sheets are optically transparent, physically light and mechanically strong. One interesting application of such CNT thin sheets was reported by Xiao et al. [155]. They found that thin CNT free-standing film could emit loud sounds, once it was fed with electric currents at sound frequency, which was attributed to a thermoacoustic effect. The ultra-small heat capacity per unit area of CNT thin film led to responses over a wide range of frequency and a high sound pressure level. With that observation, loudspeakers were made by using the CNT thin sheet. Such loudspeakers exhibited various advantages, such as nanometer thickness, high transparency, flexibility and magnet-free. Thicker CNT-based 2D structures have mainly found applications for energy conversion and storage, owing to their high accessible surface area, porous microstructures, high electrical conductivity, low mass density and good chemical stability [156–159]. Supercapacitors with CNT-based electrodes provided much higher capacitances than those ordinary dielectric-based capacitors. CNT-based supercapacitors have also possessed additional functionalities, which are flexible, stretchable and even transparent [160]. Izadi-Najafabadi et al. used aligned SWCNT arrays as electrodes to construct SWCNT-based supercapacitor, which had energy density and maximum power density of 94 Wh kg^{-1} and 210 kW kg^{-1}, respectively [161]. More importantly, CNT-sheet supercapacitors were able to retain charge in the absence of contacting electrolyte, which greatly extended their charge-storage time durations and thus could find various applications [162].

Besides the applications, such as huge oil absorption capability, as discussed in the respective sub-sections, CNT-based 3D could also find applications as membrane filters. For example, Yu et al. fabricated free-standing CNT membranes with high porosity of about 20 % by using a solvent-evaporation-based shrinking method [163]. In the CNT arrays, the diameters of both the CNTs and interstitial pores were about 3 nm. Because no polymer was incorporated, such CNT membranes had very high gas permeability, which was 4–7 orders of magnitude higher than that of the membranes made with other materials reported in the literature. As compared with 1D and 2D structures, CNT-based 3D assemblies should be further explored for more applications.

2.7 Concluding Remarks

Various carbon nanomaterials have been fabricated by using individual carbon nanotubes as building blocks. Examples include super-strong 1D CNT fibers, highly flexible 2D CNT transparent thin sheet, super-compressible 3D CNT foams and so on. These macroscaled CNT assemblies have provided possibility to handle and use under various conditions. More importantly, they have exhibited unique multiple functionalities, including high electrical conductivity, strong mechanical strength, optical transparency (in some cases) and high chemical and thermal stability. As alternative materials, they could be used as high-performance composites for aircraft and automotive industries. In electronics, they can be used as flexible electrodes in displays, antistatic coatings, organic light-emitting diodes, energy-storage devices, filters and biomimic adhesives. It is expected that more and more applications will be reported in the near future.

Although significant progress has been achieved, there are serious challenges that we need to face. For instance, there are too many parameters during the synthesis to have influence on the properties of the individual CNTs, such as SW, DW, MW, dimension (diameter, thickness of wall and length), chirality (conductivity), as so on. Also, no standard is available to guide the techniques and approaches for the large-scale production of macroscopic CNT assemblies. Additionally, more efforts should be made to explore the fabrication, characterization and application of such unique type of nanomaterials.

Acknowledgments One of the authors (LBK) would like to acknowledge the financial support of SUG (M4080845, 2012) by Nanyang Technological University and AcRF Tier 1 (RG 44/12 (M4011056.070), 2013) by MOE, Singapore.

References

1. S. Iijima, Helical microtubules of graphitic carbon. Nature **354**(6348), 56–58 (1991)
2. R.H. Baughman, A.A. Zakhidov, W.A. de Heer, Carbon nanotubes—the route toward applications. Science **297**(5582), 787–792 (2002)
3. R. Andrews et al., Multiwall carbon nanotubes: synthesis and application. Acc. Chem. Res. **35**(12), 1008–1017 (2002)
4. W.R. Yang et al., Carbon nanotubes for biological and biomedical applications. Nanotechnology **18**(41), 412001 (2007)
5. M. Endo, M.S. Strano, P.M. Ajayan, Potential applications of carbon nanotubes. Carbon Nanotubes **111**, 13–61 (2008)
6. J.M. Schnorr, T.M. Swager, Emerging applications of carbon nanotubes. Chem. Mater. **23**(3), 646–657 (2011)
7. M.F. Yu et al., Tensile loading of ropes of single wall carbon nanotubes and their mechanical properties. Phys. Rev. Lett. **84**(24), 5552–5555 (2000)
8. P.G. Collins, P. Avouris, Nanotubes for electronics. Sci. Am. **283**(6), 62–69 (2000)
9. Q.W. Li et al., Structure-dependent electrical properties of carbon nanotube fibers. Adv. Mater. **19**(20), 3358-336 (2007)
10. A.A. Balandin, Thermal properties of graphene and nanostructured carbon materials. Nat. Mater. **10**(8), 569–581 (2011)
11. N. Behabtu, M.J. Green, M. Pasquali, Carbon nanotube-based neat fibers. Nano Today **3**(5–6), 24–34 (2008)
12. L.Q. Liu, W.J. Ma, Z. Zhang, Macroscopic carbon nanotube assemblies: preparation, properties, and potential applications. Small **7**(11), 1504–1520 (2011)
13. K.L. Jiang et al., Superaligned carbon nanotube arrays, films, and yarns: a road to applications. Adv. Mater. **23**(9), 1154–1161 (2011)

14. W.B. Lu et al., State of the art of carbon nanotube fibers: opportunities and challenges. Adv. Mater. **24**(14), 1805–1833 (2012)
15. T.W. Chou et al., An assessment of the science and technology of carbon nanotube-based fibers and composites. Compos. Sci. Technol. **70**(1), 1–19 (2010)
16. A.B. Dalton et al., Super-tough carbon-nanotube fibres. Nature **423**(6941), 703-703 (2003)
17. B. Vigolo et al., Macroscopic fibers and ribbons of oriented carbon nanotubes. Science **290**(5495), 1331–1334 (2000)
18. L.M. Ericson et al., Macroscopic, neat, single-walled carbon nanotube fibers. Science **305**(5689), 1447–1450 (2004)
19. N. Behabtu et al., Strong, light, multifunctional fibers of carbon nanotubes with ultrahigh conductivity. Science **339**(6116), 182–186 (2013)
20. X.B. Zhang et al., Spinning and processing continuous yarns from 4-inch wafer scale super-aligned carbon nanotube arrays. Adv. Mater. **18**(12), 1505–1510 (2006)
21. M. Zhang, K.R. Atkinson, R.H. Baughman, Multifunctional carbon nanotube yarns by downsizing an ancient technology. Science **306**(5700), 1358–1361 (2004)
22. K.L. Jiang, Q.Q. Li, S.S. Fan, Nanotechnology: spinning continuous carbon nanotube yarns. Nature **419**(6909), 801-801 (2002)
23. S. Zhang et al., Solid-state spun fibers and yarns from 1-mm long carbon nanotube forests synthesized by water-assisted chemical vapor deposition. J. Mater. Sci. **43**(13), 4356–4362 (2008)
24. Y.L. Li, I.A. Kinloch, A.H. Windle, Direct spinning of carbon nanotube fibers from chemical vapor deposition synthesis. Science **304**(5668), 276–278 (2004)
25. M. Motta et al., High performance fibres from 'dog bone' carbon nanotubes. Adv. Mater. **19**(21), 3721–3726 (2007)
26. H.W. Zhu et al., Direct synthesis of long single-walled carbon nanotube strands. Science **296**(5569), 884–886 (2002)
27. H.W. Zhu et al., Co-synthesis of single-walled carbon nanotubes and carbon fibers. Mater. Res. Bull. **37**(1), 177–183 (2002)
28. M. Motta et al., Mechanical properties of continuously spun fibers of carbon nanotubes. Nano Lett. **5**(8), 1529–1533 (2005)
29. L. Ci et al., Multifunctional macroarchitectures of double-walled carbon nanotube fibers. Adv. Mater. **19**(13), 1719–1723 (2007)
30. L.X. Zheng et al., Carbon-nanotube cotton for large-scale fibers. Adv. Mater. **19**(18), 2567–2570 (2007)
31. X.-H. Zhong et al., Fabrication of a multifunctional carbon nanotube "cotton'" yarn by the direct chemical vapor deposition spinning process. Nanoscale **4**(18), 5614–5618 (2012)
32. J. Tang et al., Manipulation and assembly of SWNTS by dielectrophoresis. Abstracts of Papers of the American Chemical Society, vol 227 (2004), p. U1273–U1273
33. J. Tang et al., Assembly of 1D nanostructures into sub-micrometer diameter fibrils with controlled and variable length by dielectrophoresis. Adv. Mater. **15**(16), 1352–1355 (2003)
34. W.J. Ma et al., Monitoring a micromechanical process in macroscale carbon nanotube films and fibers. Adv. Mater. **21**(5), 603–608 (2009)
35. J.M. Feng et al., One-step fabrication of high quality double-walled carbon nanotube thin films by a chemical vapor deposition process. Carbon **48**(13), 3817–3824 (2010)
36. S.J. Zhang, S. Kumar, Carbon nanotubes as liquid crystals. Small **4**(9), 1270–1283 (2008)
37. M.E. Kozlov et al., Spinning solid and hollow polymer-free carbon nanotube fibers. Adv. Mater. **17**(5), 614–617 (2005)
38. J. Steinmetz et al., Production of pure nanotube fibers using a modified wet-spinning method. Carbon **43**(11), 2397–2400 (2005)
39. W. Zhou et al., Single wall carbon nanotube fibers extruded from super-acid suspensions: preferred orientation, electrical, and thermal transport. J. Appl. Phys. **95**(2), 649–655 (2004)
40. V.A. Davis et al., True solutions of single-walled carbon nanotubes for assembly into macroscopic materials. Nat. Nanotechnol. **4**(12), 830–834 (2009)

41. V.A. Davis et al., Phase behavior and rheology of SWNTs in superacids. Macromolecules **37**(1), 154–160 (2004)
42. S.J. Zhang et al., Macroscopic fibers of well-aligned carbon nanotubes by wet spinning. Small **4**(8), 1217–1222 (2008)
43. S. Ramesh et al., Dissolution of pristine single walled carbon nanotubes in superacids by direct protonation. J. Phys. Chem. B **108**(26), 8794–8798 (2004)
44. X. Lepro, M.D. Lima, R.H. Baughman, Spinnable carbon nanotube forests grown on thin, flexible metallic substrates. Carbon **48**(12), 3621–3627 (2010)
45. X.F. Zhang et al., Strong carbon-nanotube fibers spun from long carbon-nanotube arrays. Small **3**(2), 244–248 (2007)
46. J. Jia et al., A comparison of the mechanical properties of fibers spun from different carbon nanotubes. Carbon **49**(4), 1333–1339 (2011)
47. L. Zheng, G. Sun, Z. Zhan, Tuning array morphology for high-strength carbon-nanotube fibers. Small **6**(1), 132–137 (2010)
48. M. Miao et al., Poisson's ratio and porosity of carbon nanotube dry-spun yarns. Carbon **48**(10), 2802–2811 (2010)
49. Q. Zhang et al., Dry spinning yarns from vertically aligned carbon nanotube arrays produced by an improved floating catalyst chemical vapor deposition method. Carbon **48**(10), 2855–2861 (2010)
50. Q. Li et al., Sustained growth of ultralong carbon nanotube arrays for fiber spinning. Adv. Mater. **18**(23), 3160–3163 (2006)
51. M.D. Lima et al., Biscrolling nanotube sheets and functional guests into yarns. Science **331**(6013), 51–55 (2011)
52. C.P. Huynh et al., Evolution of directly-spinnable carbon nanotube growth by recycling analysis. Carbon **49**(6), 1989–1997 (2011)
53. S. Fang et al., Structure and process-dependent properties of solid-state spun carbon nanotube yarns. J. Phys. Condens. Matter. **22**(33) (2010)
54. A.A. Kuznetsov et al., Structural model for dry-drawing of sheets and yarns from carbon nanotube forests. ACS Nano **5**(2), 985–993 (2011)
55. C. Zhu et al., A self-entanglement mechanism for continuous pulling of carbon nanotube yarns. Carbon **49**(15), 4996–5001 (2011)
56. L.J. Ci et al., Preparation of carbon nanofibers by the floating catalyst method. Carbon **38**(14), 1933–1937 (2000)
57. K. Koziol et al., High-performance carbon nanotube fiber. Science **318**(5858), 1892–1895 (2007)
58. J.J. Vilatela, A.H. Windle, Yarn-like carbon nanotube fibers. Adv. Mater. **22**(44), 4959–4963 (2010)
59. X.-H. Zhong et al., Continuous multilayered carbon nanotube yarns. Adv. Mater. **22**(6), 692–696 (2010)
60. L. Song et al., Fabrication and characterization of single-walled carbon nanotube fiber for electronics applications. Carbon **50**(15), 5521–5524 (2012)
61. T. Gong et al., Connection of macro-sized double-walled carbon nanotube strands by bandaging with double-walled carbon nanotube films. Carbon **45**(11), 2235–2240 (2007)
62. H.Z. Geng et al., Effect of acid treatment on carbon nanotube-based flexible transparent conducting films. J. Am. Chem. Soc. **129**(25), 7758–7759 (2007)
63. R. Duggal, F. Hussain, M. Pasquali, Self-assembly of single-walled carbon nanotubes into a sheet by drop drying. Adv. Mater. **18**(1), 29–34 (2006)
64. X.L. Li et al., Langmuir-Blodgett assembly of densely aligned single-walled carbon nanotubes from bulk materials. J. Am. Chem. Soc. **129**(16), 4890–4891 (2007)
65. M.C. LeMieux et al., Self-sorted, aligned nanotube networks for thin-film transistors. Science **321**(5885), 101–104 (2008)
66. H.W. Zhu, B.Q. Wei, Assembly and applications of carbon nanotube thin films. J. Mater. Sci. Technol. **24**(4), 447–456 (2008)

67. Q. Cao, J.A. Rogers, Ultrathin films of single-walled carbon nanotubes for electronics and sensors: a review of fundamental and applied aspects. Adv. Mater. **21**(1), 29–53 (2009)
68. G. Gruner, Carbon nanotube films for transparent and plastic electronics. J. Mater. Chem. **16** (35), 3533–3539 (2006)
69. J. Liu et al., Fullerene pipes. Science **280**(5367), 1253–1256 (1998)
70. Z.C. Wu et al., Transparent, conductive carbon nanotube films. Science **305**(5688), 1273–1276 (2004)
71. F. Hennrich et al., Preparation, characterization and applications of free-standing single walled carbon nanotube thin films. Phys. Chem. Chem. Phys. **4**(11), 2273–2277 (2002)
72. S.M. Cooper et al., Gas permeability of a buckypaper membrane. Nano Lett. **3**(2), 189–192 (2003)
73. X.F. Zhang et al., Properties and structure of nitric acid oxidized single wall carbon nanotube films. J. Phys. Chem. B **108**(42), 16435–16440 (2004)
74. U. Vohrer et al., Carbon nanotube sheets for the use as artificial muscles. Carbon **42**(5–6), 1159–1164 (2004)
75. P.G. Whitten, G.M. Spinks, G.G. Wallace, Mechanical properties of carbon nanotube paper in ionic liquid and aqueous electrolytes. Carbon **43**(9), 1891–1896 (2005)
76. A. Kukovecz et al., Controlling the pore diameter distribution of multi-wall carbon nanotube buckypapers. Carbon **45**(8), 1696–1698 (2007)
77. R. Smajda et al., Structure and gas permeability of multi-wall carbon nanotube buckypapers. Carbon **45**(6), 1176–1184 (2007)
78. G.H. Xu et al., The feasibility of producing MWCNT paper and strong MWCNT film from VACNT array. Appl. Phys. A Mater. Sci. Process. **92**(3), 531–539 (2008)
79. L.J. Hall et al., Sign change of Poisson's ratio for carbon nanotube sheets. Science **320**(5875), 504–507 (2008)
80. J.G. Park et al., The high current-carrying capacity of various carbon nanotube-based buckypapers. Nanotechnology **19**(18), 185710 (2008)
81. J. Boge et al., The effect of preparation conditions and biopolymer dispersants on the properties of SWNT buckypapers. J. Mater. Chem. **19**(48), 9131–9140 (2009)
82. A. Anson-Casaos et al., Surfactant-free assembling of functionalized single-walled carbon nanotube buckypapers. Carbon **48**(5), 1480–1488 (2010)
83. J.W. Zhang, D.Z. Jiang, Influence of geometries of multi-walled carbon nanotubes on the pore structures of Buckypaper. Compos. Part A Appl. Sci. Manuf. **43**(3), 469–474 (2012)
84. R.L.D. Whitby et al., Geometric control and tuneable pore size distribution of buckypaper and buckydiscs. Carbon **46**(6), 949–956 (2008)
85. K. Mukai et al., Highly conductive sheets from millimeter-long single-walled carbon nanotubes and ionic liquids: Application to fast-moving, low-voltage electromechanical actuators operable in air. Adv. Mater. **21**(16), 1582–1585 (2009)
86. J.L. Rigueur et al., Buckypaper fabrication by liberation of electrophoretically deposited carbon nanotubes. Carbon **48**(14), 4090–4099 (2010)
87. M. Zhang et al., Strong, transparent, multifunctional, carbon nanotube sheets. Science **309**(5738), 1215–1219 (2005)
88. A.E. Aliev et al., Giant-stroke, superelastic carbon nanotube aerogel muscles. Science **323**(5921), 1575–1578 (2009)
89. P. Liu et al., Fast high-temperature response of carbon nanotube film and its application as an incandescent display. Adv. Mater. **21**(35), 3563–3566 (2009)
90. W. Fu et al., Super-aligned carbon nanotube films as aligning layers and transparent electrodes for liquid crystal displays. Carbon **48**(7), 1876–1879 (2010)
91. C. Feng et al., Flexible, stretchable, transparent conducting films made from superaligned carbon nanotubes. Adv. Funct. Mater. **20**(6), 885–891 (2010)
92. A.E. Aliev, Y.N. Gartstein, R.H. Baughman, Mirage effect from thermally modulated transparent carbon nanotube sheets. Nanotechnology **22**(43), 435704 (2011)
93. A.A. Kuznetzov et al., Electron field emission from transparent multiwalled carbon nanotube sheets for inverted field emission displays. Carbon **48**(1), 41–46 (2010)

94. D. Wang et al., Highly oriented carbon nanotube papers made of aligned carbon nanotubes. Nanotechnology **19**(7), 075609 (2008)
95. L. Song et al., Direct synthesis of a macroscale single-walled carbon nanotube non-woven material. Adv. Mater. **16**(17), 1529–1534 (2004)
96. W. Ma et al., Directly synthesized strong, highly conducting, transparent single-walled carbon nanotube films. Nano Lett. **7**(8), 2307–2311 (2007)
97. Q. Liu et al., In situ assembly of multi-sheeted Buckybooks from single-walled carbon nanotubes. ACS Nano **3**(3), 707–713 (2009)
98. P.X. Hou et al., Bulk storage capacity of hydrogen in purified multiwalled carbon nanotubes. J. Phys. Chem. B **106**(5), 963–966 (2002)
99. R.Z. Ma et al., Electrical conductivity and field emission characteristics of hot-pressed sintered carbon nanotubes. Mater. Res. Bull. **34**(5), 741–747 (1999)
100. J.L. Li et al., Microstructure and mechanical properties of hot-pressed carbon nanotubes compacted by spark plasma sintering. Carbon **43**(13), 2649–2653 (2005)
101. H.L. Zhang et al., Spark plasma sintering and thermal conductivity of carbon nanotube bulk materials. J. Appl. Phys. **97**(11)
102. H.-L. Zhang et al., Electrical and thermal properties of carbon nanotube bulk materials: Experimental studies for the 328-958 K temperature range. Phys. Rev. B. **75**(20)
103. C. Qin et al., High temperature electrical and thermal properties of the bulk carbon nanotube prepared by SPS. Mater. Sci. Eng. A-Struct. Mater. Prop. Microstruct. Process. **420**(1–2), 208–211 (2006)
104. J.L. Li et al., Surface graphitization and mechanical properties of hot-pressed bulk carbon nanotubes compacted by spark plasma sintering. Carbon **45**(13), 2636–2642 (2007)
105. J. Li et al., Transport properties of hot-pressed bulk carbon nanotubes compacted by spark plasma sintering. Carbon **47**(4), 1135–1140 (2009)
106. J. Li, L. Wang, W. Jiang, Super-hydrophobic surface of bulk carbon nanotubes compacted by spark plasma sintering followed by modification with polytetrofluorethylene. Carbon **48**(9), 2668–2671 (2010)
107. G. Yamamoto et al., Single-walled carbon nanotube-derived novel structural material. J. Mater. Res. **21**(6), 1537–1542 (2006)
108. G. Yamamoto et al., Preparation of single-walled carbon nanotube solids and their mechanical properties. J. Mater. Res. **20**(10), 2609–2612 (2005)
109. G. Yamamoto et al., Mechanical properties of binder-free single-walled carbon nanotube solids. Scripta Mater. **54**(2), 299–303 (2006)
110. C. Laurent et al., Spark plasma sintering of double-walled carbon nanotubes. Carbon **46**(13), 1812–1816 (2008)
111. K. Yang et al., Inter-tube bonding, graphene formation and anisotropic transport properties in spark plasma sintered multi-wall carbon nanotube arrays. Carbon **48**(3), 756–762 (2010)
112. Y. Sato et al., Influence of the structure of the nanotube on the mechanical properties of binder-free multi-walled carbon nanotube solids. Carbon **50**(1), 34–39 (2012)
113. P.D. Bradford et al., Tuning the compressive mechanical properties of carbon nanotube foam. Carbon **49**(8), 2834–2841 (2011)
114. X. Gui et al., Soft, highly conductive nanotube sponges and composites with controlled compressibility. ACS Nano **4**(4), 2320–2326 (2010)
115. X. Gui et al., Carbon nanotube sponges. Adv. Mater. **22**(5), 617–621 (2010)
116. M. Xu et al., Carbon nanotubes with temperature-invariant viscoelasticity from-196 degrees to 1000 degrees C. Science **330**(6009), 1364–1368 (2010)
117. M. Xu et al., Carbon nanotubes with temperature-invariant creep and creep-recovery from-190 to 970 degrees C. Adv. Mater. **23**(32), 3686–3691 (2011)
118. M. Xu et al., Tailoring temperature invariant viscoelasticity of carbon nanotube material. Nano Lett. **11**(8), 3279–3284 (2011)
119. X. Gui et al., Three-dimensional carbon nanotube sponge-array architectures with high energy dissipation. Adv. Mater. **26**(8), 1248–1253 (2014)

120. Z. Zeng et al., Integrated random-aligned carbon nanotube layers: deformation mechanism under compression. Nanoscale **6**(3), 1748–1755 (2014)
121. Z. Zeng et al., Carbon nanotube sponge-array tandem composites with extended energy absorption range. Adv. Mater. **25**(8), 1185–1191 (2013)
122. N. Thongprachan et al., Preparation of macroporous solid foam from multi-walled carbon nanotubes by freeze-drying technique. Mater. Chem. Phys. **112**(1), 262–269 (2008)
123. J.H. Zou et al., Ultralight multiwalled carbon nanotube aerogel. ACS Nano **4**(12), 7293–9302 (2010)
124. K.H. Kim, Y.S. Oh, M.F. Islam, Mechanical and thermal management characteristics of ultrahigh surface area single-walled carbon nanotube aerogels. Adv. Funct. Mater. **23**(3), 377–383 (2013)
125. H.Y. Sun, Z. Xu, C. Gao, Multifunctional, ultra-flyweight, synergistically assembled carbon aerogels. Adv. Mater. **25**(18), 2554–2560 (2013)
126. R. Orru et al., Consolidation/synthesis of materials by electric current activated/assisted sintering. Mater. Sci. Eng. R **63**(4–6), 127–287 (2009)
127. W. Li et al., Densification mechanisms of spark plasma sintering: multi-step pressure dilatometry. J. Mater. Sci. **47**(20), 7036–7046 (2012)
128. K. Hata et al., Water-assisted highly efficient synthesis of impurity-free single-waited carbon nanotubes. Science **306**(5700), 1362–1364 (2004)
129. T. Yamada et al., Size-selective growth of double-walled carbon nanotube forests from engineered iron catalysts. Nat. Nanotechnol. **1**(2), 131–136 (2006)
130. A.Y. Cao et al., Super-compressible foamlike carbon nanotube films. Science **310**(5752), 1307–1310 (2005)
131. J. Suhr et al., Fatigue resistance of aligned carbon nanotube arrays under cyclic compression. Nat. Nanotechnol. **2**(7), 417–421 (2007)
132. X. Gui et al., Recyclable carbon nanotube sponges for oil absorption. Acta Mater. **59**(12), 4798–4804 (2011)
133. H. Li et al., Photocatalytic, recyclable CdS nanoparticle-carbon nanotube hybrid sponges. Nano Res. **5**(4), 265–271 (2012)
134. M. Xu et al., Alignment control of carbon nanotube forest from random to nearly perfectly aligned by utilizing the crowding effect. ACS Nano **6**(7), 5837–5844 (2012)
135. X. Gui et al., Controllable synthesis of spongy carbon nanotube blocks with tunable macro- and microstructures. Nanotechnology **24**(8), 085705 (2013)
136. T. Yamada et al., Revealing the secret of water-assisted carbon nanotube synthesis by microscopic observation of the interaction of water on the catalysts. Nano Lett. **8**(12), 4288–4292 (2008)
137. R.J. Mora, J.J. Vilatela, A.H. Windle, Properties of composites of carbon nanotube fibres. Compos. Sci. Technol. **69**(10), 1558–1563 (2009)
138. A.B. Dalton et al., Continuous carbon nanotube composite fibers: properties, potential applications, and problems. J. Mater. Chem. **14**(1), 1–3 (2004)
139. Y. Gao et al., Axial compression of hierarchically structured carbon nanotube fiber embedded in epoxy. Adv. Funct. Mater. **20**(21), 3797–3803 (2010)
140. H. Zhao et al., Carbon nanotube yarn strain sensors. Nanotechnology **21**(30) (2010)
141. J.Q. Wei et al., Carbon nanotube filaments in household light bulbs. Appl. Phys. Lett. **84**(24), 4869–4871 (2004)
142. R.M. Sundaram, K.K.K. Koziol, A.H. Windle, Continuous direct spinning of fibers of single-walled carbon nanotubes with metallic chirality. Adv. Mater. **23**(43), 5064–5068 (2011)
143. Y. Zhao et al., Iodine doped carbon nanotube cables exceeding specific electrical conductivity of metals. Sci. Rep. **1**, 83 (2011)
144. Z. Zhu et al., Nano-yarn carbon nanotube fiber based enzymatic glucose biosensor. Nanotechnology **21**(16), 165501 (2010)
145. F. Cai, T. Chen, H. Peng, All carbon nanotube fiber electrode-based dye-sensitized photovoltaic wire. J. Mater. Chem. **22**(30), 14856–14860 (2012)

146. T. Chen et al., High-performance transparent and stretchable all-solid supercapacitors based on highly aligned carbon nanotube sheets. Sci. Rep. **4** (2014)
147. T. Chen et al., Polymer photovoltaic wires based on aligned carbon nanotube fibers. J. Mater. Chem. **22**(44), 23655–23658 (2012)
148. X. Chen et al., Smart, stretchable supercapacitors. Adv. Mater. **26**(26), 4444 (2014)
149. X. Fang et al., Core-sheath carbon nanostructured fibers for efficient wire-shaped dye-sensitized solar cells. Adv. Mater. **26**(11), 1694–1698 (2014)
150. Z. Yang et al., A highly stretchable, fiber-shaped-supercapacitor. Angew. Chem. Int. Ed. **52** (50), 13453–13457 (2013)
151. Y. Zhang et al., Super-stretchy lithium-ion battery based on carbon nanotube fiber. J. Mater. Chem. A **2**(29), 11054–11059 (2014)
152. R.H. Baughman et al., Carbon nanotube actuators. Science **284**(5418), 1340–1344 (1999)
153. J. Foroughi et al., Torsional carbon nanotube artificial muscles. Science **334**(6055), 494–497 (2011)
154. M.D. Lima et al., Electrically, chemically, and photonically powered torsional and tensile actuation of hybrid carbon nanotube yarn muscles. Science **338**(6109), 928–932 (2012)
155. L. Xiao et al., Flexible, stretchable, transparent carbon nanotube thin film loudspeakers. Nano Lett. **8**(12), 4539–4545 (2008)
156. G. Zheng et al., Nanostructured paper for flexible energy and electronic devices. MRS Bull. **38**(4), 320–325 (2013)
157. G. Zheng et al., Paper supercapacitors by a solvent-free drawing method. Energy Environ. Sci. **4**(9), 3368–3373 (2011)
158. K.H. An et al., Supercapacitors using single-walled carbon nanotube electrodes. Adv. Mater. **13**(7), 497–500 (2001)
159. K.H. An et al., Electrochemical properties of high-power supercapacitors using single-walled carbon nanotube electrodes. Adv. Funct. Mater. **11**(5), 387–392 (2001)
160. V.L. Pushparaj et al., Flexible energy storage devices based on nanocomposite paper. Proc. Natl. Acad. Sci. USA **104**(34), 13574–13577 (2007)
161. A. Izadi-Najafabadi et al., Extracting the full potential of single-walled carbon nanotubes as durable supercapacitor electrodes operable at 4 V with high power and energy density. Adv. Mater. **22**(35), E235–E241 (2010)
162. A.A. Zakhidov et al., Electrochemically tuned properties for electrolyte-free carbon nanotube sheets. Adv. Funct. Mater. **19**(14), 2266–2272 (2009)
163. M. Yu et al., High density, vertically-aligned carbon nanotube membranes. Nano Lett. **9**(1), 225–229 (2009)

Chapter 3
The Synthesis, Properties, and Applications of Heteroatom-Doped Graphenes

Yu-Cheng Chang and Wei-Hung Chiang

Abstract Graphene is a unique form of carbon nanomaterials, and it possesses exceptional properties including high electrical conductivity, high thermal conductivity, high surface area, high mechanical strength, and high chemical stability. However, pristine graphene is a zero-band-gap material, making it difficult to be used in many applications required materials with a band gap. There are several methods to modify the properties of graphene and heteroatom doping has been demonstrated as an effective method to create a band gap in graphene. Heteroatom doping can endow graphene with various enhanced optical, electromagnetic, structural, and physicochemical properties, making graphene become a promising material in various applications including nanoelectronics, catalysis, energy storage, functional composites, and biomedical applications. Since heteroatom-doped graphene is both of academic and technical importance, here we comprehensively discuss the properties, synthetic methods, emerging applications, and the present status of various aspects of this important issue to assist a better understanding of doped graphene materials.

Keywords Carbon nanomaterials · Graphene · Electrical conductivity · Chemical stability · Heteroatom · Doping · Energy storage

3.1 Introduction

Graphene, a sp^2-hybridized monolayer carbon atom, is one of the most interesting and promising nanomaterials with unique physical and chemical properties and has been proposed for many applications such as nanoelectronics, energy generation and storage devices, nanocomposites, catalysis support, and biomedical applications [1–4]. The unique two-dimensional (2D) structure of graphene possesses several

Y.-C. Chang · W.-H. Chiang (✉)
Department of Chemical Engineering, National Taiwan University of Science and Technology, Taipei 106, Taiwan ROC
e-mail: whchiang@mail.ntust.edu.tw

M. Husain and Z.H. Khan (eds.), *Advances in Nanomaterials*,
Advanced Structured Materials 79, DOI 10.1007/978-81-322-2668-0_3

excellent properties, for example, ultrahigh carrier mobility [5, 6], ambipolar electric field effect [7–9], massless relativistic carriers [10], anomalous quantum Hall effect (QHE) [10, 11], and over micrometer-scale spin coherence length [12], etc., it could contribute to the structure of graphene with sp^2-hybridized carbon densely packed into a honeycomb lattice structure offers the superior properties.

So far, graphene has been used to fabricate catalyst carriers [13], electrochemical biosensor [14], field-effect transistors (FETs) [9, 15, 16], supercapacitors [17, 18], transparent conducting electrodes [19–21], solar cells [22], lithium ion batteries [23, 24], and so on. However, as a zero-band-gap material, pristine graphene cannot be directly used in semiconductor industry. For example, to fabricate logic circuits for industrial applications, both p-type and n-type conductions are needed. To solve these problems and expand the area of applications, it is necessary to tune the physicochemical properties of graphenes by chemical modification [25].

3.2 Heteroatom Doping of Graphene

Graphene is made out of sp^2 hybrid carbon atoms with the s, p_x and p_y atomic orbitals on each carbon atom forming with other three surrounding atoms by three strong s bonds [26]. Overlap of the remaining p_z orbital on each carbon atom with neighboring carbon atoms produces a filled band of π orbitals, called valence band and an empty band of π^* orbitals known as the conduction band. And in what reason that makes graphene a zero-band-gap semiconductor? It may attributed to the valence and conduction bands touch at the Brillouin zone corners [27]. Figure 3.1a shows the energy spectrum of graphene and zoom-in of the energy bands at one of the Dirac point (at the K or K′ points in the Brillouin zone or the intersection of the valence band and the conduction band) [28]. Graphene can be doped by electric field (Fig. 3.1b) [2], by adsorption of metal atoms and by substrate (Fig. 3.1c) [29].

Similar to that elaborated for the silicon-based technology, chemical doping is one of the most promising ways to tailor the electronic structure by carrier injection or extraction. Doped-graphene must be synthesized with controlled methods for the first step to tailor the electrical properties and to realize practical applications. In principle, chemical doping of graphene can be realized either by graphene surface adsorption or by heteroatom substitution of carbon atom in graphene [25, 30].

Due to a partly submonolayer coverage of dopant molecules on graphene surface [25], the former is effective only for a low doping level. Also surface adsorption usually allows for simple wet chemistry approach and unlimited choices of adsorbed molecules. In addition, when interaction between surface dopants and graphene is relatively weak [25], the stability of doping effect may be strongly affected by external stimulation. In contrast, for stable doping effect, the latter is the most desirable and high doping level modulation.

Both doping ways have been extensively explored, and many techniques including chemical vapor deposition (CVD) [31–34], solvothermal chemistry

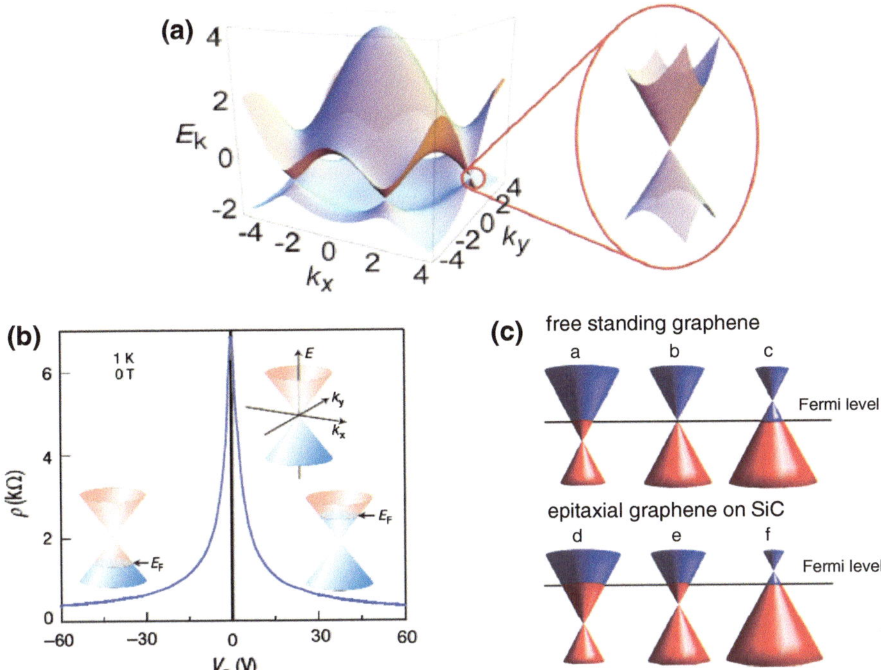

Fig. 3.1 **a** *Left* the band structure of graphene in the honeycomb lattice. *Right* zoom-in of the energy bands close to one of the dirac points. Reprinted with permission from Ref. [28]. Copyright 2009 by the American Physical Society. **b** Ambipolar electric field effect in single-layer graphene. The *insets* show the position of the dirac point and the Fermi energy EF of graphene as a function of gate voltage. Reprinted with permission from Ref. [2]. Copyright 2007, Nature Publishing Group. **c** A schematic diagram of the position of the dirac point and the Fermi level as a function of doping. The upper panel is n-type doped, pristine and p-type doped free standing graphene (**a**)–(**c**). The lower panel is n-type doped, pristine and p-type doped epitaxial graphene grown on silicon carbide (SiC) (**d**)–(**f**). Reprinted with permission from Ref. [29]. Copyright 2008, American Chemical Society

[25, 35], plasma treatment [31, 36, 37], molecule synthesis [38], molecule adsorption [39–41], graphene oxide (GO) reduction with NH_3 at high temperature [42, 43] have been used to dope graphene for various applications and the basic scientific research. In addition, doping graphene with different heteroatoms was also extensively investigated [44].

The complete controls over doping level, doping location, and the chemical bonding structure between heteroatoms and graphene and the quality of doped graphene are the challenges for graphene chemical doping and the related applications, which are the key issues [25]. Furthermore, chemical doping of graphene covers a wide research field and open wide the application sides. Therefore, it is necessary to update the status of this field to gain an insight into the future development [25, 30].

Charge transfer is determined by the relative position of density of states (DOS) of the highest occupied molecular orbital (HOMO) and lowest unoccupied molecular orbital (LUMO) of the dopant and the Fermi level of graphene [15, 40]. Dopant could act as different character, if the HOMO of a dopant is above the Fermi level of graphene, there is a charge transfer from dopant to the graphene layer, then the dopant acts as a donor; in contrast, if the LUMO is below the Fermi level of graphene, charge transfers from graphene layer to dopant, then the dopant acts as an acceptor [28, 40].

Graphene can be p-type or n-type doped via chemical doping [3, 40, 45, 46]. p-Type doping drives the Dirac points of graphene above the Fermi level by adding atoms with fewer valence electrons than carbon like boron; n-type doping drives the Dirac points below the Fermi level which achieved by adding atoms with more valence electrons than carbon like nitrogen. In general, molecules with drawing groups adsorbed on the surface of graphene will lead to p-type doping of graphene, and molecules with donating groups will lead to n-type doping. In contrast, mechanism of substitution doping still remains uncertain [15].

In the following sections, we summarize the significant development in both theory and experiment studies of doped graphenes. The doping methods, doping levels, heteroatom sources, the chemical bond structures between heteroatom and graphene, the electrical stability, the quality of doped graphene, and their applications are discussed in detail.

3.3 Properties of Heteroatom-Doped Graphene

Graphene is a semimetal or a zero-gap semiconductor [2]. Few-layer graphenes show a semiconducting nature, with the resistivity showing little change in the 100–300 K range [47]. Resistance of few-layer graphene decreases markedly if it is heated to high temperatures and also with increase in number of layers [47]. Single layer graphene showed a very high mobility of ~15,000 cm^2/V s at room temperature with ballistic transport up to sub-micrometer distances [8]. Chemical doping of heteroatoms is one of the effective methods to modify the properties of graphene [48]. The electronic structure of graphene can be tuned by doping with nitrogen and boron atoms which can form strong bonds with the carbon atoms and have similar atomic sizes [15, 49]. Nitrogen-doped (N-doped) graphene prepared under CVD conditions shows n-type behavior at a gate voltage of $V < \sim -20$ V in vacuum [43]. N-doped graphene prepared at 900 °C shows less resistance than that prepared at 700 °C since GO gets reduced effectively at higher temperatures [43]. The Dirac point of GO gets shifted to the negative side at negative gate voltages, confirming n-type doping because of the N dopants [50]. Usachov et al. [51] demonstrated that N-doping leads n-type doping by means of angle-resolved photoemission spectroscopy (ARPES). The band structure of N-doped graphene is shown in Fig. 3.2a. The Dirac cone shifted to ~0.3 eV toward higher binding energy with respect to pristine graphene, which has the Dirac point at the Fermi

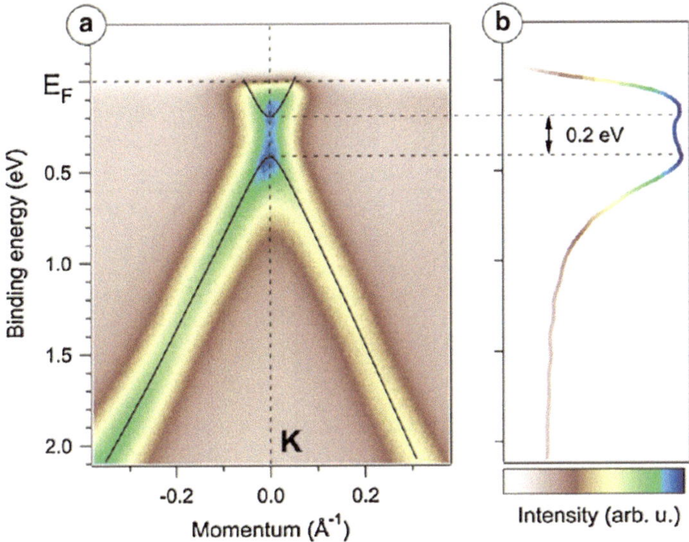

Fig. 3.2 **a** ARPES of N-graphene/Au/Ni(111)/W(110), measured at the photon energy of 35 eV, through the *K*-point, few degrees off the direction, perpendicular to the ΓK. **b** PE spectrum at the *K*-point. Reprinted with permission from Ref. [51]. Copyright 2011, American Chemical Society

energy, *Ef*. The conduction band appears below *Ef*, indicating n-type doping. The band gap of ~0.2 eV appears at the K-point (see Fig. 3.2b).

Because of the adsorption of oxygen or water in air, pristine graphene itself shows p-type behavior [52]. The dipole moment of H_2O adsorbate causes local electrostatic fields that shift the substrate's defect states with respect to graphene electrons and causes doping [53]. N-doped graphene shows hole and electron mobility of around 6000 cm^2/V s [30]. Boron-doped (B-doped) graphene exhibits p-type behavior and has lower electron mobility than N-doped graphene. The carrier mobility is around 800 cm^2/V s as confirmed by electric transport measurements [54]. Wu et al. [55] report electron mobility of B- and N-doped graphene-based back-gate FETs to be about 350–550 cm^2/V s and 450–650 cm^2/V s, respectively.

Graphene is a zero-band-gap semiconductor, for device applications, it is highly desired to induce a gap in graphene. Many methods have been proposed to open the band gap in graphene such as substrate-induced band gap [56–59], bilayer graphene [60–63] and edge effects induced band gap such as graphene quantum dots [64] and nanoribbons [65–68].

Charge-transfer doping is also known to open the band gap in graphene. Although pristine graphene is semi-metallic, it may possess a band gap from epitaxial graphene grown over SiC [69]. The symmetry of the graphene sublattice is broken down by the dissimilar interaction of the individual sublattice with the substrate, which results in a band gap of 0.26 eV [70]. Unfortunately, due to the

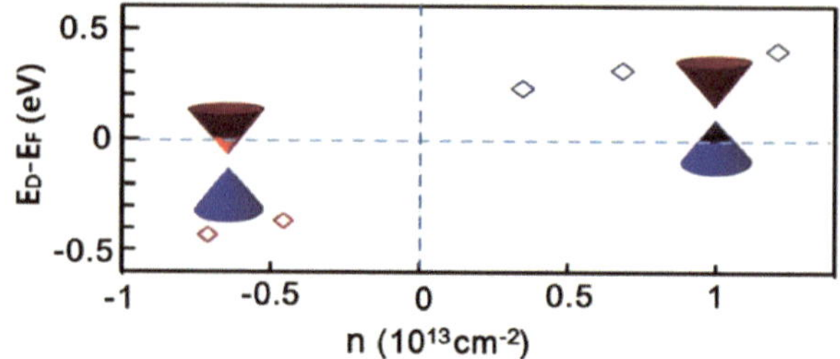

Fig. 3.3 Fermi level shift from conduction band to valence band of epitaxial monolayer graphene by charge-transfer doping. Reprinted with permission from Ref. [70]. Copyright 2014, Wiley-VCH

charge-transfer doping effect from the SiC substrate, which shifts the Fermi level far from the opened gap into the conduction band, such a substrate-induced band gap opening is hard to observe practically. Zhou et al. [57] reported that this unintentional doping from the SiC substrate can be neutralized by molecular doping via NO_2 absorption. Electron transfer from graphene to NO_2 may move the Fermi level within the band gap. They observed a reversible metal-insulator transition in single and bilayer graphenes by using the reversible adsorption and desorption of NO_2.

Figure 3.3 illustrates a quantitative measure of the Fermi energy change as a function of the carrier concentration (or degree of NO_2 absorption) [70]. Charge-transfer doping can induce charge distribution asymmetry between the two layers of the bilayer graphenes and break the inversion symmetry [70], interestingly, it can also open up band gap in bilayer graphenes. It is well-known that if the inversion symmetry of bilayer graphene is broken, a band gap is induced. For instance, Zhang et al. [71] showed that simultaneous top and bottom surface doping with an organic molecule, water/oxygenand trizine, may open up a band gap of 111 meV for bilayer graphenes.

Fluorination of graphene may induce a gap like hydrogenation. Theoretical calculation shows that only hydrogen and fluorine can completely covalently cover the basal plane of graphene [72]. The band gap for graphene fluorides may be tailored by different fluorination level [73], it could be synthesized and exhibit insulator behaviors with large band gap while the reduced graphene fluoride has low resistance [74].

By molecular doping, gaps can also be formed in graphene, such as NO_2 [57], Br_2 and I_2 adsorption [75]. There're big difference between Br_2 and I_2 adsorption. For few-layer graphene, Br_2 adsorbs and intercalates into graphene layers, in contrast I_2 only adsorbs on the surface of graphene introducing different doping levels between surface and interior layers then creating symmetric potential (from two surfaces to the center) thus opening the band gap [15]. Chemical doping is a method which more effective and simpler compared with other proposals for opening a band gap in graphene. The band gaps induced by chemical doping can be controlled by doping levels [15].

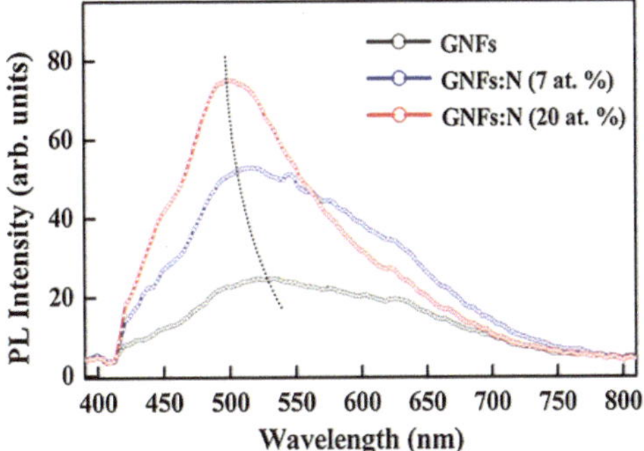

Fig. 3.4 PL spectra of graphene and N-doped graphene. Reprinted with permission from Ref. [77]. Copyright 2012, American Chemical Society

Because of its optically transparent, unique luminescent and photoluminescence (PL) quenching properties, Single-layer graphene is useful in biological applications [76]. PL is observed in chemically modified graphene. Chiou et al. [77] found out N-doping of graphene enhances the PL emission of graphene significantly. The emission increases as increasing the nitrogen content (Fig. 3.4). The PL of reduced GO is quenched by doping with nitrogen and the quenching efficiency depends on the pyridine nitrogen content [78].

The occurence of high-temperature ferromagnetism in graphite-related materials is a topic of considerable interest [4]. Yazyev et al. [79] showed that magnetism in graphene can be induced by vacancy defects or by hydrogen chemisorption. Thus interaction with electron donor and acceptor molecules as well as hydrogenation affects the magnetic properties of graphene [80]. Du et al. claimed that doping of nitrogen in reduced graphene oxide (RGO) increases the magnetization [81]. The magnetization curves (M-H) of NG-500 measured at 2 and 300 K are shown in Fig. 3.5a. No ferromagnetism was found in N-doped graphene as found in pristine graphene instead a linear, purely Curie-like paramagnetic behavior was observed at low temperatures. The magnetization was 0.312 emu/g for RGO, and 0.482 emu/g for NG-400, 0.514 emu/g for NG-500, 0.379 emu/g for NG-600, 0.246 emu/g for NG-700, and 0.173 emu/g for NG-900 as shown in Fig. 3.5c (here 400, 500, 600 and 700 refer to the synthetic temperature of the samples). The magnetic properties are attributed to the change in the proportion of pyrrolic nitrogen relative to graphitic nitrogen.

According to Li et al. [82] pyridinic and pyrrolic N-doping has significant effects on the spin distributions. For pyridinic nitrogen, the unpaired spins are mainly concentrated on edges and are localized on N atoms. Hence, it has less influence on the spin polarization of the edge states than pyrrolic nitrogen where the spin polarization on the doped edge was nearly removed. In the case of graphitic N, the opposite edges get localized and are completely canceled. Thus, increasing the pyrrolic nitrogen in graphene structures increases the magnetization of graphene.

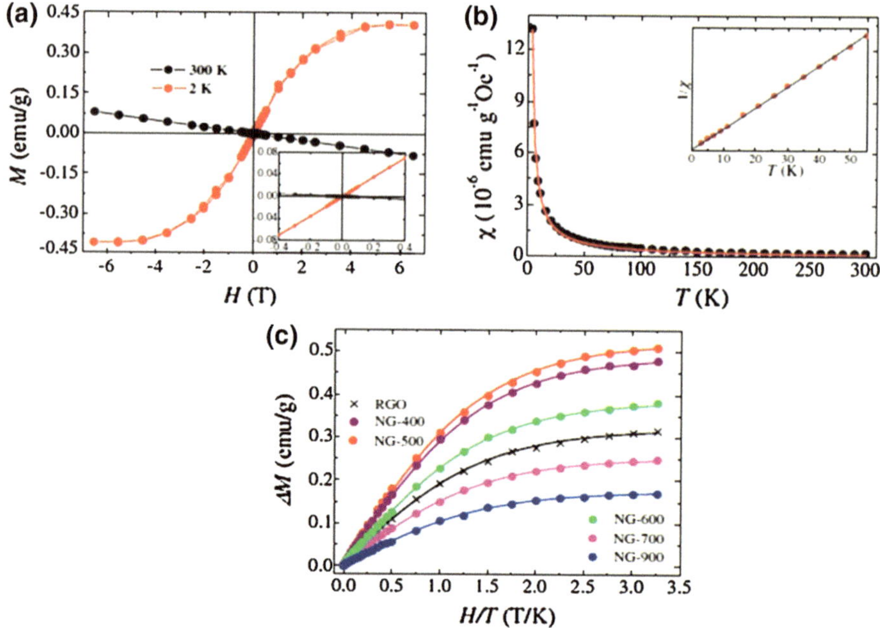

Fig. 3.5 **a** Typical M–H curves of NG-500 measured at 2 and 300 K with the, *inset* showing part of the magnetization curves. **b** Typical χ–T curve of NG-500 (applied field H = 3 kOe), *solid lines* are fitted by the Curie law gives the $1/\chi$–T curve. **c** ΔM of RGO and NG samples as a function of H/T at 2 K. *Solid curves* are fits to the brillouin function with $J = 1/2$ and $g = 2$. Reprinted with permission from Ref. [81]. Copyright 2013, Elsevier B.V.

Theoretical calculations claim that the magnetism in B-doped graphene systems to be generally weak. The presence of boron clusters weakens the edge magnetism in zigzag graphene edges [83].

3.4 Synthesis of Heteroatom-Doped Graphene

To date, several techniques have been established for graphene synthesis. However, thermal CVD [20], arc discharge approach [78], graphite oxide post treatment [37], and plasma treatment synthesis [84] are the most commonly used methods today. In the upcoming sections, a few of the graphene synthesis methods and their scientific and technological importance are described in details.

CVD is a chemical process by which a substrate is exposed to thermally decompose precursors and the desired product deposited onto the substrate surface at high temperature. There are numerous advantages of the CVD process. The process yields high quality and high purity final products in a large scale. Moreover, by controlling the CVD process parameters, control over the morphology,

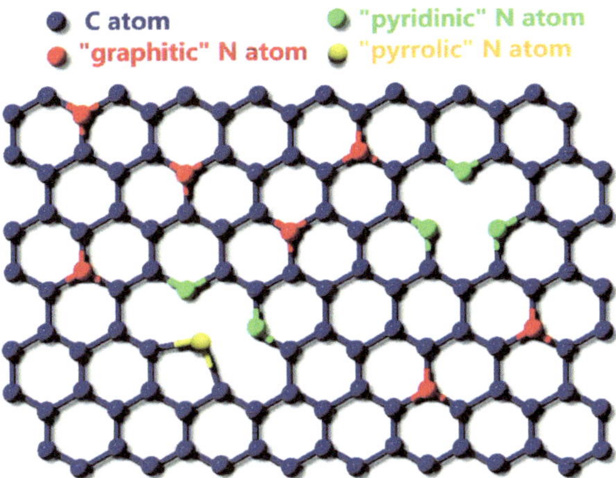

Fig. 3.6 Schematic representation of the N-doped graphene. The *blue*, *red*, *green*, and *yellow spheres* represent the C, "graphitic" N, "pyridinic" N, and "pyrrolic" N atoms in the N-doped graphene, respectively. Reprinted with permission from Ref. [32]. Copyright 2009, American Chemical Society

crystallinity, shape, and size of the desired product is possible. However, tailoring high precision atomic-level properties using CVD method is still under investigation [85].

Few layer N-doped graphene were produced by using NH_3 and CH_4 as the nitrogen and carbon sources, respectively, and in 2009 by Wei et al. [32] (Fig. 3.6), who the first reported using copper as the catalyst by CVD method. From X-ray photoelectron spectroscopy (XPS) analysis, they discovered that the nitrogen atoms were combined with carbon atoms by covalent bond structure in the forms of "pyrolic", "pyridinic" and "graphitic" nitrogen bonding configurations and the dominated form was the content of "graphitic" nitrogen. The doping level of N could be controlled by the ratio of NH_3 and CH_4 flow rates and which reached as high as 8.9 at.%. In addition, N-doped graphene showed n-type behavior with the mobility in the range of 200–450 $cm^2\ V^{-1}\ s^{-1}$, which measured by fabricating FETs. Imamura and Saiki [34] also synthesized N-doped graphene with pyridine molecules by using CH_4 and NH_3 as the carbon and nitrogen sources, but they reduced the growth temperature down to 500 °C on Pt (111) surface. The nitrogen concentration of the N-doped graphene sample was estimated to be 4.0 at.%, which indicated that the chemical bonds of the hexahydric ring in a pyridine molecule were broken and decomposed to single atoms before forming graphene.

Not only NH_3 or gases could be the source of nitrogen element, worthless biomass wastes contained N element were also used to prepared N-doped graphene. Wu et al. [86] synthesized large area, single or few layer N-doped graphene with the chitosan as the sole C and N sources on arbitrary substrates including silicon, quartz and glasses. Wu et al. [55] took polystyrene and boric acid as the solid precursor to synthesize monolayer B-doped graphene by using CVD method with. By adjusting the dopant elemental precursors, the boron concentration could be tailored from 0.7 to 4.3 at.% and among different kinds of bonding, the BC_3 bonding was a remarkable type in the B-doped graphene films, which played an important role in increasing the density of the states near the Fermi level and altering the valence band structure of graphene. For oxygen reduction reaction (ORR), the electrochemical measurements of the 3D B-doped graphene showed high electrocatalytic activities with better stability, higher current generation capability and superior tolerance than pristine graphene (Fig. 3.7d, e) [87]. Wang et al. [88] developed a

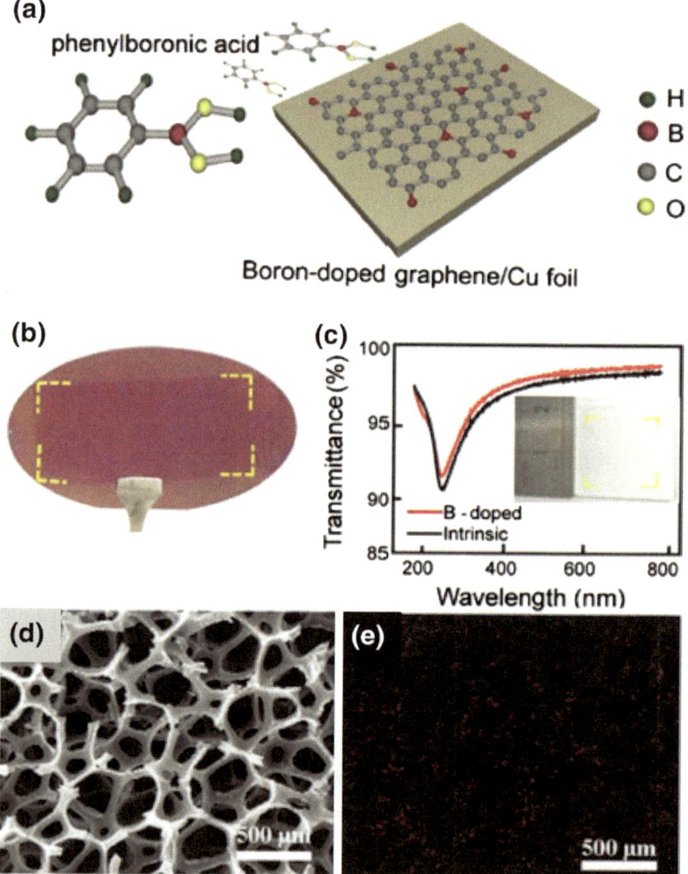

Fig. 3.7 **a** Schematic diagram of CVD growth of B-doped graphene on Cu surface with phenylboronic acid as the carbon and boron sources. The *red*, *grey*, *yellow*, and *green spheres* represent boron, carbon, oxygen and hydrogen atoms, respectively. **b** Contrast enhanced photograph of the B-doped graphene sample on 4-in. Si/SiO_2 substrate. **c** UV–vis transmittance spectra of the B-doped graphene and the reference intrinsic graphene on quartz substrate. Reprinted with permission from Ref. [88]. Copyright 2013, Wiley-VCH. **d** SEM image of BN-GF and the corresponding EDX mapping of (**e**) B. Reprinted with permission from Ref. [87]. Copyright 2013, Royal Society of Chemistry

method to synthesized B-doped graphene with phenyl boronic acid ($C_6H_7BO_2$) as the only source of carbon and boron (Fig. 3.7a–c). The sole precursor was beneficial for the large scale growth of uniform and high quality B-doped graphene under controllable manner and the concentration of boron was estimated to be 1.5 at.%. Some et al. [89] synthesized phosphorus and N-doped bilayer graphene, they had demonstrated the lone pair electron donating ability of the phosphorus atom is much stronger than nitrogen due to its higher nucleophilicity, and air-stable n-type performance in molecularly doped FETs.

The electric arc reactor for synthesis of graphene mainly comprises two electrodes and a steel chamber which cooled by water. A direct-current arc voltage is applied across two graphite electrodes immersed in an inert gas such as H_2, NH_3 and He, air. As the rods are brought close together, discharge occurs resulting in the formation of plasma. When the discharge ends, the soot generated is collected under ambient conditions. Only the soot deposited on the inner wall of the chamber is collected, avoiding the substance at the bottom of the chamber, for the latter tends to contain other graphitic particles [90].

Arc-discharge method has been widely used for high quality, high purity, large scale and controllable synthesis of carbon nanotubes and fullerene, which had many similarities to graphene preparations [25]. Huang et al. [91] demonstrated gram-scale and high-quality N-doped graphene flakes by arc-discharge method with flake graphite as carbon source, ZnO or ZnS as catalyst, melamine as the nitrogen source and the content of N reached 4.92 at.%. Using cyanuric chloride and trinitrophenol as reactants, Feng et al. [92] reported a facile method for large scale preparation of N-doped graphene in gram scale by a detonation technique at low temperature and the doping level reached 12.5 at.%, which was higher than most of the other methods. Panchakarla et al. [48] synthesized N-doped few-layer graphene by performing arc-discharge between two graphite electrodes in the presence of H_2, He and NH_3 or pyridine vapors. The graphene obtained contain few-layers (~3–4 layers) with nitrogen content of 1–1.4 at.% as found by electron-energy loss spectroscopy(EELS). Figure 3.8 shows AFM and TEM images of N-doped graphene obtained in this manner. B-doped graphene is obtained by carrying out the discharge in the presence of a mixture of a H_2 and B_2H_6 [93].

Reduction of GO is one of the established procedures to produce graphene in large quantities [94]. There are several methods to reduce GO, by thermal annealing, plasma treatment, wet chemical reaction, etc. It has been investigated for more than 50 years for GO synthesized through strong oxidants, and this method could fabricate GO flakes in large quantity of gram or even kilogram. There are many kinds of oxygen functional groups including –O–, –OH, –CHO, –COOH–, etc. in or on the GO surface. If these oxygen functional groups could be replaced with heteroatoms, we can not only obtain high level doped graphene but also reduce the GO at the same time [25].

Li and co-authors [43] reported 5 at.% nitrogen containing graphene by thermal annealing of GO with NH_3 under different thermal conditions ranging from 300 to 1100 °C. The degree of doping depends upon the temperature and oxygen

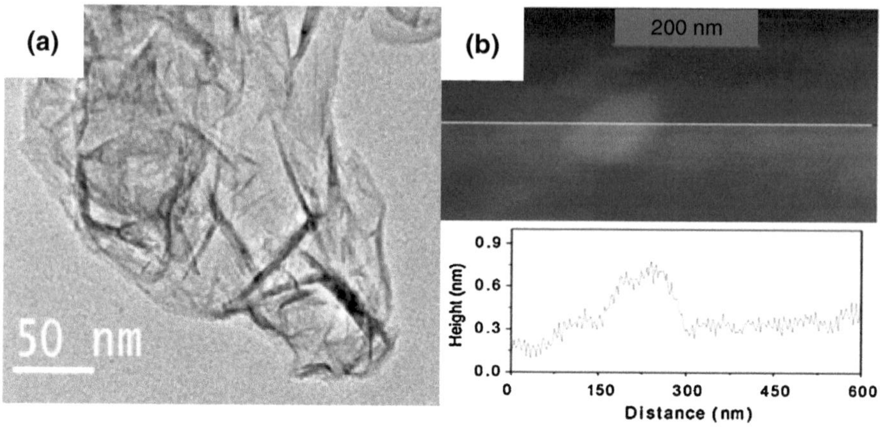

Fig. 3.8 **a** TEM and **b** AFM images of N-doped graphene. Reprinted with permission from Ref. [48]. Copyright 2009, Wiley-VCH

functionalities at graphene edges. Wen et al. [95] synthesized N-doped graphene by thermal annealing of GO mixed with cyanamide and the concentration of N reached up to 9.96 at.% at 900 °C (Fig. 3.9). The pore volume of the obtained N-doped graphene was about 3.42 $cm^3\ g^{-1}$ with highly crumpled morphology. It was used to fabricate electrode for supercapacitors with significant improvement of several performance parameters, including high capacity, excellent rate capability and long term stability (5000 cycles with 96.1 % maintained). Khai et al. [96] prepared B-doped graphene by annealing films obtained from suspensions of GO and H_3BO_3 in N,N-dimethylformamide. Not only N-, B-doped graphene, but also sulfur-doped (S-doped), phosphorus-doped (P-doped) now been extensively studied. Yang et al. [97] reported that the S-doped and Selenium-doped (Se-doped) graphene could also possess excellent properties. They prepared doped grapheme by annealing GO with benzyl disulfide and diphenyl diselenide in argon atmosphere, S-doped and Se-doped exhibit excellent catalytic activity, long-term stability and high methanol tolerance in alkaline media for ORR, even better than the commercial Pt/C. Zhang et al. [98] fabricated P-doped graphene with 1.81 at.% (Fig. 3.10) phosphorus by thermal annealing of GO and triphenyl phosphine (TPP). The as-synthesized P-doped graphene exhibits outstanding ORR activity as well as excellent stability and selectivity, which endow it great potential to be an efficient metal-free electrocatalyst.

Wu et al. [99] prepared 3D B and N co-doped monolithic graphene aerogels through hydrothermally treated GO and ammonia boron trifluoride at 180 °C for 12 h (Fig. 3.11). The as-synthesized B and N co-doped graphene aerogels based supercapacitors exhibited several excellent performances, including high specific capacitance, enhanced energy density or power density, good rate capability, and the minimized device thickness.

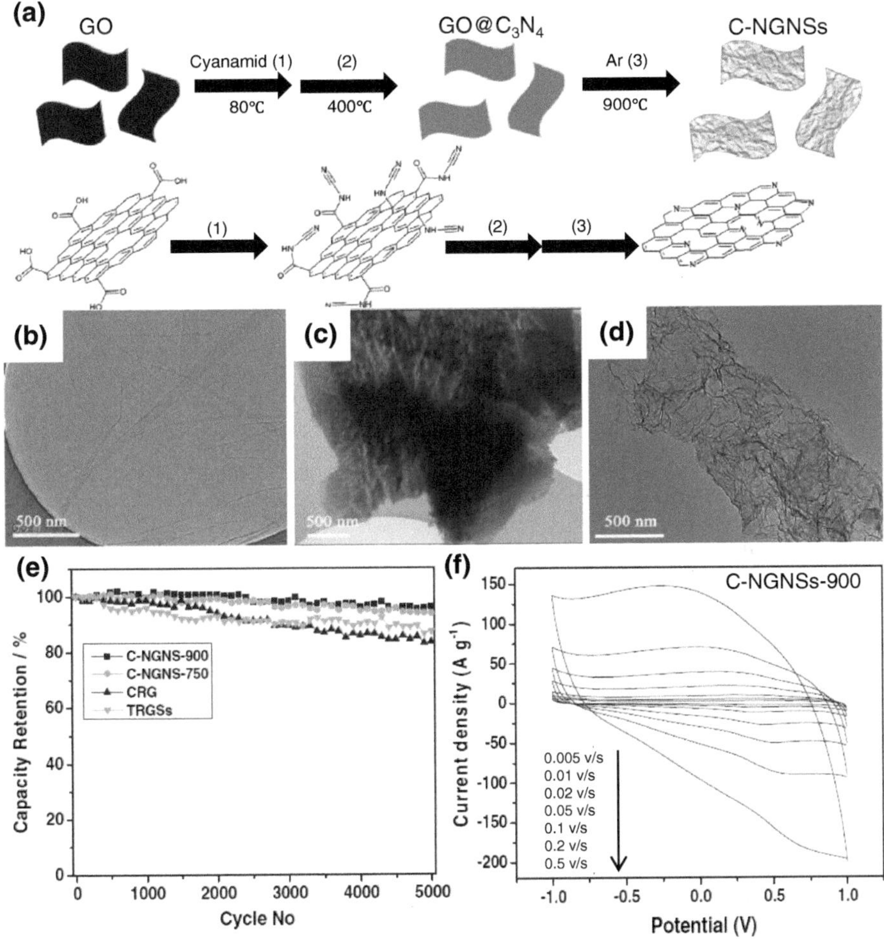

Fig. 3.9 **a** Schematic illustration for fabricating crumpled N-doped graphene. TEM images of the **b** GO, **c** GO@p-C3N4, **d** C-NGNSs-900. **e** Capacitance retention ratio as a function of the potential sweep rates for different graphene supercapacitors up to 5000 cycles. **f** CVs of the C-NGNSs-900 supercapacitor in 6 M KOH aqueous solution between −1.0 and 1.0 V at different scan rates. Reprinted with permission from Ref. [95]. Copyright 2012, Wiley-VCH

Plasma treatment can be used for the introduction of heteroatoms, structures, or groups onto bulk scaffold surfaces and it is also a simple way to modify material surface [100]. Wang et al. [37] reported an efficient approach for simple, tunable and fast reduction for preparation of N-doped graphene at the same time by treating GO with nitrogen plasma. The content of N could be changed in a range of 0.11–1.35 at.% by controlling the exposure time. The obtained N-doped graphene

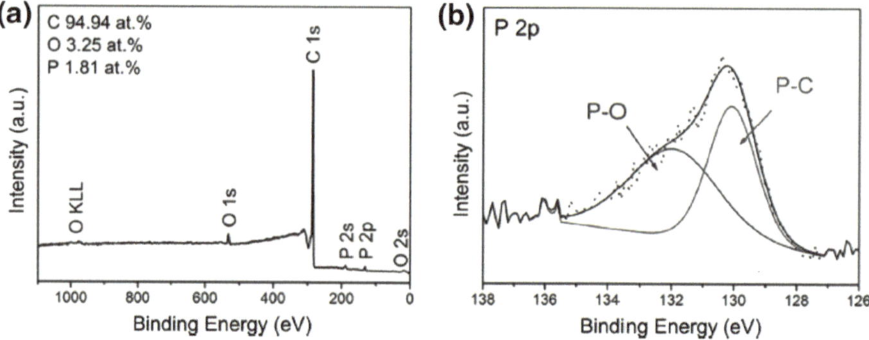

Fig. 3.10 **a** XPS survey spectrum and **b** the high-resolution P 2p XPS spectrum of P-doped graphene. Reprinted with permission from Ref. [98]. Copyright 2013, Wiley-VCH

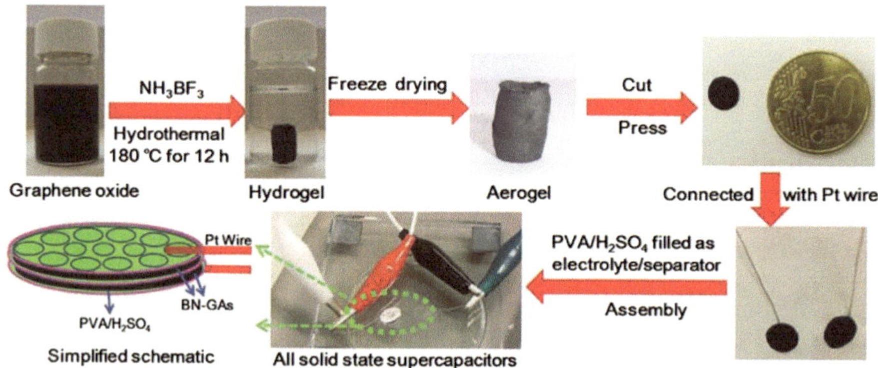

Fig. 3.11 Fabrication illustration of all-solid-state supercapacitors (ASSSs) based on BN-GAs that were involved by a combined hydrothermal process and freeze-drying process. The as-fabricated supercapacitors with a diameter of 7 mm indicated by the *dotted green ring* and a simplified schematic of ASSSs based on aerogels are shown (below *left*). Reprinted with permission from Ref. [99]. Copyright 2012, Wiley-VCH

not only showed excellent electrocatalytic activity for reduction of H_2O_2 but also fast electron transfer kinetics of glucose biosensing to concentrations as low as 0.01 mM in the presence of interferences. Tang et al. [101] achieved the tunable band gaps and transport properties of B-doped graphene through controllable doping with the ion atmosphere of trimethylboron decomposed by microwave plasma (Fig. 3.12a–e). The boron doping level can be changed from 0 to 13.85 at.% by controlling the ion reaction time and the band gaps were tuned from 0 to 0.54 eV as the doping level increasing, leading to a series of modulated transport properties. In addition, the obtained B-doped graphene showed a typical p-type conductivity with a current on/off ratio higher than 100 from electrical measurements [25].

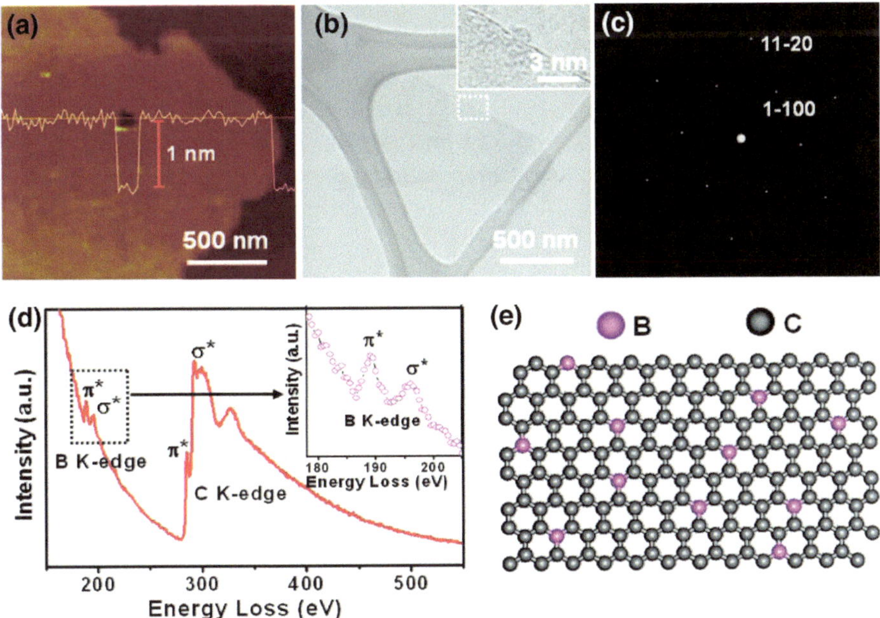

Fig. 3.12 Structure and composition characterizations of the graphenes reacted with 5 min of the B ion atmosphere decomposed from TMB plasma. **a** Typical AFM image of the reacted graphene; **b** TEM image and HRTEM image (*inset*) of the reacted graphene, and the corresponding SAED pattern; **c**, **d** typical EELS spectrum taken from the doped graphenes; **e** schematic structure of a B-doped graphene with B content of ~6.8 at.%, as detected in the synthesized graphenes. Reprinted with permission from Ref. [101]. Copyright 2012, American Chemical Society

3.5 Applications of Heteroatom-Doped Graphene

The following section emphasizes how to optimize the material properties and the performance of device via modification of doping, also lists relevant applications of doped graphene for advanced nanomaterials and devices.

Fuel cells are electrochemical energy conversion device that oxidize fuel at the anode and reduce oxygen from air at the cathode to produce electricity [102]. Commonly, platinum (Pt) is used as the catalyst, however, the commercial application of Pt catalysts is limited by its scarcity, time-dependent drift, and CO poisoning [102]. While Pt-based catalysts [103, 104] and other metal catalysts, such as Au and Pd [105, 106], have been developed, nowadays graphitic carbon-based metal-free catalysts have emerged as a promising alternative with low cost, long-term durability, and high ORR activity [70]. Whereas pristine graphene do not show considerable electrocatalytic activity, with performance optimization, heteroatom doping can provide excellent ORR catalytic activity [31, 97, 107–110], which is comparable to conventional Pt catalysts.

Pristine graphene showed a two-step, two-electron ORR pathway with a low onset potential. Zhang et al. [111] proposed a four-electron ORR pathway on N-doped graphene in an acidic environment. Qu et al. [109] also reported that N-doped graphene exhibited a typical one-step, four-electron ORR pathway, which is similar to the ORR pathway of a Pt catalyst (Fig. 3.13a). In another work [112], the H_2O_2 formed in the ORR process was about 10 % at −0.5 V in the diffusion-controlled region. Here the low proportion of H_2O_2 indicated the four electron reduction process dominated in ORR.

B-doped graphene (BG) also shows good electrocatalytic activity toward ORR in alkaline conditions similar to the performance of Pt catalysts [47]. In addition, the non-metallic BG catalyst shows good CO tolerance superior and long-term stability to that of Pt-based catalysts [113]. Three-dimensional B,N-doped graphene foam (BN-GF) performed as excellent metal free catalyst for ORR [87].

To date, the exact doping configuration that contributes to the ORR activity is still under debate [114–116]. Recent theoretical [117] and experimental [115] investigations have reported that quaternary N is more active than pyridinic N. The lone pair electrons of pyridinic N may cause a repulsive interaction with O_2 molecules possessing many lone pair electrons. This repulsion reduces the effective

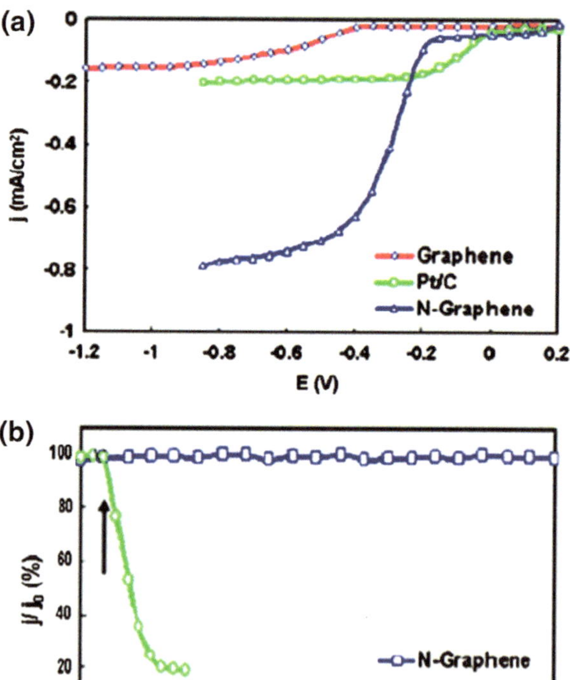

Fig. 3.13 **a** Ring rotating disc electrode voltammograms for ORR in air-saturated 0.1 M KOH on the electrodes graphene (*red*), Pt/C (*green*), and N-graphene (*blue*) with a scan rate of 0.01 V s^{-1}. Electrode rotating rate: 1000 rpm. **b** Current–time chronoamperometric response of Pt/C and NG to CO. The *arrow* indicates the addition of 10 % (v/v) CO at −0.4 V. Reprinted with permission from Ref. [109]. Copyright 2010, American Chemical Society

polarized interaction at C–O_2 [115]. Doping with other elements with lower electronegativity than carbon, such as B (2.04), P (2.19), and S (2.58), can also bring a high ORR activity via similar bond polarization [113, 118–120]. On account of those heteroatoms take positive charges, oxygen chemisorption occurs over them [121].

Lithium-ion batteries (LIBs) have become predominant in battery technology in recent years due to their high power density, high energy density, high reversible capacity good cycle life and excellent storage characteristics. The carbon electrode plays an important role in the battery performance [47]. Although graphene-based materials can reach a high reversible capacity at a low charge rate [122], it is still rate-limited at a high charge/discharge rate (≥500 mA/g) [123]. Thus, doped graphene based device is proposed with the intent to achieve a high reversible capacity at the high charge/discharge rate.

There are calculations which show that B-doping in grapheme creates electron deficient areas distributed around boron (Fig. 3.14a) and depresses its Fermi level into the valence band [124]. This modified electronic configuration results in a higher lithium ion absorption efficiency of B-doped graphene and the most stable site for lithium is slightly shifted towards boron (Fig. 3.14b) [124].

Reddy and co-authors [125] deposited N-doped graphene on Cu foil by the CVD process and show edits double reversible discharge capacity. The increase in the reversible discharge capacity can be ascribed from the defects created by N-doping. Ai et al. [126] reported a novel approach for the syntheses of N- and S-codoped graphene (NS-G) by means of covalent functionalization of GO followed by thermal treatment. The unique structural properties and synergistic effects of N and S codoping make NS-G a superior anode material for LIBs in terms of high

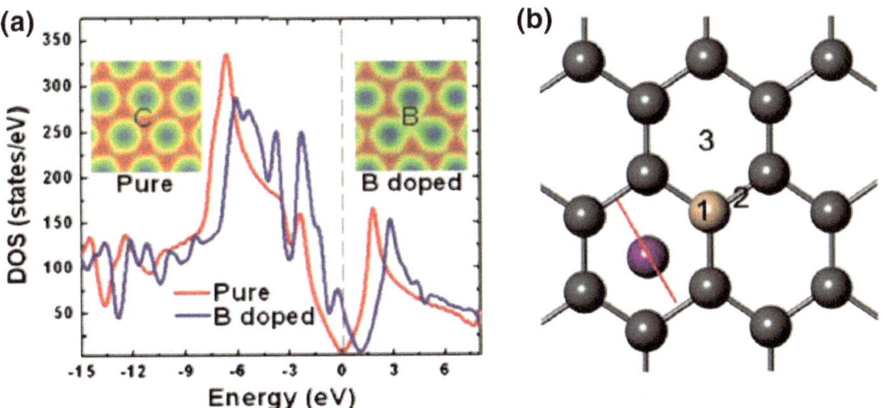

Fig. 3.14 **a** The DOS of pristine and B-doped graphene. Contour plots of the electron density along the graphene plane for pristine and B-doped graphene are shown in the *left* and *right insets*. **b** Schematic for Li^+ adsorption on B-doped graphene. *1*, *2*, and *3* represent the three high symmetry sites at top, bridge, and hollow, respectively. Reprinted with permission from Ref. [124]. Copyright 2009, American Institute of Physics

reversible capacity, long-term cycling performance and excellent rate capability (Fig. 3.15). The N-doped graphene exhibits 452 and 684 mA h/g in the 100th cycle and 501st cycle respectively, with high electrochemical performance than pristine graphene or the commercialized graphite anode. N-doped graphene containing 3 at. % N exhibits a reversible capacity of 1043 mA h/g in the first cycle and 872 mA h/g in the 30th cycle, which is higher than that of pristine graphene, which shows 955 and 638 mh A/g in the first and 30th cycles respectively [127].

Electrochemical capacitors are potential applications in many fields such as mobile electronics, hybrid vehicles and power supply devices due to their energy density and high power, long cycle life and cost effectiveness [128]. Carbon-based supercapacitors show excellent capacitance behavior because of their high surface

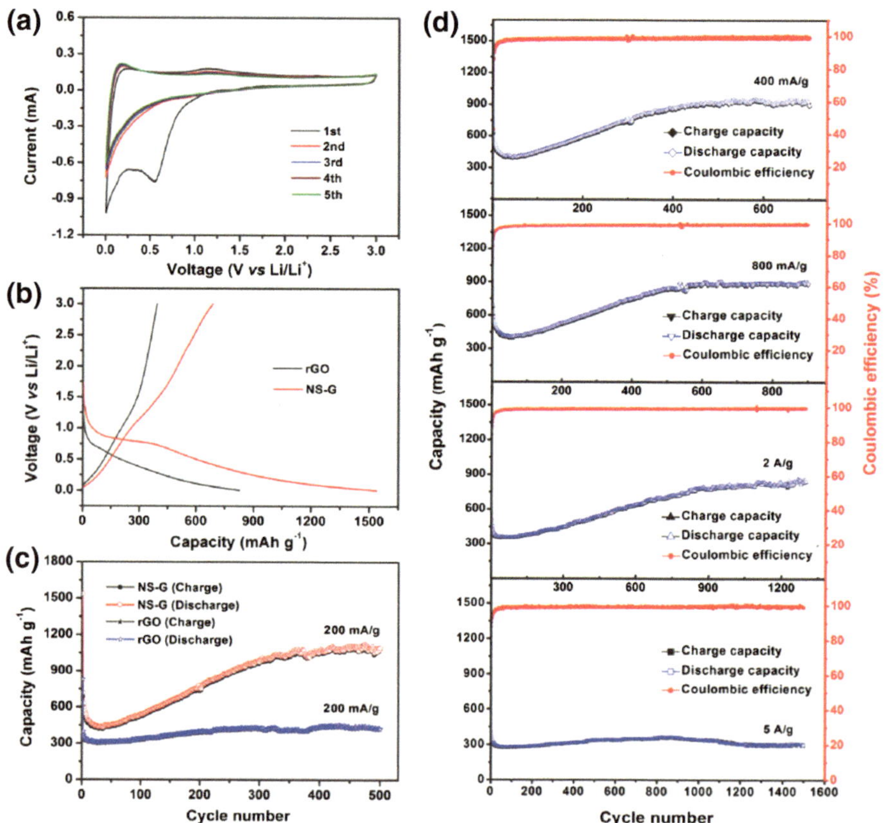

Fig. 3.15 Electrochemical Li storage performance of RGO and NS-G. **a** CV curves of the NS-G electrode at a sweep rate of 0.5 mV s^{-1}. **b** Initial galvanostatic charge–discharge voltage versus capacity profiles of the RGO and NS-G electrodes at a current density of 200 mA g^{-1}. **c** Cycling performance of the RGO and NS-G electrodes at a current density of 200 mA g^{-1}. **d** Rate capability of the NS-G electrode at different rates. Reprinted with permission from Ref. [126]. Copyright 2014, Wiley-VCH

area and high electrical conductivity [129]. The reason why graphene is a good base material for supercapacitors is because of the electron mobility around room temperature and also its high surface area. Owing to the modification of the electronic structure and surface energy, heteroatom doping can greatly influence the charge-storage capacity of graphene [84, 130]. Lee et al. [130] reported that N-doping improves the specific capacity of macroporous graphene film, which is attributed to the enhanced electrical conductivity and wettability [70]. The energy density of supercapacitors can be increased using organic solvents or ionic liquids as electrolytes [131]. N-doped graphene prepared under plasma conditions shows capacitance 4 times higher than that of pristine graphene in an organic electrolyte (1 M tetraethyl ammonium tetrafluoroborate in acetonitrile) and showed superb cyclic stability of 100,000 cycles [84]. Lei and co-authors [132] obtained a specific capacitance of 255 F/g at 0.5 A/g in an aqueous electrolyte (6 M KOH). The graphene electrode showed 43 % decrease in the initial capacitance at a current density of 30 A/g. Density functional theory (DFT) calculations by Jeong et al. [84] demonstrated that N-doping with the contribution of pyridinic N have the highest proportion could increase the binding energy of electrolyte ions at the graphene surface. Because of this enhanced binding, N-doped surfaces can accommodate more ions and thus have a higher charge-storage capacity. Based on this theoretical prediction, they fabricated wearable supercapacitors consisting of N-doped graphene (Fig. 3.16a). Irrespective of the electrolytes, the specific capacity of N-doped graphene was about four times higher than pristine graphene (Fig. 3.16b) [70]. N-doped graphene hydrogels showed a specific capacitance of 308 F/g at 3 A/g [133]. It has been found recently that increase in the nitrogen content in graphene increases the specific capacitance [115]. Qiu et al. [134] reduced GO by hydrazine hydrate and annealed subsequently in NH_3 atmosphere to produce N-graphene sheets. N-graphene shows a maximum capacitance of 144.9 F/g at a current density of 0.5 A/g in conventional organic solvent-based electrolyte. The maximum energy density obtained was 80.5 W h/kg at a power density of 558 W/kg. B-doped

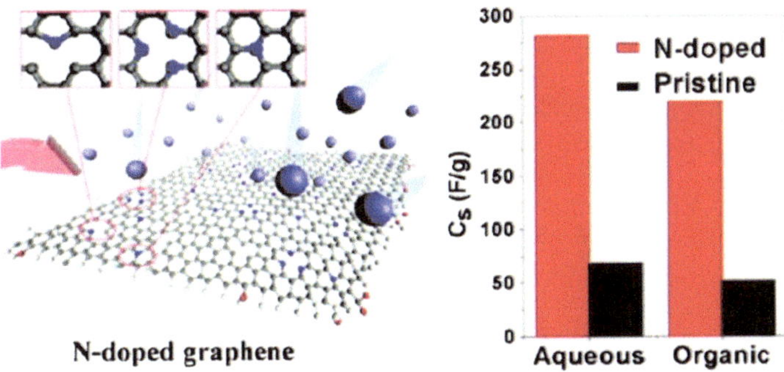

Fig. 3.16 N-doped graphene prepared by N_2 plasma immersion and specific capacity of N-doped graphene and pristine graphene based capacitors. Reprinted with permission from Ref. [84]. Copyright 2011, American Chemical Society

graphene (BG) made by the 'fried ice' method gives a specific capacitance of 281 F/g in aqueous electrolyte (2 M H_2SO_4) [135]. Han et al. [136] produced B-graphene by reduced GO (1.1 at.% B) has a high surface area of 466 m^2/g and showed a specific capacitance of 200 F/g in aqueous electrolyte with good cyclic stability of more than 4500 cycles.

Field emission (FE) has been extensively explored from various exotic low dimensional carbon nanomaterials [137], such as single and multi-walled carbon nanotubes (CNTs) [138], vertically aligned nanowalls [139], tubular graphitic cones [140], amorphous carbon films [141], few-layered grapheme nanoflakes [142, 143], and, more recently, from doped and pristine graphene [144, 145]. Though FE in CNTs is highly efficient, it has been shown that heteroatom doping by elements, could reduce the turn-on field and significantly increasing the electron emission current due to doped graphene reduce the effective tunneling potential barrier, such as nitrogen [146, 147]. Palnitkar et al. [145] studied on undoped as well as N- and B-doped graphene samples forming by arc-discharge method in a hydrogen atmosphere for FE studies (Fig. 3.17). They discovered containing nitrogen and boron as dopants showed low turn-on fields of 0.6 and 0.8 V/μm respectively at an emission current density of 10 $\mu A/cm^2$. Sharma et al. [148] prepared doped materials which show excellent FE current stability for a long-range of 3 h (current value = 1 μA). The number of emission spots and the number of tiny spots do not change during current-time measurements. Kashida and co-authors [149] have measured FE characteristics of N-doped graphene using in situ transmission electron microscopy. The turn-on voltage of N-doped graphene was found to be less than that of pristine graphene. To draw 1 nA current, the turn-on voltage required is 230 V for pristine and 110 V for N-doped graphene. This is due to the improved electrical conductivity of the N-doped sample.

Composites are made from more than two components and demonstrate synergistic characteristics differentiated from the individual components [70]. Due to the molecular scale dimension and large shape anisotropy of CNTs and graphene, a percolation threshold is expected with a small amount of fillers [150–153].

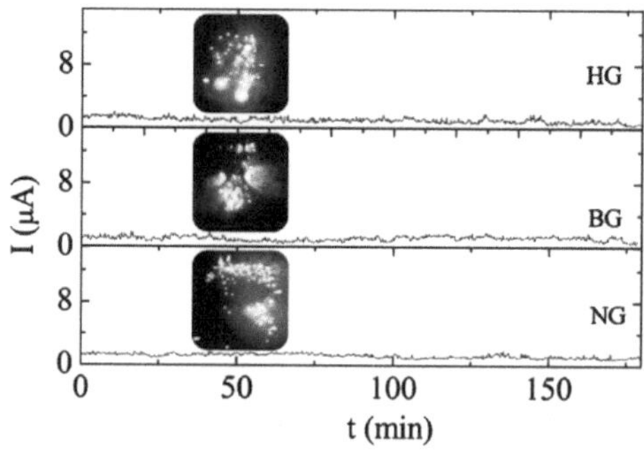

Fig. 3.17 Current stability of undoped (HG), B-doped (BG), and N-doped (NG) graphenes (*inset* FE pattern corresponding to HG, BG, and NG). Reprinted with permission from Ref. [145]. Copyright 2010, American Institute of Physics

Furthermore, CNTs and graphene may incorporate new characteristics into composite systems, such as high electrical/thermal conductivity [154–158] and optical transparency [7, 159–161], which can be exploited for electromagnetic shielding [162, 163], transparent conductive electrodes [164–166], and heat-management system [167–169]. Nevertheless, the molecular-scale dimensions of CNTs and graphene make it difficult to disperse them homogeneously within a matrix. Indeed, one of the initial motivations for the chemical modification of CNTs and graphene [170–173] is molecular level dispersion in a polymer or other metrics.

Gopalakrishnan et al. [174] composited of TiO_2 nanoparticles with B- and N-doped graphenes show selectivity in photo degradation of dyes. The mechanism is likely to involve the photo excitation of electrons in the TiO_2 nanoparticles followed by their transfer to graphene. The adsorbed dye molecules interact with graphene using π-π and charge transfer interaction (Fig. 3.18). Accordingly, a dye which is a good electron donor interacts more strongly with B-doped graphene (B = 3 at.%), while a dye which is a poor electron donor interacts more strongly with N-doped graphene (N = 1.2 at.%). The electronic structure of graphene, therefore, plays a significant role in the photo degradation kinetics [47].

Pt is the widely used counter electrode in dye-sensitized solar cells (DSSCs) due to their high electrical conductivity but the high cost and risk of corrosion of Pt through the redox species in the electrolyte hindered the development of the solar cells. Recently, much effort has been made to reduce or replace Pt-based electrodes in DSSCs [175–184]. In particular, carbon black [177], carbon nanotubes [179, 182], carbon nanoparticles [181], and grapheme nanosheets [176, 180, 183–185] have been studied as the counter electrodes in DSSCs. However, their electrical conductivities and reduction catalytic activities still cannot match up to those of Pt [186]. It is important to balance its electrocatalytic activity and the electrical conductivity [183–185] for improving the device performance of DSSCs with a carbon-based counter electrode, Heteroatom-doped graphene has been used as a

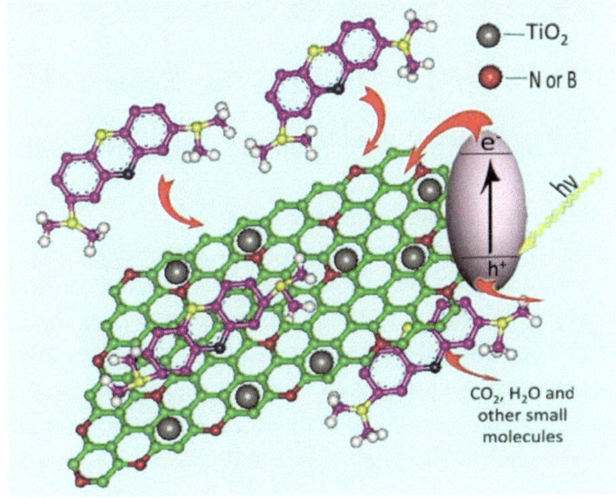

Fig. 3.18 The schematic shows structure of TiO_2-doped graphene and tentative processes of the photodegradation of methylene blue over TiO_2-doped graphene. Reprinted with permission from Ref. [174]. Copyright 2011, Elsevier B.V.

counter electrode in DSSCs. Modification of doping improves the electrochemical activity of graphene by increasing the conductivity which helps to reduce the total internal resistance and improves the photovoltaic performance [47]. Doping graphene with heteroatoms, for example, nitrogen, boron, sulfur, phosphorus, etc., can introduce electrocatalytic active sites with a minimized change of the conjugation length [187]. Furthermore, heteroatom doping has also been demonstrated to enhance the electrical conductivity and surface hydrophilicity to facilitate charge-transfer and electrolyte–electrode interactions, respectively, and even impart electrocatalytic activities [187, 188]. Park et al. [189] observed that $AuCl_3$ doping significantly improves the graphene OPV device performance, but also improves the surface wetting of the graphene electrodes, thus the device success rates are improved. Xue et al. [186] prepared N-doped 3D graphene foam (N-GF) by freeze drying method and demonstrated its application as a metal-free electrocatalyst for the reduction of triiodide to replace the Pt cathode in DSSCs, leading to power conversion efficiency up to 7.07 %. The increase in the performance is due to the presence of nitrogen in graphene which acts as active sites for the interaction of iodide ions which in turn gets reduced.

Hydrogen produced from water using solar energy is clearly the ultimate source of clean renewable energy. Inspired by natural photosynthesis, artificial solar water-splitting materials are now being designed and tested [47]. Maitra et al. [190] have reported that a nanocomposite of few-layer MoS_2 with heavily nitrogenated graphene performs as an excellent hydrogen evolution reaction (HER) catalyst. Nitrogen incorporation in graphene improved the catalytic activity of the composite with 2H–MoS_2 layers and enhances the electron donating ability of the graphene. The composite of exfoliated graphene (EG)-MoS_2 exhibited much better H_2 evolution (0.54 mol/g h) compared to 2H–MoS_2 alone (0.05 mol/g h).

Graphene-based biosensors for detecting bacteria, glucose, pH values and proteins have also been fabricated. Fan et al. [191] produced N-doped graphene sheets and chitosan (CS) to prepare electrochemical bisphenol A (BPA), which could disrupt endocrine system and cause cancer, has been considered as an endocrine disruptor sensor. They compared with graphene; N-doped graphene has favorable electron transfer ability and electrocatalytic property, which could enhance the response signal towards BPA. They also found CS exhibits excellent film forming ability and improved the electrochemical behavior of N-doped graphene modified electrode. The sensor exhibited a sensitive response to BPA in the range of 1.0×10^{-8}–1.3×10^{-6} mol L^{-1} with a low detection limit of 5.0×10^{-9} mol L^{-1} under the optimal conditions.

Environmental remediation is the process to reduce hazardous pollutants or contaminants from the environment. Environmental sensing is another significant field of environmental remediation for which the potential of CNTs and graphene has been intensively explored [192, 193]. To date, chemical modification of graphene has been reported to show higher potential for better selectivity and sensitivity than pristine graphene [194]. For instance, such as Cd (II) [195], Co

(II) [195], Hg (II) [196], Cr (VI) [197], Pb (II) [198], these heavy metal ions are toxic in aqueous solution, here heteroatom-doped graphene has been utilized to remediate aqueous solution, and even more capture gaseous molecules, like CO_2 [199].

3.6 Perspective

To sum up, different kinds of chemical modification schemes for graphene have already widened their application window; however, it is still just the tip of the iceberg of prospects. There are still many challenges at present. First of all, we should pay more attention to improve the doping level for changing properties of graphene. Also focus on band gap opening, one of the high priority goals in development of graphene electronics. And even same dopant has different doping types, how to control the doping types further exist different properties still remain questions. Enhancing the stability of electrical properties of doped graphene as environment has large influence to their stability, and improving the doping quality often results in defects in grapheme framework. And because of most doped graphene are difficult to reproduce, to explore new methods to improve the reproducibility of doped graphene is also an important issue [25].

Heteroatom doping can greatly extend the arsenal of graphene materials and their potential for a spectrum of applications. There is reason to believe that some of the applications may find practical value in the near future. To fully explore the potential of chemical modification, several principal remaining issues need to be resolved. The foremost challenge is functionalization without major sacrifice of the intrinsic excellent properties of graphene [70].

References

1. F. Schwierz, Graphene transistors. Nat. Nanotechnol. **5**(7), 487–496 (2010)
2. A.K. Geim, K.S. Novoselov, The rise of graphene. Nat. Mater. **6**(3), 183–191 (2007)
3. T. Wehling et al., Molecular doping of graphene. Nano Lett. **8**(1), 173–177 (2008)
4. C.N.R. Rao, A.K. Sood, *Graphene: Synthesis, Properties, and Phenomena* (Wiley, NJ, 2013)
5. X. Du et al., Approaching ballistic transport in suspended graphene. Nat. Nanotechnol. **3**(8), 491–495 (2008)
6. K.I. Bolotin et al., Ultrahigh electron mobility in suspended graphene. Solid State Commun. **146**(9), 351–355 (2008)
7. K.S. Kim et al., Large-scale pattern growth of graphene films for stretchable transparent electrodes. Nature **457**(7230), 706–710 (2009)
8. K.S. Novoselov et al., Electric field effect in atomically thin carbon films. Science, **306** (5696), pp. 666–669 (2004)
9. X. Li et al., Large-area synthesis of high-quality and uniform graphene films on copper foils. Science **324**(5932), 1312–1314 (2009)

10. K. Novoselov et al., Two-dimensional gas of massless dirac fermions in graphene. Nature **438**(7065), 197–200 (2005)
11. K. Novoselov et al., Unconventional quantum Hall effect and Berry's phase of 2π in bilayer graphene. Nat. Phys. **2**(3), 177–180 (2006)
12. N. Tombros et al., Electronic spin transport and spin precession in single graphene layers at room temperature. Nature **448**(7153), 571–574 (2007)
13. R. Nie et al., Platinum supported on reduced graphene oxide as a catalyst for hydrogenation of nitroarenes. Carbon **50**(2), 586–596 (2012)
14. F. Yavari et al., High sensitivity gas detection using a macroscopic three-dimensional graphene foam network. Sci. Rep. **1**, (2011)
15. H. Liu, Y. Liu, D. Zhu, Chemical doping of graphene. J. Mater. Chem. **21**(10), 3335–3345 (2011)
16. Y. Xue et al., Synthesis of large-area, few-layer graphene on iron foil by chemical vapor deposition. Nano Res. **4**(12), 1208–1214 (2011)
17. H. Wang et al., Ni (OH) 2 nanoplates grown on graphene as advanced electrochemical pseudocapacitor materials. J. Am. Chem. Soc. **132**(21), 7472–7477 (2010)
18. Y. Zhu et al., Carbon-based supercapacitors produced by activation of graphene. Science **332** (6037), 1537–1541 (2011)
19. X. Li et al., Transfer of large-area graphene films for high-performance transparent conductive electrodes. Nano Lett. **9**(12), 4359–4363 (2009)
20. A. Reina et al., Layer area, few-layer graphene films on arbitrary substrates by chemical vapor deposition. Nano Lett. **9**(8), 3087 (2009)
21. S. Bae et al., Roll-to-roll production of 30-inch graphene films for transparent electrodes. Nat. Nanotechnol. **5**(8), 574–578 (2010)
22. Z. Liu et al., Graphene doping of P3HT: PCBM photovoltaic devices. Synth. Met. **160**(9), 1036–1039 (2010)
23. J.K. Lee et al., Silicon nanoparticles–graphene paper composites for Li ion battery anodes. Chem. Commun. **46**(12), 2025–2027 (2010)
24. H. Wang et al., Mn3O4− graphene hybrid as a high-capacity anode material for lithium ion batteries. J. Am. Chem. Soc. **132**(40), 13978–13980 (2010)
25. Y.Z. Xue et al., Controllable synthesis of doped graphene and its applications. Small **10**(15), 2975–2991 (2014)
26. S. Russo et al., Electronic transport properties of few-layer graphene materials. arXiv preprint arXiv:1105.1479 (2011)
27. Z. Berger, C. Naud, D. Mayo, T.B. Li, J. Hass, A.N. Marchenkov, E.H. Conrad, P.N. First, W.A. de Heer et al., Science **312**, 1191 (2006)
28. A.C. Neto et al., The electronic properties of graphene. Rev. Mod. Phys. **81**(1), 109 (2009)
29. I. Gierz et al., Atomic hole doping of graphene. Nano Lett. **8**(12), 4603–4607 (2008)
30. B. Guo et al., Controllable N-doping of graphene. Nano Lett. **10**(12), 4975–4980 (2010)
31. Y. Shao et al., Nitrogen-doped graphene and its electrochemical applications. J. Mater. Chem. **20**(35), 7491–7496 (2010)
32. D. Wei et al., Synthesis of N-doped graphene by chemical vapor deposition and its electrical properties. Nano Lett. **9**(5), 1752–1758 (2009)
33. Z. Jin et al., Large-scale growth and characterizations of nitrogen-doped monolayer graphene sheets. ACS Nano **5**(5), 4112–4117 (2011)
34. G. Imamura, K. Saiki, Synthesis of nitrogen-doped graphene on Pt (111) by chemical vapor deposition. J. Phys. Chem. C **115**(20), 10000–10005 (2011)
35. D. Deng et al., Toward N-doped graphene via solvothermal synthesis. Chem. Mater. **23**(5), 1188–1193 (2011)
36. R.I. Jafri, N. Rajalakshmi, S. Ramaprabhu, Nitrogen doped graphene nanoplatelets as catalyst support for oxygen reduction reaction in proton exchange membrane fuel cell. J. Mater. Chem. **20**(34), 7114–7117 (2010)
37. Y. Wang et al., Nitrogen-doped graphene and its application in electrochemical biosensing. ACS Nano **4**(4), 1790–1798 (2010)

38. Y. Zhu et al., Graphene and graphene oxide: synthesis, properties, and applications. Adv. Mater. **22**(35), 3906–3924 (2010)
39. J. Dai, J. Yuan, P. Giannozzi, Gas adsorption on graphene doped with B, N, Al, and S: A theoretical study. Appl. Phys. Lett. **95**(23), 232105 (2009)
40. O. Leenaerts, B. Partoens, F. Peeters, Adsorption of H_2O, NH_3, CO, NO_2, and NO on graphene: a first-principles study. Phys. Rev. B **77**(12), 125416 (2008)
41. J. Dai, J. Yuan, Adsorption of molecular oxygen on doped graphene: atomic, electronic, and magnetic properties. Phys. Rev. B **81**(16), 165414 (2010)
42. B. Dai et al., High-quality single-layer graphene via reparative reduction of graphene oxide. Nano Res. **4**(5), 434–439 (2011)
43. X. Li et al., Simultaneous nitrogen doping and reduction of graphene oxide. J. Am. Chem. Soc. **131**(43), 15939–15944 (2009)
44. X. Wang et al., Heteroatom-doped graphene materials: syntheses, properties and applications. Chem. Soc. Rev. **43**(20), 7067–7098 (2014)
45. S. Adam et al., A self-consistent theory for graphene transport. Proc. Natl. Acad. Sci. **104** (47), 18392–18397 (2007)
46. A. Lherbier et al., Charge transport in chemically doped 2D graphene. Phys. Rev. Lett. **101** (3), 036808 (2008)
47. C.N.R. Rao, K. Gopalakrishnan, A. Govindaraj, Synthesis, properties and applications of graphene doped with boron, nitrogen and other elements. Nano Today **9**(3), 324–343 (2014)
48. L. Panchakarla et al., Synthesis, structure, and properties of boron-and nitrogen-doped graphene. Adv. Mater. **21**(46), 4726–4730 (2009)
49. C. Coletti et al., Charge neutrality and band-gap tuning of epitaxial graphene on SiC by molecular doping. Phys. Rev. B **81**(23), 235401 (2010)
50. X. Wang et al., N-doping of graphene through electrothermal reactions with ammonia. Science **324**(5928), 768–771 (2009)
51. D. Usachov et al., Nitrogen-doped graphene: efficient growth, structure, and electronic properties. Nano Lett. **11**(12), 5401–5407 (2011)
52. X. Wang et al., Science **324**, 768 (2009)
53. S.J. Goncher et al., Substrate level control of the local doping in graphene. Nano Lett. **13**(4), 1386–1392 (2013)
54. J. Gebhardt et al., Growth and electronic structure of boron-doped graphene. Phys. Rev. B, **87**(15), p. 155437 (2013)
55. T. Wu et al., Nitrogen and boron doped monolayer graphene by chemical vapor deposition using polystyrene, urea and boric acid. New J. Chem. **36**(6), 1385–1391 (2012)
56. S. Zhou et al., Substrate-induced bandgap opening in epitaxial graphene. Nat. Mater. **6**(10), 770–775 (2007)
57. S. Zhou et al., Metal to insulator transition in epitaxial graphene induced by molecular doping. Phys. Rev. Lett. **101**(8), 086402 (2008)
58. S. Kim et al., Origin of anomalous electronic structures of epitaxial graphene on silicon carbide. Phys. Rev. Lett. **100**(17), 176802 (2008)
59. S.-Y. Kwon et al., Growth of semiconducting graphene on palladium. Nano Lett. **9**(12), 3985–3990 (2009)
60. E. McCann, Asymmetry gap in the electronic band structure of bilayer graphene. Phys. Rev. B **74**(16), 161403 (2006)
61. E.V. Castro et al., Biased bilayer graphene: semiconductor with a gap tunable by the electric field effect. Phys. Rev. Lett. **99**(21), 216802 (2007)
62. Y. Zhang et al., Direct observation of a widely tunable bandgap in bilayer graphene. Nature **459**(7248), 820–823 (2009)
63. J. Park et al., Single-gate bandgap opening of bilayer graphene by dual molecular doping. Adv. Mater. **24**(3), 407–411 (2012)
64. L. Ponomarenko et al., Chaotic dirac billiard in graphene quantum dots. Science **320**(5874), 356–358 (2008)

65. K. Nakada et al., Edge state in graphene ribbons: nanometer size effect and edge shape dependence. Phys. Rev. B **54**(24), 17954 (1996)
66. V. Barone, O. Hod, G.E. Scuseria, Electronic structure and stability of semiconducting graphene nanoribbons. Nano Lett. **6**(12), 2748–2754 (2006)
67. Y.-W. Son, M.L. Cohen, S.G. Louie, Energy gaps in graphene nanoribbons. Phys. Rev. Lett. **97**(21), 216803 (2006)
68. X. Li et al., Chemically derived, ultrasmooth graphene nanoribbon semiconductors. Science **319**(5867), 1229–1232 (2008)
69. S. Zhou et al., Origin of the energy bandgap in epitaxial graphene. Nat. Mater. **7**(4), 259–260 (2008)
70. U.N. Maiti et al., 25th anniversary article: chemically modified/doped carbon nanotubes and graphene for optimized nanostructures & nanodevices. Adv. Mater. **26**(1), 40–67 (2014)
71. W. Zhang et al., Opening an electrical band gap of bilayer graphene with molecular doping. ACS Nano **5**(9), 7517–7524 (2011)
72. D. Boukhvalov, M. Katsnelson, Chemical functionalization of graphene. J. Phys. Condens. Matter **21**(34), 344205 (2009)
73. J.T. Robinson et al., Properties of fluorinated graphene films. Nano Lett. **10**(8), 3001–3005 (2010)
74. S.-H. Cheng et al., Reversible fluorination of graphene: evidence of a two-dimensional wide bandgap semiconductor. Phys. Rev. B **81**(20), 205435 (2010)
75. N. Jung et al., Charge transfer chemical doping of few layer graphenes: charge distribution and band gap formation. Nano Lett. **9**(12), 4133–4137 (2009)
76. C. Shan et al., Direct electrochemistry of glucose oxidase and biosensing for glucose based on graphene. Anal. Chem. **81**(6), 2378–2382 (2009)
77. J. Chiou et al., Nitrogen-Functionalized graphene nanoflakes (GNFs: N): tunable photoluminescence and electronic structures. J. Phys. Chem. C **116**(30), 16251–16258 (2012)
78. S. Gao et al., Density functional theory prediction for diffusion of lithium on boron-doped graphene surface. Appl. Surf. Sci. **257**(17), 7443–7446 (2011)
79. O.V. Yazyev, L. Helm, Defect-induced magnetism in graphene. Phys. Rev. B **75**(12), 125408 (2007)
80. K. Gopalakrishnan et al., A Raman study of the interaction of electron-donor and-acceptor molecules with chemically doped graphene. J. Mol. Struct. **1023**, 2–6 (2012)
81. Y. Liu et al., Increased magnetization of reduced graphene oxide by nitrogen-doping. Carbon **60**, 549–551 (2013)
82. Y. Li et al., Spin gapless semiconductor—metal—half-metal properties in nitrogen-doped zigzag graphene nanoribbons. ACS Nano **3**(7), 1952–1958 (2009)
83. C. Özdoğan, J. Kunstmann, A. Quandt, Localization of metallicity and magnetic properties of graphene and of graphene nanoribbons doped with boron clusters. Phil. Mag. **94**(16), 1841–1858 (2014)
84. H.M. Jeong et al., Nitrogen-doped graphene for high-performance ultracapacitors and the importance of nitrogen-doped sites at basal planes. Nano Lett. **11**(6), 2472–2477 (2011)
85. W. Choi, J.-W. Lee, *Graphene: Synthesis and Applications* (CRC Press, USA, 2011)
86. G. Wu et al., Nitrogen-doped graphene-rich catalysts derived from heteroatom polymers for oxygen reduction in nonaqueous lithium–O_2 battery cathodes. ACS Nano **6**(11), 9764–9776 (2012)
87. Y. Xue et al., Three-dimensional B, N-doped graphene foam as a metal-free catalyst for oxygen reduction reaction. Phys. Chem. Chem. Phys. **15**(29), 12220–12226 (2013)
88. H. Wang et al., Synthesis of boron-doped graphene monolayers using the sole solid feedstock by chemical vapor deposition. Small **9**(8), pp. 1316–1320 (2013)
89. S. Some et al., Highly air-stable phosphorus-doped n-type graphene field-effect transistors. Adv. Mater. **24**(40), 5481–5486 (2012)
90. N. Li, Z. Wang, Z. Shi, Synthesis of Graphenes with Arc-Discharge Method (InTech, European Union, 2011)

91. L. Huang et al., Gram-scale synthesis of graphene sheets by a catalytic arc-discharge method. Small **9**(8), pp. 1330–1335 (2013)
92. L. Feng, Y. Chen, L. Chen, Easy-to-operate and low-temperature synthesis of gram-scale nitrogen-doped graphene and its application as cathode catalyst in microbial fuel cells. ACS Nano **5**(12), 9611–9618 (2011)
93. K. Subrahmanyam et al., Chemical storage of hydrogen in few-layer graphene. Proc. Natl. Acad. Sci. **108**(7), 2674–2677 (2011)
94. W.S. Hummers Jr, R.E. Offeman, Preparation of graphitic oxide. J. Am. Chem. Soc. **80**(6), 1339 (1958)
95. Z. Wen et al., Crumpled nitrogen-doped graphene nanosheets with ultrahigh pore volume for high-performance supercapacitor. Adv. Mater. **24**(41), 5610–5616 (2012)
96. T.V. Khai et al., Comparison study of structural and optical properties of boron-doped and undoped graphene oxide films. Chem. Eng. J. **211**, 369–377 (2012)
97. Z. Yang et al., Sulfur-doped graphene as an efficient metal-free cathode catalyst for oxygen reduction. ACS Nano **6**(1), 205–211 (2011)
98. C. Zhang et al., Synthesis of phosphorus-doped graphene and its multifunctional applications for oxygen reduction reaction and lithium ion batteries. Adv. Mater. **25**(35), 4932–4937 (2013)
99. Z.S. Wu et al., Three-dimensional nitrogen and boron co-doped graphene for high-performance all-solid-state supercapacitors. Adv. Mater. **24**(37), 5130–5135 (2012)
100. Q. Chen et al., Plasma activation of carbon nanotubes for chemical modification. J. Phys. Chem. B **105**(3), 618–622 (2001)
101. Y.-B. Tang et al., Tunable band gaps and p-type transport properties of boron-doped graphenes by controllable ion doping using reactive microwave plasma. ACS Nano **6**(3), 1970–1978 (2012)
102. Y. Yu et al., The study of Pt@ Au electrocatalyst based on Cu underpotential deposition and Pt redox replacement. Electrochim. Acta **54**(11), 3092–3097 (2009)
103. S. Sun et al., Monodisperse FePt nanoparticles and ferromagnetic FePt nanocrystal superlattices. Science **287**(5460), 1989–1992 (2000)
104. N. Kristian, Y. Yan, X. Wang, Highly efficient submonolayer Pt-decorated Au nano-catalysts for formic acid oxidation. Chem. Commun. **3**, 353–355 (2008)
105. N. Kristian et al., Controlled synthesis of Pt-decorated Au nanostructure and its promoted activity toward formic acid electro-oxidation. Electrochim. Acta **54**(21), 4916–4924 (2009)
106. Y. Wang et al., Novel palladium–lead (Pd–Pb/C) bimetallic catalysts for electrooxidation of ethanol in alkaline media. J. Power Sour. **195**(9), 2619–2622 (2010)
107. P. Serp, M. Corrias, P. Kalck, Carbon nanotubes and nanofibers in catalysis. Appl. Catal. A **253**(2), 337–358 (2003)
108. S. Maldonado, K.J. Stevenson, Influence of nitrogen doping on oxygen reduction electrocatalysis at carbon nanofiber electrodes. J. Phys. Chem. B **109**(10), 4707–4716 (2005)
109. L. Qu et al., Nitrogen-doped graphene as efficient metal-free electrocatalyst for oxygen reduction in fuel cells. ACS Nano **4**(3), 1321–1326 (2010)
110. H. Wang, T. Maiyalagan, X. Wang, Review on recent progress in nitrogen-doped graphene: synthesis, characterization, and its potential applications. ACS Catalysis **2**(5), 781–794 (2012)
111. L. Zhang, Z. Xia, Mechanisms of oxygen reduction reaction on nitrogen-doped graphene for fuel cells. J. Phys. Chem. C **115**(22), 11170–11176 (2011)
112. D. Geng et al., High oxygen-reduction activity and durability of nitrogen-doped graphene. Energy Environ. Sci. **4**(3), 760–764 (2011)
113. Z.-H. Sheng et al., Synthesis of boron doped graphene for oxygen reduction reaction in fuel cells. J. Mater. Chem. **22**(2), 390–395 (2012)
114. P. Matter, U. Ozkan, Non-metal catalysts for dioxygen reduction in an acidic electrolyte. Catal. Lett. **109**(3–4), 115–123 (2006)
115. Z. Luo et al., Pyridinic N doped graphene: synthesis, electronic structure, and electrocatalytic property. J. Mater. Chem. **21**(22), 8038–8044 (2011)

116. N.P. Subramanian et al., Nitrogen-modified carbon-based catalysts for oxygen reduction reaction in polymer electrolyte membrane fuel cells. J. Power Sour. **188**(1), 38–44 (2009)
117. X. Bao et al., A first-principles study of the role of quaternary-N doping on the oxygen reduction reaction activity and selectivity of graphene edge sites. Top. Catal. **56**(18–20), 1623–1633 (2013)
118. Z. Yang et al., Investigation of homologous series as precursory hydrocarbons for aligned carbon nanotube formation by the spray pyrolysis method. Nano **06**(03), 205–213 (2011)
119. S. Yang et al., Efficient synthesis of heteroatom (N or S)-doped graphene based on ultrathin graphene oxide-porous silica sheets for oxygen reduction reactions. Adv. Funct. Mater. **22** (17), 3634–3640 (2012)
120. R. Li et al., Phosphorus-doped graphene nanosheets as efficient metal-free oxygen reduction electrocatalysts. RSC Adv. **3**(25), 9978–9984 (2013)
121. L. Yang et al., Boron-doped carbon nanotubes as metal-free electrocatalysts for the oxygen reduction reaction. Angew. Chem. **123**(31), 7270–7273 (2011)
122. D. Pan et al., Li storage properties of disordered graphene nanosheets. Chem. Mater. **21**(14), 3136–3142 (2009)
123. Z.-S. Wu et al., Doped graphene sheets as anode materials with superhigh rate and large capacity for lithium ion batteries. ACS Nano **5**(7), 5463–5471 (2011)
124. X. Wang et al., First-principles study on the enhancement of lithium storage capacity in boron doped graphene. Appl. Phys. Lett. **95**(18), 183103 (2009)
125. A.L.M. Reddy et al., Synthesis of nitrogen-doped graphene films for lithium battery application. ACS Nano **4**(11), 6337–6342 (2010)
126. W. Ai et al., Nitrogen and sulfur codoped graphene: multifunctional electrode materials for high-performance Li-ion batteries and oxygen reduction reaction. Adv. Mater. (2014)
127. P. Wu et al., Facile synthesis of nitrogen-doped graphene for measuring the releasing process of hydrogen peroxide from living cells. J. Mater. Chem. **22**(13), 6402–6412 (2012)
128. B.E. Conway, Transition from "supercapacitor" to "battery" behavior in electrochemical energy storage. J. Electrochem. Soc. **138**(6), 1539–1548 (1991)
129. K.H. An et al., Supercapacitors using single-walled carbon nanotube electrodes. Adv. Mater. **13**(7), 497–500 (2001)
130. S.H. Lee et al., Three-. Angew. Chem. **122**(52), 10282–10286 (2010)
131. T.Y. Kim et al., High-performance supercapacitors based on poly (ionic liquid)-modified graphene electrodes. ACS Nano **5**(1), 436–442 (2010)
132. Z. Lei, L. Lu, X. Zhao, The electrocapacitive properties of graphene oxide reduced by urea. Energy Environ. Sci. **5**(4), 6391–6399 (2012)
133. H.-L. Guo et al., Synthesis and characterization of nitrogen-doped graphene hydrogels by hydrothermal route with urea as reducing-doping agents. J. Mater. Chem. A **1**(6), 2248–2255 (2013)
134. Y. Qiu, X. Zhang, S. Yang, High performance supercapacitors based on highly conductive nitrogen-doped graphene sheets. Phys. Chem. Chem. Phys. **13**(27), 12554–12558 (2011)
135. Z. Zuo, Z. Jiang, A. Manthiram, Porous B-doped graphene inspired by fried-ice for supercapacitors and metal-free catalysts. J. Mater. Chem. A **1**(43), 13476–13483 (2013)
136. J. Han et al., Generation of B-doped graphene nanoplatelets using a solution process and their supercapacitor applications. ACS Nano **7**(1), 19–26 (2012)
137. N. Soin et al., Enhanced and stable field emission from in situ nitrogen-doped few-layered graphene nanoflakes. J. Phys. Chem. C **115**(13), 5366–5372 (2011)
138. W. Zhu et al., Large current density from carbon nanotube field emitters. Appl. Phys. Lett. **75** (6), 873–875 (1999)
139. Y. Wu et al., Carbon nanowalls and related materials. J. Mater. Chem. **14**(4), 469–477 (2004)
140. J. Li et al., Field emission from high aspect ratio tubular carbon cones grown on gold wire. Appl. Phys. Lett. **87**(14), 143107 (2005)
141. G.A. Amaratunga, S. Silva, Nitrogen containing hydrogenated amorphous carbon for thin-film field emission cathodes. Appl. Phys. Lett. **68**(18), 2529–2531 (1996)

142. A. Malesevic et al., Field emission from vertically aligned few-layer graphene. J. Appl. Phys. **104**(8), 084301 (2008)
143. J. Qi et al., Ar plasma treatment on few layer graphene sheets for enhancing their field emission properties. J. Phys. D Appl. Phys. **43**(5), 055302 (2010)
144. Z.S. Wu et al., Field emission of single-layer graphene films prepared by electrophoretic deposition. Adv. Mater. **21**(17), 1756–1760 (2009)
145. U. Palnitkar et al., Remarkably low turn-on field emission in undoped, nitrogen-doped, and boron-doped graphene. Appl. Phys. Lett. **97**(6), 063102 (2010)
146. L. Chan et al., Role of extrinsic atoms on the morphology and field emission properties of carbon nanotubes. Appl. Phys. Lett. **82**(24), 4334–4336 (2003)
147. H.-S. Ahn et al., Field emission of doped carbon nanotubes. Appl. Phys. Lett. **88**(9), 093122 (2006)
148. R. Sharma et al., Field emission properties of boron and nitrogen doped carbon nanotubes. Chem. Phys. Lett. **428**(1), 102–108 (2006)
149. R.V. Kashid et al., Field emission characteristics of pristine and N-doped graphene measured by in-situ transmission electron microscopy. J. Appl. Phys. **113**(21), 214311 (2013)
150. J. Li et al., Correlations between percolation threshold, dispersion state, and aspect ratio of carbon nanotubes. Adv. Funct. Mater. **17**(16), 3207–3215 (2007)
151. W. Bauhofer, J.Z. Kovacs, A review and analysis of electrical percolation in carbon nanotube polymer composites. Compos. Sci. Technol. **69**(10), 1486–1498 (2009)
152. T. Kuilla et al., Recent advances in graphene based polymer composites. Prog. Polym. Sci. **35**(11), 1350–1375 (2010)
153. S.D. Sarma et al., Electronic transport in two-dimensional graphene. Rev. Mod. Phys. **83**(2), 407 (2011)
154. S. Huang, L. Dai, A.W.H. Mau, J. Phys. Chem. B **103**, 4223 (1999)
155. Q. Wang et al., The effects of CNT alignment on electrical conductivity and mechanical properties of SWNT/epoxy nanocomposites. Compos. Sci. Technol. **68**(7–8), 1644–1648 (2008)
156. A.A. Balandin et al., Superior thermal conductivity of single-layer graphene. Nano Lett. **8**(3), 902–907 (2008)
157. J.H. Seol et al., Two-dimensional phonon transport in supported graphene. Science **328** (5975), 213–216 (2010)
158. S. Berber, Y.-K. Kwon, D. Tomanek, Unusually high thermal conductivity of carbon nanotubes. Phys. Rev. Lett. **84**(20), 4613 (2000)
159. M. Zhang et al., Strong, transparent, multifunctional, carbon nanotube sheets. Science **309** (5738), 1215–1219 (2005)
160. X.M. Liu et al., Transparent boron-doped carbon nanotube films. Nano Lett. **8**(9), 2613–2619 (2008)
161. G. Eda, G. Fanchini, M. Chhowalla, Large-area ultrathin films of reduced graphene oxide as a transparent and flexible electronic material. Nat Nano **3**(5), 270–274 (2008)
162. J. Liang et al., Electromagnetic interference shielding of graphene/epoxy composites. Carbon **47**(3), 922–925 (2009)
163. H.-B. Zhang et al., Tough graphene–polymer microcellular foams for electromagnetic interference shielding. ACS Appl. Mater. Interfaces **3**(3), 918–924 (2011)
164. X. Wang, L. Zhi, K. Müllen, Transparent, conductive graphene electrodes for dye-sensitized solar cells. Nano Lett. **8**(1), 323–327 (2007)
165. A.K. Geim, Graphene: status and prospects. Science **324**(5934), 1530–1534 (2009)
166. J.O. Hwang et al., Workfunction-tunable, N-doped reduced graphene transparent electrodes for high-performance polymer light-emitting diodes. ACS Nano **6**(1), 159–167 (2011)
167. H. Huang et al., Aligned carbon nanotube composite films for thermal management. Adv. Mater. **17**(13), 1652–1656 (2005)
168. S. Subrina, D. Kotchetkov, A.A. Balandin, Heat removal in silicon-on-insulator integrated circuits with graphene lateral heat spreaders. Electron Device Lett. IEEE **30**(12), 1281–1283 (2009)

169. M.F. De Volder et al., Carbon nanotubes: present and future commercial applications. Science **339**(6119), 535–539 (2013)
170. J. Lee, T. Cocke, F. Gonzalez, Regional vs "tight" heparinization in hemodialysis. in *Proceedings of the Clinical Dialysis and Transplant Forum* (1974)
171. J. Chen et al., Solution properties of single-walled carbon nanotubes. Science **282**(5386), 95–98 (1998)
172. S. Stankovich et al., Graphene-based composite materials. Nature **442**(7100), 282–286 (2006)
173. T. Ramanathan et al., Functionalized graphene sheets for polymer nanocomposites. Nat Nano **3**(6), pp. 327–331 (2008)
174. K. Gopalakrishnan et al., Selectivity in the photocatalytic properties of the composites of TiO_2 nanoparticles with B- and N-doped graphenes. Chem. Phys. Lett. **511**(4–6), 304–308 (2011)
175. M. McCune, W. Zhang, Y. Deng, High efficiency dye-sensitized solar cells based on three-dimensional multilayered ZnO nanowire arrays with "caterpillar-like" structure. Nano Lett. **12**(7), 3656–3662 (2012)
176. W. Hong et al., Transparent graphene/PEDOT–PSS composite films as counter electrodes of dye-sensitized solar cells. Electrochem. Commun. **10**(10), 1555–1558 (2008)
177. T.N. Murakami et al., Highly efficient dye-sensitized solar cells based on carbon black counter electrodes. J. Electrochem. Soc. **153**(12), A2255–A2261 (2006)
178. H. Han et al., A design for monolithic all-solid-state dye-sensitized solar cells with a platinized carbon counterelectrode. Appl. Phys. Lett. **94**(10), 103102 (2009)
179. Z. Yang et al., Aligned carbon nanotube sheets for the electrodes of organic solar cells. Adv. Mater. **23**(45), 5436–5439 (2011)
180. H. Choi et al., Graphene counter electrodes for dye-sensitized solar cells prepared by electrophoretic deposition. J. Mater. Chem. **21**(21), 7548–7551 (2011)
181. R. Jia et al., Synthesis of highly nitrogen-doped hollow carbon nanoparticles and their excellent electrocatalytic properties in dye-sensitized solar cells. J. Mater. Chem. **20**(48), 10829–10834 (2010)
182. J. Han et al., Water-soluble polyelectrolyte-grafted multiwalled carbon nanotube thin films for efficient counter electrode of dye-sensitized solar cells. ACS Nano **4**(6), 3503–3509 (2010)
183. J.D. Roy-Mayhew et al., Functionalized graphene as a catalytic counter electrode in dye-sensitized solar cells. ACS Nano **4**(10), 6203–6211 (2010)
184. D.W. Zhang et al., Graphene-based counter electrode for dye-sensitized solar cells. Carbon **49**(15), 5382–5388 (2011)
185. C. Mattevi et al., Evolution of electrical, chemical, and structural properties of transparent and conducting chemically derived graphene thin films. Adv. Funct. Mater. **19**(16), 2577–2583 (2009)
186. Y. Xue et al., Nitrogen-doped graphene foams as metal-free counter electrodes in high-performance dye-sensitized solar cells. Angew. Chem. Int. Ed. **51**(48), 12124–12127 (2012)
187. D. Yu et al., Metal-free carbon nanomaterials become more active than metal catalysts and last longer. J. Phys. Chem. Lett. **1**(14), 2165–2173 (2010)
188. S. Yang et al., Graphene-based carbon nitride nanosheets as efficient metal-free electrocatalysts for oxygen reduction reactions. Angew. Chem. Int. Ed. **50**(23), 5339–5343 (2011)
189. H. Park et al., Doped graphene electrodes for organic solar cells. Nanotechnology **21**(50), 505204 (2010)
190. U. Maitra et al., Highly effective visible-light-induced H_2 generation by single-layer 1T-MoS_2 and a nanocomposite of few-layer 2H-MoS_2 with heavily nitrogenated graphene. Angew. Chem. Int. Ed. **52**(49), 13057–13061 (2013)
191. H. Fan et al., Electrochemical bisphenol a sensor based on N-doped graphene sheets. Anal. Chim. Acta **711**, 24–28 (2012)

192. B.L. Allen, P.D. Kichambare, A. Star, Carbon nanotube field-effect-transistor-based biosensors. Adv. Mater. **19**(11), 1439–1451 (2007)
193. S.N. Kim, J.F. Rusling, F. Papadimitrakopoulos, Carbon nanotubes for electronic and electrochemical detection of biomolecules. Adv. Mater. **19**(20), 3214–3228 (2007)
194. Q. He et al., Graphene-based electronic sensors. Chem. Sci. **3**(6), 1764–1772 (2012)
195. G. Zhao et al., Few-layered graphene oxide nanosheets as superior sorbents for heavy metal ion pollution management. Environ. Sci. Technol. **45**(24), 10454–10462 (2011)
196. V. Chandra, K.S. Kim, Highly selective adsorption of Hg2+ by a polypyrrole-reduced graphene oxide composite. Chem. Commun. **47**(13), 3942–3944 (2011)
197. H. Jabeen et al., Enhanced Cr(vi) removal using iron nanoparticle decorated graphene. Nanoscale **3**(9), 3583–3585 (2011)
198. G. Zhao et al., Removal of Pb(ii) ions from aqueous solutions on few-layered graphene oxide nanosheets. Dalton Trans. **40**(41), 10945–10952 (2011)
199. A.K. Mishra, S. Ramaprabhu, Nanostructured polyaniline decorated graphene sheets for reversible CO_2 capture. J. Mater. Chem. **22**(9), 3708–3712 (2012)

Chapter 4
Chalcogenides to Nanochalcogenides; Exploring Possibilities for Future R&D

Zishan H. Khan, Shamshad A. Khan, Faisal A. Agel, Numan A. Salah and M. Husain

Abstract Chalcogenides (Se, Te and S) are one of the interesting classes of materials studied so far. They have potential applications in phase change recording, memory and switching and various other solid state electronic devices. These materials got a great deal of attention of the scientists worldwide due to their low phonon energy, infrared transparency, large value of refractive index, high photosensitivity, reversible phase transformation etc. There are several techniques for the synthesis of chalcogenide materials, which include melt quenching, thermal evaporation, sputtering, chemical vapor deposition etc. Among all these techniques, melt quenching is one of the simplest and popular techniques for producing chalcogenide glasses. Recently, a lot of work is focussed on production of chalcogenides at nanoscale. The understanding of electrical, optical and thermal properties of these chalcogenides at nanoscale is of great interest both from fundamental and technological point of view. Due to their interesting physical properties, these nanochalcogenides has raised considerable deal of research interest followed by technological applications in the field of micro/optoelectronics. The structure of chalcogenides is disordered at the atomic scale. Therefore, the nanostructures of these materials can easily be tailored and may yield a greater variety than that of crystalline nanostructures. The synthesis of chalcogenide nanostructures in the form

Z.H. Khan (✉)
Department of Applied Sciences and Humanities, Jamia Millia Islamia, New Delhi, India
e-mail: zishan_hk@yahoo.co.in

S.A. Khan
Department of Physics, St. Andrews PG College, Gorakhpur, India

F.A. Agel
Department of Physics, University of Hail, Hail, Kingdom of Saudi Arabia

N.A. Salah
Center of Nanotechnology, King Abdulaziz University, Jeddah, Kingdom of Saudi Arabia

M. Husain
Department of Physics, Jamia Millia Islamia, New Delhi, India

M. Husain
MJP Rohilkhand University, Bareilly, UP, India

M. Husain and Z.H. Khan (eds.), *Advances in Nanomaterials*,
Advanced Structured Materials 79, DOI 10.1007/978-81-322-2668-0_4

of nanoparticles, nanobelts, nanorods, and nanowires has stimulated intense research activity due to their improved properties at nanoscale. With these interesting results, the nano-chalcogenides have become the focus of attention and are expected to present interesting properties. A dramatic change in the physical and chemical properties of these materials is observed due to size reduction. Moreover, the work on the synthesis and characterization of nano-chalcogenides is still in the primarily stages and accordingly, overall features have not been explored so far. Therefore, more research work on these nanochalcogenides is needed for complete understanding of the mechanism responsible for change in properties in these materials at nanoscale. This chapter provides a comprehensive review of chalcogenides and nanochalcogenides, their synthesis and applications.

Keywords Chalcogenides · Nanochalcogenides · DC conductivity · Optical band gap · Crystallization kinetics

4.1 Introduction

Amorphous materials were insignificant in the past. In fact, the use of glass in decorative and packaging applications has a history of more than 10,000 years. It was a long believed that these materials might not be semiconductor as the theory of semiconductor is based on periodicity of atom i.e. long-range order. But in 1950, Kolomiets [1] showed that glasses containing large proportional of chalcogen elements (S, Se, Te) may act as a semiconductor. These workers did not, however, find that these materials are very interesting because their electrical conductivity could not be increased by adding the conventional doping elements of III and V groups of periodic table. In 1951, Spear [2] reported the first drift mobility measurement in vitreous Se and Tauc et al. [3] reported the first studies on amorphous Ge. Much stimulating work on amorphous semiconductors started after the Ovshinsky [4] reported threshold and memory switching in chalcogenide glasses in 1968. He also demonstrated the use of chalcogenide thin films for computer memory. A further interest in these semiconductors arose when W.E. Spear and P.G. Le Comber (1975) at the University of Dundee, U.K., developed a technology to produce n and p type amorphous silicon similar to their crystalline counterpart. Since then, these materials have been utilized in various solid state devices. It also presents several surprising features that stimulated pure academic interest. Therefore, a serious investigation of the properties of amorphous semiconductors requires not only familiarity with one particular area of science or engineering, but also a detailed knowledge of results from all the fields including physics, chemistry, metallurgy and electrical engineering. Although a lot of work on amorphous semiconductors has been reported, but it is not enough to completely predict the behavior of these materials as compared with the crystalline materials. Therefore, much fundamental experimental and theoretical work remains is still needed to completely understand these materials.

One of the most widespread current applications of amorphous semiconductors is in the field of xerography in document copying machines. Perhaps, the most technologically developed and potentially the most important application of the amorphous semiconductor (a-Se:H) is in the direct conversion of sunlight to electrical power. The use of an amorphous material as an electrochemical sensor is in the so-called 'glass electrode' commonly used to monitor proton activity, i.e. pH, of the aqueous solutions. Glassy metallic alloys have important advantage over their crystalline counterparts, namely the important magnetic properties, such as saturation magnetic moment (or magnetization), Curie temperature and magnetostriction, can be varied smoothly simply by changing the composition. Being magnetic, amorphous alloys are also found useful as magnetic sensors. The features that differentiate them from other soft magnetic materials in this application include: high permeability. The greatest impact of the use of glassy metallic alloys has been in transformer-core applications, for operation both at low frequencies (50–60 Hz) in distribution transformers and at high frequencies in a variety of special applications.

There are many optical properties of these materials, which are common with the classical semiconductors. The optical absorption spectra of the semiconductors exhibit a threshold, below the threshold frequency, the light can pass through practically without losses, while above it, the light is strongly absorbed. Moreover, a good luminescence property in the visible and infrared spectral range is also characteristic of many semiconductors. Thus, a semiconductor material should possess several characteristic properties, not just the one of the moderately good electrical conduction. Therefore the material exhibiting the above mentioned properties falls in the category of semiconductors. This criterion excludes ionic conductors, for example, which exhibit conductivity values of the same order of magnitude as that of pure semiconductor, but do not display the characteristic temperature dependence. The atomic structure of amorphous solids is characterized by the same parameters, which are used for crystalline materials: coordination number, valency, bond length and bond angle. However, the atomic structure of amorphous solids is different for each atom and the whole structure of the disorder material is a superposition of local structural elements. The valency of atoms and stoichiometry also vary in the amorphous state. Such variable structure units are the topological elements in the isotropic random network structure of amorphous solids. Efforts are being made to dope these glasses so that they can be used for other solid state devices, where n or p semiconductor is required. In some Ge–Se based glasses, Bi doping has shown n type conduction. This has opened the possibility of doping in chalcogenides glasses and based on doping, the devices using n and p type semiconductors have been developed. However, more research work is required to get efficient doping in chalcogenides glasses to obtain good quality devices.

The arduous search for materials with molecular scale properties that can satisfy the demands of the 21st century has led to the development of nanochalcogenides. Every type of traditional material has its nanostructured counterpart. The application of nanomaterials in nanoelectronics, nanodevices and systems, nanocomposite materials, alternative energy resources and nanobiotechnology are remarkable. The interface of nanoscience and technology with biological and therapeutic sciences is expected to radically improve the ability to provide efficient treatments in otherwise impossible situations. From the innovation of fullerene, carbon nanotubes and graphene; there have been numerous reports that discussed the fundamental and technological importance of novel nanostructured materials. The characteristic structural features in between the isolated atoms and the bulk macroscopic materials result in significantly different physical properties of this special class of materials. The most popular term in the nano world is "Quantum confinement", which is due to the changes in the atomic structure. The main implication of such confinement is the change in the system total energy; and hence the overall thermodynamic stability.

Nano-chalcogenides continues as an attraction by researchers and engineers and are viewed as large group of interesting solids in which unusual physical and chemical phenomena are revealed and as the materials, it opens new ways in science and technology. Due to the large optical nonlinearity and short response time of nano-chalcogenides, the studies on nonlinear optical properties are also important. The optical properties of these nanoscale chalcogenides are also strongly depend on the size, shape and surface characteristics. Therefore, much attention has been paid to control these parameters to manipulate the optical properties of nanomaterials. Nano-chalcogenide possesses unique characteristics which are different from those in oxide and halide glasses, i.e. molecular structures and semiconductor properties. Due to the potential applications in nanoscale devices, a lot of work is focused on the synthesis and characterization of the nanochalcogenides. One of the present challenges in materials science is the production of materials at nanometric scale with controlled composition and structure due to the observation of novel properties at nanometric scales. An extensive investigation of varieties of phase change optical recording materials have been made, but the exploration of these materials at nano-scale is still needed. Development of optical materials at nanoscale for the use in optical disk, is expected to be cost effective and more stable, which is the main requirement for advanced data storage devices. It is also understood that the reduction in size (nano-particles) will change the properties of these materials dramatically (enhanced band gap, large absorption coefficient and highly compositional dependence of reflection, fast crystallization). Therefore, the chalcogenides produced at nano-scale i.e. nanochalcogenides may find interesting applications in future nano-optoelectronic and other semiconducting nanodevices.

Recently, efforts have been dedicated to prepare nano-chalcogenide materials for various applications. Shen et al. [5] reported the preparation of dextran/Se nanocomposites for different nanomedicine applications. Lee et al. [6] prepared semiconducting thermo-optic material for potential application to super-resolution optical data storage. Pinkas et al. [7] prepared amorphous nanoscopic iron (III) oxide from Fe(acac). Concha et al. [8] studied Monte Carlo simulation of Fe–Co amorphous nanoparticles magnetization. Xi and Lam [9] synthesized and characterized CdSe nanorods using a novel microemulsion method at moderate temperature. Rajamathi and Seshadri [10] studied oxide and chalcogenide nanoparticles from hydrothermal/solvothermal reactions. The work on nanoparticles layers of CdSe buried in oxide and chalcogenide thin film matrices reported by Nesheva et al. [11].

4.2 Synthesis of Chalcogenide Glasses

The formation of glass by cooling a liquid requires the prevention of the nucleation and growth process responsible for crystallization. Both the rate of homogeneous nucleation and the subsequent rate of crystal growth go through a maximum temperature as the temperature is lowered below the thermodynamic melting point. Thus, if temperature of a liquid can be lowered below these maximum quickly enough as to avoid nucleation or the growth of detectable crystals, further cooling will not be needed and the liquid will eventually form a glass. In this sense, glass formation is a kinetics process but there are features of the phenomenon, which can be described within framework of thermodynamics. Whether or not a particular liquid can be cooled to form a glass will obviously depend upon the rates of the atomic and/or molecular transport process involved in nucleation and crystal growth. Viscosity is often regarded as the determining factor and it is true that many liquid are comparatively viscous at their melting points, this is not always an infallible guide. For example, in the homologous series of primary alcohol, the lowest member (methyl) crystallizes vary rapidly on cooling, as do the higher members with log chains, while intermediate members tends to form glasses. Viscosity on the other hand increases steadily with increasing chain length. There are clearly steric factors involved, which are not simply related to the flow processes. Nevertheless, it is possible to roughly relate the glass batches that can exhibit high vapor pressure, and the melts tend to have a high viscosity. They are susceptible to oxidation and hydrolysis. To avoid these problems, melting is preferably carried out inside a vitreous ampoule sealed under vacuum and the melt is agitated, purification of the batch and ampoule must be completed prior to sealing and it is important to minimize hydride, hydro oxide and oxide impurities, which impair the transmission window of the finished glass.

Both the amorphous and polycrystalline chalcogenides in bulk form can be prepared by using diffusion/melt quenching method. The highly pure materials (99.999 %) are used for preparing the amorphous and polycrystalline chalcogenides materials. Initially, materials are weighed according to their atomic percentages and are sealed in quartz ampoules under a vacuum of 10^{-5} Torr. The sealed ampoules are then placed in a Microprocessor-Controlled Programmable Muffle Furnace, where the temperature increases at a certain fixed heating rate up to a temperature higher than melting points of the constituents elements and are kept at that temperature for 12–16 h with frequent rocking to ensure the homogenization of the melt. As a requirement of homogeneity, heating and shaking of the elements in the ampoules are performed simultaneously. The melt are then rapidly quenched ice water. The X-ray diffraction technique is used to verify the amorphous nature of

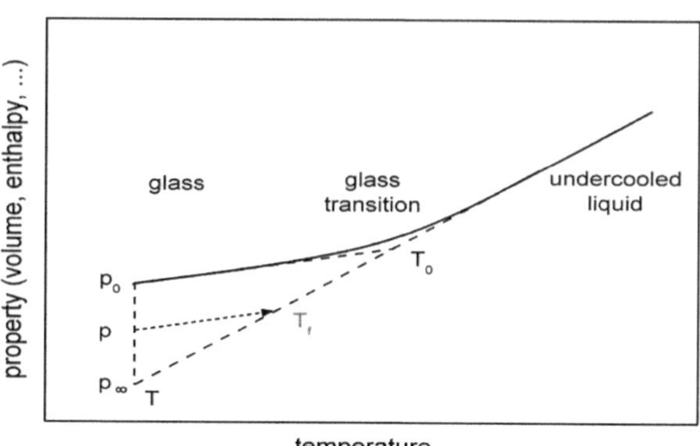

Fig. 4.1 Temperature dependence property of quenched materials at constant pressure

as-prepared bulk sample. Figure 4.1 shows the temperature dependence property of quenched materials at constant pressure.

4.2.1 Preparation of Chalcogenides Thin Films

There are several methods for preparation of chalcogenide thin films. Some of them are physical vapour condensation, sputtering, pulsed-laser deposition and chemical vapour deposition method. These techniques are most commonly employed to fabricate the nanostructures of chalcogenide glasses. The brief discussion about these methods is as follows;

4.2.1.1 Physical Vapour Condensation

Over the last few decades, amorphous thin films have predominantly been prepared by physical vapour deposition techniques. Physical vapour condensation (Fig. 4.2) is one the simplest techniques employed so far. The vapour is formed by heating of a solid. With the introduction of more versatile deposition techniques such as sputtering and chemical vapour deposition, tradition thermal evaporation is now

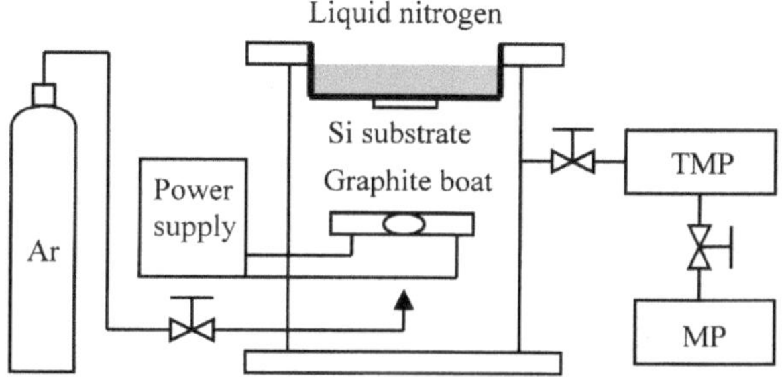

Fig. 4. 2 Schematic diagram of physical vapour condensation system

less widely used. This technique is replaced by physical vapour condensation technique, which one of simplest techniques to produce the nano-chalcogenides. The physical vapour condensation remains attractive for good glass-forming binary chalcogenides, because it is low-cost, relatively simple, and reproducible, and capable of deposition over a large substrate area. This is one of the favoured techniques for deposition of nanostructures of amorphous selenium (a-Se) for flat-panel X-ray detectors, nano-ionic memories, thin-film waveguides, hard and soft stamp-imprinting techniques. In the nanochalcogenide films, it can be observed that the refractive index is non-uniform through the thickness: the bottom part of the film is enriched with the more volatile species and therefore, it has a lower refractive index than nearer to the surface.

Nanochalcogenide thin films can be prepared in the presence of ambient argon gas atmosphere in the chamber. Initially, a small quantity of powder of these alloys is kept in a molybdenum boat and the chamber is evacuated to a vacuum of 10^{-6} Torr. An inert-gas (argon) is purged into the chamber after attaining a vacuum of 10^{-6} Torr. The bulk chalcogenide materials are then evaporated in the presence of ambient argon gas atmosphere in the chamber to deposit the thin film. The substrates is cooled with liquid nitrogen and the evaporated material is deposited on a glass substrate pasted on the LN_2 cooled substrate.

4.2.1.2 Sputtering

Sputtering is the most versatile technique for depositing chalcogenide thin films (Fig. 4.3). It can be used for deposition of any binary, ternary, multi-component or rare-earth doped glasses. Sputtering is the dominant industrial technique in depositing chalcogenide films for optical (DVD, Blu-ray) and electrical data-storage (FLASH, RAM) applications, but is now being challenged by promising new techniques.

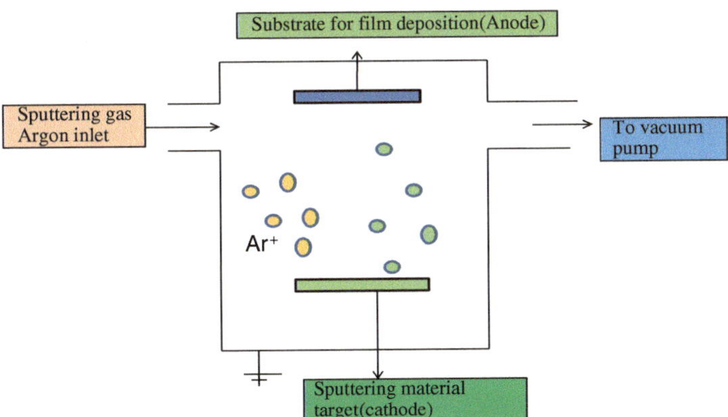

Fig. 4.3 Schematic diagram of RF sputtering system

In most forms of sputtering, there is an electrical potential difference (can be of the order of 1500 V) between the target and the substrate, that (via generation of plasma) is the source of the high-energy ions. Sputtering is a complex process where the material in a solid target is vaporized by bombardment with high-energy ions. There are many variants of sputtering, and here we limit ourselves to in between the target and the substrate, that (via generation of plasma) is the source of the high-energy ions. It is a process where the material in a solid target is vaporized by bombardment with high-energy ions. The target material is in the form of elemental (multi-target deposition) or multi-component amorphous/crystalline discs made from high-purity elements. When high-energy ions arrive at a target surface, ablation of target atoms is just one of many possible events which can happen to the target material. Based on time-scale, these can be classified as prompt effects (<10–12 s), e.g. lattice collisions, physical sputtering, reflection from the surface; cooling effects (from >10–12 to <10–10 s), e.g. thermal spikes along collision cascades; delayed effects (>10–10 s), e.g. atomic diffusion in the target material, strain-induced diffusion, segregation, etc.; and persistent effects, e.g. gas incorporation, compressive stress due to recoil implantation, etc.

An important quantity in sputtering is the sputtering yield, which is the number of atoms leaving a target per incident ion. The momentum transfer from an incident ion to the atoms in the target is most efficient when the ions and the atoms have the same atomic mass. Argon ions (40 amu) have high sputtering yields for targets atoms of roughly similar mass. For heavier target atoms, heavier incident ions give better yields. However, argon is still the most common sputtering gas because of the high cost of the heavier noble gases. The sputtering yield is dependent on the angle of incidence (of the Ar^+ ions on the target); it increases with off-normal incidence, and peaks at 50–60°, where the yield is increased by 20–50 %.

4.2.1.3 Pulsed-Laser Deposition

Pulsed-laser deposition is a relatively new physical vapour deposition technique (Fig. 4.4), which has been used for deposition of a vast range of materials including chalcogenide thin films. In pulsed-laser deposition, laser pulses of high-energy density are focused onto a target, the material of which is vaporized after the laser power reaches the ablation threshold and forms a plasma plume. This plume of ablated species is projected towards substrate, where the species condense and form a thin film. Interaction of the laser pulse with the target is a highly non-equilibrium process, where the absorption of the laser energy is confined to very small volume. The stoichiometry of the target is then well preserved in the deposited film even for multi-component or rare-earth doped compositions. Chalcogenide films deposited by pulsed-laser deposition can often be obtained containing structural units similar to those in the corresponding bulk glass. Pulsed-laser deposition is a complex deposition technique with many technological aspects reviewed in detail elsewhere.

There are two pulsed-laser deposition approaches for chalcogenide films: conventional pulsed-laser deposition and ultrafast pulsed-laser deposition. The

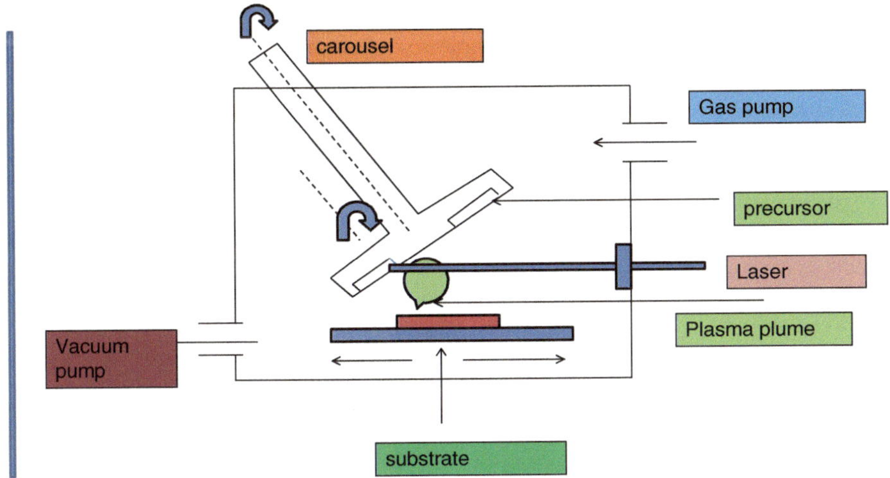

Fig. 4.4 Schematic diagram of pulsed-laser deposition

techniques make use of different pulsed-laser sources, resulting in different predominant ablation mechanisms. In conventional pulsed-laser deposition, ultraviolet excimer gas lasers provide nanosecond laser pulses of average energy ~0.01–10 J, intensity ~108–109 W/cm^2 and repetition rates ~1–50 Hz. The laser spot area on the target is typically 0.5–5 mm^2 but can be larger and non-circular in shape. Penetration depths are up to several micrometres, and depend strongly on laser power density and material absorption coefficient. Similarly, deposition rates can reach up to tens of nm per pulse (10–24 nm per pulse) and are likely difficult to control. These conditions are associated with the thermal ablation mechanism, in which species are ejected from the target mostly by local thermal overheating, i.e. melting and vaporization.

In contrast, in ultrafast pulsed-laser deposition, electrostatic ablation dominates, in which species (ions) are ejected from the target due to the strong electric fields. These fields are formed through charge separation by ionization, associated with electrons being emitted from the target. Ultrafast pulsed-laser deposition uses ps or fs pulsed lasers, e.g. higher harmonic generation in Nd:YAG laser (natural frequency at $\lambda = 1064.1$ nm) at 532, 355, 266 nm. The repetition rates reach ~50–100 MHz, the intensities above ~1013 W/cm^2 and the average energy is in the mJ range. The penetration depths in the target, up to 1–2 mm, are typically smaller than in conventional pulsed-laser deposition. Deposition rates can be controlled easily and are 1–15 nm per pulse. Ultrafast pulsed-laser deposition has the advantages of: (1) forming of a continuous flow of ablated species, (2) changing the ablation mechanism from thermal to electrostatic, and (3) smoothing the effective intensity distribution, allowing the creation of a single homogeneous phase (suppressing preferential ablation which can occur in conventional pulsed-laser deposition) in the ablated plume.

Pulsed-laser deposition has been used in the deposition of a variety of chalcogenide thin films. Pulsed-laser deposition of chalcogenide films also has several technological issues which curtail its wider application. It is still difficult to control film deposition rates precisely, especially in conventional pulsed-laser deposition. To achieve good thickness uniformity (this could be an issue even for small samples) and composition homogeneity of films over a large substrate area, optimization of the laser power and optical path are necessary, as well as control of the process geometry and target-to-substrate positioning (by substrate rotation/ translation).

4.2.1.4 Chemical Vapour Deposition

Chemical vapour deposition is an important technique, particularly promising for conformal deposition of chalcogenide thin films into embedded 1D, 2D and 3D structures. Chemical vapour deposition is a method where gaseous precursors react either in gas phase or at the substrate gas interface, producing thin film on the substrate (Fig. 4.5). The composition of the thin film is different from the gases being used. The thickness of the conformal coatings can be more uniform and controlled more precisely than in physical vapour deposition. Chemical vapour deposition is well known for the deposition of crystalline films (ceramics), but it is rather new for thin films of amorphous chalcogenides. The key advantages of chemical vapour deposition using gaseous sources are: (1) uniform thickness (and composition) can be achieved in coatings of very complex spatial features, reaching remote areas of substrates; (2) the correct purification processes can give high purity precursors; and (3) high deposition rates ($\sim$1 mm/min). Optimizing the deposition of chalcogenides involves a strong and complex competition between the various techniques, and is emerging as a preferred option.

Chemical vapour deposition reactions typically start and proceed only at quite high temperatures of 400–600 °C, too high for pre-deposited temperature-sensitive structures (memory, sensors, etc.) to be used as substrates. High temperatures are

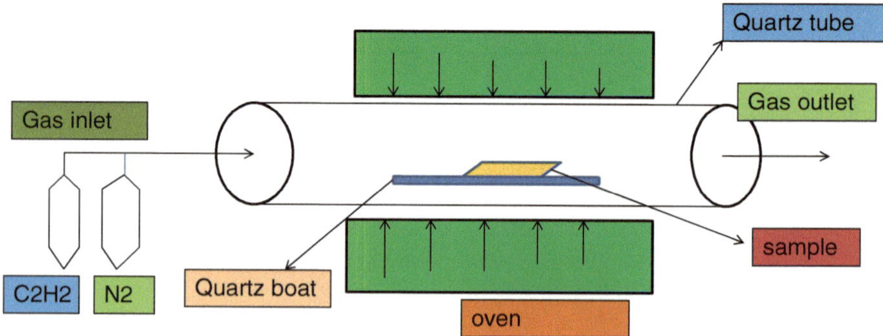

Fig. 4.5 Schematic diagram of CVD system

necessary to achieve large values of equilibrium constants and a good conversion rate of precursor's reactions at the substrate. The required temperature can be reduced by employing a plasma in the reaction chamber to enhance (i.e., accelerate) the chemical reactions, or by using metal-organic precursors. The pressure in the deposition chamber may vary from atmospheric-pressure chemical vapour deposition to low-pressure chemical vapour deposition depending on the required deposition conditions. Chemical vapour deposition has already been demonstrated to be promising for the deposition of amorphous and crystalline nano-chalcogenide thin films for waveguides, phase-change RAM and photovoltaic applications. It is still challenging to find appropriate precursors for rare-earth metals in the deposition of rare-earth doped nano-chalcogenide films.

4.3 Electrical, Optical and Thermal Properties

4.3.1 Electrical Transport Properties

During last several years, the mechanism of charge carrier transport in amorphous semiconductors has been the subject of intensive theoretical and experimental investigation. These studies have motivated the use of structural disorder in amorphous materials for the development of cheaper, reliable and more efficient solid state devices.

It has long been believed that the chalcogenide glasses show only p-type conduction and it is impossible to control the conduction type simply by doping technique because the valence of doped atoms is always satisfied in the glasses. This is explained by the pinning of Fermi energy due to the equilibrium between negatively D^- and positively D^+ charged dangling bonds on the basis of the band model of amorphous semiconductor proposed by Street and Mott [12], Mott et al. [13] and Kastner et al. [14].

Various models have been proposed for the energy distribution of the density of states and these models are mainly concerned with the important question of the existence of localized states in the tails of the valence and conduction band and the mobility edge separating extended from localized states.

4.3.1.1 The CFO Model

Cohen et al. [15] supposed that the non-crystalline structures would lead to Over-lapping of band tails of localized states. In the overlap region, they would be charged, leading to centers with unpaired spins. Such overlapping states would lead to the pinning of Fermi level. The tail from the conduction band would consist of donors and tail from the valence band will contain acceptors. Thus, the higher acceptor states would give up an electron to the lower donors, producing positively charged donors and negative acceptors. The other principal feature of this model was the existence of mobility edges at energies in the band tails. These are identified with the critical energies separating localized from extended states introduced earlier [16]. This model is sometimes called the Mott-CFO model. The difference between the energies of the mobility edges in the valence and conduction bands gap is called the mobility gap. Although, there is considerable evidence about the concept of mobility edges, the concept of overlapping tails is now considered unlikely to apply to amorphous semiconductors and insulators that are transparent in the visible or infrared. This model predicts that at high temperature the material will behave like an intrinsic semiconductor. Since, the density of states is infinite, electrons at low temperature and at the Fermi level will carry current and the $T^{1/4}$ law is also expected.

4.3.1.2 The Davis-Mott Model

According to Davis and Mott [17] for an ideal amorphous semiconductor in which all bonds are saturated and there are no long-range fluctuations, should have the density of states. Deep tails should be arising only from gross density or bond angle fluctuations. However, strongly overlapping tails as envisaged by Cohen et al. [15], which probably exist in some liquids such as expended fluid mercury. Real non-crystalline materials, however, are thought to contain imperfections such as impurities, or dangling bonds at point defects or micro voids and these just as in crystal may lead to levels within the band gap. In evaporated films of Ge and Si and some of their alloys, the conductivity, particularly at low temperatures, is due to the hopping conduction between such defect states by electron with energies near the Fermi level. The density of such states depends on the condition of deposition. Due to this, the Fermi level is located near mid gap and appears to be pinned there over a wide temperature range. This model was based on several experimental results and implied a finite density of states at Fermi level as long as the total density of states in the gap is not large. The model allowed optical transparency with the need for assumption about the magnitude of the matrix elements. However no explanation was offered as to why the controlling states should lie near mid-gap. Mott [18] suggested that if the states arise from a defected centre, e.g. a dangling bond, then they could act both as deep donors and acceptors single and double occupancy conditions lading to two bands separated by appropriate correlation energy. Even without additional compensation centers, on this model Fermi level should lie between the two bands if they do not overlap or be pinned within them if they do. If they overlap strongly, as would be expected if the density of centers is high, then the model becomes essentially indistinguishable.

4.3.1.3 The Charged Dangling Band Model

This model was put forwarded by Street and Mott [12]. This model is preferred for the systems where the density vanishes, and in which deep donors and acceptors locate the Fermi energy, simply because it is successful in explaining a wide variety of phenomena. According to this model the defects can have three charged states D^+, D^- and D^o. Since each of them are associated with a different local atomic configuration, they can be considered as different defects. The negatively charged defects D is a dangling bond associated with an under-co-ordinate atom, for example a chalcogen (say Se) bonded to another atom (As) bonded to two other atoms. When an electron is removed from the dangling bond (forming D^o), it is assumed that there is an attraction of the atom in question towards a fully co-ordinated neighboring chalcogen atom, one of the lone pair electrons on the latter being used to form a bonding orbital and the other an intending orbital. However that this bond is as strong as when a second electron is removed for then both lone pair electrons from the neighboring chains are used in bonding and the former singly co-ordinate chalcogen becomes essentially three-fold co-ordinate the D^+ centre. It assumed that the reaction

$$2D^o \rightarrow D^+ + D^-$$

is exothermic, that is the total energy (electrons plus lattice) associated with the pair of charged defects D^+ and D^- (both without spin) is lower than that of two neutral defects D_o (both with spin). The coulomb repulsive energy between the two electrons at D^- is more than compensated by lattice energy (U) for the defect.

Kastner et al. [14] and Kastner and Fritzsche [19] provided insight into the above processes using chemical-bond arguments. In their notation D^+ and D^- are denoted by C_{3^+} and C_{1^-}; C stands for chalcogen and the subscript indicating the co-ordination. In contrast to Street and Mott [12] and Mott et al. [13] description of D^o, they consider the neutral centre to be three fold co-ordinated with the anti-bonding electron residing symmetrically at the defect, which they therefore, designate C_{3^0}. The creation of a valence alternation pair, by C_{3^+} and C_{1^-} is considered to occur in two stages. First, a neutral dangling bond C_{10} interacts with the lone pair of a neighboring chalcogen forming a three-fold chalcogen, according to the reaction $C_{1^o} + C_{2^o} \rightarrow C_{2^o} + C_{3^o}$. During this reaction, (i) one of the three electrons in lone pair orbital at C_{10} is transferred to a lower lying bonding orbital and (ii) one of the two lone pair electrons at C_{20} is transferred to a bonding orbital and the other anti bonding orbital. Secondly, two C_{30} defects convert to C_{3^+} and C_{3^-} by transferring two electrons in antibonding orbital and two in bonding orbital into lone pair orbital's an exothermic reaction. The above mentioned make the important observation that the valence-alternation pair is associated with the same number of bonds as the continuous/random network, so that the energy to form it may be quite low.

4.3.1.4 The QMT Model

A model for the frequency-dependent conductivity was first proposed by Pollak and Geballe [20]. They assumed that the condition in doped and compensated semiconductors arises from tunneling between neighboring donor impurity states, randomly situated in space and energy: this has been called the quantum-mechanical tunneling (QMT) model. According to this model the motion of carrier occurs through quantum-mechanical tunneling between localized (defects) states near the Fermi level [21]. It thus appears that this model is inappropriate for the case of these materials.

4.3.1.5 Elliot's Theory

Elliot [21] proposed an alternative theory for ac conduction in chalcogenide glasses that is based on the model for charged defect in chalcogenide glasses [12–14]. This model assumes that carrier separating the two defect centers. For simple activation energy over a barrier whose height (W) is a random variable, the exponent (s) of the frequency dependence of σ_{ac} is predicted to be exactly unity [22–24] in the present case. The experimentally observed variation of s with temperature is accounted for by assuming a columbic correlation between the charged defects centers [21], resulting in a correlation between W and the intersite separation [25], i.e. correlated barrier hopping (CBH).

In case of amorphous semiconductors, at higher temperature, the conductivity is due to thermally activated tunnelling of the charge carriers in the localized states near the hand edges, and is given by [26]

$$\sigma = \sigma_o \exp(-\Delta E/KT) \tag{4.1}$$

where σ_o is the pre-exponential factor and ΔE is the activation energy. At low temperatures, the conduction process in both the cases involves localized states near the Fermi level and the conduction is due to variable range hopping in accordance with Mort's relation [27–29]

$$\sigma = \sigma_o \exp\left[-(T_o/T)^{1/4}\right] \tag{4.2}$$

where,

$$T_o = 16\alpha^3/KN(E_F) \tag{4.3}$$

$N(E_F)$ is the density of states near the Fermi level and α is a measure of the spatial extent ion of the wave function $\exp(-\alpha\gamma)$ associated with the localized states. The value of (T_o/T) which represents the ratio of the characteristic disorder energy to the thermal energy is radically different for the two cases. For amorphous semiconductors (T_o/T) is in the range of 10^{-5}–10^{-8} while for polycrystalline semiconductors this ratio is generally in the range 10–10^{-3} [30].

4.3.1.6 Conduction in Band Tails

If the wave functions are localized, so that σ(E) = 0, conduction occurs by thermally activated hopping. Every time, an electron moves from one localized state to another. It will exchange energy with photon. It is most likely that the mobility will have a thermally activated nature.

$$\mu_{hop} = \mu_o.\exp[-W(E)/kT] \tag{4.4}$$

The pre-exponential μ_o has the form

$$\mu_o = (1/6)\nu_{ph}eR^2/kT \tag{4.5}$$

where ν_{ph} is the phonon frequency and R the distance covered in one hop. For typical phonon frequency $\nu_{ph} = 10^{13}\,S^{-1}$ and W = KT. Equation (4.5) yields a mobility of the order 10^{-2} cm^{-2} V^{-1} s^{-1} at room temperature. As postulated by Mott, comparison of this value with the one calculated for conduction in the extended states suggests that the mobility may drop by a factor of at least 100 at the energy which separates the localized and non localized states.

4.3.1.7 Conduction in Localized States

If the Fermi level lies in a band of localized states as predicted by the Davis-Mott model, the carriers can move between the states via a phonon assisted tunnelling process. This is the transport analogous top impurity conduction observed in heavily doped and highly compensated semiconductors at low temperatures. An estimate for the temperature dependence of the hopping conductivity has been given by Mott [27]. We shall follow here his original derivation. Let us consider an electron that is scattered by phonons from one localized state to another. The energy difference between the states is denoted by W. The probability that an electron will jump from one state to another is determined by three factors which are following:

1. The probability of finding a photon with an excitation energy equal to W given by Boltzmann expression exp (−W/kT).
2. An attempt frequency ν_{ph}, which can not be greater than the maximum photon frequency (in the range 10^{12}–10^{13} s^{-1}).
3. The probability of an electron transfer from one state to another. This factor depends on the overlapping of the wave functions and should be given by exp (−2αR). Here R is the jumping distance which at higher temperatures equals the interatomic spacing and α is a quantity which is a representative for rate of

fall-off of the wave-function at a site. If overlapping of the wave-function is important, the factor exp(−2αR) becomes of the order of one.

The probability p that an electron jumps will then be expressed by

$$P = \nu_{ph}.\exp(-2\alpha R - W/kT) \tag{4.6}$$

The conductivity may be written as

$$\sigma = (1/6)e^2R^2\nu_{ph}N(E_F)\exp(-2\{\alpha R)\exp(-W/kT) \tag{4.7}$$

where, $N(E_F)$ is the density of states at Fermi level.

As the temperature is lower, the number and the energy of phonons decrease and the more energetic phonon-assisted hops will progressively become less favorable. Carriers will tend to hop to larger distances in order to find sites.

Many experimental studies, especially the tetrahedral materials, dealt with variable range hopping. In many cases, unreasonably high values were calculated for $N(E_F)$ by using the pre-exponential factor $\sigma_0(T)$. Although the $T^{-1/4}$ dependence of lnσ is widely observed, these findings seem to indicate that the above expression do not describe in an exact quantitative way. It must be mentioned here that Mott's derivation of variable range hopping implies to a large number of simplifying assumption of which the most important are: (i) energy independence of density of states at Fermi level (E_F), neglection of correction effects in the tunneling process. (ii) omission of multi-phonon process and (iii) neglection of electron-phonon interaction. Several authors have investigated the effect of departure from a constant density of states and have clearly demonstrated that the energy distribution of the density of states is of major importance in the theory of variable range hopping.

4.3.2 *Optical Properties*

For the measurement of the intensity of reflected or transmitted light as a function of wavelength and angle of incidence, there are several techniques available. Normally, the reflectance and transmittance of light are taken as the periodic functions of path length difference between waves reflected from the air-film and film-substrate interfaces respectively. The path length difference is associated with the angle of incidence and the wavelength of incident light as well as the film thickness. Therefore, the monitoring reflectance and transmittance as function of wavelength or angle of incidence is not only important, but their variation with film thickness is also an important factor [31]. In case, where the film thickness changes slowly but the optical parameters change rapidly; a rapid monitoring is not possible by such means.

Many methods have been developed for the complete analysis of thin films by measurement of reflectance and transmittance measured as a function of wavelength or angle of incidence [32, 33]. For the calculation of optical parameters of weakly absorbing films from the reflectance and transmittance data, there are many methods available. However, it is convenient and simpler to measure experimentally either the reflectance or the transmittance of the chalcogenides thin films and many methods have been developed for complete optical analysis using reflectance/transmittance measurements alone. Explicitly formulas valid for extreme in reflectance and transmittance help us to calculate the spectral dependence of the optical constants [34, 35].

4.3.2.1 Optical Absorption and Optical Gap

The optical absorption and optical gap depend on the short range order in the amorphous structure and defects associated with it. The variation of the optical gap and properties of thin films deserve a comprehensive investigation. The mechanism of disorder and defect interaction in amorphous chalcogenides is expected to be understood from the studies of the variation of the optical gap and the short range order as a function of thickness for the films of different compositions [35]. The decrease in the disorder and detect states in the structural bonding is known to increase the optical gap.

4.3.2.2 Absorption Process

These properties span over a wide range of phenomena, and help us greatly in understanding the basic physical properties of semiconductors. These properties are also used in the development of optical devices widely used in research and industry. The measurement of absorption spectra is the direct method for probing the band structure of semiconductors. In the absorption process, a photon of known energy excites an electron from a lower to higher energy state. In a different manner, we can say that absorption is also a result for interaction between atoms and electromagnetic radiation. In semiconductors, a number of distinct optical absorption processes take place independently. These processes are given below:

Fundamental Absorption Process

The most important process involves the transition of electrons from the valence to the conduction band. Due to its importance, the process is referred to as fundamental absorption. In fundamental absorption, an electron absorbs a photon (from incident beam) and jumps from the valence into the conduction band. The photon energy must be equal to the energy gap or larger. The frequency must therefore be;

$$\nu \geq (E_g/h) \tag{4.8}$$

The frequency $\nu_o = E_g/h$ is referred to as the absorption edge. In transition process (photon absorption), the total energy and momentum of the electron-photon system must be conserved.

Calculating the absorption coefficient for fundamental absorption requires quantum manipulations. Essentially, these consist of treating the incident radiation as a perturbation, which couples the electron state in the valence band to its counterpart in the conduction band, and using the technique of quantum perturbation theory. One then finds that the absorption coefficient has the form [36].

$$\alpha_d = A(h\nu - E_g)'' \tag{4.9}$$

where A is a constant involving the properties of the bands and E_g is the energy gap. The absorption coefficient increases parabolically with the frequency above the fundamental edge. A useful application of these results is their use in measuring energy gap in semiconductors. Thus, E_g is directly related to the frequency edge $E_g = h\nu$. This is now the standard procedure of determining the gap. Due to its accuracy and convenience, the optical method also reveals more details about the band structure than the conductivity method.

The absorption coefficient associated with fundamental absorption is large ($\sim 10^4\ cm^{-1}$ for chalcogenides glasses). Thus, absorption is readily observable even in thin samples. Since, the energy gaps in semiconductors are small (1 eV or less), the fundamental edge usually occurs in the infrared region. The development of a large variety of reliable infrared detectors has been one of many benefits, which have accrued from this work. The absorption process occurs in the so-called direct-gap semiconductors. Here the bottom of the conduction band lies at k = 0, and hence directly above the top of the valence band. Electrons near the top of the valence band are able to make transitions to states near the bottom of the conduction band, consistent with the selection rule. Examples of such substances are GaAs, InSb and many other III–V compounds.

There are also indirect-gap semiconductors, in which the bottom of the conduction band does not lie at the origin. In this case, the electron can not make a direct transition from the top of the valence band to the bottom of the conduction band because this would violet the momentum selection rule. Such a transition may still take place but as a two step process. The electron absorbs both a photon and a phonon simultaneously. The photon supplies the needed energy, while the phonon supplies the require momentum.

Calculation of the indirect-gap absorption coefficient, which is more involved than that of direct absorption, is given by the formula [36]

$$\alpha_i = A'(T)(h\nu - E_g)^{1/2} \tag{4.10}$$

where A′(T) is a constant containing parameters pertaining to the bands and the temperature. Here α_i increases as the second power of $(h\nu - E_g)$, much faster than the half power of this energy difference as in direct transition. So, we may use the optical method to discriminate between direct and indirect gap semiconductors, an improvement over the conductivity method.

Exciton Absorption

In discussing fundamental absorption, we assumed that the excited electron becomes a free particle in the conduction band, and similarly, that the hole left in the valence band is also free. The electron and hole attract each other and may possibly form a bound state in which the two particles revolve around each other. Such a state is referred as an exciton. The binding energy of the exciton is small, about 0.01 eV, and hence the excitation level falls very slightly below the edge of the conduction band.

The energy of the photon involved in exciton absorption is given by.

$$h\nu = E_g - E_{ex} \tag{4.11}$$

where E_{ex} is the exciton binding energy. The exciton spectrum therefore, consists of a sharp line, falling slightly below the fundamental edge. This line is often broadened by interaction of the exciton with impurities or other similar effects, and may well merge with the fundamental absorption band, although often the peak of the exciton line remains clearly discernible. This illustrates a fact which is often observed: absorption of an exciton introduces complications into the fundamental absorption spectrum, particularly near the edge, and renders the determination of the energy gap in semiconductors more difficult. However exciton absorption is important in the discussion of optical properties of insulators in the ultraviolet region of the spectrum.

Free Carrier Absorption

Free carriers both (electrons and holes) absorb radiation without becoming excited into the other band. In absorbing a photon, the electron (or hole) in this case makes a transition to another states in the same band. Such process is usually referred to as an interband transition. For concreteness, take the substance to be n-type, so that only electrons are present. The real and imaginary parts of the dielectric constant are given by

$$\varepsilon_r' = \varepsilon_{L,r} - \sigma_o t/\varepsilon_o(1 + \omega^2\tau^2) = n_o^2 - k^2 \tag{4.12}$$

$$\varepsilon_r'' = \sigma_o/\varepsilon_o\omega(1 + \omega^2\tau^2) = 2n_o k \tag{4.13}$$

At low frequency and small conductivity (low concentration), the lattice contribution $\varepsilon_{L,r}$ dominates the dielectric polarization. Thus, the substance acts as a normal dielectric. There is, however, a slight absorption associated with ε_r'' which represents the absorption of radiation by free carriers.

In the region of low frequency and high conductivity, the free-carrier term, in Eq. (4.12) dominates. Thus $\varepsilon_r' < 0$ and the substance exhibits total reflection such as a metal does. This is to be expected, since the electron concentration is very high, approaching the electron concentration in metals. In the high-frequency (short-wavelength) region, $\omega\tau \gg 1$ (but small conductivity), the materials acts like a normal dielectric with $n_o = \varepsilon_{L,r}^{1/2}$ and the absorption coefficient is

$$A = \sigma_0/\varepsilon_o c n_o \omega^2 \tau^2 \tag{4.14}$$

The free carrier absorption take place even when $h\nu < E_g$ and frequently this absorption dominates the spectrum below the fundamental edge. For $h\nu > E_g$, both types of absorption-fundamental and free-carrier occur simultaneously.

Absorption Process Involving Impurities

Absorption process involving impurities often take place in semiconductors. The type and degree of absorption depend on the type of impurity present in the concentration. The transition to higher impurity levels appear as sharp lines in the absorption spectrum, which is analogous to the donor-conduction band transition above. For shallow impurities, the absorption lines associated with donors and acceptors fall in the far infra-red region. Such processes may serve in principle as a basis for detectors in this rather difficult region of the spectrum. The spectrum may also serve as a diagnostic technique for determining the type of impurity present. Such processes lead to absorption which is close to the fundamental absorption, and are seldom resolved from it.

The energy of the photon in this case is

$$h\nu = E_g - E_d - E_a \tag{4.15}$$

This leads to a discrete structure in the absorption curve. But this is often difficult to resolve because of its proximity to the fundamental edge. Impurities may also affect the absorption spectrum in other direct ways. For instance, an exciton is often found to be trapped by an impurity. This may happen as follows: The impurity first traps an electron and once this happens the impurity—now charged—attracts a hole through the Coulomb force. Thus both an electron and a hole are trapped by the impurity. The spectrum of this exciton is different from that of a free exciton because of the interaction with the impurity.

4.3.2.3 Optical Constants of Thin Films

The optical behaviour of a material is generally utilized to determine its optical constants, refractive index (n) and coefficient of extinction (k). Films are ideal specimen for reflectance and transmittance types of measurements. Separate determinations of n and k can be made by measuring reflectance and transmittance of the same specimen. To calculate these optical constants, the reflectance and transmittance data obtained from spectrophotometric measurements is used.

The absorption coefficient (α) can also be calculated directly from the absorbance versus wave number curves using the relation [36]

$$OD = \log I_B/I_T = \alpha\, t \tag{4.16}$$

$$\alpha = OD/t \tag{4.17}$$

OD is the optical density measured at a given layer thickness t (nm). I_B and I_T are the intensities by light irradiation and transmittance respectively.

Chalcogenide glasses exhibit highly reproducible optical edges, which are relatively insensitive to preparation conditions and only the observable absorption [37] with a gap under equilibrium conditions account for the first process. In the second process, the absorption edge depends exponentially on the photon energy according to the Urbach relation [38]. In crystalline materials the fundamental edge is directly related to the conduction and valance band, i.e. direct and indirect band gaps, while in the case of amorphous materials a different type of optical absorption edge is observed. In these materials, the absorption coefficient (α) increases exponentially with the photon energy near the energy gap. Generally, this type of behavior is observed in chalcogenides [38]. This optical absorption edge is known as the Urbach edge and is given by,

$$\alpha \sim \exp\,.\,[A(h\nu - h\nu_0)]/kT \tag{4.18}$$

where A is a constant of the order of unity and ν_0 is the constant corresponding to the lowest excitonic frequency.

The relationship between the absorption coefficient α and the energy hν of the incident photon is given by [39, 40],

$$(\alpha\,.\,h\nu) \propto (h\nu - Eg)^m \tag{4.19}$$

For direct optical band gap, we take $m < 1$, whereas for indirect transition, $m > 1$ is used. The value of optical band gap (E_g) is calculated from the plot of $(\alpha\ .\ h\nu)^{1/m}$ versus photon energy (hν) by taking the intercept on the X-axis. The values of refractive index (n) and extinction coefficient (k) will be calculated by using the theory of reflectivity of light.

On the basis of this theory, the reflectance of light measured for the thin film can be expressed in term of Fresnel's coefficient. The reflectivity on an interface can be given by;

$$R = [(n-1)^2 + k^2]/[(n+1)^2 + k^2] \tag{4.20}$$

and

$$\alpha = 4\pi k/\lambda \tag{4.21}$$

where λ is the wavelength.

Amorphous chalcogenide nano-particles have been under intensive investigation in the past few years because of their size-dependent properties and the possibility of arranging them in micro and nano-assemblies. Chalcogenide possesses unique characteristics which are different from those in oxide and halide glasses, i.e. molecular structures and semiconductor properties. Photo-induced phenomena are observed in chalcogenide thin films. These changes are accompanied by changes in the optical constants, i.e., changes in the electronic band gap, refractive index and optical absorption coefficient. During the last few decades, many studies on ion-selective electrodes using chalcogenide glass materials have been undertaken. A large number of chalcogenide glasses for the detection of heavy metals in aqueous media were used. The main advantage of the sensors based on chalcogenide glasses are their high chemical stability and the low detection limit as compared to their crystalline counterpart [41–46]. However, to keep pact with modern technology, the fabrication of fast responsive, cost effective and miniaturized sensor devices based nano-chalcogenides are needed. For the fabrication of multi-sensor systems, this is important and can be employed for the simultaneous measurement of different ions in solutions.

A lot of work is reported on the thin films of chalcogenides. Here, we are presenting some of the reported work on thin films of chalcogenide glasses both at micro and nanoscale in amorphous as well as in crystalline form.

4.3.3 *Studies on Chalcogenides Films*

The electrical charge transport in As–Se–Te glasses by correlating the dc and ac conductivity measurements performed at different temperatures was studied by El-Den et al. [47]. They reported that the dc conductivity measurements did not reflect necessarily the bulk properties but can be completely dominated by the charge-depleted contact region effects. Using these measurements in combination with ac admittance spectroscopy, it was possible to separate the effects of these two reasons and to describe them quantitatively. The observed ac behaviour stems from the spatial distribution of the charge density with in the contact region. From the relaxation time distribution entering microscopic conductivity, a single-channel hopping of charges strongly localized through finite charge-lattice interaction was deduced for the given system.

The dc electrical conductivity as a function of temperature (294–383 K) for as grown evaporated thin films of $Se_{75}S_{25-x}Ag_x$ was studied by Khan et al. [48]. The dc conductivity decreased at all the temperatures with the increase of silver content in the binary system. They suggested that the conduction was due to thermally assisted tunneling of the carriers in the localized states near the band edges. Compositional dependence of the optical properties of as deposited $Se_{75}S_{25-x}Ag_x$ thin films of thickness 300 nm were also studied in the spectral range from 400 to 1000 nm. The optical band gap increased on incorporation of silver contents in Se–S system. The results were interpreted in terms of the change in concentration of localized states due to the shift in Fermi level. The results on electrical properties of vacuum evaporated thin films of a-$(Se_{70}Te_{30})_{100-x}(Se_{98}Bi_2)_x$ system in the temperature range (308–355 K) suggested that the dc conductivity and activation energy depend on the Bi concentration [49]. Photocurrent dependence on incident radiation followed the power law (Iph/F^{γ}) and the transient photocurrent exhibited the non-exponential decay time. On the basis of the calculated parameters, it is suggested that the recombination within the localized states was predominant. To study the crystallization kinetics, they studied the heating rate dependence of glass transition and crystallization temperatures and the activation energy for thermal relaxation and activation energy for crystallization were calculated. Finally, the composition dependence of the activation energy for thermal relaxation and activation energy for crystallization was interpreted in terms of the structure of Se–Te–Bi glassy system. The absorbance, reflectance and transmittance of as-deposited thin films of a-$Se_{80}Te_{20-x}Cu_x$ (where x = 2, 6, 8 and 10) were measured in the wavelength region 400–1000 nm [50]. They studied the optical band gap and optical constants of amorphous thin films as a function of photon energy. The optical band gap increased on adding the copper in $Se_{80}Te_{20-x}Cu_x$ system. The value of refractive index (n) was found to decrease where as, the value of the extinction coefficient (k) showed an increasing trend with the increase in photon energy. The results were discussed in terms of concentration of localized states [50]. An investigation of the electrical and optical properties of vacuum evaporated $(Ge_2S_3)_1(Sb_2Te_3)_1$ thin films as a function of the film thickness (118.9–200 nm) and annealing temperatures (373, 423, and 473 K) was carried out El-Wahabb and Farid [51]. The X-ray diffraction pattern revealed the formation of amorphous films. On the basis of electrical conductivity measurements, the film thickness and annealing temperature dependence of electrical conductivity was found. Two types of conduction channels were observed, which suggested two conduction mechanisms. The measurements of optical absorption suggested that the fundamental absorption edge depended on annealing temperatures. An allowed indirect transition was responsible for the optical absorption. A decrease in the optical energy gap (E_{opt}) was observed with the increase in film thickness and annealing temperatures below T_g. The analysis of transmission spectra of thin films of $Se_{85-x}Te_{15}Bi_x$ (x = 0, 1, 2, 3, 4, 5) glassy alloys, measured at normal incidence, in the spectral range 400–1500 nm were studied [52]. To determine the refractive index (n) and film thickness (d), the authors employed Swanepoel's method. An increase in refractive index (n) with the increase in Bi concentration was associated with the increased polarizability of the larger Bi atom (atomic radius 1.46 Å) as compared with the Se atom (atomic radius ~1.16 Å). Wemple–Di Domenico model (WDD) was applied to estimate the dispersion energy (E_d), average energy gap (E_0) and static refractive index (n_o). On the basis of Tauc's extrapolation, the optical band gap (E_g) was calculated and its calculated value decreased from 1.46 to 1.24 eV with the increase in Bi content. On the basis the electronegativity difference of the atoms involved and cohesive energy of the system, the variation of optical band gap with Bi content was discussed. Nearly stoichiometric thin films of $In_{49}Se_{48}Sn_3$ at room temperature, by conventional thermal evaporation of the pre-synthesized materials were deposited by Salem et al. [53]. Using transmission electron microscopy and diffraction (TEMD), the microstructure of as-deposited and annealed films was studied and it was found that the as-deposited films were amorphous in nature, while those annealed at 498 K were crystalline. An analysis of the optical absorption spectra revealed a non-direct energy gap characterizing the amorphous films, while both allowed and forbidden direct energy gaps characterized the crystalline films. Temperature dependence of the resistance (ln(R) vs. 1000/T) for crystalline films showed two straight lines, which was related to the extrinsic and intrinsic conduction. Using similar Ag metal electrodes, the room temperature I–V characteristics of the as-deposited films sandwiched showed an ohmic behavior, where as the use of dissimilar Ag/Al metal electrode suggested non-ohmic behavior, which was attributed to space charge limited conduction. Khan et al. [54] reported the measurements of optical constants (absorption coefficient, refractive index, extinction coefficient, real and imaginary part of the dielectric constant) of $Se_{75}S_{25-x}Cd_x$ (where x = 0, 2, 4, 6 and 8) thin films of thickness 3000 Å as a function of photon energy in the wave length range 400–1000 nm. An in the optical band gap and extinction coefficient was observed where as the value refractive index showed a decreasing trend on the addition of cadmium in Se–S system. The results were discussed in terms of the change in concentration of localized states, which may be due to the shift in Fermi level. CdSe thin films were characterized by various techniques such as X-ray diffraction, scanning electron microscopy and UV–Vis–NIR double beam spectrophotometer [55]. The optical band gap energy (E_g) was reported to be 1.7 eV on the basis of interference transmission spectra T(λ) at normal incidence in the wavelength range 400–2500 nm for $Ge_{20}Se_{75}M_5$ thin films [56]. Orava et al. [57] studied the optical and structural properties of $Ge_{20}Se_{80}$, $Ge_{25}Se_{75}$, $Ge_{30}Se_{70}$ and $Ag_x(Ge_{0.20}Se_{0.80})_{100-x}$ thin films. All samples were confirmed as amorphous according to XRD. The Raman spectra showed increase in the intensity of 260 and 237 cm^{-1} and decrease in the intensity of 198 and 216 cm^{-1} bands with different Se content in the bulk samples, whereas the intensity of peaks at 260 and 216 cm^{-1} decreased with increasing Ag content. The significant red shift of band gap energy occurred upon different Ag content. The optical band gap of bulk samples decreased (2.17–2.08 eV) and refractive index increased (2.389–2.426 at 1550 nm) with increasing Se content in bulk glasses, whereas the optically induced dissolution and diffusion resulted in graded refractive index profile along the film thickness caused by different Ag concentration.

Excellent nano-multilayers (NML) chalcogenide were developed using pulsed laser deposition (PLD) method based on KrF excimer laser operating at 248 nm with constant output energy of 250 mJ per pulse, with pulse duration of 30 ns and with repetition rate of 10 Hz [58]. They mentioned that in NMLs more important was the stimulated mass transport (interdiffusion) across the interfaces by mains, the rate of interdiffusion increases by an order of magnitude if the a-Se/As2S3 chalcogenide–chalcogenide NML was irradiated by laser light. The stimulated interdiffusion resulted in the bleaching of the NML (blue shift of the absorption edge) since the intermixing of the wide- and narrow band gap materials gave a solid solution with a wider band gap in comparison with an initial narrow-gap, active sub-layer. The resulting amplitude relief of optical transmission may be tuned by the selection of the certain pairs of components, some of which were patented. The influence of laser-irradiation on structural and optical properties of phase change $Ga_{25}Se_{75-x}Te_x$ thin films is studied by Agel et al. [59]. This amorphous to crystalline phase change in $Ga_{25}Se_{75-x}Te_x$ thin films due to laser-irradiation suggested that this material is promising materials for optical recording. Alvi et al. [60] studied the photo-induced effects on structural and optical properties of $Ga_{15}Se_{81}Ag_4$ chalcogenide thin films. Analysis of the optical absorption data showed that the rule of non-direct transition predominated. The optical band gap decreased with the increase in the illumination time. The value of absorption and extinction coefficients increased while the refractive index decreased by increasing the illumination time from 0 to 150 min. The photo-induced effects on electrical properties of $Ga_{15}Se_{81}Ag_4$ chalcogenide thin films were studied [61]. The activation energy in $Ga_{15}Se_{81}Ag_4$ chalcogenide thin films decreased with increasing the exposure time whereas the dc conductivity increased at each temperature by increasing the illumination time. Khan et al. [62] measured the dc electrical conductivity of as deposited thin films of a-$Se_xTe_{(100-x)}$ (x = 3, 6, 9 and 12) as a function of temperature ranging from 298 to 383 K. The value of activation energy was found to decrease on incorporation of dopant (Se) content in the Te system. The optical absorption measurements showed an indirect optical band gap in this system and it decreased on increasing Se concentration.

Shaheen et al. [63] reported the effect of gamma irradiation on the optical properties of a-$Se_{90}In_{10-x}Sn_x$ chalcogenide thin films. Al-Agel [64] studied the optical and electrical properties of $Ga_{15}Se_{77}In_8$ chalcogenide thin films before and after annealing. El-Sebaii et al. [65] reported the role of heat treatment on structural and optical properties of thermally evaporated $Ga_{10}Se_{81}Pb_9$ chalcogenide thin films. Al-Ghamdi et al. [66] studied the effects of laser irradiation on optical properties of amorphous and annealed $Ga_{15}Se_{81}In_4$ and $Ga_{15}Se_{79}In_6$ chalcogenide thin films, Khan et al. [67] reported the thermal annealing effect of on optical constants of vacuum evaporated $Se_{75}S_{25-x}Cd_x$ chalcogenide thin films, Al-Hazmi [68] studied the effect of annealing on optical constants of $Se_{75}S_{25-x}Cd_x$ chalcogenide thin films, Chauhan et al. [69] studied the photo-induced optical changes in $Ge_xAs_{40}Se_{60-x}$ thin films, Farid et al. [70] studied the compositional effects on the optical properties of Ge_xSb_{40-x} Se_{60} thin films, Chauhan et al. [71] studied the linear and nonlinear optical changes in amorphous As_2Se_3 thin film upon UV exposure, Imran et al. [72] reported the effect of thermal annealing on some electrical properties and optical band gap of vacuum evaporated $Se_{65}Ga_{30}In_5$ thin films.

The work on optical properties of amorphous thin film of Se–Te–Ag system prepared by using thermal evaporation technique by Madhu et al. [73], phase change and optical band gap behavior of $Se_{0.8}S_{0.2}$ chalcogenide glass films by Rafea and Farid [74], influence of indium content on the optical, electrical and crystallization kinetics of $Se_{100-x}In_x$ thin films deposited by flash evaporation technique by Ammar et al. [75], Structural and optical properties of $In_{35}Sb_{45}Se_{20-x}Te_x$ phase-change thin films by Diab et al. [76], temperature effects on the optoelectronic properties of $AgIn_5S_8$ thin films by Qasrawi [77], optical band gap and refractive index dispersion parameters of In–Se–Te amorphous films by Aly et al. [78], photo-induced effects on electrical properties of $Ga_{15}Se_{81}Ag_4$ chalcogenide thin films by Alvi et al. [79] are also worth mentioning.

4.3.4 Studies on Nanostructures of Chalcogenides

Recently, a lot of work is focused on the synthesis and characterization of nanoscale materials due to their potential applications. The major challenge in this field is the mass production of materials at nanometric scale with a controlled composition and structure as the materials produced at nanoscale provide novel properties. Due to their small grain size, these materials are characterized by the rather high number of atoms located in the grain boundary. Many studies on phase change optical recording materials have been reported, but the studies based on nanoscale materials are still limited. It is expected that the use of these chalcogenides materials at in nano size, which may be useful for making nano-electronic and nano-memory devices. Optical materials produced at nanoscale for the applications in optical disk are expected to be cost effective. This is one of necessities for the development of advanced data storage devices. The studies on nanochalcogenides are still limited, therefore more studies on these wonderful materials are needed for their utilization in various R&D applications [7–81].

Liu and Zhu [82] reported the synthesis of polycrystalline CdS nanowires assembled by nanocrystals. They proposed that this simple route can also be used for the synthesis of other metal nanochalcogenides. Amorphous CdS nanoparticles capped with cetyltrimethyl ammonium bromide (CTAB) were also prepared under various conditions with the help of co-precipitation method [83]. A blue shift in the band gap in the UV–visible absorption spectra was observed. On the basis of the obtained results, they reported the transformation of amorphous CdS nanoparticles into CdS nanocrystals having a zinc blende or a wurtzite structure, depending on the heat treatment. Wageh et al. [84] reported the preparation of strong-confined PbSe Quantum Dots in P_2O_5–Na_2O–ZnO–Li_2O phosphate glass. This confinement was confirmed by the observation of a high shift in the absorption spectra of PbSe QDs.

Tintu et al. [85] has reported the nanocomposite thin films of $Ga_5Sb_{10}Ge_{25}Se_{60}$ chalcogenide glass for optical limiting applications. They reported a low-cost, scalable method to fabricate optical grade composite thin films of $Ga_5Sb_{10}Ge_{25}Se_{60}$/PVA composite films for nonlinear optical applications. The synthesis of polycrystalline nanotubular of Bi_2Te_3 via a high-temperature solution process decomposed from trioctylphosphine oxide (TOPO) extracted tellurium species (Te-TOPO), as sacrificial template was reported by Chai et al. [86]. The formation of such tubular structure was believed to be the result of outward diffusion of Te during the alloying process. On the basis of the measurements electrical parameters (Seebeck coefficient and electrical conductivity) of the polycrystalline nanotubular Bi_2Te_3, it was observed that the electrical conductivity was reduced approximately by three orders of magnitude as compared to that of the bulk bismuth telluride materials. High-quality monodisperse Cu_2S nanocrystals (sizes from 2 to 20 nm) by the reaction of copper stearate ($CuSt_2$) and dodecanethiol (DDT) in 1-octadecene (ODE) were synthesized by Li et al. [87]. X-ray diffraction (XRD), X-ray photoelectron spectrum (XPS), and transmission electron microscopy (TEM) techniques were employed to characterize the nanocrystals. For certain particle size, these as-prepared Cu_2S nanocrystals showed good self-assembly behaviors, and were easily assembled into two-dimensional and three-dimensional superlattice structures. With the hep of a simple and environmentally benign chemical bath deposition (CBD) method, CdS nanocrystalline films were deposited onto flexible plastic and titanium substrates at room temperature. As-despised films contained clusters of CdS nanoparticles, which were successfully converted into nanowire (NW) networks using chemical etching process [88]. The width of these nanowires was in the range of 50–150 nm and lengths was of the order of a few micrometers. Optical absorption measurements revealed a band gap of 2.48 eV for CdS nanowires, whereas for CdS nanoparticles, the value was found to be 2.58 eV. CdSe nanocrystals were successfully synthesized by mechanical alloying Cd and Se elemental powders [89]. HRTEM images as well as Fourier transformation in reciprocal space provided a good pathway to identify the structure of individual CdSe nanocrystal. An amorphous thin oxide layer covering on the surface of the as-milled CdSe nanocrystals was observed on HRTEM images, which was assigned to be SeO_2 and SeO_3 by X-ray photoelectronic spectroscopy (XPS). Subsequent capping the surface of as-milled CdSe nanocrystals with long chain TOP/TOPO molecules was achieved colorful dispersion colloid solution, which showed similar optical properties to those CdSe nanocrystals prepared by wet chemical process. Yang et al. [90] reported the controlled synthesis of hexagonal flower-like Zn-doped CdSe ($Cd_{1-x}Zn_xSe$, x = 0.1) microstructures from the solid air-stable precursor by two facile steps. X-ray powder diffraction (XRD), field-emission scanning electron microscopy (FE-SEM), energy dispersive spectrum (EDS) and UV–vis absorption spectroscopy techniques were used to characterize this alloy. A shift toward shorter wavelength in the optical band gap energies of the as-synthesized flower-like $Cd_{0.9}Zn_{0.1}Se$ microstructures was observed as compared with that of CdSe. Ingole et al. [91] synthesized 1-Hexanethiolate capped copper (I) selenide quantum dots at room temperature by the method of arrested precipitation in Triton X-100, water-in-oil microemulsions. X-ray and electron diffraction analysis revealed the formation of nanocrystallites of cubic $Cu_{2-x}Se_x$ (Berzelianite) phase having a lattice constant of 5.70 ± 0.07 Å. Using Debye–Scherrer equation, the average particle size was estimated to be ca. 7.3 nm. Two major peaks at 114 and 619 cm^{-1} in the FT-IR spectra, were fitted to Cu–S stretching and bending, respectively, which suggested the formation of stable Cu–S bond between surface copper ions and the adsorbed 1-hexanethiolate. Commercially available polysaccharides, agarose and gellan, as morphology-directing agents were used for the synthesis of t-Se nanowires in water at room temperature in the presence of ascorbic acid as reducing agent [92]. The diameter of the nanowires prepared in the presence of agarose varied from 100 to 208 nm, where as the diameter of the nanowires prepared from gellan was varied from 51 to 145 nm. Han et al. [93] synthesized peanut-shaped Sb_2S_3 superstructures via a hydrothermal process at 120 °C for 8 h using hydrochloric acid and antimony O-benzyl dithiocarbonate (benzylxanthate, $Sb(S_2COC_7H_7)_3$) as starting materials. Transmission electron microscopy (TEM) and scanning electron microscopy (SEM) studies reveal that the peanut-shaped Sb_2S_3 superstructures are aggregated by nanorods. The possible mechanism for the formation of these structures were interpreted with the help of obtained results. Liu et al. [94] synthesized single phase $SnSe_2$ at 180 °C by hydrothermal co-reduction method from $SnCl_2 \cdot 2H_2O$ and SeO_2, its morphology and growth direction were investigated. Experimental results showed that the $SnSe_2$ powder consisted of regular and homogenous hexagonal nanoflakes which growed along (0001) crystal plane and the nanoflakes were about 600–700 nm in side length and 30–40 nm in thickness. Li et al. [95] described a facile and rapid method for preparing In_2S_3 nanoparticles via ultrasound dispersion. This method allowed them to prepare In_2S_3 nanoparticles from bulk indium and sulfur with ease and without using expensive agents and in a short time. The possible growing mechanism of the In_2S_3 nanoparticles was presented. In addition, they provided detailed characterizations including TEM, XRD, TG–DTA, and XPS to study the shape, composition and structure of In_2S_3 nanoparticles. ZnS nanoparticles were synthesized in the pores of the mesoporous silica (MS) particles coated with two bilayers of poly (allylamine hydrochloride) (PAH)/poly (styrene sulfonate) (PSS) via the layer-by-layer (LbL) self-assembly technique. XRD and TEM results confirmed that these nanocrystals were inserted into the pores of the MS spheres [96].

Khan et al. [97] reported the electrical and optical studies of thin films of a-Ga_xSe_{100-x} nanorods (x = 3, 6, 9 and 12) (Fig. 4.6). DC electrical conductivity of as deposited thin films of a-Ga_xSe_{100-x} nanorods was measured as a function of temperature range from 298 to 383 K. Thermally assisted tunneling of the carriers were responsible for the electrical conduction in the localized states near the band edges. Optical absorption measurements suggested an indirect optical band gap in this system (Fig. 4.7), which showed a decreasing trend with the increase in Ga content. Khan et al. [98] studied the optical constants of thin films of Ga_xSe_{100-x} nanoparticles (Fig. 4.8) with x = 3, 6, 9 and 12 in a wavelength region 400–900 nm. The experimental result showed that the optical absorption follows the rule of

Fig. 4.6 SEM images of a-$Ga_{12}Se_{88}$ nanorods [97]

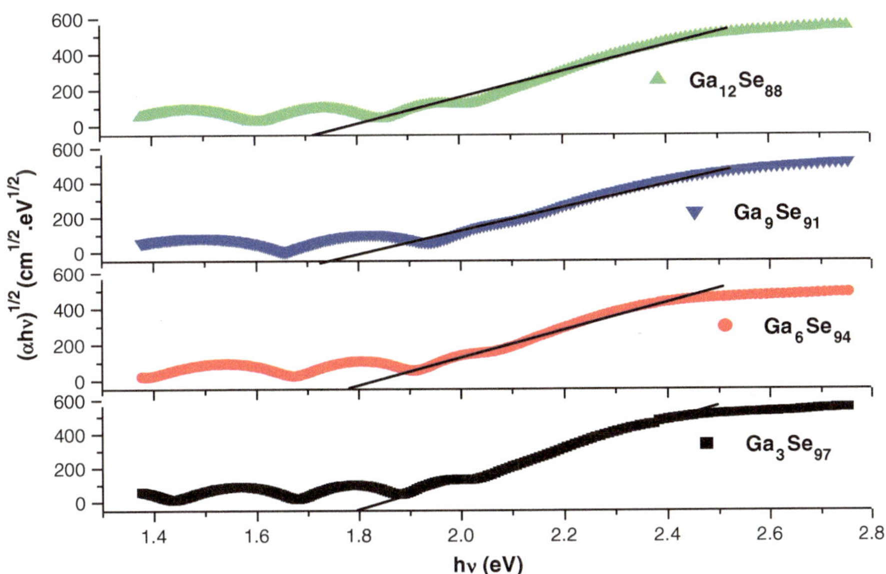

Fig. 4.7 $(\alpha h\nu)^{1/2}$ against photon energy (hν) of thin films of a-Ga_xSe_{100-x} nanorods [97]

indirect transition. The optical band gap and optical constants of thin films were studied as a function of photon energy (Fig. 4.9).

Thermal evaporation method was used for synthesis of thin films of Ga_xTe_{100-x} (x = 3, 6, 9 and 12) [99]. SEM images suggest that the films contained nanoparticles of size varying from 100 to 200 nm (Fig. 4.10). XRD pattern shown in Fig. 4.11 predicts the amorphous nature of these as-synthesized nanoparticles. The dc

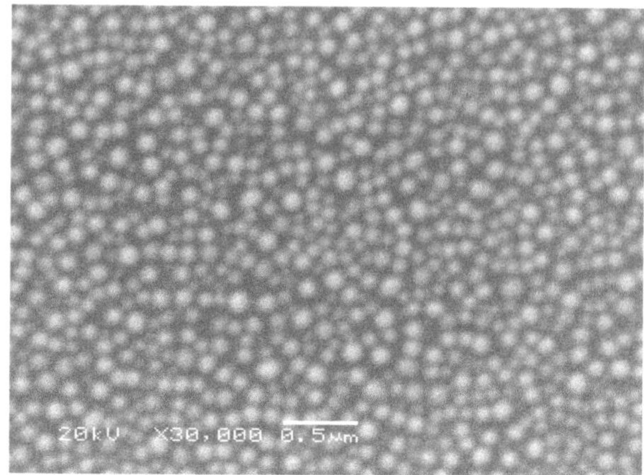

Fig. 4.8 SEM image of as-prepared thin films of a-Ga_xSe_{100-x} nanoparticles [98]

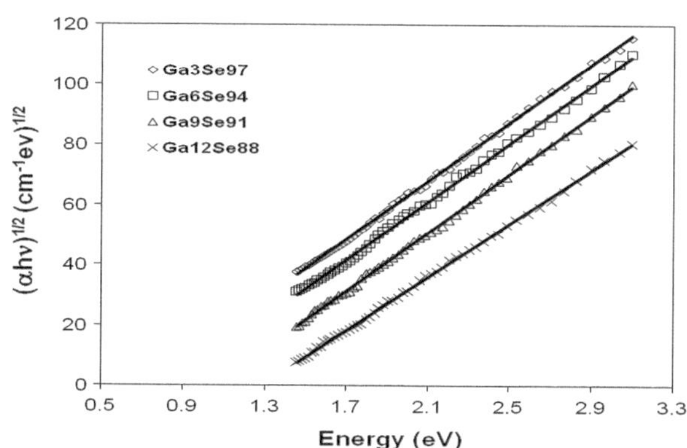

Fig. 4.9 Variation of $(\alpha h\nu)^{1/2}$ with incident photon energy ($h\nu$) for thin films of Ga_xSe_{100-x} nanoparticles [98]

Fig. 4.10 SEM image of a-$Ga_{12}Te_{88}$ nanoparticles [99]

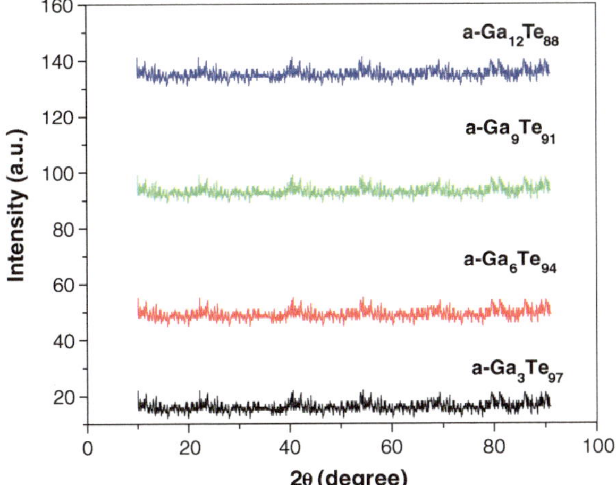

Fig. 4.11 XRD pattern of a-Ga_xTe_{100-x} [99]

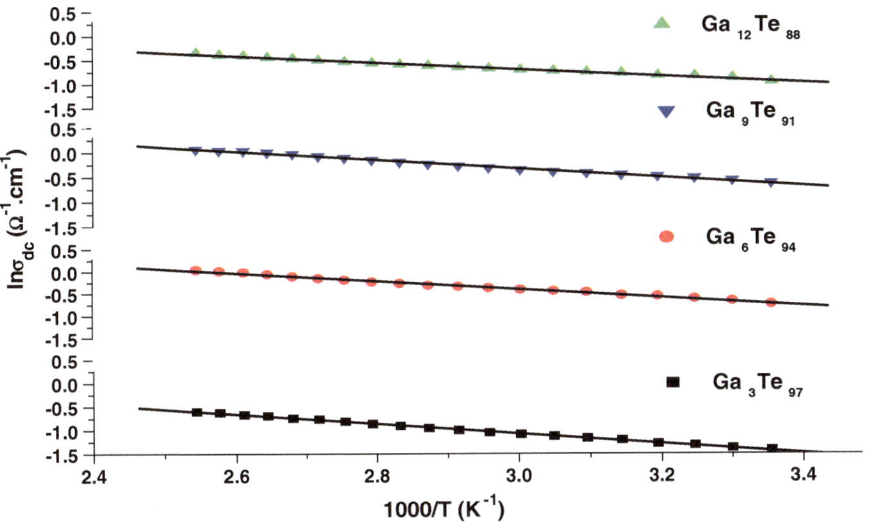

Fig. 4.12 Temperature dependence of dc conductivity in the temperature range 298–383 K of a-Ga_xTe_{100-x} nanoparticles [99]

electrical conductivity of the as-deposited films of Ga_xTe_{100-x} nanoparticles was measured as a function of temperature range from 298 to 383 K (Fig. 4.12), which increased exponentially with temperature. The optical measurements suggested an indirect optical band gap in this system. The value of the optical band gap decreased on increasing the Ga concentration.

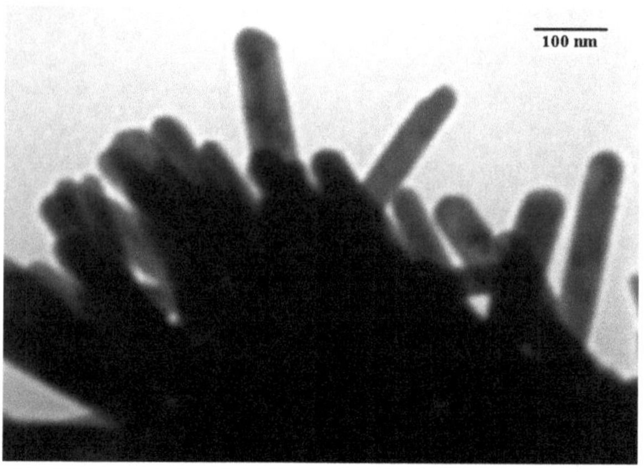

Fig. 4.13 TEM image of as grown a-$Se_{87}Te_{13}$ nanorods [100]

Electrical transport properties of a-$S_{87}Te_{13}$ nanorod film were studied by Khan et al. [100]. Using physical vapour condensation system, thin film consisting of a-$S_{87}Te_{13}$ nanorods were deposited on a glass substrate under an ambient gas (Ar) atmosphere. With the help of scanning electron microscopy (SEM) and transmission electron microscopy (TEM), the morphology and microstructure of these nanorods were studied (Fig. 4.13). Temperature dependence of dc conductivity for these nanorods over a temperature range of 500–100 K was also reported.

The structural and optical studies of $Se_{88}Te_{12}$ chalcogenide nanoparticles prepared by ball milling were analyzed [101]. The ball milling was performed in a Laboratory 8000 M-Mixer/Mill (SPEX) mill using hardened steel balls and a vial with a ball-to-powder weight ratio of 10:1. Optical absorption measurements indicated that the absorption mechanism is due to direct transition. The optical band gap increased with the increase in milling time. The results were interpreted in terms of the change in concentration of localized states due to the shift in Fermi level. Synthesis and optical characterization of nanocrystalline CdTe thin films were reported by Al-Ghamdi et al. [102]. A wet chemical route at pH $\approx$ 11.2 using cadmium chloride and potassium telluride as starting materials was used to prepare the nanocrystalline CdTe bulk powder. Using transmission electron microscope (TEM), X-ray diffraction (XRD) and scanning electron microscopy (SEM), the as-prepared sample was characterized. The optical constants which includes absorption coefficient, optical band gap, refractive index, extinction coefficient, real and imaginary part of dielectric constant was studied as a function of photon energy in the wavelength region 400–2000 nm. Analysis of the optical absorption data showed that the rule of direct transitions was predominated. The optical characterization of nanocrystalline $Se_{85}Te_{10}Pb_5$ and $Se_{80}Te_{10}Pb_{10}$ chalcogenides was studied by Khan et al. [103]. The polycrystalline $Se_{85}Te_{10}Pb_5$ and $Se_{80}Te_{10}Pb_{10}$ chalcogenides were used as a starting material for the milling process. The experimental result showed that the optical absorption follows the rules of direct transition. TEM measurements showed that after 60 h of milling, nanoparticles with a typical diameter of about 5–20 nm were produced. It was observed that the absorption coefficient, optical band gap and extinction coefficient increased while the refractive index decreased with an increase in milling time.

Thin films of $Te_{94}Se_6$ nanoparticles deposited by physical vapor condensation technique at different argon (Ar) pressures were studied by Salah et al. [104]. As-grown films showed a polycrystalline structure as revealed by XRD pattern. SEM images showed that the nanoparticles are uniformly distributed whose size varied from 12 to 60 nm with decrease in Ar pressure from 667 to 267 Pa. A direct band gap in these as-grown thin films was reported and its value was decreased with increasing particle size. The studies on optical and structural phenomenon in as-deposited thin films composed of aligned nanorods of a-Se_xTe_{100-x} (x = 3, 6, 9 and 12) were reported by Agel [105]. FESEM images suggest that these thin films contain high yield of aligned nanorods (Fig. 4.14). Optical absorption measurement suggested a direct optical band gap in this alloy system (Fig. 4.15). Alvi and Khan

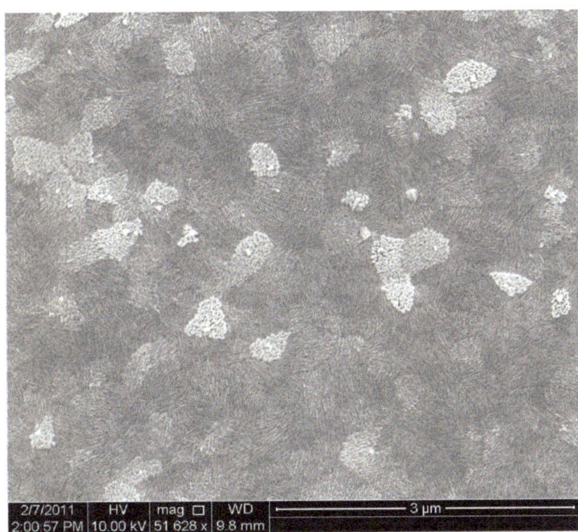

Fig. 4.14 SEM image of aligned nanorods of a-Se_xTe_{100-x} [105]

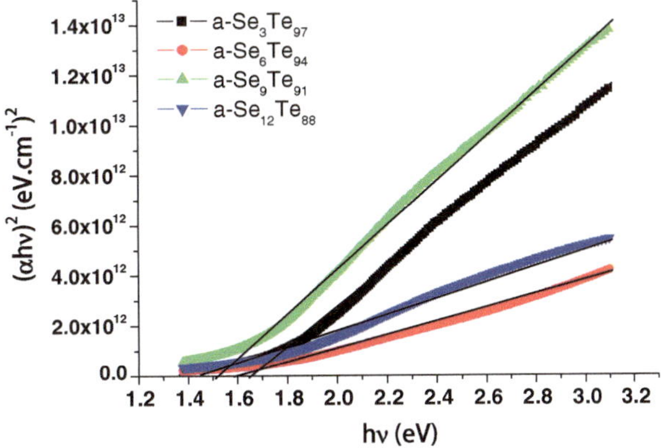

Fig. 4.15 $(\alpha h\nu)^2$ against photon energy ($h\nu$) of thin films of aligned nanorods of a-Se_xTe_{100-x} [105]

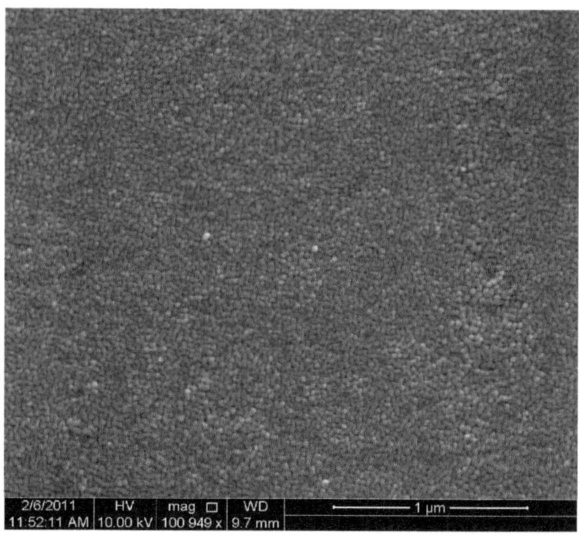

Fig. 4.16 FESEM image of thin film of a-$(PbSe)_{90}Cd_{10}$ nanoparticles [106]

[106] synthesized the a-$(PbSe)_{100-x}Cd_x$ (Fig. 4.16) (x = 5, 10, 15 and 20) nanoparticles using a thermal evaporation method. They reported the optical properties of these nanoparticles (Figs. 4.17 and 4.18). Salah et al. [107] synthesized thin films containing nanoparticles of $Se_{35}Te_{65-x}Ge_x$ (x = 0, 3, 6, 9, 12). X-ray diffraction (XRD), scanning electron microscope (SEM), and UV–Vis spectroscopy were used to characterize the as-synthesized nanoparticles. The amorphous nature of these films was confirmed by XRD patterns. SEM images suggested the growth of nanoparticles with average particle size around 30 nm. Changing concentrations of Te and Ge in the alloy system did not affect the morphology of the films. The optical behavior of these films was studied using the absorption and transmission spectra in the spectral region 400–1100 nm. The optical band gap (E_g) values were also determined and were found to decrease from 0.83 to 0.69 eV by increasing the concentration of Ge from 0 to 12. The electrical and optical properties of thin film of a-$Se_{70}Te_{30}$ nanorods were studied by Khan and Husain [108]. X-ray diffraction technique is used to verify the amorphous nature of this alloy. Transmission electron microscopy (TEM) suggested that the film contains nanorods of diameters varying from 30 to 80 nm and length of the order of few hundreds of nanometers. On the basis of the temperature dependence of dc conductivity, the conduction mechanism in this film of a-$Se_{70}Te_{30}$ nanorods was elucidated. In optical properties, optical absorption measurements of the thin film of a-$Se_{70}Te_{30}$ nanorods indicated that the absorption mechanism is due to indirect transition.

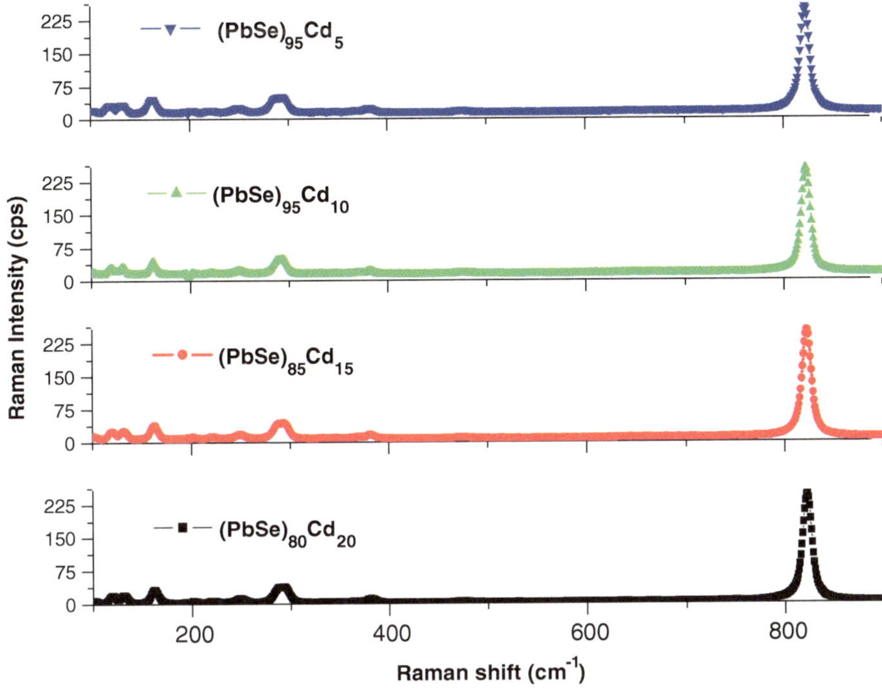

Fig. 4.17 Raman Spectra of a-$(PbSe)_{100-x}Cd_x$ nanoparticles [106]

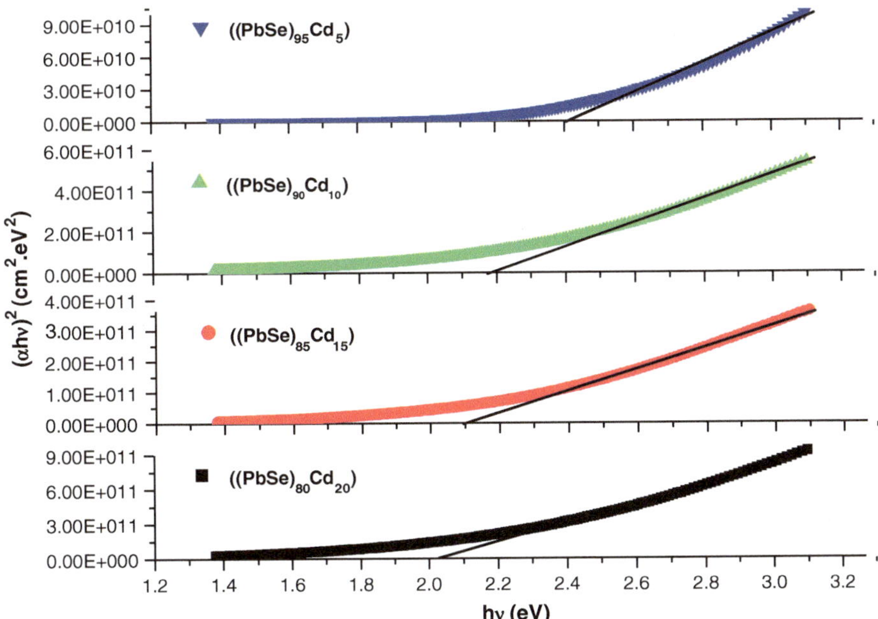

Fig. 4.18 $(\alpha h\nu)^2$ against photon energy $(h\nu)$ for thin films of a-$(PbSe)_{100-x}Cd_x$ nanoparticles [106]

4.3.5 Thermal Properties

A glass is an amorphous solid, which exhibits a glass transition. The glass transition is the phenomena in which a solid amorphous phase exhibits a more or less abrupt change in derivative thermodynamic properties (e.g. heat capacity or thermal expansively) from crystal—like to liquid—values with change of temperature. The term glassy is confined to materials, which can be obtained in a reproducible state (even after temperature cycling), since the material can be in a state of internal equilibrium above the glass transition. Glassy materials therefore need not be prepared solely by quenching from the melt; the glassy solids are a special sub set of amorphous materials. To express it in a different way, all glasses are amorphous but not all amorphous solids are necessarily glasses. To study the crystallization kinetics in these glasses, differential scanning calorimetry is a one of the powerful tool. Using this system, we can measure the glass transition, glass crystallization temperatures, enthalpy, order parameter and many other thermodynamical. These parameters are the most important parameters and one can easily explain the crystallization kinetics in glasses with the help of these parameters.

When a liquid is cooled, one of two events may occur. Either crystalline may take place at the melting point T_m or else the liquid will super cooled for temperature below T_m, becoming more viscous with decreasing temperature and may ultimately form a glass. The crystallization process is manifested by an abrupt change in volume at T_m, where as glass formation is characterized by a gradual break in slope. The region over which the change of slope occurs is termed the glass transition temperature T_g. It is found that the slower the rate of cooling, the larger is the region for which the liquid may be super cooled and hence the lower is the glass transition temperature. Thus the glass transition temperature of a particular material partly depends on its thermal history. Glass transition is a thermodynamic process common to all amorphous solids. To calculate the activation energy for relaxation time (ΔE_t), the knowledge of thermal relaxation in chalcogenide glasses and its kinetics interpretation is required. When a glass following an instantaneous change in temperature (during the quenching) relaxes from a state of higher enthalpy towards an equilibrium state of lower enthalpy, the thermal relaxation occurs. The annealing temperature plays a crucial role for this type of thermal relaxation and the process becomes faster near the glass transition temperature.

The glass transitions appear as an endothermic peak or a shift in the base line in DSC as result of change in specific heat. However, such an endothermic peak in chalcogenide glasses can be seen due to fast change in enthalpy when the glassy system relaxes quickly due to decrease in viscosity at the glass transition temperature.

Several parameters such as band gap, co-ordination numbers, bond energy, effective molecular weight, the type and fraction of various structural units formed are important for the glass transition temperature (T_g) of a multi component glass. In chalcogenide glasses, the glass tradition temperature (T_g) varies with the heating rare (β). This variation takes the form of a power law behavior and may be represented by the following expression;

$$T_g = [\beta]^y \quad (4.29)$$

Normalization has been done with respect to a heating rate of 1 K/min. Using Eq. (2.1) the exponent y is given by,

$$y = \log_{10}\left[(T_g)_{10}/(T_g)_1\right] \quad (4.30)$$

where $(T_g)_{10}$ stands for the T_g values at a heating rate of 10 K/min Eq. (4.1) gives an excellent description of the dependence of T_g, which increases with increasing heating rate and hence the glass transition temperature increases.

Different material has their different thermal properties. In the theory of solids, the specimen is taken at rest as background medium and only the propagation of various excitations is considered, which are either waves or particles. The presence of other excitations is also considered, which are either waves or particles. The presence of these excitations in the materials gives rise to the thermodynamically properties. The propagation of these excitations depends mainly on the internal structure of the materials. Specifically, thermal energy in a solid is transported by lattice wave (whose quanta is known as phonon) and by electron. In addition to this, there are also other processes i.e. convection and radiation. Solids can be divided into two categories as ordered system and disordered system. As name suggested, the ordered systems are perfect crystals with defects, liquids, amorphous substances, glasses, alloys ceramics constitute disordered system, powered systems, polymers or the systems, which are genetic in nature also form the disordered system. The theory of ordered system, therefore, cannot directly be applied to disordered system without substantial modification. Even in disordered system, there exist wide varieties of structures, groups like amorphous substances, glasses etc.,. the studies of thermal properties like crystallization kinetics, phase transformation etc. have been undertaken by using so many techniques. Common experimental techniques used to study these transformations include optical or electron microscopy, X-ray diffraction, electrical conductivity, thermal analysis and dilatometry.

Another method, which is also quite convenient for crystallization measurements, because of its simplicity and accuracy of continuous measurement, is conductivity measurement. The electrical properties are influenced by the structural effect associated with the thermal effects and can be related to the thermally induced transition. During a first order transformation, a latent heat is evolved and the transformation obeys the classical clausius-clapeyron equation. Second order transitions do not have accompanying latent heats, but like first order changes, can be detected by abrupt variations in compressibility, heat capacity, thermal expansion coefficients and the like. It is these variations that reveal phase transformations using thermal analysis techniques. The study of phase transformation, which is encountered in glassy materials, is of great importance as the temperatures at which the transformations take place, control their industrial applications. These transformation temperatures can be monitored using thermal analysis techniques. Amorphous materials do not have a well-defined melting point. The phase transition from the solid to the liquid state takes place gradually at a temperature of $T > T_g$, the temperature of the transition of glass.

The physical properties of chalcogenide glasses are important and on the basis these properties, these glasses find potential applications [109]. The physical properties of chalcogenide semiconducting glasses also strongly depend on the alloy compositions and doping [110, 111]. These materials have become a multipurpose materials, used for the fabrication of technologically important devices such as IR detector [112], electronic and optical switches [113, 114], inorganic resist [115] and optical recording media [116]. Due to the observation of unique phenomena like photo induced structural transformations [117], photo darkening/bleaching [118] and columnar collapsing [119], they have wide range applications. Recently, efforts are dedicated to develop chalcogenide based erasable optical storage media. For studying crystallization in semiconducting chalcogenide glasses, thermal processes are important. The crystallization under thermal annealing of chalcogenide thin films have been studied a lot [120–124]. In last few years, many reports on the use of chalcogenide thin films for phase change optical storage were appeared [125–127]. Large optical reflectivity and optical absorption changes in some semiconductors-semimetal thin films produced by heat treatment or laser irradiation, are useful for optical storage based on the amorphous-crystalline phase transition [127]. The significant understanding of the material properties required by the technological aspects of phase change storage has been produced in past years but still the engineering and development aspects of the field surpass the underlying basic scientific understanding.

When a materiel or system is given heat energy, then the propagation of the energy from one end of the system to other is indeed a challenging task to convert the picture of individual atoms vibrating about their lattice sites into a model, which would correctly predict the experimental values of some of the thermal properties like thermal conductivity, specific heat, phase transformation etc. The term phase transition is not only important for academic interest but also for the selection of the particular material for specific purpose e.g. in the industries, fabrication, research, space, technology etc. The study of phase transformation, which is encountered in glassy materials, is of great importance as the temperatures at which the transformations take place, control their industrial applications. These transformation temperatures can be monitored using thermal analysis techniques. The concept of the activation energy has always been important to study the kinetics of chalcogenides. During glass crystallization phenomena, the parameters related crystallization kinetics give the information about the nucleation and the growth processes, and dominate the devitrification of most glassy solids. It is important to identify separate activation energies with individual nucleation and growth steps in a transformation, although they have been combined into activation energy, which represents the overall crystallization process. When a liquid is cooled below its freezing point the onus is on the crystalline form to establish itself and if it fails to do this, glass will ultimately result by default. The difference in free energy between the stable (crystalline) phase and the metastable (supercooled liquid) phase is the crucial parameter for studying the phase change. The amount of material that is transformed in a given period of time will depend upon both the rate of formation of crystalline nuclei and their rate of growth. Therefore, crystallization involves both nucleation and diffusion of the crystallizable units to the crystal front (growth). Nucleation occurs more readily at lower crystallization temperatures because of the lower critical nucleus size and lower free energy barrier associated with the processes. Short-range diffusion occurs at higher temperatures. All diffusive motions are completely frozen-in at temperatures below the glass transition temperature. Studies on the compositional dependence and the effect of addicting a small quantity of material on crystallization is important as the write and erase time is closely related to the melting temperature (T_m) and the speed of crystallization.

4.3.5.1 Differential Scanning Calorimetric Studies

Thermal analysis encompasses a wide verity of techniques. Differential Scanning Calorimetry (DSC) is the more sensitive technique among them. It is used for measuring the energy necessary to establish a nearly zero temperature difference between a substance and an inert reference material, as the two specimens are subjected to identical temperature regimes in an environment heated or cooled at a controlled rate. Such measurements provide qualitative and quantitative information about physical and chemical changes that involve endothermic or exothermic or changes in heat capacity. Basically the DSC measurement is a function of differential heat flow with temperature for compounds that exhibits thermal transitions. These transitions are typically melting, crystallization and the glass transition.

The thermal behavior is normally investigated using Differential Scanning Calorimeter. The DSC is calibrated prior to the measurement using high purity standard Pb, Sn and In with well-known melting points. The crystallization experiments are carried out through continuous heating rates. Using the microprocessor of thermal analyzer, the glass transition (T_g), crystallization (T_c), melting temperature (T_m) are measured. The crystallized fraction (α) is calculated with the help of partial area analysis. At any temperature (T), the fraction of volume crystallized (α) is given as, $\alpha = A_t/A$, where A represents the total area of the exotherm between temperature (T_1) at which crystallization just begins and temperature (T_2) where the crystallization is completed. A_t represents the area between T_1 and T (Fig. 4.19).

Thermal properties of chalcogenides play an important role in determining the transport mechanisms, thermal stability of these glasses for the practical applications [128–131]. Using Differential Scanning Calorimetry, the crystallization kinetics in amorphous materials has been widely studied [132–137]. A large variety of theoretical models and functions have been proposed to explain the crystallization kinetics. The application of each depends on the type of amorphous

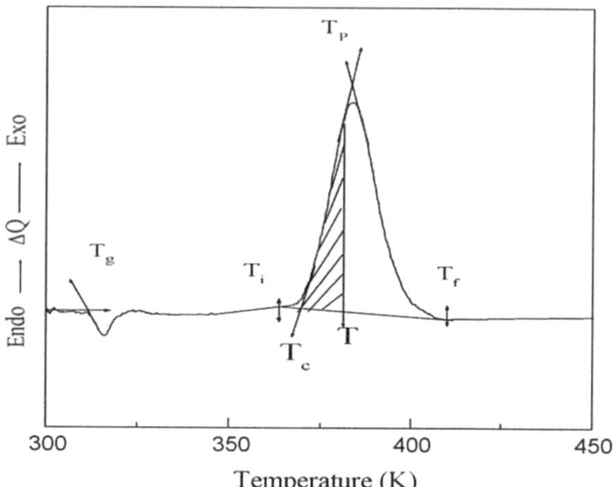

Fig. 4.19 DSC curve indicating the estimation of volume fraction crystallized

material studied. Normally, we study isothermal and non-isothermal crystallization kinetics in chalcogenides [135–137]. In isothermal method, the sample is heated quickly to a temperature above the glass transition temperature (T_g) and the heat evolved during the crystallization process is measured as a function of time. Whereas in non-isothermal method, the sample is heated at a fixed heating rate and the heat evolved is measured as a function of temperature or time. The experiments conducted at constant heating rate are a rapid way of studying the glass transformation. On the other hand, the isothermal experiments are time-consuming. Another disadvantage of isothermal analysis is the impossibility of reaching a test temperature instantaneously and during the time, which the system needs to stabilize, no measurements are possible. The non-isothermal method involving constant heating rate experiments do not have this drawback [128, 137]. Therefore, most of the researchers prefer non-isothermal method to study the crystallization kinetics.

4.3.5.2 Theoretical Consideration of Crystallization Kinetics

Generally, DSC results are interpreted with the help of theoretical basis given by the theory of transformation kinetics as the volume fraction (α) crystallized in time (t) are related by Johnson and Mehl [138] and Avrami's equation [139–141], which is given as;

$$\alpha(t) = 1 - \exp . [-(kt)^n] \tag{4.22}$$

where $\alpha(t)$ is the volume fraction crystallized after time t, n is the Avrami exponent, which is associated with the nucleation and growth mechanisms and k is the reaction rate constant.

In the thermally activated process the reaction rate constant k is related to temperature T through the equation

$$k = \nu \exp(-\Delta E/RT) \tag{4.23}$$

where ΔE represents the effective activation energy describing the overall crystallization process, and ν is the rate constant depending on the operating nucleation and growth modes [142, 143]. Generally, the crystallization process is determined by nucleation and growth, which will have different activation energies [144]. At different degrees of crystallization, different growth mechanisms may come into picture, which gives temperature dependent activation energy. Using the model given by Matusita et al. [145] for non-isothermal experiments, the activation energy for crystallization and the Avrami exponent can be obtained. The volume fraction of crystallites (α) precipitated in a glass heated at constant heating rate (β) is related to the effective activation energy for crystallization (ΔE_c) using the following equation;

$$\ln[-\ln(1-\alpha)] = -n\ln(\beta) - 1.052\,m(\Delta E_c)/RT + \text{constant} \tag{4.24}$$

where, m is an integer, which depends upon the dimensionality of the crystal. When the nuclei formed during the heating at a constant rate are dominant, n is equal to (m + 1) and when nuclei formed during any previous heat treatment prior to thermal analysis are dominant, n is equal to m. According to Eq. (4.24), a plot of ln [−ln (1 − α)] versus ln β yield a straight line with slope equal to n (order parameter).

The experimental crystallization data can be interrelated with the help of Kissinger's, Matusita's and modified Ozawa's equations for non-isothermal crystallization. Using Kissinger's equation [133], the activation energy (ΔE_c) for crystallization may evaluated using the equation given below;

$$\ln(\beta/T_c^2) = -\Delta E_c/RT_c + D \tag{4.25}$$

The plot of ln (β/T_c^2) versus 1000/T_c, which comes out to be a straight line and the slope of the line will give the value of E_c.

Using the variation of the onset crystallization temperature with heating rate by given by Ozawa's relation [146], the activation energy of crystallization can also be determined using the following relation;

$$\ln\beta = -\Delta E_c/RT_c + C \tag{4.26}$$

where, C is a constant.

The glass transition activation energy (ΔE_g) can be estimated using Kissinger's equation [113] which is given as;

$$\ln(\beta/T_g^2) = -\Delta E_g/RT_g + \text{constant} \tag{4.27}$$

It is evident from this equation that a plot of ln (β/T_g^2) against 1000/T_g should be a straight line and the activation energy involved in the molecular motions and rearrangements around T_g can be calculated from the slope of this plot.

For the chalcogenide glasses, the dependence of heating rate with the glass transition temperature can be discussed using thermal relaxation phenomena and the activation energy of structural relaxation (ΔE_g) can also be related to T_g and β by the following equation [147];

$$\ln\beta = -(\Delta E_g/RT_g) + \text{constant} \tag{4.28}$$

From this equation, a plot of ln β against 1000/T_g should also be straight line and slope gives the activation energy involved in the molecular motions and rearrangements around T_g.

4.3.6 *Studies on Bulk Chalcogenides*

The phase separation in $Ga_{25}Se_{75-x}Te_x$ chalcogenide thin films was reported by Al-Agel et al. [148]. They observed that the thermally evaporated amorphous $Ga_{25}Se_{75-x}Te_x$ thin films undergone structural transformations when annealed at a temperature in between glass transition and crystallization temperature. Thermal annealing caused a decrease in the indirect optical band gap. This behavior was attributed to the production of surface dangling bonds around the formed crystallites during the process of crystallization.

The phase transformation kinetics and optical properties of Ga–Se–Sb phase-change thin films were studied [149]. A systematic investigation on of the heating-rate dependence of the glass transition and crystallization temperature in $Ga_{25}Se_{75-x}Sb_x$ chalcogenides glasses was carried out using DSC technique. The crystallization kinetics under non-isothermal conditions and effect of thermal annealing on optical constants in a-$Ge_{25}Se_{75-x}Te_x$ glasses were studied by Al-Agel [150]. With the increase in heating rate and incorporation of Te in Ge–Se system, the glass transition temperature (T_g) and crystallization temperature (T_c) was found to shift to higher temperatures (Figs. 4.20 and 4.21). With increasing Te content, the activation energy of the amorphous–crystalline phase transformation (ΔE_c) and the activation energy of glass transition (ΔE_g) showed an increasing trend. On thermal annealing, the crystallization of $Ge_{25}Se_{75-x}Te_x$ thin films resulted in a decrease in the optical band gap. This decrease in the band gap was due to crystallization of $Ge_{25}Se_{75-x}Te_x$ thin films. Bulk sample of $Ga_{15}Se_{76}Pb_9$ chalcogenide glass was prepared by melt quenching technique [151]. A study of crystallization kinetics of this glass was investigated using non-isothermal DSC measurements. The values of different crystallization parameters such as the activation energy for structural

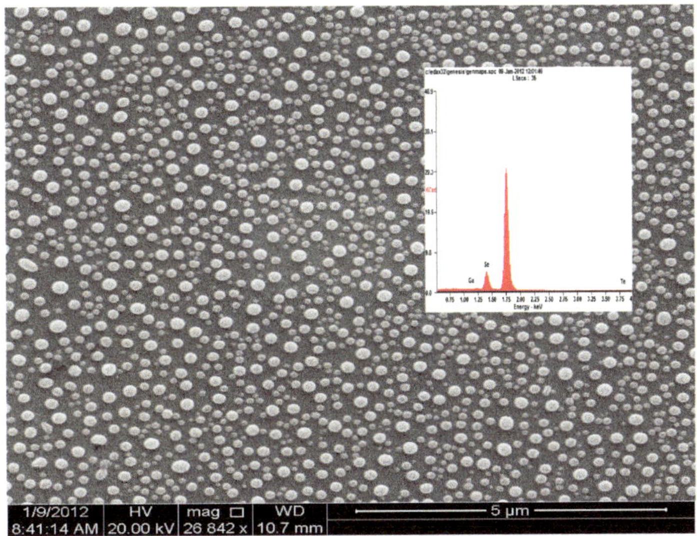

Fig. 4.20 FESEM image of as-prepared thin film of $Ga_{25}Se_{66}Te_9$ glass and the *inset shows* the energy dispersive X-ray spectroscopy (EDAX) of $Ga_{25}Se_{66}Te_9$ thin films [150]

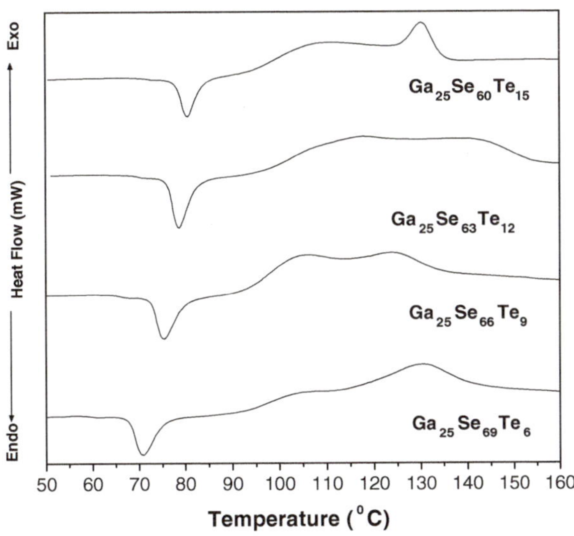

Fig. 4.21 DSC trace for the powdered $Ga_{25}Se_{75-x}Te_x$ chalcogenide glasses at 20 K/min [150]

relaxation (ΔE_t) and activation energy of crystallization (ΔE_c) are estimated using the T_g and T_c dependence with heating rate (β). The order parameter (n) and the crystallization enthalpy (ΔH_c) were determined using the area analysis of crystallization peaks. The results of crystallization were discussed on the basis of different models. The non-isothermal crystallization in Ga–Se–Ag chalcogenide glass was reported by Khan et al. [152]. On the basis of value of local Avrami exponents, the volume nucleation with two dimensional growths was found to be responsible for the transformation process. The glass transition and crystallization behavior in $Ga_{15}Se_{85-x}Pb_x$ were investigated by Khan et al. [153]. The crystallization kinetics of $Se_{90}Bi_{10}$ and $Se_{85}Bi_5Te_{10}$ chalcogenide glasses was studied and discussed using Differential Thermal Analysis (DTA) under a non-isothermal condition at different heating rates [154].

Differential scanning calorimetry (DSC) and X-ray diffraction (XRD) analysis were carried out to study crystallization in $Ge_2Sb_2Se_4Te$ glass under non-isothermal conditions as a function of the particle size [155]. With help of autocatalytic Šesták–Berggren model, the crystallization kinetics was discussed. All aspects of a full-scale kinetic study of a crystallization process were discussed extensively. The effect of replacement of antimony by silver on the physical parameters of new quaternary $Ge_{20}Se_{60}Sb_{20-x}Ag_x$ (where x = 0, 5, 10, 15 and 20 at.%) glasses were studied [156]. The average coordination number and the glass transition temperature decreased while the density, the compactness, the average heat of atomization, the cohesive energy and the overall mean bond energy increased with increasing Ag content. The observed dependence of the physical parameters on composition for the $Ge_{20}Se_{60}Sb_{20-x}Ag_x$ (where x = 0, 5, 10, 15 and 20 at.%) glasses can be explained on the basis of a chemically ordered network arrangement. The structural characterization of $Se_{70}Te_{30}$ chalcogenide glass for the as-prepared and thermal annealed samples were identified using the X-ray diffraction (XRD) and scanning electron microscopy (SEM) [157]. The isothermal thermal annealing process resulted in a transformation from the amorphous to the crystalline state. The thermal analyses of $Se_{70}Te_{30}$ glass were investigated using a differential thermal analyzer (DTA) at five different heating rates under non-isothermal conditions. Ternary $Se_{90}Te_{10-x}Sn_x$ ($x = 2, 4, 6$, and 8) chalcogenide glassy alloys were prepared by melt quenching technique [158, 159]. Various crystallization parameters, such as onset (T_c) and peak (T_p) crystallization temperatures, activation energy of crystallization (E_c) and Avrami exponent (n) were determined for these alloys. Differential scanning calorimetric (DSC) technique was employed to determine the thermal properties of Se-rich $Se_{85-x}Te_{15}Bi_x$ glassy alloys [159, 160]. The glass transition temperature (T_g) and peak crystallization temperature (T_p) were found to shift to a higher temperature with increasing heating rate. Optical band gap was estimated using Tauc's extrapolation and was found to decrease from 1.46 to 1.24 eV with the Bi addition.

Abu El-Oyoun [161] has studied the DSC studies on the transformation kinetics of two separated crystallization peaks of $Si_{12.5}Te_{87.5}$ chalcogenide glass; an application of the theoretical method developed and isoconversional method, Patial et al. [162] has studied the crystallization kinetics of In additive Se–Te chalcogenide glasses, Kumar and Singh [163] has studied the effect of indium additive on crystallization kinetics and thermal stability of Se–Te–Sn chalcogenide glasses, Kozmidis-Petrović et al. [164] reported the non-isothermal crystallization parameters for the $Cu_{15}(As_2Se_3)_{85}$ metal-chalcogenide glass, Lafi [165] has studied the glass transition kinetics and crystallization mechanism in $Se_{90}Cd_8Bi_2$ and $Se_{90}Cd_6Bi_4$ chalcogenide glasses, Abu El-Oyoun [166] has reported the evaluation of the transformation kinetics of $Ga_{7.5}Se_{92.5}$ chalcogenide glass using the theoretical method developed and isoconversional analyses, Al-Agel et al. [167] reported the kinetics of non-isothermal crystallization and glass transition phenomena in $Ga_{10}Se_{87}Pb_3$ and $Ga_{10}Se_{84}Pb_6$ chalcogenide glasses by DSC, Elnaeim et al. [168] has studied the glass transition and crystallization kinetics of $In_x(Se_{0.75}Te_{0.25})_{100-x}$ chalcogenide glasses.

The work on thermal characterization of $Se_{85-x}Sb_{15}Sn_x$ chalcogenide glasses by Imran [169], crystallization kinetics determination of $Pb_{15}Ge_{27}Se_{58}$ chalcogenide glass by using the various heating rates (VHR) method by Abd El-Raheem and Ali [170], crystallisation kinetics, glass forming ability and thermal stability in glassy $Se_{100-x}In_x$ chalcogenide alloys by Muiva et al. [171], non-isothermal crystallization kinetic study on $Ga_{15}Se_{85-x}Ag_x$ chalcogenide glasses by using differential scanning calorimetry by Al-Ghamdi et al. [172], a study of DSC non-isothermal pre-crystallization kinetics of $Pb_{10}Se_{90}$ glass using isoconversional kinetic analysis by Soltan [173], observation of phase separation in some Se–Te–Sn chalcogenide glasses by Abdel-Wahab [174], an investigation of the kinetic transformation mechanism of $Ge_{12.5}Te_{87.5}$ chalcogenide glass under non-isothermal regime by Abu El-Oyoun [175], studying the crystallization behavior of the $Se_{85}S_{10}Sb_5$ chalcogenide semiconducting glass by DSC and X-ray diffraction by Shapaan and Shaaban [176], Glass transition, thermal stability and glass-forming tendency of $Se_{90-x}Te_5Sn_5In_x$ multi-component chalcogenide glasses by Kumar and Singh [177], effect of addition of gallium on the thermal stability and crystallization kinetic parameters of Ga_xSe_{100-x} glass system by M. Abu El-Oyoun [178], kinetics of the glass transition in $As_{22}S_{78}$ chalcogenide glass: Activation energy and fragility index by Abu-Sehly [179], crystallization kinetics, glass transition kinetics, and thermal stability of $Se_{70-x}Ga_{30}In_x$ semiconducting glasses by Imran [180], calorimetric studies of $Se_{75}Te_{15}Cd_{10}$ and $Se_{75}Te_{10}Cd_{10}In_5$ multicomponent chalcogenide glasses by Kumar and Singh [181], isochronal and isothermal crystallization in $Zr_{55}Cu_{30}Ni_5Al_{10}$ bulk metallic glass by Qiao and Pelletier [182] are also worth mentioning.

Recently, a lot of research work is focused on the phase change chalcogenide glasses for phase change memory devices. Zhang et al. [183] has studied the phase change $Si_2Sb_2Te_5$ material and its application in chalcogenide random access memory. Jialin et al. [184] reported the effects of Ge doping on the properties of Sb_2Te_3 phase-change thin films. Ge-doped Sb_2Te_3 films were deposited by magnetron sputtering of Ge and Sb_2Te_3 targets on SiO_2/Si (1 0 0) substrates. X-ray diffraction, differential scanning calorimetry and X-ray photoelectron spectroscopy measurements were employed to understand the effect of Ge doping on the structure. These measurements indicated that substitution of Ge atoms for Sb/Te in lattice sites and form Ge–Te bonds, moreover, a metastable phase was seen in Ge-doped specimens. After Ge doping, the crystallization temperature and resistivity of amorphous Sb_2Te_3 were found to increase. This is useful for improvement of room temperature stability of the amorphous state and the SET current of chalcogenide random access memory may also be reduced. Kolobov et al. [185] has studied the thermal component of amorphisation in phase-change alloys and chalcogenide glasses, Lankhorst [186] has studied the modelling glass transition temperatures of chalcogenide glasses applied to phase-change optical recording materials, Abrutis et al. [187] reported the chemical vapor deposition of chalcogenide materials for phase-change memories, Takata et al. [188] has studied the thermal strain imaging of chalcogenide in a phase change memory, Mehta and Kumar [189] reported the observation of phase separation in some Se–Te–Ag chalcogenide glasses, Saxena [190] has studied the phase transformation kinetics and related thermodynamic and optical properties in chalcogenide glasses, Hafiz et al. [191] reported the reversible phase change in Bi_xSe_{100-x} chalcogenide thin films for optical recording medium. Kozicki et al. [192] has studied nanoscale phase separation in Ag–Ge–Se glasses, Kumar et al. [193] has studied the characterization of phase transition in silver photo-diffused $Ge_2Sb_2Te_5$ thin films. The review articles on status and challenges of phase change memory modeling by Lacaita et al. [194] and why are chalcogenide glasses the materials of choice for Ovonic switching devices? by Fritzsche [195] are also worth mentioning.

4.3.7 Studies on Nano-chalcogenides

The glass transition kinetics in ball milled amorphous Ga_xTe_{100-x} nanoparticles were studied Khan [196]. The ball milled samples of a-Ga_xTe_{100-x} contained high yield of nanoparticles with their diameter in the range of 20–40 nm (Fig. 4.22). The studies of glass transition kinetics suggested an increase in the value of glass transition temperature with the increase in composition and heating rate (Fig. 4.23). A good agreement is observed in the values of activation energy for structural relaxation calculated using two methods. The calculated value showed an

Fig. 4.22 SEM image of as-prepared a-Ga_xTe_{100-x} nanoparticles [196]

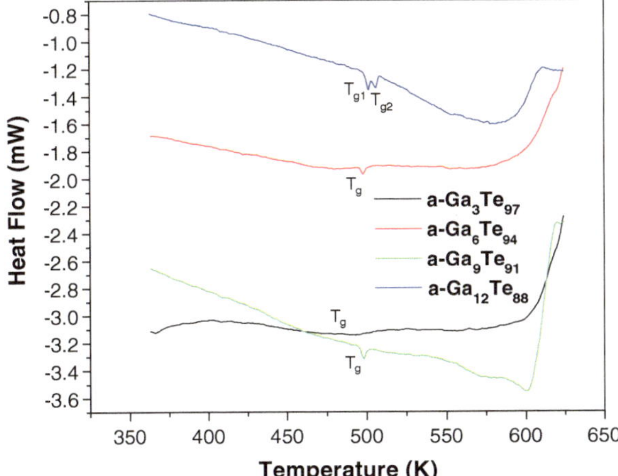

Fig. 4.23 DSC scans for a-Ga_xTe_{100-x} (x = 3, 6, 9 and 12) nanoparticles at the heating rate of 5 K/min [196]

increasing trend with increase in gallium content. The results were can be interpreted in terms of change in chemical bonding at higher concentrations.

The crystallization kinetics in as-synthesis high yield of a-$Se_{100-x}Te_x$ nanorods were studied [197]. SEM investigation suggested that the samples of a-Se_xTe_{100-x} contained high yield of nanorods and their diameter was of the order of several hundred nanometers (Fig. 4.24). An amorphous nature is confirmed by the XRD pattern of these nanorods (Fig. 4.25). The activation energy of crystallization (ΔE_c) and the order parameter (n) were calculated (Fig. 4.26) using the heating rate dependence of glass transition temperature and crystallization temperature, the activation energy for structural relaxation (ΔE_t). Thermal properties of amorphous Ga_xSe_{100-x} nanorods (Fig. 4.27) were studied at different heating rates (5, 10, 15, and 20 K/min) using differential scanning calorimetry [198]. With the help of the heating rate dependence of glass transition temperature and crystallization

Fig. 4.24 FESEM image of as grown a-$Se_{100-x}Te_x$ nanorods [198]

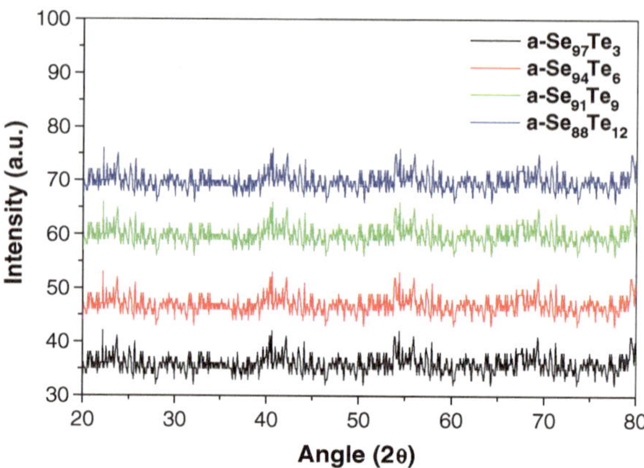

Fig. 4.25 XRD pattern of a-$Se_{100-x}Te_x$ nanorods [198]

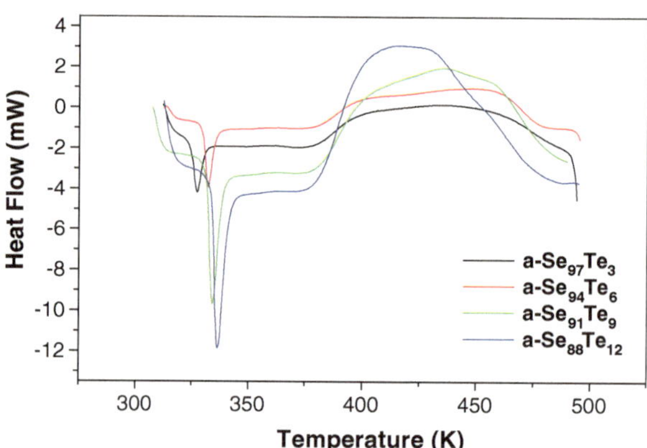

Fig. 4.26 DSC scans for a-$Se_{100-x}Te_x$ (x = 3, 6, 9 and 12) nanorods at the heating rate of 20 K/min [198]

Fig. 4.27 SEM image of as-prepared a-$Ga_{12}Se_{88}$ nanorods [198]

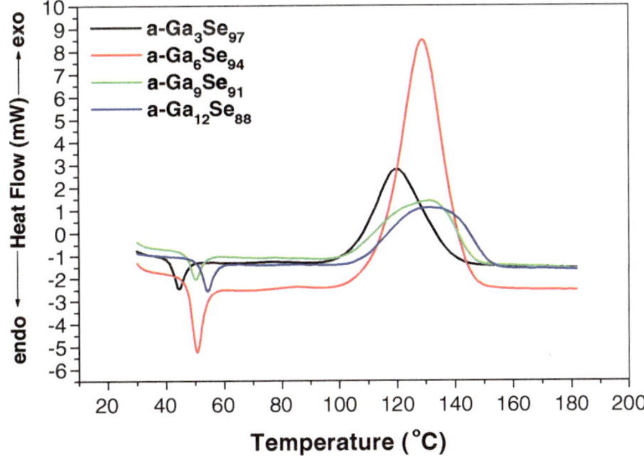

Fig. 4.28 DSC scans for a-Ga_xSe_{100-x} (x = 3, 6, 9 and 12) nanorods at the heating rate of 20 K/min [198]

temperature (Fig. 4.28), the activation energy for structural relaxation (ΔE_t), the activation energy of crystallization (ΔE_c) and the order parameter (n) were calculated. To study the thermal stability of these nanorods, several parameters such as thermal stability parameter (S), the crystallization enthalpy (ΔH_c), entropy (ΔS) changes during crystallization process and crystallization rate factor (Kp) were calculated. The glass forming ability (GFA) was studied using the calculated values of Hruby number (H_R) and relaxation time. Further, the calculated value of fragility index (F_i) was used to confirm whether these glasses were obtained from kinetically strong-glass-forming (KS) or kinetically fragile-glass forming (KF) liquids.

Nanoparticles of a-Se_xTe_{100-x} were prepared by a melt-quenching technique and their crystallization kinetics was studied [199]. The diameter of these nanoparticles varied from 100 to 200 nm (Fig. 4.29). On the basis of the obtained results, it is suggested that the glass a-$Se_{12}Te_{88}$ with lowest activation energy value and highest

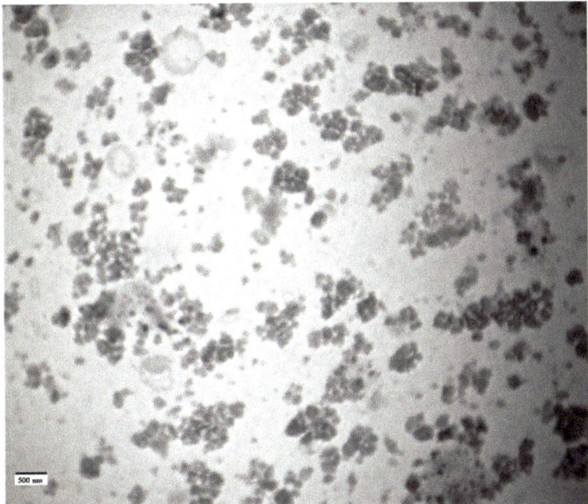

Fig. 4.29 TEM image of as-prepared a-$Se_{12}Te_{88}$ nanoparticles [199]

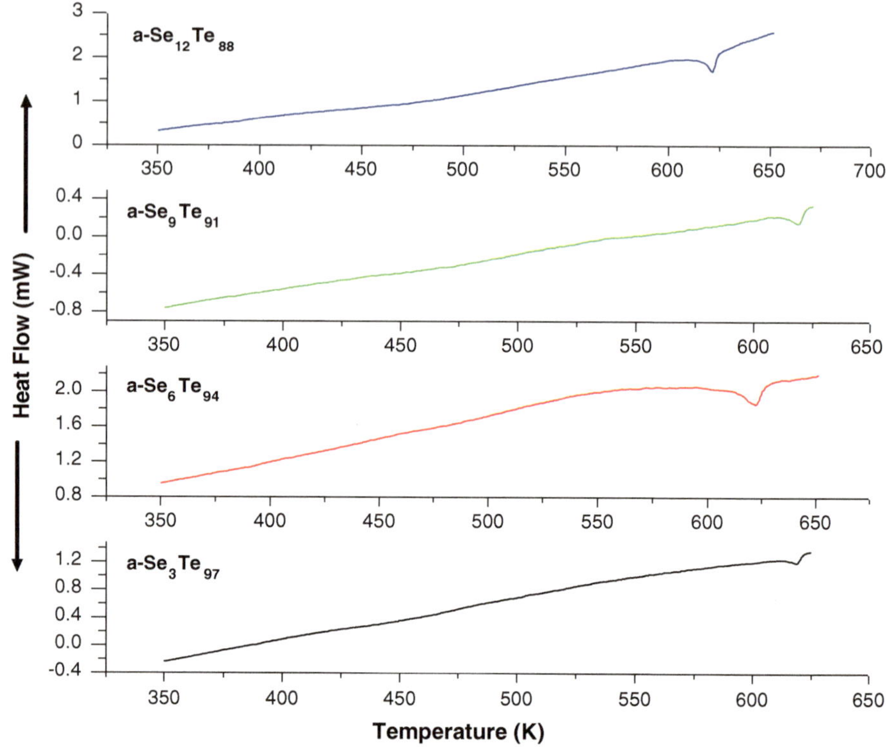

Fig. 4.30 DSC scans at the heating rate of 10 K/min for a-Se_xTe_{100-x} nanoparticles [199]

value of total relaxation time, was the most stable glass with good glass-forming ability among all of the compositions studied (Fig. 4.30). The composition dependence of the activation energy for thermal relaxation and activation energy for crystallization has been discussed in terms of the structure of Te–Se glassy system.

A lot of research work is presently concentrated on the thermal, electrical, structural and optical properties of amorphous semiconductors. However, the scientist feels that the field is highly intriguing. The basic physics of these materials has not reached to a level, which is required to develop devices. Thus the scientific interest in this field is rapidly expanding and a lot of works on thermal properties are available in the literature. However, many questions and quarries are yet to be explained satisfactory viz. to prepare doped semiconductors which are useful for optical data storage, phase change optical data storage media on increasing the erase speed, the electrical switching ability, to develop chalcogenide based erasable optical storage media, the glass transition temperature, crystallization temperature, melting temperature, the activation energy for structural relaxation, activation energy of crystallization, Avrami exponent, order parameter, enthalpy released, thermal stability of the material, the structural and chemical bonding in amorphous semiconductors.

4.4 Applications of Chalcogenides

4.4.1 Memories Based on Phase-Change in Chalcogenide Glasses

Recently, a lot of efforts are dedicated to scale the non-volatile-memory technology beyond the flash memory architecture. Chalcogenide glasses are thought to ideal candidate for phase-change memories [200]. In these glassy systems, a reversible switching between the amorphous and crystalline state is observed. Due to this interesting phenomenon, they have potential to be used for rewritable optical recording [201–203] and electrically programmable non-volatile memories [204]. In glass science, the phase separation effects are important. It is a phenomenon used to describe a homogeneous system such as a liquid, where two or more finely mixed chemically or structurally different, component or phases are separated. Usually, a significant change in glass physical properties is observed due to such type of structural effects. The change in the properties includes a decrease in the glass transition temperature, a decrease in the optical band gap and an enhancement in the molar volumes [205]. In last few years, the chalcogenides have been developed as a new phase change material with demanding applications. Based on the reversible switching between the amorphous and crystalline states, phase change materials have become very popular and have been used for optical rewritable data storage media [206–208]. It has been observed recently that these materials have great potential for future electronic nonvolatile memory devices, the so-called phase change random access memory (PRAM) [209–211]. For applications of optical data storage, a focused laser beam is used as a heat source and the active layer of PRAM devices are heated by a current pulse. Using the recorded data based on difference in reflectivity or electrical conductivity, the amorphous and crystalline states of the phase change recording layer are distinguished. During last few decades, a lot of work on the synthesis and optical properties of alloys with a focus on the performance of optical disks is reported. Therefore, the reversible transformation of amorphous-to-polycrystalline phases is useful for optical data storage devices [212–214].

4.4.2 Memories Based on Electrical Switching in Chalcogenide Glasses

For the development of next generation nonvolatile memories (NVMs) providing excellent speed, superior density with the consumption of low power and good endurance at a low cost compared to the existing flash/dynamic/static random access memories (RAMs), most commonly available technologies involves the use of magneto-resistance (MRAM), ferroelectricity (FRAM), phase transition (PRAM) and electrical-resistance (RRAM) [215–230]. Out of these technologies, phase transition (PRAM) and electrical-resistance (RRAM) uses the phenomenon of electrical resistance switching. Recently, it has become more popular due to high potential for the next generation memories. PRAM involves the use switching of resistances due to the amorphous-crystalline phase transition of a chalcogenide material. This switching is triggered by an electric pulse irrespective of its polarity. Whereas, PRAM uses the switching of resistance due to change in the polarity of the applied electric field and it is not related o the change in phase. Although this type of switching behaviour is commonly observed, full understanding of the RRAM mechanism is still needed.

The trap-controlled space-charge-limited current [219], charge-trapping-defect states inside the band gap [222], charge-trap states at metal/oxide interface with a change of Schottky-like barrier [223], electric-pulse-induced vacancy/ion motion [222, 223], and filamentary conduction [224–230] have been explained with the help of different models. A variety of materials has been utilized for RRAM, in particular transition metal oxides [219, 231–238], perovskite-based oxides [222, 223, 239–243] and (Ag or Cu saturated) chalcogenides [231–238, 244, 245]. The mechanism of polarity-dependent resistance (PDR) switching with metal-saturated chalcogenides depends on the nature of the solid-state electrolytic and the ionic conductivities of amorphous materials. Due to this, it is also named as ionic/electrolytic switching. In such materials, the medium becomes electrically conductive for one polarity due to the formation of conducting filamentary pathways, whereas it behaves as an insulating medium for the opposite polarity.

Among the various choices of active materials, chalcogenides are promising and versatile as they exhibit more than one type of memory switching. However, the phase-change and polarity-dependent resistance (PDR, non-phase-change) switching were thought to be independent in these materials and no single system is reported showing these two switching mechanisms together [246]. There are many compositions of GeSbTe system, which exhibited amorphous-crystalline switching [247]. Many others such as AgS, CuS, AgGeSe, AgGeTe and AgInSbTe [229–234] presented PDR switching.

Amorphous materials are the interesting class of materials and have been the focus of attention during last several years due to their interesting properties. In these materials, the long range order is missing, which is responsible for the creation of the localized states in the mobility gap. These localized states are commonly observed in these materials, which may also be due to other inherent defects. During last few years, the studied on electrical properties in such amorphous systems has suggested the existence of gap states in these materials. These gas states significantly affect electrical phenomena in semiconductors [248, 249]. It is reported that most of the chalcogenides exhibit interesting electrical effects due to a change in resistance. This is known as switching. Later on, it is observed that the chalcogenide glasses are the typical materials, showing these properties [247]. In chalcogenides, this type of memory switching occurs due to reversible change in structure from crystalline to amorphous state and vice versa. The amorphous sate is of high resistance state, whereas, the crystalline state is of low resistance state. The change in the structure from amorphous to crystalline takes place in the ON state in the current filament. The cause of such crystallization is Joule heating, which gives rise to the high field and increased carrier concentration. Boer and Ovshinsky [250] reported such type of crystallization by Joule heating of the current channel. The thermally stabilized high electric field induced in these materials is responsible for the switching transition. The switching behavior in chalcogenides is well studied with the help of different models [251, 252]. Last few years have seen many papers on switching in chalcogenides, but the explanation of switching mechanism in these materials is still not clear and requires serious efforts in this direction [253].

Intense research on amorphous materials started after the Ovshinsky's discovery of memory switching in these glasses [254]. After this discovery, many technologies for digital optical random access memory discs and the new non-volatile electrical memory could be realized [254–256]. Later on in the late 1970 [257] and early 1980, the physics governing this electronic switching effect was discussed [258]. There are normally two kinds of chalcogenide switches classified as threshold which is mono-stable and reversible whereas, the memory switch is bi-stable and irreversible. In case of chalcogenides especially selenium and tellurium rich chalcogenide glasses, electrical switching occurs on applying an appropriate voltage, which is called as the threshold or switching voltage (V_T) and on the application of such voltage, glass switches to a high conducting state, known as ON state from a low conducting state, known as OFF state. Being reversible in nature, the threshold switching glasses revert to the OFF state and the memory switches remain in the ON state after removing the switching field due to irreversible nature. Both in threshold as well as memory glasses, the process of switching is primarily electronic in nature and it is observed due to filling of charged defect states by the field-injected charge carriers. However, the process of forming a conducting crystalline channel takes place due to additional thermal effects in memory materials [259]. Since these glassy materials show interesting phenomenon of electrical switching, the phase change memories (PCMs) are thought to be the best candidate for future Non-Volatile Random Access Memories. With the discovery of the phenomena of phase change memories, the chalcogenide glasses can be used in memory switching. The research of non-crystalline semiconductors started intensively after the discovery of electronic switching effect in chalcogenide glasses [260]. Due to this, the technologies of digital optical random access memory discs [255] and the non-volatile electrical memory were developed [254, 256–258]. In present semiconductor memory technology, the memory components used different specialized features. There is not a single technology providing all the important features such as good speed, cheap, lowest power consumption, long cycle and non-volatility. Presently, DRAM, SRAM and FLASH are the commonly used memory technologies. Therefore, we need a memory technology that should include only the attractive features excluding any of the unwanted properties. Recently, the development of a new memory technology named Ovonic Unified Memory (OUM) is under progress. These memories will be based on chalcogenide materials. We expect that the development of memory chips based this technology will have all important feature such as SRAM speed, DRAM cost, power characteristics and non-volatility. Presently, many high volume portable products such as hand-held cell phones, Personal Digital Assistants (PDAs), electronic digital cameras are available in the market. All of these need non-volatility. We also need radiation resistant nonvolatile memory for space applications. Currently, the alloys used in large-scale production of re-writable CD disks are also used as the storage medium for the OUM chips [261, 262]. It is understood that the chalcogenides based memory technology has the potential for excellent memory devices. The chalcogenide family includes S, Se and Te. All of these are Group VI elements. Normally, the alloys used for memory devices contain at least one of the group-VI elements. The alloys containing Ge, Sb and Te are being used by Energy Conversion Devices Inc. to develop the phase-change memory technology [259, 262]. The re-writeable CD and DVD disks are using this technology for the commercial production. The data storage in such disks involves a thermally activated, rapid, reversible change in the alloy structure. This process is inherently non-volatile as it uses two different phases of material to represent the binary information. The rewriteable CD and DVD disk technology uses the difference in reflectivity to read the state of each memory bit. This is done by throwing a low-power laser at the material and collecting the reflected light. The chalcogenide alloys has two phases, i.e., crystalline and amorphous. These two phases have different electrical properties. The resistivity of amorphous state has been shown to four times higher than that of polycrystalline state. These two states switch to a long resistance state independent of the phase. The structural phase of the material remains unaffected during this type of switching and resistance can be restored when the applied voltage is brought below the holding voltage V_H. The low resistance state allows high current to be flown into the material on relatively low voltages. These extraordinary electrical properties are very interesting when we consider these materials to be used as the in semiconducting memory devices. One can read a particular state of the device by charging a capacitive load line, the variable resistance chalcogenide element above V_T [263], rate of charging depends upon the resistance and can be recorded and converted into a logic state. Chalcogenide elements can be programmed by applying high current (for the voltage above V_T) for heating the material up to the crystallization or the melting temperature. The resistance of the device behaves differently at different currents. At low currents, the resistance of the device is independent of current and is only a function of initial state. At moderate currents, the temperature of the device is high enough to reach at the crystallization temperature of the material and consequently the resistance drops. High current increases the temperature of the material to the melting temperature, which increases the resistance. The materials with one or more chalcogen elements i.e., sulphur, selenium and tellurium form a special class of semiconducting materials, called chalcogenide materials [264]. The chalcogenide materials can be fabricated in the form of glasses either as thin films, or in the form of bulk samples. The chalcogenide materials can be obtained in different composition, including non-stoichiometric compounds and mixtures. This has given a new direction to chalcogenides. Scientists, all over the world are working on various new chalcogenide materials and their applications.

The chalcogenide glasses have high resistivity, which limits their use in opto-electronic deices [265, 266]. When electric field is applied to the high resistive materials, the material either switches to sate of low resistance or enters in the regime of current-controlled negative resistance (CCNR). Since the first report on the electrical phenomena by Ovshinsky [267], a lot of work is being carried out using amorphous semiconductors, which includes chalcogenide semiconductors [267], oxides [268], and organic semiconductors [269]. There are two types of switching states: "threshold switching" and 'memory switching'. For threshold switching, the only stable state is the low resistance above some minimum holding Voltage, and "memory switching", occurs at zero bias. There are different methods to achieve the low resistance states including pure electronic [257] and thermal [270] breakdown or via the fusion of polycrystalline conducting filaments [271]. For thin films, the electronic mechanism is applicable, whereas, the electrothermal breakdown mechanism controls the bulk material [272]. Glassy semiconductors have a number of properties important for device application. Their physical properties change with their chemical compositions. A lot of workers investigated the effects of impurity on conductivity and optical properties [273, 274]. There are many factors such as method of synthesis, melt cooling rate, purity of materials and thermal treatment, which affect the electrical conductivity and switching behavior of the chalcogenide material [275–277].

The relation of switching parameters with other properties of chalcogenide glasses has been one of the rigorously focused research field. Several investigators are working towards figuring out the relation between the switching voltages and glass transition temperature [270, 278], crystallization temperature [279], electrical resistivity [280], ambient temperature and conductivity activation energy [281], etc., There are two network topological thresholds, namely the rigidity percolation threshold (RPT) and chemical threshold (CT). Some recent studies are dedicated to understand the effect of these thresholds on the switching behavior of chalcogenide glasses [282]. There are many reports on electrical switching in chalcogenides available in the literature [283]. Atyia and Bekheet [283] fabricated $Se_{85}Te_{10}Bi_5$ films of different thicknesses (126–512 nm) and investigated the I–V characteristics, which revealed that it is typical for the application in memory switches. The switching phenomenon was explained using electrothermal model for the switching mechanism. Bang and Lee [284] investigated on the impurity-doped $Ge_2Sb_2Te_5$ (GST) chalcogenide alloy exhibiting threshold switching properties. The doped GST thin film prepared by the doping of indium (In) and phosphorous (P) into GST was not crystallized, even at post-annealing temperatures. The specific crystallization behavior in the doped GST thin film was attributed to the stabilization of the amorphous phase of GST due to In and P doping. Lee et al. [285] fabricated threshold switching device based on As–Ge–Te–Si. Das et al. [286] studied the electrical switching in a-$Al_{23}Te_{77}$ thin film. They observed the memory type electrical switching in their samples. Madhu et al. [287] prepared the bulk $Ge_7Se_{93-x}Sb_x$ glasses by melt quenching. They further studied the electrical switching in the as-prepared alloy to examine the switching behaviour. Gunti and Asokan [288] have undertaken alternating differential scanning calorimetric (ADSC) and electrical switching studies on $Ge_{20}Se_{80-x}Bi_x$ glasses. They elucidated the role of topological thresholds on thermal properties and electrical switching behavior. Madhu et al. [289] studied electrical switching in bulk $Ge_{10}Se_{90-x}Tl_x$ glasses. They explained the mechanism of switching and the effect of composition and thickness dependence on switching voltages. Thermal studies and electrical-switching behavior of $Sb_xSe_{55-x}Te_{45}$ ($2 \leq x \leq 9$) glasses have been studied by Bhanu Prashanth and Asokan [290]. They reported memory type of electrical switching in $Sb_xSe_{55-x}Te_{45}$ glasses ($2 \leq x \leq 9$), which was explained with the help of polymeric network and high devitrifying ability. Kumar et al. [291] studied the electrical switching, structural, optical and photo-acoustic properties of chalcogenide $GeSe_{1.5}S_{0.5}$ thin films of different prepared by vacuum evaporation technique. The decrease of band gap energy with increase in film thickness was explained using the 'density of states' theory. All thin films samples exhibited memory-type electrical switching behavior. An enhancement in the threshold voltages of $GeSe_{1.5}S_{0.5}$ thin films were observed with increase in film thickness. The thickness dependence of switching voltages was explained by the Joule heating effect. Verma et al. [292] reported that the bulk $Ge_{15}Te_{85-x}In_5Ag_x$ exhibited electrical switching with switching/threshold voltages in the range of 70–120 V for 0.3 mm thickness sample, their samples exhibited threshold or memory behavior depending on the ON state current. In their work, SET-RESET studies were also performed. Raman studies of the samples after the SET and RESET operations revealed that the SET state is a crystalline phase, obtained by thermal annealing and the RESET state is the glassy state, similar to the as-quenched samples.

Although the above-mentioned studies on electrical switching in amorphous semiconductors showed significant results, but the electrical switching in nano-chalcogenides is not studied very well. It is therefore, suggested that the switching in the nanostructures of chalcogenides may be future research direction. The size effect may result in a dramatic change in switching behaviour of these materials and the nanodevices based electrical switching in these nanochalcogenides shall provide the most powerful and valuable alternate.

4.5 Conclusion

During last few years, a lot of emphasis is focused on the production of chalcogenides at nano-scale. Due to unusual physical and chemical phenomena are revealed in these nanochalcogenides, they continue to attract the attention of researchers and engineers as a very large group of interesting solids, thereby opening new ways in science and technology. Nanostructures formation has been explored for many kinds of materials, and this becomes an interesting topic also for chalcogenides. We know that producing atomically flat surfaces in single crystalline materials is possible by manipulating atoms, whereas this is meaningless, in case of glassy systems. The structure of amorphous materials is disordered at the atomic level. Therefore, these nano-materials can easily be tailored and may be produced in greater variety than that of crystalline nanostructures. Due to large optical nonlinearity and short response time observed in these materials, studying nonlinear optical properties of these materials at nanoscale is also interesting. Optical properties of nanocrystals are strongly affected by size reduction, shape and surface characteristics. Therefore, many reports based on effect these parameters on the optical properties of Nanomaterials are available in the literature. From the above literature, it may be suggested that the production of nanostructures of chalcogenides is interesting and results in the improvement in many physical properties. Normally, the bulk chalcogenides show an indirect band gap and they could not become very popular for the devices due to this reason. The observation of direct optical band gap in these nanostructures is one of the interesting phenomena and will open-up new directions for the application of these materials in semiconducting devices. From the above, it is finally concluded that the nanochalcogenides is one of important directions of future research on chalcogenides and the exploration of new chalcogenides at nanoscale is expected to provide a breakthrough in this field.

References

1. B.T. Kolomiets, Vitreous semiconductors (I), (II). in Proceedings of International Conference on Semiconductor Physics, 1960, (Czechoslovak Academy of Science) p. 884, Phys. Stat. Solidi, 1964, 7, p. 359
2. W.E. Spear, Amorphous and liquid semiconductors. Proc. Phys. Soc. (London) **870**, 1139 (1957)
3. J. Tauc, R. Grigorovici, A. Vancu, Optical properties and electronic structure of amorphous germanium. Phys. Stat. Solidi **15**, 627 (1966)
4. S.R. Ovshinisky, Reversible electrical switching phenomena in disordered structures. Phys. Rev. Lett. **21**, 1450 (1968)
5. Y. Shen, X. Wang, A. Xie, L. Huang, J. Zhu, L. Chen, Synthesis of dextran/Se nanocomposites for nanomedicine application. Mater. Chem. Phys. **109**, 534–540 (2008)
6. H.S. Lee, B. Cheong, T.S. Lee, K.S. Lee, W.M. Kim, J.Y. Huh, Optic material for potential application to super-resolution optical data storage. Surf. Coat. Technol. **193**, 335–339 (2005)
7. J. Pinkas, V. Reichlova, R. Zboril, Z. Moravec, P. Bezdicka, Sonochemical synthesis of amorphous nanoscopic iron(III) oxide from Fe(acac)$_3$. Ultrason. Sonochem. **15**, 257–264 (2008)
8. B. Molina Concha, E. De Biasi, R.D. Zysler, Monte Carlo simulation of Fe–Co amorphous nanoparticles magnetization. Phys. B **403**, 390–393 (2008)
9. L.F. Xi, Y.M. Lam, Synthesis and characterization of CdSe nanorods using a novel microemulsion method at moderate temperature. J. Colloid Interface Sci. **316**, 771–778 (2007)
10. M. Rajamathi, R. Seshadri, Oxide and chalcogenide nanoparticles from hydrothermal/solvothermal reactions. Curr. Opin. Solid State Mater. Sci. **6**, 337–345 (2002)
11. D. Nesheva, H. Hofmeister, Z. Levi, Z. Aneva, Nanoparticle layers of CdSe buried in oxide and chalcogenide thin film matrices. Vacuum **65**, 109–113 (2002)
12. R.A. Street, N.F. Mott, States in the gap in glassy semiconductors. Phys. Rev. Iett. **35**, 1293 (1975)
13. N.F. Mott, E.A. Davis, R.A. Street, States in a gap and recombination in amorphous semiconductors. Phil. Mag. **32**, 961 (1975)

14. M. Kastner, D. Adler, H. Fritzsche, Valence-alternation model for localized gap states in lone-pair semiconductors. Phys. Rev. Lett. **37**, 1504 (1976)
15. M.H. Cohen, H. Fritzsche, S.R. Ovshinsky, Simple band model for amorphous semiconducting alloys. Phys. Rev. Lett. **22**, 1065 (1969)
16. N.F. Mott, Introduction to the electron theory of metals. Phil. Mag. **13**, 989 (1966)
17. E.A. Davis, N.F. Mott, Conduction in non-crystalline systems V. conductivity, optical absorption and photoconductivity in amorphous semiconductors. Phil. Mag. **22**, 903 (1970)
18. N.F. Mott, Tetrahedrally-bonded amorphous semiconductors. Phil. Mag. **26**, 505 (1972)
19. A.V. Kastner, H. Fritzsche, Photo-induced metastability in amorphous semiconductors. Phil. Mag. **37**, 199 (1978)
20. M. Pollak, T.H. Geballe, Low-frequency conductivity due to hopping processes in silicon. Phys. Rev. **122**, 1742 (1961)
21. S.R. Elliot, A theory of AC conduction in chalcogenide glasses. Phil. Mag. **36**, 1291 (1977)
22. M. Pollak, Coherence and energy transfer in glasses. Phil. Mag. **23**, 519 (1976)
23. M. Grevers, F. Du Pre, Discuss Faraday Soc. A. **42**, 47 (1946)
24. H. Frohlich, *Theory of dielectrics; dielectric constant and dielectric loss* (Clarendon Press, Oxford, 1949)
25. G.E. Pike, AC conductivity of scandium oxide and a new hopping model for conductivity. Phys. Rev. **8**(6), 1572 (1972)
26. H. Fritzsche, in *Amorphous and Liquid Semiconductors*, ed. by J. Tau (Plenum, London, 1974), p. 232
27. N.F. Mott, Localized states in a pseudogap and near extremities of conduction and valence bands. Phil. Mag. **19**, 3 (1969)
28. N.F. Mott, Introductory talk; conduction in non-crystalline materials. J. Non-Crys. Solids **1** (72), p. 8 (1993)
29. V. Ambegokar, B.I. Helperin, J.S. Langer, Hopping conductivity in disordered systems. Phys. Rev. B, **4**, p. 2162 (1971)
30. A.M. Phak, Electrical properties of thermally evaporated tellurium films. Thin Solid Films **41**, 235 (1977)
31. S.K. Srivastava, Krishna K. Srivastava, Shiveom Srivastava, Narayan P. Srivastava, Optical study of thin film of $Ge_{10}Se_{90-x}bi_x$. Int. J. Emerg. Technol. **3**(2), 4 (2012)
32. J.P. Borgogro, B. Lazarides, E. Pelletier, Automatic determination of the optical constants of inhomogeneous thin films. Appl. Optics. **21**, 4020 (1982)
33. S.V. Babu, M. David. R.C. Patel, Improved hybrid solar cells via in situ UV polymerization. Appl. Optics. **30**(7), p. 839 (1991)
34. F.J. Biatt, *Physics of Electron Conduction in Solids* (McGraw-Hill, New York, 1968)
35. K.L. Chopra, in *Thin Film Phenomena* (Mcgraw-Hill, New York, 1969)
36. K.A. Rubin, M. Chen, Progress and issues of phase-change erasable optical recording media. Thin Solid Films **181**, 129 (1989)
37. J. Tauc, in *Amorphous and Liquid Semiconductors*, ed. by J. Tauc (Plenum Press, New York, 2012) 197, p. 159
38. F. Urbach, The long-wavelength edge of photographic sensitivity and of the electronic absorption of solids. Phys. Rev. **92**, 1324 (1953)
39. T.T. Nang, M. Okuda, T. Matsushita, S. Yokota, A. Suzuki, Electrical and optical properties of gese amorphous thin films. Jpn. J. Appl. Phys. **14**, 849 (1976)
40. M.L. Theye, in *Proceedings 5th International Conference on Amorphous and Liquid Semiconductors*, vol. 1, p. 479 (1973)
41. YuG Vlasov, E.A. Bychkov, Ion-selective chalcogenide glass electrodes. Ion-sel. Electrode Rev. **9**, 5 (1987)
42. Y.G. Vlasov, New solid-state ion-selective electrodes—sensors for chemical analysis of solution. Fres. Z. Anal. Chem. **335**, 92 (1989)
43. Y.G. Vlasov, E.A. Bychkov, A.V. Legin, Chalcogenide glass chemical sensors: research and analytical applications. Talanta **41**, 1059 (1994)

44. J.H. Christensen, P. Clemmesen, G.H. H, K. Liltorp, J. Mortensen, Response characteristics and application of chalcogenide glass Cr(VI) selective electrode. Sens. Actuators B **45**(3), 2393 (1997)
45. R. Todorov, T. Iliev, K. Prtkev, Light-induced changes in the optical properties of thin films of Ge–S–Bi(Tl, In) chalcogenides. J. Non-Crys. Solid **263**, 326 (2003)
46. A.A. Alnajjar, The role of thermal treatment on the optical properties of $Ge_{0.15}Se_{0.85}$ system. Renew. Energy **34**, 71 (2009)
47. M.B. El-Den, N.B. Olsen, I.H. Pedersen, P. Viscor, Dc and ac electrical transport in assete systems. J. Non-Crys. Solids **92**, 20 (1987)
48. S.A. Khan, F.S. Al-Hazmi, A.M. Al-Sanosi, A.S. Faidah, S.J. Yaghmour, A.A. Al-Ghamdi, Effect of Ag incorporation on electrical and optical properties of Se–S chalcogenide thin films. Phys. B **404**, 1415 (2009)
49. Z.H. Khan, M. Zulfequar, M. Ilyas, M. Husain, K.S. Begum, Electrical and thermal properties of a-$(Se_{70}Te_{30})_{100-x}(Se_{98}Bi_2)_x$ alloys. Current Appl. Phys. **2**, 167 (2002)
50. A. Ahmad, S.A. Khan, K. Sinha, L. Kumar, Z.H. Khan, M. Zulfequar, M. Husain, Optical characterization of vacuum evaporated a-$Se_{80}Te_{20-x}$ Cu_x thin films. Vacuum **82**, 608 (2008)
51. E. Abd El-Wahabb, A.M. Farid, Electrical conductivity and optical absorption of $(Ge_2S_3)_1(Sb_2Te_3)_1$ amorphous thin films. J. Alloys Comp. **472**, 352 (2009)
52. Ambika P.B. Barman, An optical study of vacuum evaporated Se_{85-x} Te_{15} Bi_x chalcogenide thin films. Phys. B Condense Matter **405**, 822 (2010)
53. A.M. Salem, Y.A. El-Gendy, E.A. El-Sayad, Optical and electrical properties of thermally evaporated $In_{49}Se_{48}Sn_3$ films. Phys. B **404**, 2425 (2009)
54. S.A. Khan, F.S. Al-Hazmi, S. Al-Heniti, A.S. Faidah, A.A. Al-Ghamdi, Effect of cadmium addition on the optical constants of thermally evaporated amorphous Se–S–Cd thin films. Current Appl. Phys. **10**, 145 (2010)
55. A.S. Khomane, P.P. Hankare, Structural, optical and electrical characterization of chemically deposited cdSe thin films. J. Alloys Comp. **489**, 605 (2010)
56. K.A. Aly, N. Afify, A.M. Aboushly, Incorporation of Bi, Cd and Zn on the optical properties of $Ge_{20}Se_{80}$ thin films. Phys. B Condense Matter. **405**, 1846 (2010)
57. J. Orava, T. Kohoutek, T. Wagner, Z. Cerna, M. Vlcek, L. Benes, B. Frumarova, M. Frumar, Optical and structural properties of Ge–Se bulk glasses and Ag–Ge–Se thin films. J. Non-Crys. Solids **355**, 1951 (2009)
58. V. Takats, P. Nemec, A. Csik, S. Kokenyesi, Photo-and thermally induced interdiffusion in Se/As_2S_3 nanomultilayers prepared by pulsed laser deposition and thermal evaporation. J. Phys. Chem. Solids **68**, 948 (2007)
59. F.A. Al-Agel, S.A. Khan, E.A. Al-Arfaj, F.M. Al-Marzouki, A.A. Al-Ghamdi, Z.H. Khan, M. Zulfequar, Influence of laser-irradiation on structural and optical properties of phase change $Ga_{25}Se_{75-x}Te_x$ thin films. Mater. Lett. **92**, 424 (2013)
60. M.A. Alvi, S.A. Khan, A.A. Al-Ghamdi, Photo-induced effects on structural and optical properties of $Ga_{15}Se_{81}Ag_4$ chalcogenide thin films. J. Luminescence **132**, 1237 (2012)
61. F.A. Al-Agel, Structural and optical properties of Te doped Ge–Se phase-change thin films: a material for optical storage. Mater. Sci. Semicond. Process. **18**, 36 (2014)
62. Z.H. Khan, N. Salah, S. Habib, A.A. Al-Ghamdi, S.A. Khan, Electrical and optical properties of a-Se_xTe_{100-x} thin films. Opt. Laser Technol. **44**, 6 (2012)
63. A.A. Shaheen, M.M.A. Imran, O.A. Lafi, M.I. Awadallah, M.K. Abdullah, Optical properties of a-$Se_{90}In_{10-x}Sn_x$ chalcogenide thin films before and after gamma irradiation. Radiat. Phys. Chem. **79**(9), 923 (2010)
64. F.A. Al-Agel, Effects of annealing temperatures on optical and electrical properties of vacuum evaporated Ga $_{15}Se_{77}In_8$ chalcogenide thin films. Vacuum **85**(9), 892 (2011)
65. A.A. El-Sebaii, S.A. Khan, F.M. Al-Marzouki, A.S. Faidah, A.A. Al-Ghamdi, Role of heat treatment on structural and optical properties of thermally evaporated $Ga_{10}Se_{81}Pb_9$ chalcogenide thin films. J. Luminescence **132**(8), 2082–2087 (2012)

66. A.A. Al-Ghamdi, S.A. Khan, S. Al-Heniti, F.A. Al-Agel, T. Al-Harbi, M. Zulfequar, Effects of laser irradiation on optical properties of amorphous and annealed $Ga_{15}Se_{81}In_4$ and $Ga_{15}Se_{79}In_6$ chalcogenide thin films. J. Alloys Compd. **505**(1), 229 (2010)
67. S.A. Khan, J.K. Lal, A.A. Al-Ghamdi, Thermal annealing effect of on optical constants of vacuum evaporated $Se_{75}S_{25-x}Cd_x$ chalcogenide thin films, Opt. Laser Technol. **42**(5), p. 839 (2010)
68. F.S. Al-Hazmi, Effect of annealing on optical constants of $Se_{75}S_{25-x}Cd_x$ chalcogenide thin films. Phys. B **404**(8–11), 1354 (2009)
69. R. Chauhan, A.K. Srivastava, A. Tripathi, K.K. Srivastava, Photo-induced optical changes in $Ge_xAs_{40}Se_{60-x}$ thin films. Prog. Nat. Sci. Mater. Int. **20**, 54 (2010)
70. A.M. Farid, I.K. El-Zawawi, A.H. Ammar, Compositional effects on the optical properties of $Ge_xSb_{40-x}Se_{60}$ thin films. Vacuum **86**(9), 1255 (2012)
71. R. Chauhan, A.K. Srivastava, A. Tripathi, K.K. Srivastava, Linear and nonlinear optical changes in amorphous As_2Se_3 thin film upon UV exposure. Prog. Nat. Sci. Matet. Int. **21**(3), 205 (2011)
72. M.M.A. Imran, O.A. Lafi, M. Abu-Samak, Effect of thermal annealing on some electrical properties and optical band gap of vacuum evaporated $Se_{65}Ga_{30}In_5$ thin films. Vacuum **86** (10), 1589 (2012)
73. M. Mishra, R. Cauhan, A. Katiyar, K.K. Srivastava, Optical properties of amorphous thin film of Se–Te–Ag system prepared by using thermal evaporation technique. Prog. Nat. Sci. Mater. Int. **21**(1), 36–39 (2011)
74. M. Abdel Rafea, H. Farid, Mater. Chem. Phys. **113**, 268 (2009)
75. A.H. Ammar, N.M. Abdel-Moniem, M. Farag, Influence of indium content on the optical, electrical and crystallization kinetics of $Se_{100-x}In_x$ thin films deposited by flash evaporation technique. Phys. B **407**(3), 356 (2012)
76. A.K. Diab, M.M. Wakkad, EKh Shokr, W.S. Mohamed, Structural and optical properties of $In_{35}Sb_{45}Se_{20-x}Te_x$ phase-change thin films. J. Phys. Chem. Solids **71**(9), 1381 (2010)
77. A.F. Qasrawi, Temperature effects on the optoelectronic properties of agin $_5S_8$ thin films. Thin Solid Films **519**(11), p. 3768 (2011)
78. K.A. Aly, N. Afify, A.M. Abousehlly, A.M. Abd Elnaeim, Optical band gap and refractive index dispersion parameters of In–Se–Te amorphous films. J. Non-Crys. Solids **357**(10), 2029 (2011)
79. M.A. Alvi, S.A. Khan, A.A. Al-Ghamdi, Photo-induced effects on electrical properties of $Ga_{15}Se_{81}Ag_4$ chalcogenide thin films. Mater. Lett. **66**(1), 273 (2012)
80. Zong-Hong Lin and R. Chris, "Evidence on the size-dependent absorption spectral evolution of selenium nanoparticles" Maters. Chem. Phys., 2005, 92(2–3), p.591
81. Xueyun Gao, Tao Gao, Lide Zhang, Solution-solid growth of α-monoclinic selenium nanowires at room temperature. J. Mater. Chem. **13**, 6 (2003)
82. X.L. Liu, Y.J. Zhu, A precursor nanowire templated route to CdS nanowires. Mater. Lett. **63**, 1085 (2009)
83. M.F. Kotkata, A.E. Masoud, M.B. Mohamed, E.A. Mahmoud, Synthesis and structural characterization of CdS nanoparticles. Phys. E **41**, 640–645 (2009)
84. S. Wageh, M.H. Badr, M.H. Khalil, A.S. Eid, Strong confinement of PbSe nanocrystals in phosphate glass. Phys. E **41**, 1157–1163 (2009)
85. R. Todorov, A. Paneva, K. Petkov, Optical characterization of thin chalcogenide films by multiple-angle-of-incidence ellipsometry. Thin Solid Films **518**(12), 3280 (2010)
86. Z. Chaia, Z. Penga, C. Wanga, H. Zhang, Synthesis of polycrystalline nanotubular Bi_2Te_3. Mater. Chem. Phys. **113**, 664 (2009)
87. S. Li, H.Z. Wang, W.W. Xu, H. Si, X. Tao, S. Lou, Z. Du, L.S. Li, Synthesis and assembly of monodisperse spherical Cu_2S nanocrystals. J. Colloid. Interface Sci. **330**, 483 (2009)
88. J.K. Dongre, V. Nogriya, M. Ramrakhiani, Structural, optical and photoelectrochemical characterization of CdS nanowire synthesized by chemical bath deposition and wet chemical etching. Appl. Surf. Sci. **255**, 6115 (2009)

89. G.L. Tan, J.H. Du, Q.J. Zhang, Structural evolution and optical properties of CdSe nanocrystals prepared by mechanical alloying. J. Alloys Compd. **468**, 421 (2009)
90. Y. Yang, Y. Chai, D. Fanglin, Controllable synthesis of flower-like $Cd_{1-x}Zn_xSe$ microstructures from the self-prepared precursor. J. Alloys Compd. **478**, 513 (2009)
91. P.P. Ingole, P.M. Joshi, S.K. Haram, Room temperature synthesis of 1-hexanethiolate capped $Cu_{2-x}Se$ quantum dots in Triton $_{X\text{-}100}$ water-in-oil microemulsions. Colloids Surf. A Physicochem. Eng. Aspects **337**, 136 (2009)
92. S. Lee, S. Hong, B. Park, S.R. Paik, S. Jung, Agarose and gellan as morphology-directing agents for the preparation of selenium nanowires in water. Carbohydr. Res. **344**, 260 (2009)
93. Q. Han, L. Chen, W. Zhu, M. Wang, X. Wang, X. Yang, L. Lu, Synthesis of Sb_2S_3 peanut-shaped superstructures. Mater. Lett. **63**, 1030 (2009)
94. K. Liu, H. Liu, J. Wang, L. Feng, Synthesis and characterization of $SnSe_2$ hexagonal nanoflakes. Mater. Lett. **63**, 512 (2009)
95. Z. Li, X. Tao, W. Zhishen, P. Zhang, Z. Zhang, Preparation of In_2S_3 nanopraricle by ultrasonic dispersion and its tribology property. Ultrason. Sonochem. **16**, 221 (2009)
96. Y. Li, Y. Zhu, C. Li, X. Yang, C. Li, Synthesis of ZnS nanoparticles into the pore of mesoporous silica spheres. Mater. Lett. **63**, p. 1068 (2009)
97. Z.H. Khan, S.A. Khan, N. Salah, S. Habib, S.M. Abdallah El-Hamidy, A.A. Al-Ghamdi, Effect of composition on electrical and optical properties of thin films of amorphous Ga_xSe_{100-x} Nanorods. Nano Res. Letts. **5**, 1512 (2010)
98. Z.H. Khan, A.A. Al-Ghamdi, S.A. Khan, S. Habib, N. Salah, Morphology and optical properties of thin films of Ga_xSe_{100-x} nanoparticles. Nanosci. Naotech. Lett. **3**, 1 (2010)
99. Z.H. Khan, S.A. Khan, N. Salah, A.A. Al-Ghamdi, S. Habib, Electrical properties of thin films of a-Ga_xTe_{100-x} composed of nanoparticles. Phil. Mag. Lett. **93**(7), 207 (2010)
100. Z.H. Khan, S.A. Khan, N. Salah, A.A. Al-Ghamdi, S. Habib, Electrical transport properties of a-$Se_{87}Te_{13}$ nanorods. J. Expt. Nanosci. **6**, 337 (2010)
101. S.A. Khan, F.A. Al-Agel, A.S. Faidah, S.J. Yaghmour, A.A. Al-Ghamdi, Characterization of $Se_{88}Te_{12}$ nanostructured chalcogenide prepared by ball milling. Mater. Lett. **64**, 1391 (2010)
102. A.A. Al-Ghamdi, S.A. Khan, A. Nagat, M.S. Abd El-Sadek, Synthesis and Optical characterization of nanocrystalline cdte thin films. Opt. Laser Technol. **42**, 1181 (2010)
103. S.A. Khan, F.A. Al-Agel, A.A. Al-Ghamdi, Optical characterization of nanocrystalline and chalcogenides. Superlattices Microstruct. **47**, 695 (2010)
104. N. Salah, S.S. Habib, A. Memic, N.D. Alharbi, S.S. Babkair, Z.H. Khan, Syntheses and characterization of thin films of $Te_{94}Se_6$ nanoparticles for semiconducting and optical devices. Thin Solid Films **531**, 70 (2013)
105. F.A. Agel, Optical and structural properties of a-Se_xTe_{100-x} aligned nanorods. Nanoscale Res. Lett. **8**, 520 (2013)
106. M.A. Alvi, Z.H. Khan, Synthesis and characterization of nanoparticle thin films of a-$(PbSe)_{100-x}Cd_x$ lead chalcogenides. Nanoscale Res. Lett. **8**, 148 (2013)
107. N. Salah, S.S. Habib, Z.H. Khan, E. Alarfaj, S.A. Khan, Synthesis and characterization of $Se_{35}Te_{65-x}ge_x$ nanoparticle films and their optical properties. J. Nanomater. **2012**, p. 393084 (2012)
108. Z.H. Khan, M. Husain, Electrical and optical properties of thin film of a-$Se_{70}Te_{30}$ nanorods. J. Alloy. Compd. **486**, 774 (2009)
109. R.M. Mehra, P.C. Mathur, Analysis of single polaron hopping in ac conductivity of amorphous $Ge_{20}Sb_xSe_{80-x}$ glasses. Thin Solid Films **170**, 15 (1989)
110. N.F. Mott, E.A. Davis, *Electronic Processes in Non-crystalline Materials* (Clarendon, Oxford, 1979), p. 428
111. E.A. Davis, *Electronic and Structural Properties of Amorphous Semiconductors* (Academic Press, London, 1973), p. 425
112. J. Nishi, S. Morimoto, I. Ingawa, R. Iizuka, T. Yamashita, Recent advances and trends in chalcogenide glass fiber technology: a review. J. Non-Crys. Solids **140**, 199 (1992)
113. J.A. Savage, *Infrared Opical Materials and their Anti-reflection Coatings* (Adam Hilger, Bristol, 1985)

114. A.B. Seddon, M.A. Hemingway, Thermal characterisation of infrared-transmitting gesi glasses. J. Non-Crys. Solids **161**, 323 (1993)
115. W. Leung, N.W. Cheung, Studies of Ag photodoping in Ge_xSe_{1-x} glass using microlithography techniques. Appl. Phys. Lett. **46**, 481 (1985)
116. J. Feinleib, J.P. De Neufville, S.C. Moss, S.R. Ovshinsky, Rapid reversible light-induced crystallization of amorphous semiconductors. Appl. Phys. Lett. **18**, 254 (1971)
117. J.P. Deneufville, S.C. Moss, S.R. Ovshinsky, Photostructural transformations in amorphous As_2Se_3 and As_2S_3 films. J. Non-Crys. Solids **13**, 191 (1973)
118. B. Singh, S. Rajagopalan, P.K. Bhatt, D.K. Pandey, K.L. Chopra, Photocontraction effect in amorphous in $Se_{1-X}Ge_x$ films. Solid State Commun. **29**, 167 (1979)
119. K.L. Chopra, S. Harshvardhan, S. Rajagopalan, L.K. Malhotra, On the origin of photocontraction effect in amorphous chalcogenide films. Solid State Commun. **40**, 387 (1981)
120. T. Okabe, S. Endu, S. Saito, Simultaneous crystallization of both elements in amorphous GeSb and GeAl eutectic alloys. J. Non-Cryst. Solids **222**, 117 (1990)
121. J.M. Del Pozo, M.P. Herrero, Crystallization behavior of amorphous $Ge_{(1-x)}Sb_{1-x}$ thin-films. J. Non-Crys. Solids **185**, 183 (1995)
122. T. Rajagopalen, G.B. Reddy, Effect of annealing rate on the crystallization process in $Ge_5Bi_{18}Se_{77}$ films. Thin Solid Films **353**, 254 (1999)
123. J.M. Del Pozo, L. Diaz, Optical study of $Ge_{(1-x)}Sb_{1-x}$ crystallization. J. Non-Crys. Solids **243**, 45 (1999)
124. A.H. Moharram, M.S. Rasheedy, A simple method for crystallization kinetics determination and its application to $Ge_{10}Te_{35}As_{55}$ glass. Phys. Stat. Sol. (A) **169**, 33 (1998)
125. M. Chen, K.A. Rubin, R.W. Barton, Compound materials for reversible, phase-change optical data storage. Appl. Phys. Lett. **9**, 502 (1986)
126. Z.H. Khan, M. Zulfequar, M. Husain, Effect on Sb on Transport Properties of a-$Se_{80-x}Ga_{20}Sb_x$ thin films. Jpn. J. Appl. Phys. **37**, 23 (1998)
127. M. Abkowitz, G.M.T. Foley, J.M. Morkovics, A.C. Palumbo, Etastable photoenhanced thermal generation in a-SeTe alloys. AIP Conf. Proc. **120**, 117 (1984)
128. N. Afify, Calorimetric study on the crystallization of a $Se_{0.8}Te_{0.2}$ chalcogenide glass. J. Non-Crys. Solids **142**, 247 (1992)
129. N. Afify, Kinetics study of non-isothermal crystallization in $Se_{0.7}Te_{0.3}$ chalcogenide glass. J. Non-Crys. Solids **136**, 67 (1991)
130. A.K. Agnihotri, A. Kumar, A.N. Nigam, The X-ray K-absorption studies in glassy $Se_{80}Te_{20}$ and $Se_{80}Te_{10}Sb_{10}$. J. Non-Crys. Solids **101**, 127 (1988)
131. V. Damodara Das, P. Jansi Lakshmi, "Explosive" crystallization of amorphous $Se_{80}Te_{20}$ alloy thin films and oriented growth of crystallites. Phys. Rev. B **37**, 720 (1988)
132. S. Mahadevan, A. Giridhar, A.K. Singh, Calorimetric measurements on As–Sb–Se glasses. J. Non-Cryst. Solids **88**, 11 (1986)
133. H.E. Kissinger, Reaction kinetics in differential thermal analysis. Anal. Chem. **29**, 1702 (1957)
134. M.A. Abdel-Rahim, M. Abu El-Oyoun, A.A. Abu-Sehly, Calorimetric study of the chalcogenide $Se_{72.5}Te_{20}Sb_{7.5}$ glass. J. Phys. D Appl. Phys. **34**, 2541 (2001)
135. J. Vazquez, P.L. Lopez-Alemany, P. Villares, R. Jimenez-Garay, Generalization of the Avrami equation for the analysis of non-isothermal transformation kinetics. Application to the crystallization of the $Cu_{0.20}As_{0.30}Se_{0.50}$ alloy. J. Phys. Chem. Solids **61**, 493 (2000)
136. A. El-Salam, M. Abousehly, Activation energy of $Se_2Ge_{0.2}Sb_{0.8}$ chalcogenide glass by differential scanning calorimetry. J Thermal Anal **46**, 177–186 (1996)
137. M.J. Strink, A.M. Zahra, Determination of the transformation exponent *s* from experiments at constant heating rate. Thermochim. Acta **298**, 179 (1997)
138. W.A. Johnson, K.F. Mehl, Crystallization kinetics of the chalcogenide $Bi_{10}Se_{90}$ glass. Trans. Inst. Mining Met. Eng. **135**, 315 (1981)
139. M. Avrami, Interfacial electrochemistry: theory: experiment, and applications. J. Chem. Phys. **7**, 103 (1939)

140. M. Avrami, Kinetics of phase change. II transformation-time relations for random distribution of nuclei. J. Chem. Phys. **8**, 212 (1940)
141. M. Avrami, Kinetics of phase change. III. granulation, phase change, and microstructure. J Chem. Phys **9**, 177 (1941)
142. F. Liu, F. Sommer, E.J. Mittemeijerr, An analytical model for isothermal and isochronal transformation kinetics. J. Mater. Sci. **39**, 1621 (2004)
143. A.A. Abou-Sehly, S.N. Alamri, A.A. Joraid, Measurements of DSC isothermal crystallization kinetics in amorphous selenium bulk samples. J. Alloys Compd. **476**, 348 (2009)
144. S. Vyazovkin, On the phenomenon of variable activation energy for condensed phase reactions. New J. Chem. **24**, 913 (2000)
145. K. Matusita, T. Konatsu, R. Yokota, Kinetics of non-isothermal crystallization process and activation energy for crystal growth in amorphous materials. J. Mater. Sci. **19**, 291 (1984)
146. T. Ozawa, A new method of analyzing thermogravimetric data. Bull. Chem. Soc. Jpn **38**, 188 (1965)
147. C.T. Moynihan, A.J. Easteal, J. Wilder, J. Tucker, Dependence of the fictive temperature of glass on cooling rate. J. Phys. Chem. **78**, 2673 (1974)
148. F.A. Al-Agel, E.A. Al-Arfaj, F.M. Al-Marzouki, Shamshad A. Khan, A.A. Al-Ghamdi, Study of phase separation in $Ga_{25}Se_{75-x}Te_x$ chalcogenide thin films. Prog. Nat. Sci. Mater. Int. **23**, 139 (2013)
149. F.A. Al-Agel, E.A. Al-Arfaj, F.M. Al-Marzouki, S.A. Khan, Z.H. Khan, A.A. Al-Ghamdi, Phase transformation kinetics and optical properties of Ga–Se–Sb phase-change thin films. Mater. Sci. Semicond. Process. **16**, 884 (2013)
150. F.A. Al-Agel, Crystallization kinetics and effect of thermal annealing on optical constants in a-$Ge_{25}Se_{75-x}te_x$ glasses. J. Alloys Compd. **568**, 92 (2013)
151. Zishan H. Khan, N. Salah, Sami Habib, S.A. Khan, Kinetics of non-isothermal crystallization in $Ga_{15}Se_{76}Pb_9$ chalcogenide glasses by differential scanning calorimeter (DSC). Chalcogenide Lett. **8**, 615 (2011)
152. Shamshad A. Khan, J.K. Lal, F.A. Al-Agel, M.A. Alvi, Non-isothermal crystallization in Ga–Se–Ag chalcogenide glass by differential scanning calorimetry. J. Alloy. Compd. **554**, 227 (2013)
153. Z.H. Khan, S.A. Khan, M.A. Alvi, Study of glass transition and crystallization behaviour in $Ga_{15}Se_{80-x}pb_x$ ($0 \leq x \leq 6$). Acta Phys. Pol. A, **201**, 123(1)
154. A.S. Farid, H.E. Atiya, Glass transition and crystallization study of Te additive Se Bi chalcogenide glass. J. Non-Cryst. Solids **408**, 123 (2015)
155. R. Svoboda, P. Bezdička, J. Gutwirth, J. Málek, Crystallization processes in $Ge_2Sb_2Se_4Te$ glass. Maters. Res. Bull. **61**, 207 (2015)
156. A. Dahshan, K.A. Aly, Characterization of new quaternary $Ge_{20}Se_{60}Sb_{20-x}Ag_x$ ($0 \leq x \leq 20$ at.%) glasses. J. Non-Cryst. Solids **408**, 62 (2015)
157. M.I. Abd-Elrahman, R.M. Khafagy, S.A. Zaki, M.M. Hafiz, Characterization of optical constants of $Se_{30}Te_{70}$ thin film: effect of the thickness. Thermochim. Acta **575**, 285 (2014)
158. O.A. Lafi, M.M.A. Imran, N.I. Abu-Shaweesh, F.M. Al-Kurdi, I.K. Khatatbeh, Effect of chemical ordering on the crystallization behavior of $Se_{90}Te_{10-x}Sn_x$ (x = 2, 4, 6, and 8) chalcogenide glasses. J. Phys. Chem. Solids **75**(6), 790 (2014)
159. B. Bhoi, V. Srinivas, V. Singh, Evolution of microstructure and magnetic properties of nanocrystalline $Fe_{70-x}Cu_xCo_{30}$ alloy prepared by mechanical alloying. J. Alloy. Compd. **496**, 423 (2010)
160. F.A. Al-Agel, Z.H. Khan, F.M. Al-Marzouki, S.A. Khan, A.A. Al-Ghamdi, Kinetics of phase transformation in nanostructured GaSeTe glasses. J. Nanosci. Nanotechnol. **13**, 2001 (2013)
161. M. Abu El-Oyoun, DSC studies on the transformation kinetics of two separated crystallization peaks of $Si_{12.5}Te_{87.5}$ chalcogenide glass: an application of the theoretical method developed and isoconversional method. Mater. Chem. Phys. **131**(1–2), 495 (2011)
162. B.S. Patial, N. Thakur, S.K. Tripathi, On the crystallization kinetics of In additive Se–Te chalcogenide glasses. Thermochim. Acta **513**(1–2), 1 (2011)

163. S. Kumar, K. Singh, The effect of Indium additives on crystallization kinetics and thermal stability of Se–Te–Sn chalcogenide glasses. Phys. B **406**(8), 1519 (2011)
164. A.F. Kozmidis-Petrović, S.R. Lukić, G.R. Štrbac, Calculation of non-isothermal crystallization parameters for the Cu_{15} $(As_2Se_3)_{85}$ metal-chalcogenide glass. J. Non-Cryst. Solids **356**(41–42), 2151 (2010)
165. O.A. Lafi, Glass transition kinetics and crystallization mechanism in $Se_{90}Cd_8Bi_2$ and $Se_{90}Cd_6Bi_4$ chalcogenide glasses. J. Alloy. Compd. **519**, 123 (2012)
166. M. Abu El-Oyoun, Evaluation of the transformation kinetics of $Ga_{7.5}Se_{92.5}$ chalcogenide glass using the theoretical method developed and isoconversional analyses. J. Alloy. Compd. **507**(1), 6 (2010)
167. F.A. Al-Agel, S.A. Khan, E.A. Al-Arfaj, A.A. Al-Ghamdi, Kinetics of non-isothermal crystallization and glass transition phenomena in $Ga_{10}Se_{87}Pb_3$ and $Ga_{10}Se_{84}Pb_6$ chalcogenide glasses by DSC. J. Non-Cryst. Solids **358**, 564–570 (2012)
168. A.M. Abd Elnaeim, K.A. Aly, N. Afify, A.M. Abousehlly, Glass transition and crystallization kinetics of $In_x(Se_{0.75}Te_{0.25})_{100-x}$ chalcogenide glasses. J. Alloy. Compd. **491**(1–2), 85 (2010)
169. M.M.A. Imran, Thermal characterization of $Se_{85-x}Sb_{15}Sn_x$ ($10 \leq x \leq 13$) chalcogenide glasses. Phys. B **406**(22), 4289 (2011)
170. M.M. Abd El-Raheem, H.M. Ali, Crystallization kinetics determination of $Pb_{15}Ge_{27}Se_{58}$ chalcogenide glass by using the various heating rates (VHR) method. J. Non-Cryst. Solids **356**(2), 77 (2010)
171. C.M. Muiva, S.T. Sathiaraj, J.M. Mwabora, Crystallization kinetics, glass forming ability and thermal stability in some glassy $Se_{100-x}In_x$ chalcogenide alloys. J. Non-Cryst. Solids **357**(22–23), 3726 (2011)
172. A.A. Al-Ghamdi, M.A. Alvi, S.A. Khan, Non-isothermal crystallization kinetic study on $Ga_{15}Se_{85-x}Ag_x$ chalcogenide glasses by using differential scanning calorimetry. J. Alloy. Compd. **509**(5), 2087 (2011)
173. A.S. Soltan, A study of DSC non-isothermal pre-crystallization kinetics of $Pb_{10}Se_{90}$ glass using isoconversional kinetic analysis. Phys. B **405**(3), 965 (2010)
174. F. Abdel-Wahab, Observation of phase separation in some Se–Te–Sn chalcogenide glasses. Physica B **406**(5), 1053 (2011)
175. M. Abu El-Oyoun, The effect of addition of gallium on the thermal stability and crystallization kinetic parameters of Ga_xSe_{100-x} glass system. J. Non-Cryst. Solids **357**(7), 1729 (2011)
176. M. Shapaan, E.R. Shaaban, Studying the crystallization behavior of the $Se_{85}S_{10}Sb_5$ chalcogenide semiconducting glass by DSC and X-ray diffraction. J. Phys. Chem. Solids **71** (9), 1301 (2010)
177. S. Kumar, K. Singh, Glass transition, thermal stability and glass-forming tendency of $Se_{90-x}Te_5Sn_5In_x$ multicomponent chalcogenide glasses. Thermochim. Acta **528**, 32 (2012)
178. M. Abu El-Oyoun, The effect of addition of gallium on the thermal stability and crystallization kinetic parameters of Ga_xSe_{100-x} glass system. Phys. B **406**, 125 (2011)
179. A.A. Abu-Sehly, Kinetics of the glass transition in $As_{22}S_{78}$ chalcogenide glass: activation energy and fragility index. Mater. Chem. Phys. **125**(3), 672 (2011)
180. M.M.A. Imran, Crystallization kinetics, glass transition kinetics, and thermal stability of $Se_{70-x}Ga_{30}In_x$ (x = 5, 10, 15, and 20) semiconducting glasses. Phys. B **406**(3), 482 (2011)
181. S. Kumar, K. Singh, Composition dependence UV-visible and MID-FTIR properties of $Se_{98-x}Zn_2In_x$ (X = 0, 2, 4, 6 and 10) chalcogenide glasses. Phys. B **405**(15), 3135 (2010)
182. J.C. Qiao, J.M. Pelletier, Isochronal and isothermal crystallization in $Zr_{55}Cu_{30}Ni_5Al_{10}$ bulk metallic glass. Trans. Nonferrous Met. Soc. China **22**, 577 (2012)
183. T. Zhang, Z. Song, B. Liu, S.F. Bomy, Investigation of phase change $Si_2Sb_2Te_5$ material and its application in chalcogenide random access memory. Chem. Solid-State Electron. **51**(6), 950 (2007)
184. Y. Jialin, B. Liu, T. Zhang, Z. Song, S. Feng, B. Chen, Effects of Ge doping on the properties of Sb_2Te_3 phase-change thin films. Appl. Surf. Sci. **253**(14), 6125 (2007)

185. A.V. Kolobov, P. Fons, M. Krbal, J. Tominaga, A thermal component of amorphisation in phase-change alloys and chalcogenide glasses. J. Non-Cryst. Solids **17**, 358 (2012). doi:10.1016/j.jnoncrysol.2011.10.024
186. M.H.R. Lankhorst, Modelling glass transition temperatures of chalcogenide glasses. Applied to phase-change optical recording materials. J. Non-Cryst. Solids **297**(2–3), 210 (2002)
187. A. Abrutis, V. Plausinaitiene, M. Skapas, C. Wiemer, O. Salicio, M. Longo, A. Pirovano, J. Siegel, W. Gawelda, S. Rushworth, C. Giesen, Chemical vapor deposition of chalcogenide materials for phase-change memories. Microelectron. Eng. **85**(12), 2338 (2008)
188. K. Takata, H. Maekawa, H. Endo, Thermal strain imaging of chalcogenide in a phase change memory. Cur. App. Phys. **11**(3), 731 (2011)
189. N. Mehta, A. Kumar, Observation of phase separation in some Se–Te–Ag chalcogenide glasses. Mater. Chem. Phys. **96**(1), 73 (2006)
190. N.S. Saxena, Phase transformation kinetics and related thermodynamic and optical properties in chalcogenide glasses. J. Non-Cryst. Solids **345–346**, 161 (2004)
191. M.M. Hafiz, O. El-Shazly, N. Kinawy, Reversible phase change in Bi_XSe_{100-X} chalcogenide thin films for using as optical recording medium. Appl. Surf. Sci. **171**(3–4), 231 (2001)
192. M.N. Kozicki, M. Mitkova, J. Zhu, M. Park, Nanoscale phase separation in Ag–Ge–Se glasses. Eng. **63**(1–3), 155 (2002)
193. S. Kumar, D. Singh, S. Sandhu, R. Thangaraj, Characterization of phase transition in silver photo-diffused $Ge_2Sb_2Te_5$ thin films. Vacuum **86**(10), 1443 (2012)
194. A.L. Lacaita, D. Ielmini, D. Mantegazza, Status and challenges of phase change memory modeling. Solid-State Elect. **52**, 1443 (2008)
195. H. Fritzsche, Why are chalcogenide glasses the materials of choice for Ovonic switching devices? J. Phys. Chem. Solids **68**, 878 (2007)
196. Z.H. Khan, Glass transition kinetics in ball milled amorphous Ga_xTe_{100-x} nanoparticles. J. Non-Cryst. Solids **380**, 109–113 (2013)
197. Z.H. Khan, A.A. Al-Ghamdi, F.A. Al-Agel, Crystallization kinetics in as-synthesis high yield of a-$Se_{100-x}Te_x$ nanorods. Mater. Chem. Phys. **134**, 260 (2012)
198. Z.H. Khan, Non-isothermal crystallization in amorphous $GaxSe_{100-x}$ nanorods. Jpn. J. Appl. Phys. **50**, 105603 (2011)
199. Z.H. Khan, Glass transition kinetics of a-Se_xTe_{100-x} nanoparticles. Sci. Adv. Maters. **4**, 1 (2012)
200. S. Lay, Tech. Dig. Int. Electron Devices Meet p. 255 (2003)
201. D.J. Gravesteijn, C.J. van der Poel, P.M.L.O. Scholte, C.M.J. van Uijen, Phase change optical recording. Philips Tech. Rev. **44**, 250 (1989)
202. N. Yamada, Erasable phase-change optical materials. MRS Bull. **21**, 48 (1996)
203. H.J. Borg, R. Van Woudenberg, Trends in optical recording. J. Magn. Magn. Mater. **193**, 521 (1999)
204. S. Ovshinsky, Amorphous materials—the key to new devices. IEEE Proc. CAS **1**, 33 (1998)
205. P. Boolchand, D.G. Georgiev, T. Qu, F. Wang, L. Chai, Nanoscale phase separation effects near r = 2.4 and 2.67, and rigidity transitions in chalcogenide glasses. C.R. Chem. **5**, 713 (2002)
206. N. Yamada, E. Ohno, K. Nishiuchi, M. Takao, Rapid-phase transitions of GeTe–Sb_2, Te_3, pseudobinary amorphous thin films for an optical disk memory. J. Appl. Phys. **69**, 2849 (1991)
207. J.H. Coombs, A.P.J.M. Jongenelis, W. Van Es-Spiekman, B.A.J. Jacobs, Laser-induced crystallization phenomena in GeTe-based alloys. … of nucleation and growth. J. Appl. Phys. **78**, 4906 (1995)
208. C. Peng, L. Cheng, M. Mansuripur, Experimental and theoretical investigations of laser-induced crystallization and amorphization in phase-change optical recording media. J. Appl. Phys. **82**, 4183 (1997)
209. K. Nakayama, K. Kojima, F. Hayakawa, Y. Imai, A. Kitagawa, M. Suzuki, Reversible and irreversible … Based on phase transitions in chalcogenide thin films. Jpn. J. Appl. Phys **39**, 6157 (2000)

210. S. Lai, T. Lowrey, Tech. Dig. Int. Electron Dev. Meeting **1**, 803 (2001)
211. S.M. Yoon, N.Y. Lee, S.O. Ryu, K.J. Chio, Y.S. Park, S.Y. Lee, B.G. Yu, M.J. Kang, S.Y. Chio, M. Wuttig, Se-based phase-change memory device with lower power and higher speed operations. IEEE Electron Device Lett. **27**, 445 (2006)
212. P. Arun, A.G. Vedeshwar, Mater. Res. Bull. **34**, 203 (1999)
213. H. Tashiro, M. Harigaya, Y. Kageyama, K. Ito, M. Shinotsuka, K. Tani, A. Watada, N. Yiwata, Y. Nakata, S. Emura, Structural Analysis of Ag–In–Sb–Te phase-change material. Jpn. J. Appl. Phys. **41**, 3758 (2002)
214. J. Li, F. Gan, Optical properties of $Ag_8In_{14}Sb_{55}Te_{23}$ phase-change films. Thin Solid Films **40**, 232 (2002)
215. G.A. Prinz, Magnetoelectronics. Science **282**, 1660 (1998)
216. G.R. Fox, F. Chu, T. Davenport, Current and future ferroelectric nonvolatile memory technology. J. Vac. Sci. Technol., B **19**, 1967 (2001)
217. F. Pellizzer et al., Phase-change memory technology for embedded applications. in Proceedings of the IEEE Symposium on VLSI Technology, Digest of Technical Papers (unpublished), vol. 3.1, p. 18 (2004)
218. S.H. Lee et al., in Proceeding of the IEEE Symposium on VLSI Technology, Digest of Technical Papers (unpublished), p. 20 (2004)
219. K.L. Chopra, Dielectric properties of zns films. J. Appl. Phys. **36**, 184 (1965)
220. A. Beck, J.G. Bednorz, Ch. Gerber, C. Rossel, D. Widmer, Reproducible switching effect in thin oxide films for memory applications. Appl. Phys. Lett. **77**, 139 (2000)
221. A. Sawa, T. Fujii, M. Kawasaki, Y. Tokura, Hysteretic current-voltage characteristics and resistance switching at a rectifying $Ti/Pr_{0.7}Ca_{0.3}MnO_3$ interface. Appl. Phys. Lett. **85**, 4073 (2004)
222. X. Chen, N. Wu, J. Strozier, A. Ignatiev, Spatially extended nature of resistance switching in perovskite oxide thin films. Appl. Phys. Lett. **89**, 063507 (2006)
223. A. Ignatiev, N.J. Wu, X. Chen, S.Q. Liu, C. Papagianni, J. Strozier, Resistance switching in perovskite thin films. Phys. Status Solidi B **243**, 2089 (2006)
224. K. Szot, R. Dittmann, W. Speier, R. Waser, Nanoscale resistive switching in $SrTio_3$ thin films. Phys. Status Solidi (RRL) **1**, 86 (2007)
225. Y. Hirose, H. Hirose, Polarity-dependent memory switching and behavior of Ag dendrite in Ag-photodoped amorphous As_2S_3 films. J. Appl. Phys. **47**, 2767 (1976)
226. K. Terabe, T. Hasegawa, T. Nakayama, M. Aono, Quantized conductance atomic switch. Nature **433**, 47 (2005)
227. T. Sakamoto, NEC J. Adv. Tech. **2**, 260 (2005)
228. M.N. Kozicki, M. Park, M. Mitkova, Nanoscale memory elements based on solid-state electrolytes. IEEE Trans. Nanotechnol. **4**, 331 (2005)
229. C.J. Kim, S.G. Yoon, J. Vac. Sci. Technol., B **24**, 721 (2006)
230. Y. Yin, H. Sone, S. Hosaka, Memory effect in metal–chalcogenide–metal structures for ultrahigh-density nonvolatile memories. Jpn. J. Appl. Phys. **1**(45), 4951 (2006)
231. J.F. Gibbons, W.E. Beadle, Switching properties of thin Nio films. Solid-State Electron. **7**, 785 (1964)
232. W.R. Hiatt, T.W. Hickmott, Appl. Phys. Lett. **6**, 106 (1965)
233. F. Argall, Switching phenomena in titanium oxide thin films. Solid-State Electron. **11**, 535 (1968)
234. T.W. Hickmott, J. Vac. Sci. Technol. **6**, 828 (1969)
235. J.C. Bruyere, B.K. Chakraverty, Switching and negative resistance in thin films of nickel oxide. Appl. Phys. Lett. **16**, 40 (1970)
236. Y. Ogimoto, Y. Tamai, M. Kawasaki, Y. Tokura, Resistance switching memory device with a nanoscale confined current path. Appl. Phys. Lett. **90**, 143515 (2007)
237. Y. Watanabe, J.G. Bednorz, A. Bietsch, Ch. Gerber, D. Widmer, A. Beck, Appl. Phys. Lett. **78**, 3738 (2001)
238. C.J. Kim, I.W. Chen, Resistance switching of Al/(Pr, Ca)Mno_3 thin films. Jpn. J. Appl. Phys. Part 2 **44**, 525 (2005)

239. K. Szot, W. Speier, G. Bihlmayer, R. Waser, Switching the electrical resistance of individual dislocations in single crystalline $SrTiO_3$. Nat. Mater. **5**, 312 (2006)
240. S.Q. Liu, N.J. Wu, A. Ignatiev, Electric-pulse-induced reversible resistance change effect in magnetoresistive films. Appl. Phys. Lett. **76**, 2749 (2000)
241. W.W. Zhuang et al., Tech. Dig.—Int. Electron Devices Meet **193** (2002)
242. K. Terabe, T. Nakayama, T. Hasegawa, M. Aono, Formation and disappearance of a nanoscale silver cluster realized by solid electrochemical reaction. J. Appl. Phys. **91**, 10110 (2002)
243. M.N. Kozicki, M. Mitkova, M. Park, M. Balakrishnan, C. Gopalan, Information storage using nanoscale electrodeposition of metal in solid electrolytes. Superlattices Microstruct. **34**, 459 (2003)
244. A.L. Greer, N. Mathur, Changing face of the chameleon. Nature (News and Views) **437**, 1246 (2005)
245. S.K.M. Dehaldhar, S.P. Sengupta, Ind. J. Pure Appl. Phys. **17**, 422 (1979)
246. W. Beyer, H. Mell, J. Phys. Status Solidi B **45**, 153 (1971)
247. N. Klein, Switching and breakdown in films. Thin Solid Films **7**, 149 (1971)
248. S.R. Ovshinsky, in *Disordered Materials: Science and Technology*, ed. by D. Alder (Amorphous Institute Press, New York, 1982)
249. K.W. Boer, S.R. Ovshinsky, Electrothermal Initiation of an electronic switching mechanism in semiconducting glasses. J. Appl. Phys. **41**, 2675 (1970)
250. D. Alder, M.S. Shur, M. Silver, S.R. Ovshinsky, Threshold switching in chalcogenide-glass thin films. J. Appl. Phys. **51**, 3289 (1980)
251. K.S. Hong, R.F. Speyer, J. Non-Cryst. Solids **116**, 191 (1990)
252. T. Ohta, S.R. Ovshinsky, in *Photo-Induced Metastability in Amorphous Semiconductors, (Chapter 18)*, ed. by A.V. Kolobov (Wiley, Berlin, 2003), pp. 310–326
253. S.R. Ovshinsky, D. Strand, J. Optoelectron. Adv. Mater. **7**, 1679 (2005)
254. D. Adler, H.K. Henisch, N.F. Mott, Rev. Mod. Phys. **50**, 209 (1978)
255. T. Lowrey, S. Hudgens, W. Czubatyj, C. Dennison, S. Kostylev, G. Wicker, Mater. Res. Soc. Symp. Proc. **803**, 101 (2004)
256. R. Aravinda Narayanan, S. Asokan, A. Kumar, Influence of chemical disorder on electrical switching in chalcogenide glasses. Phys. Rev. B **63**, 092203 (2001)
257. S.R. Ovshinsky, H. Fritzsche, Amorphous semiconductors for switching, memory, and imaging applications. IEEE Trans. Elect. Dev. **20**(2), 91 (1973)
258. H. Fritzsche, Electronic phenomena in amorphous semiconductors. Annu. Rev. Mater. Sci. **2**, 697 (1972)
259. S. Tyson, G. Wicker, T. Lowrey, S. Hudgens, K. Hunt, Nonvolatile, high density, high performance phase change memory. in *Proceedings 2000 IEEE Aerospace Conference*, Big Sky MT, Mar 2000, p.18
260. A. Pirovano, A.L. Lacaita, A. Benvenuti, Electronic switching in phase-change memories. IEEE Trans. Electron Devices **53**, 452 (2004)
261. S.R. Ovshinsky, S.J. Hudgens, W. Czubatyj, D.A. Strand, G.C. Wicker, Electrically erasable memory elements having improved set resistance stability. US Patent 5, p. 414, 1995
262. J. Hernandez, B. Chao, D. Strand, S.R. Ovshinsky, D. Pawlik, P. Gasiorowski, The relationship between crystal structure and performance as 58 optical recording media in Te–Ge–Sb thin films. Appl. Phys. Comm. **11**(4), 557 (1992)
263. D. Adler, B.H. Schwartz, M.C. Steele (eds.), *Physical Properties of Amorphous Materials* (Plenum, New York, 1985)
264. A.D. Pearson, Memory and switching in semiconducting glasses: a review. J. Non-Cryst. Solids **2**, 1 (1970)
265. C. Mattheck, Phys. Stat. Sol. A **11**, 117 (1971)
266. H.K. Henisch, W.R. Smith, Switching in organic polymer films. Appl. Phys. Lett. **24**, 589 (1974)
267. H.K. Henisch, E.A. Fagen, S.R. Ovshinsky, A qualitative theory of electrical switching processes in monostable amorphous structures. J. Non-Cryst. Solids **4**, 538 (1970)

268. N.F. Mott, Conduction in non-crystalline systems VII. Non-ohmic behaviour and switching. Philos. Mag. **24**, 911 (1971)
269. A. Alegria, A. Arruabarrena, F. Sanz, J. Non-Cryst, Solids **58**, 17 (1983)
270. S. Lukić, D. Petrović, *Complex noncrystaline chalcogenides* (University of Novi Sad,Faculty of Sciences, Novi Sad, 2002), p. 8
271. M. Popescu, Disordered Chalcogenide Optoelectronic Materials: Phenomena and Applications. J. Optoelectron. Adv. Mater. **7**, 2189 (2005)
272. B. Stričić, M. Živanov, M. Slankamenac, 8th International Symposium Young People and Multidisciplinary Research. Timisoara, Romania **2**, 257 (2006)
273. M. Slankamenac, M. Živanov, Research people and actual tasks on multidisciplinary sciences. Loznec, Bulgaria **2**, 304 (2007)
274. T. Ivetić, M.V. Nikolić, M. Slankamenac, M. Živanov, D. Minić, P.M. Nikolić, M.M. Ristić, Influence of Bi_2O_3 on microstructure and electrical properties of $ZnO–SnO_2$ ceramics. Sci. Sinter. **39**, 229 (2007)
275. S.R. Lukić, D.M. Petrović, A.F. Petrović, Effect of copper on conductivity of amorphous asse yi z. J. Non-Cryst. Solids **75**, 241–245 (1998)
276. M. Slankamenac, S.R. Lukić, INDEL. Banja Luka **6**, 20 (2006)
277. S. Prakash, S. Asokan, D.B. Ghare, Electrical switching behaviour of semiconducting aluminium telluride glasses. Semicond. Sci. Technol. **9**, 1484 (1994)
278. S.S.K. Titus, R. Chatterjee, S. Asokan, A. Kumar, Electrical switching and short-range order in As–Te glasses. Phys. Rev. B **48**, 1650 (1993)
279. N.A. Hegab, M. Fadel, K.A. Sharaf, Switching effects $InSe_{90-x}Sb_xBi_{10}$ thin-films. Vacuum **46**, 1351 (1995)
280. M.F. Kotkata, M.A. Afifi, H.H. Habib, N.A. Hegab, M.M. Abdel-Aziz, Memory switching in amorphous GeSeTe chalcogenide semiconductor films. Thin Solid Films **240**, 143 (1994)
281. R. Aravinda, S. Narayanan, A.K. Asokan, Influence of chemical disorder on electrical switching in chalcogenide glasses. Phys. Rev. B **63**, 2203 (2001)
282. C. Das, R. Lokesh, G.M. Rao, S. Asokan, Electrical switching behavior of amorphous $Al_{23}Te_{77}$ thin film sample. J. Non-Cryst. Solids **356**, 2203 (2010)
283. H.E. Atyia, A.E. Bekheet, Switching phenomenon and optical properties of $Se_{85}Te_{10}Bi_5$ films. Phys. B **403**, 3130 (2008)
284. K.S. Bang, S.-Y. Lee, 한국진공학회지 **23.1**, 34–39 (2014)
285. M.J. Lee, D. Lee, S.-H. Cho, J.-H. Hur, S.-M. Lee, D.H. Seo, D.-S. Kim, M.-S. Yang, S. Lee, E. Hwang, Nat. Commun. **4**, 3629 (2013)
286. C. Das, G.M. Rao, S. Asokan, Electrical switching behavior of amorphous $Ge_{15}Te_{85-x}Si_x$ thin films with phase change memory applications. Mater. Res. Bull. **49**, 388 (2014)
287. B.J. Madhu a, H.S. Jayanna a, S. Asokan, The composition dependence of electrical switching behavior of $Ge_7Se_{93-x}Sb_x$ glasses. J. Non-Cryst. Solids **355**, 2630 (2009)
288. S.R. Gunti, S. Asokan, Thermal and electrical switching studies on $Ge_{20}Se_{80-x}Bi_x$ ($1 \leq x \leq 13$) ternary chalcogenide glassy system. J. Non-Cryst. Solids **356**, 1637 (2010)
289. B.J. Madhu, H.S. Jayanna, S. Asokan, The composition dependence of electrical switching behavior of $Ge_7Se_{93-x}Sb_x$ glasses. J. Non-Cryst. Solids **355**, 459 (2009)
290. S.B.B. Prashanth, S. Asokan, Effect of antimony addition on the thermal and electrical-switching behavior of bulk Se–Te glasses. J. Non-Cryst. Solids **355**, 164–368 (2009)
291. R.T.A. Kumar, C. Das, P.C. Lekha, S. Asokan, C. Sanjeeviraja, P. Padiyan, Enhancement in threshold voltage with thickness in memory switch fabricated using $GeSe_{1.5}S_{0.5}$ thin films. J. Alloys Compd. **615**, 629–635 (2014)
292. G. Sreevidya Varma, D.V.S. Muthu, A.K. Sood, S. Asokan, Electrical switching, SET-RESET, and Raman scattering studies on $Ge_{15}Te_{80-x}In_5Ag_x$ glasses. J. Appl. Phys. **115**, 164505 (2014)

Chapter 5
Metal Oxide Nanostructures: Growth and Applications

Mukesh Kumar

Abstract This chapter provides a comprehensive review of metal oxide nanostructures, their growth and applications. These metal oxides are important transparent conducting materials that have drawn great attention in the scientific community owing to their high optical transparency ($\sim$85–95 %) in the visible region of spectrum along with high electrical conductivity (10^3–10^4 Ω^{-1} cm^{-1}). The combination of these properties make these materials suitable for many technological applications in solar cells, heat mirrors, transparent conducting electrodes, display devices and light emitting diodes. The metal oxide nanostructures, particularly nanowires and nanotubes, have also drawn considerable research interest as sensor material because of their high surface-to-volume ratio and suitable surface chemistry for verities of sensor applications. This has opened up potential applications of these materials in the area of environment control devices with high sensing response, low detection limit, low power consumption and high compatibility with microelectronic processing for IC's. Thus, 1-D metal oxide nanostructures have proven to be the most promising interface for communicating with the outer world. Different growth techniques and various growth models have been discussed for the growth of nanowires, metal-filled nanotubes, octahedrons and nanoflute structures. The article concludes with a glimpse of some recent work based on metal oxide photodetector especially in deep ultraviolet region (DUV) (<280 nm) which would provide futuristic application in optical switching, single photon detection and communication.

Keywords Metal oxide nanostructures · Nanowires · Nanotubes growth techniques · Growth models and applications

M. Kumar (✉)
Functional and Renewable Energy Materials Laboratory, Department of Physics, Indian Institute of Technology Ropar, Rupnagar 140001, Punjab, India
e-mail: mkumar@iitrpr.ac.in

M. Husain and Z.H. Khan (eds.), *Advances in Nanomaterials*,
Advanced Structured Materials 79, DOI 10.1007/978-81-322-2668-0_5

5.1 Introduction to Metal Oxides

Metal oxides are an important class of functional materials because of their unique combination of high electrical conductivity (10^3–10^4 S/cm) and high transparency in the visible region (400–800 nm) [1, 2]. The capability of transparent conducting oxide (TCO) material to flow and collect the current and pass the light through makes these material suitable for display devices [3]. Metal oxides have adjustable oxygen vacancies which enable the material to possess unique electronic, optical and chemical properties [4, 5]. Due to their special shape, composition, chemical and physical properties metal oxide nanostructures are the focus of current research.

Two main intrinsic properties of transparent conducting oxide materials are their high optical transparency with metal like high electrical conductivity. In general, the electrical conductivity (σ) of the material is defined as [6]:

$$\sigma = \mathrm{ne}\mu \tag{5.1}$$

where, n is the number of free carriers per unit volume and μ is the carrier mobility. It is evident from Eq. (5.1) that the conductivity is a function of two parameters: carrier concentration and carrier mobility. Crystalline tin doped indium oxide (ITO) deposited at moderate substarte temperature ($\sim$300 °C) can have electrical conductivity as high as $\sim$5000 S/cm. For comparison, the electrical conductivity of pure metallic copper and insulating quartz are 10^5 and 10^{-18} S/cm, respectively. The carrier concentration in metal oxides is generally produced by metal interstitials and oxygen vacancies. Carriers can also be generated by external doping with a substitutional dopant. Typical carrier concentrations in TCO are $\sim 10^{19}$–10^{22} cm^{-3}. The mobility (μ) can be described as the easiness of carrier to travel in the material. It is defined in term of scattering time (τ) and effective mass of carrier (m*):

$$\mu = \frac{e\tau}{m^*} \tag{5.2}$$

From above equation it is clear that to design higher electron mobility TCO, τ should be increased and m* should be reduced.

The optical transmittance of TCO thin films is intimately related to its optical band gap and carrier concentration. Typical optical spectra of ZnO thin film is shown in Fig. 5.1 [1]. The spectra consists of three regions: (i) High transmittance region in 400–750 nm (corresponding to 3.1–1.7 eV), (ii) Low transmittance at λ_{min} or lower wavelength region that corresponds to the optical band edge transition in the material and (iii) Low transmission and high reflection at λ_{max} or higher wavelength region. The low transmittance in λ_{max} side corresponds to collective oscillations of free electrons in the TCO materials.

For TCO materials, there is always a tradeoff between high electrical conductivity and high transparency in the visible region. The transmittance window from λ_{min} to λ_{max} depends on the direct band gap (E_g) ($\lambda_{min} \propto 1/E_g$) and free carrier concentration (N) in the material (λ_{max} or $\lambda_p \propto 1/N^{1/2}$). If, N is increased to get high electrical

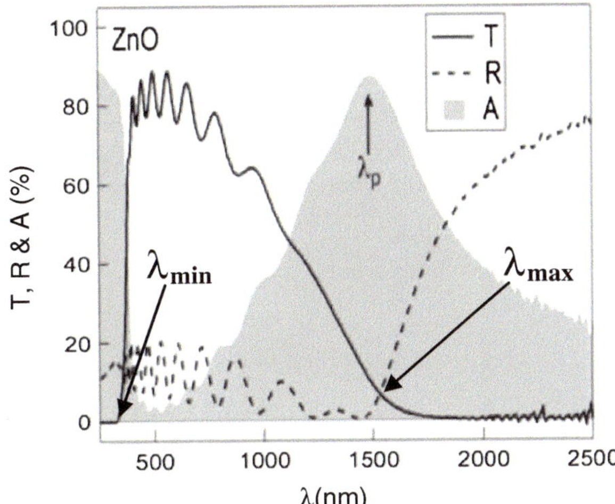

Fig. 5.1 Optical spectra of ZnO thin film in the range of 300–2500 nm [1]

conductivity (σ = Neμ), the TCO film transmission especially in the near-infrared (NIR) region reduces as a result of absorption by free carriers or by metal-like reflection. Therefore, high mobility TCO got attention to overcome this problem.

Figure 5.2 illustrates metal oxide compositional space that is well known for n-type transparent conducting electrodes.

The corners of the hexahedron shows the individual oxides, edges denote binary oxide systems and faces sights represent ternary oxide systems. Tin doped indium oxide (ITO), zinc oxide doped indium oxide (IZO) and other binary metal oxide nanostructures are reported in the literature [7]. Indium oxide (IO) is the host material for industrial standard ITO. Indium oxide is the n-type semiconductor with a band gap of 3.7 eV. At normal temperature and pressure, In_2O_3 exhibits cubic bixbyite structure (a = 10.117 Å) with 80 atoms or 16 formula units per unit cell. The elementary cell of IO lattice consists of 32 cations and 48 anions arranged in eight cubes as shown in Fig. 5.3. Each cube consists of two nonequivalent sites of indium referred to as In1 and In2, respectively [8].

The selected physical properties of IO are listed in Table 5.1. The combination of these properties make IO thin films and nanostructures suitable for a number of

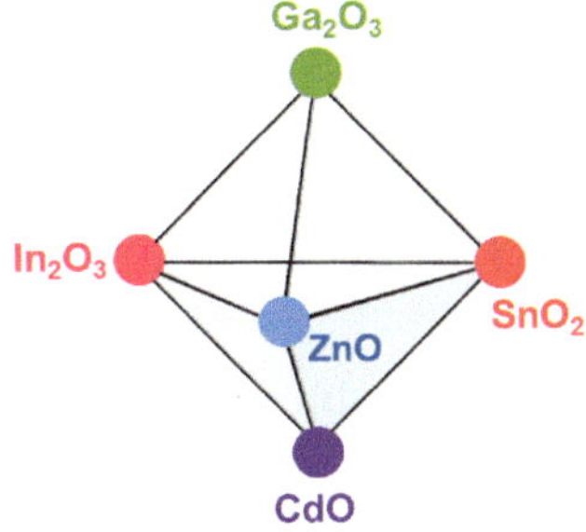

Fig. 5.2 Conventional transparent conducting oxide composition space [1]

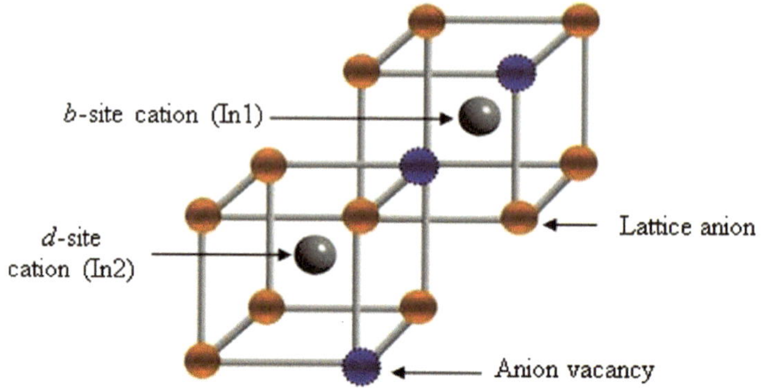

Fig. 5.3 Nonequivalent cation sites in In_2O_3 structure [8]

Table 5.1 Physical properties of bulk indium oxide at room temperature

Properties	Value/parameters
Crystal structures	Bixbyite cubic, a = 10.117 Å
Semiconductor	n-type
Melting temperature	1910 °C
Band gap	3.7 eV
Electrical conductivity	$\sim 10^3\ \Omega^{-1}\ cm^{-1}$
Donor activation energy	10 meV
Mass density	7.17 gm cm^{-3}

applications in solar cell, liquid crystal display, high-sensitive sensor, nanoelectronic and optoelectronic devices [1, 2]. In particular, IO nanowires have been directly integrated on sapphire substrate to fabricate the vertical field effect transistor. The seamless and direct integration of IO nanowires was proposed for it's potential use in terabyte ultra-high density nanoscale electronic and optoelectronic devices [9]. IO nanowires have shown room temperature ultraviolet emission at 398 nm (upon excitation by 274 nm laser) which was proposed to be a source for the next generation compact disc writers [10]. Besides, IO is the host material for ITO which has highest electrical conductivity and highest plasma frequency among all the doped TCO materials [11]. Recently, IO nanostructures have also shown applications in devices, such as, multi-level memory devices, carbon nanotubes-IO composite based super capacitors and IO nanofibers based optical waveguides [12–14].

5.1.1 *Metal Oxide Nanostructures*

The nanostructures, having at least one dimension between 1 and 100 nm, exhibit size-dependent physical, electronic and chemical properties which bridges atomistic scale properties and the extended states of the bulk [15–18]. Depending on their degrees of freedom, nanostructures are of three types: 0-D, 1-D and 2-D [19]. Quantum wells/thin films are classified as 2-D nanostructures, nanorods and nanowires are classified as 1-D nanostructures and small clusters or quantum dots are considered as 0-D nanostructures. The charge carriers in quantum dots are confined in all three directions. In a quantum wire, the charge carriers are confined in two dimensions and free to move in third dimension while in a quantum well the charge carriers are confined only in one dimension and free to move in the remaining two dimensions. Due to the confinement of electrons in nano dimension, the density of energy states is no longer continuous like bulk counterparts, but discrete that begins to dictate material's properties.

Figure 5.4 depicts the schematic representation of all nanostructures along with their density of state curves.

$$E = \frac{\hbar^2(k_x^2 + k_y^2 + k_z^2)}{2m} \quad \text{(Bulk)} \tag{5.3}$$

$$E_l = \frac{\pi^2\hbar^2 l}{2mL_x^2} + \frac{\hbar^2(k_y^2 + k_z^2)}{2m} \quad \text{(Quantum well)} \tag{5.4}$$

$$E_{l,m} = \frac{\pi^2\hbar^2}{2m}\left(\frac{l^2}{L_{x^2}} + \frac{m^2}{L_y^2}\right) + \frac{\hbar^2 k_z^2}{2m} \quad \text{(Quantum wire)} \tag{5.5}$$

$$E_{l,m,n} = \frac{\pi^2\hbar^2}{2m}\left(\frac{l^2}{L_{x^2}} + \frac{m^2}{L_y^2} + \frac{n^2}{L_z^2}\right) \quad \text{(Quantum dot)} \tag{5.6}$$

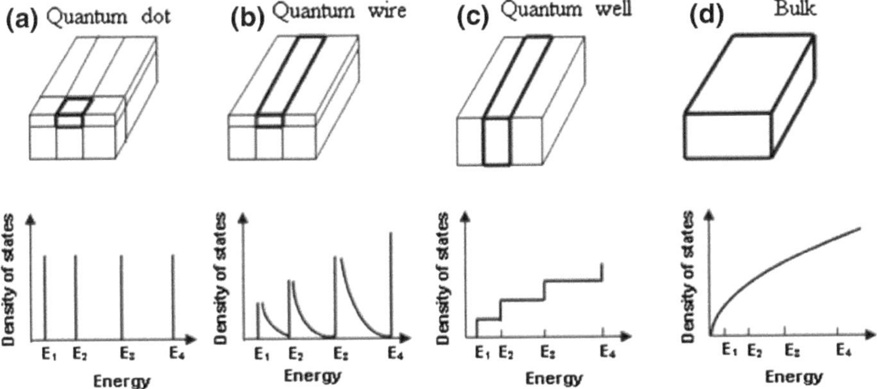

Fig. 5.4 Schematic representations of **a** quantum dot, **b** quantum wire and **c** quantum well nanostructures in contrast to **d** bulk counterparts with their respective energy density [20, 21]

where l, m, n = 1, 2, 3… are the quantum number of the energy levels due to the carrier confinement in x, y and z direction, respectively. The $k_{x,y,z}$ are the wave vectors and $\hbar^2 k_{x,y,z}^2/2m$ represent kinetic energies in unconfined direction. A large number of interesting and new phenomena is associated with nanometer-sized structures and their fundamental properties like band gap, melting point, conductivity, mechanical and catalytic are size-dependent [20, 21].

1-D nanomaterials have gained attention because of their unique properties and potential applications in nanodevices. Although the size-dependence of different properties is significantly higher for 0-D nanostructures as compared to other nanostructures, the 1-D nanostructures have advantage over the 0-D nanostructures. For example, electrical contact can be easily made on 1-D nanostructures due to their high aspect ratio. Also, 1-D nanostructures can serve as the interconnects for different devices and are thus ideal for investigating the size dependent electrical, optical, thermal and catalytic properties, etc.

Each morphology of 1-D nanostructures has specific applications. The nanotubes filled with metal are exciting because nanocavity at the center provides possible applications in different nanofluidic devices like nanothermometer, nanorobot spot welding and frictionless mass transport of liquids at nanoscale [22]. In general, 1-D nanostructures have wide applications as field emission sources [23], storage materials [24], light emitting devices [25], sensors [26], lasing action [27] and optical applications [28].

5.2 Growth of Metal Oxide Nanostructures

Controlled growth of metal oxide nanostructures is essential for the development of different devices based on them. A number of synthesis techniques were employed for the growth of metal oxide nanostructures which include solution based chemical synthesis methods, physical vapor deposition (PVD), and chemical vapor deposition (CVD) techniques. Here, we focus on the chemical vapor deposition technique. Table 5.2 summarizes different techniques used to grow indium oxide nanostructures as reported in the literature.

5.2.1 *Chemical Vapor Deposition*

Chemical vapor deposition (CVD) technique involves chemical growth process. The thermal CVD and metalorganic CVD have been used for growing In_2O_3 nanostructures. In thermal CVD, a solid source is placed upstream in a high-temperature region while the substrates are placed in the downstream direction in lower temperature regions. The solid is vaporized through heating, which is transported by a carrier gas that feeds the catalyst and source material for nanowire

Table 5.2 Growth techniques reported for indium oxide nanostructures along with growth parameters

S. No.	Synthesis technique	Processing parameters	Morphology [length (L) and diameter (D)]	Ref.
1.	AAM/electrodeposition	Source: $InCl_3 + Na_3C_6H_5O_6$, substrate: Au/AAM, air annealing temperature: 973–1073 K for 10 h	Nanowires D = 60 nm, L ~ 1–2 μm	[33]
2.	AAM/sol-gel	Source: 0.4 M $In(NO_3)_3$, air annealing: 973 K for 12 h, substrate: AAM	Nanotube ID = 100 nm, OD = 200 nm L ~ 50 μm	[34]
3.	AAM/sputtering	Target: In_2O_3, substrate: Porous alumina membrane/Si, substrate temperature—453 K, deposition pressure—0.003 Torr	Nanorods D = 13 nm, L = 150 nm	[35]
4.	Physical evaporation	Source: In, substrate: quartz tube, substrate temperature: 873–1173 K, N_2 gas flow: 500 sccm through DI water at 353 K	Nanobelt Thickness = 20–200 nm, L ~ 100–300 μm	[36]
5.	Molecular beam epitaxy	Source: In:Sn (90:10), substrate: Glass and hydrogenated Si, substrate temperature: 573–923 K	Nanowires D = 8–20 nm, L = 40–500 nm	[37]
6.	Laser ablation	Target: InAs, substrate: Au/Si, deposition pressure: 200 Torr, gas flow rate: 150 sccm (Ar + 0.002 %O_2), substrate temperature: 973 K, Nd:YAG laser (λ = 532 nm, power = 1 W)	Nanowires D = 10 nm, 20 nm and 30 nm, L ~ 1–2 μm	[38]
7.	Metalorganic chemical vapor deposition	Source: triethylindium vapors at 308 K, Substrate: Si, substrate temperature: 623 K, gas flow rate: Ar + O_2	Nanobumps decorated rectangular rods: 200–700 nm, L ~ 1–2 μm	[39]
8.	Thermal chemical vapor deposition	Source: In + In_2O_3 (1:4), substrate: alumina crucible, source temperature: 1573 K, substrate temperature: 973–1073 K, vacuum heating for 30 min	Indium-filled nanotubes, D = 30–150 nm	[31, 40]

growth. Thermal CVD has been utilized for the growth of a number of IO nanostructures like nanopyramids, nanorods, nanowires, nanobelts and nanotubes filled with indium metal, with good crystalline nature and controlled morphology [29, 30]. Li et al. [31] has reported the growth of IO nanostructures filled with metallic indium by the heating of In + In_2O_3 precursor at 1573 K for 30 min, while the Si substrate was held at 973–1073 K. Different morphologies can be obtained by varying the deposition parameters like temperature, pressure, precursor etc. Yan and co-workers have reported the growth of IO nanostructures (nanowires, nanoarrows, nanorods) by selecting three different deposition conditions [32]. Although solution based chemical methods like electrodeposition and sol-gel are suitable for low temperature growth of TCO nanostructures, the crystalline nature of grown TCO nanostructures and purity was observed to be poor in these techniques which dramatically hinder their applications for many critical applications. The techniques utilizing the PAM template have the advantage in terms of control over the diameter and length of grown nanostructures. On the other hand, the complete removal of the template from nanostructures still remains a problem.

5.3 Growth of Indium Oxide Nanostructures

5.3.1 Effect of Growth Ambient: Tunable Growth of Nanowires, Nanotubes and Octahedrons

Kumar et al. [41] have reported the vapor phase transport and condensation in the presence of oxidizing and reducing ambients for the growth of IO nanostructures. Three different conditions were used for the growth of different IO nanostructure samples: (i) IO + C (sample IO-1) (ii) same as (i) but in the presence of water (sample IO-W) and (iii) same as in (i) but in the presence of ethanol vapors (sample IO-E). In all the three conditions, temperature (960 °C), Ar gas flow rate (200 ml min^{-1}), substrate (5 nm Au coated silicon) and deposition time (1 h), source material, indium oxide and activated charcoal were kept identical.

Figure 5.5 showed the TEM image of IO octahedrons, nanowires and nanotube structures. The HRETM was carried out on these structures which revealed that the IO nanostructures were crystalline in nature. The nanostructures grown under ethanol were observed to have the nanocavity filled with indium metal, as shown in Fig. 5.5d. Figure 5.6 shows the precursor stage, nucleation stage, growth stage and finally morphology for the growth of different IO nanostructures under three different conditions. The growth of octahedra, under condition (i) is initiated by carbon assisted reduction of In_2O_3 into In_xO (x = 1, 2) vapors and shown in Fig. 5.6a. The reduced indium oxide started to deposit on the gold coated silicon substrate. The stability of the catalyst assisted growth or self-catalytic growth critically decided by the initial supersaturating of reaction species [42, 43, 44]. In the initial stage the incoming vapors condensed and forms the nuclei of In_xO on

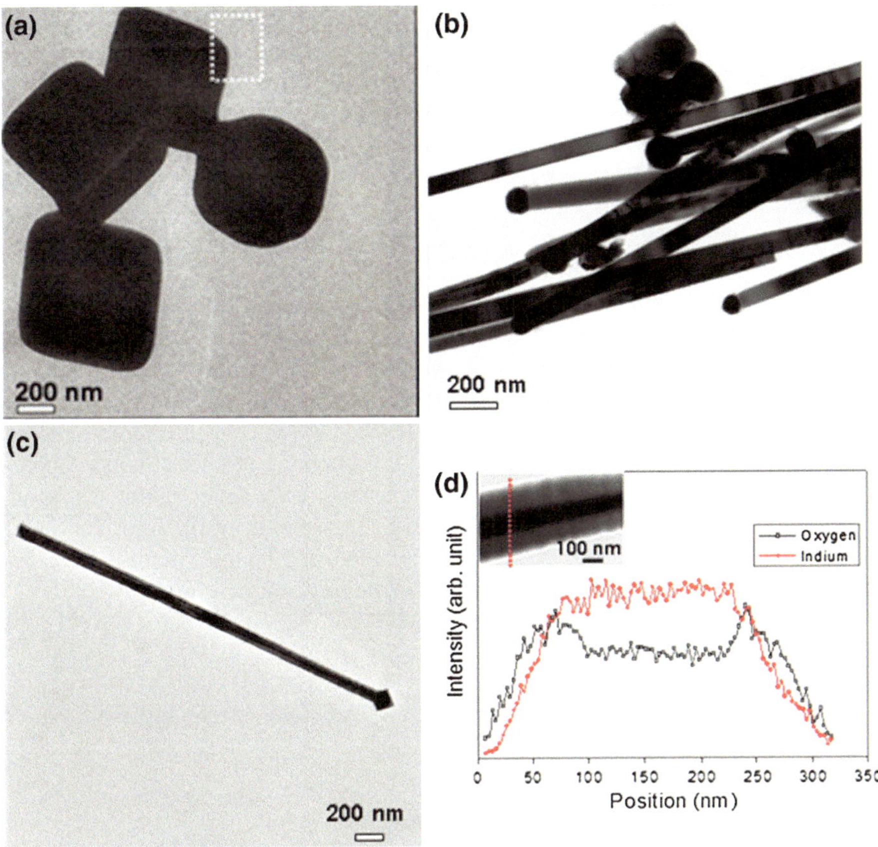

Fig. 5.5 TEM image of IO octahedrons (**a**), nanowires (**b**) nanotubes (**c**) and STEM-EDX along the radial direction of IO tubular structure (**d**). Inset shows the STEM image of IO tubular structure [41]

silicon substrate. The nuclei grow according to the surface energy of different crystallographic planes. Although, indium oxide is cubic crystalline material but the energy of different crystallographic planes may be different and may decide the growth direction of nanostructures. Hao et al. [30] have reported that the growth rate of different crystallographic planes in cubic material depends on their surface energy, $\gamma_{\{111\}} < \gamma_{\{100\}} < \gamma_{\{110\}}$. Since in growth condition (i) the reaction species are reduced indium oxide therefore it does not support Au assisted growth and resulted in formation of IO octahedrons structures. According to the surface energy of different planes {111} is the slowest growing plane and have the tendency to grow as the facets on IO octahedrons. For the growth of IO nanowires (growth condition ii), all the experimental conditions remain unchanged from those used for IO octahedra except introduction oxidizing agent water. The In_xO vapors may have a tendency to form In_2O_3 vapor species in the presence of extra oxygen source at growth temperature. Therefore, the reaction species in growth condition (ii) are like stoichiometric In_2O_3 and these species have higher chance to mix with Au and forms an alloy which supports the catalyst assisted growth of indium oxide nanowires [45]. Precursor stage in Fig. 5.6b shows a spherical gold droplet on a silicon substrate, which acts as the preferential adsorbing site for the incoming In_2O_3 vapor species. Depending on the surface energies of different crystallographic planes, as discussed above, the nanowires grow in [100] direction [30]. The reagent species adsorbed on the surface of a grown nanowire, low surface energy planes, also reach to the growth front, resulting in the continuous growth of the nanowire.

Due to the different stoichiometry of incoming reaction species in the presence of water the structural defects like oxygen vacancies, indium interstitial and other defects may be absent during the growth. This results in the growth of long and smooth surface IO nanowires [45]. In growth condition (iii), the grown structures are tubular IO filled with metal indium. The reaction species are reduced indium oxide but due to the presence of ethanol, a reducing agent, the reduced indium oxide further reduced to form indium rich growth ambient. Now the growth ambient is indium rich, as shown in Fig. 5.6c. The growth of indium-filled indium oxide tubular nanostructures was govern by modified bottom-vapor-solid growth mechanism [43, 46]. The flowing chemical reactions are possible:

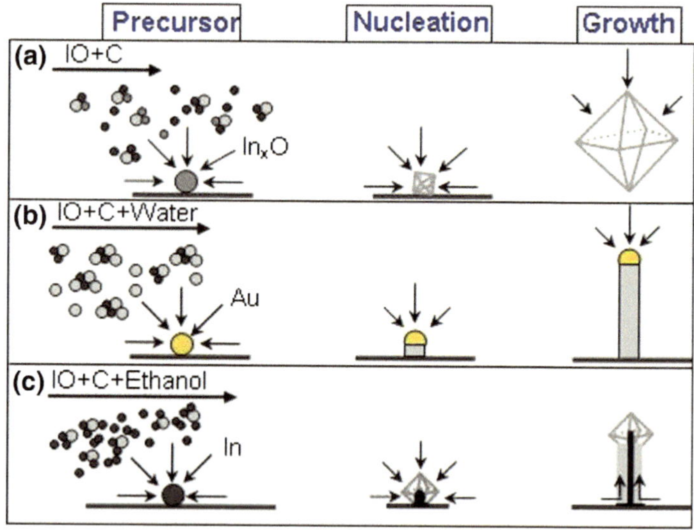

Fig. 5.6 **a–c** Illustrate the precursor stage, nucleation stage, growth stage and final morphology during the growth of octahedra, nanowires and indium-metal-filled nanoarrow structures. The different filled spheres in the precursor stage are symbolic representations of different reaction species. [•] indium; [○] oxygen; [•○] oxygen deficient IO; [••○] IO. For simplicity, the mechanism is discussed assuming that IO nanostructures are vertical to the substrate [41]

Step 1: Carbothermal reduction of In_2O_3

$$In_2O_3(s) + C(s) \xrightarrow{1000\,^\circ C} In_2O(v) + CO_2(v) \tag{5.7}$$

Step 2: Further reduction of reaction species into indium rich ambient

$$In_2O(v) + C_2H_5OH\ (v) \xrightarrow{1000\,^\circ C} CO_2(v) + 3H_2(v) + 2In(v) \tag{5.8}$$

The growth condition is identical to the growth in condition (i) except the presence of ethanol which leads to the indium rich ambient during growth. Therefore, Au assisted growth is not supported. In the initial stage, indium rich nuclei is formed which oxidized at high temperature except at the vicinity of the silicon substrate that leads to the growth of truncated IO octahedrons grown with metal indium sealed inside more thermally stable indium oxide shell, as shown in Fig. 5.6c. As the growth continues, the excess indium inside indium oxide gets phase separated and formed indium filled tubular structures. The capillarity induced filling of other metals have also been observed in carbon nanotubes [47]. Due to the low melting point of metal indium in nanometer size cavity, the large vapor pressure may be exerted during the growth. As per the literature, the vapor pressure of indium heated at 923 °C is 1 Pa. The inverse dependence of vapor pressure with microcavity size may critically increase the vapor pressure of indium at growth temperature [48]. Therefore, the large upward thrust of melted indium in nanosize cavity may result in the growth of indium-filled IO tubular nanoarrow structures.

5.3.2 *Effect of Growth Time: Regular to Irregular Nanostructures*

In our group, we have demonstrated the effect of growth time in the inert ambient [49]. The morphological analysis of nanostructures grown for different time was carried out using SEM is shown in Fig. 5.7. The IO-3 (deposition time 3 h) and IO-6 (deposition time 6 h) samples contain octahedra with truncated tip while IO octahedron structures in IO-12 (deposition time 12 h) sample are more complex

Fig. 5.7 SEM micrograph of IO microstructures deposited for **a** 3 h (IO-3), **b** 6 h (IO-6) and **c** 12 h (IO-12). The inset in **a–c** show higher magnification images of single IO octahedral [49]

in nature. The size of octahedron structures in IO-3, IO-6 and IO-12 sample are about 1, 1.5–2 and 2–3 μm, respectively. The octahedra have sharp facets with truncated tips, as shown in the inset of Fig. 5.7.

The growth of IO octahedra in an inert ambient is attributed to the self-catalytic growth of IO on silicon the substrate [43]. The sharp tip and nearly perfect octahedra were observed when deposition time was kept 1 h. The sharp tip octahedra occurs if the ratio, $r_{\{100\}}/r_{\{111\}} \approx 1.73$ and the growth of the truncated octahedron takes place if the ratio, $r_{\{100\}}/r_{\{111\}} \leq 1.73$ [30]. As the deposition time increases from 3 to 12 h the octahedron becomes more irregular in shape. The large deposition time may lead to second nucleation due to the presence of structural defects like, oxygen vacancies, indium vacancies, oxygen interstitial and anti-site defects on the regular structures resulting in the more irregular shaped octahedra.

These structural defects on the facets of octahedra may affect surface energies of different crystallographic planes and change the growth rate perpendicular to these planes, and thus form different morphologies. With longer reaction durations, these inappropriate factors will affect the end-product morphologies more. When the growth time is 3 h, the IO octahedra were grown with truncated apex. In addition, the inter-penetrated octahedron and octahedron with pyramid on facets were also observed when the deposition time increases to 12 h.

5.3.3 *Effect of Growth Pressure: Nanoflute to Nanotube*

Kumar et al. have demonstrated the growth of IO nanoflute in the presence of ethanol and low growth pressure [50]. The presence of nano size cavity was clearly seen in the Fig. 5.8a. HRTEM of a single IO nanowire clearly revealed the presence

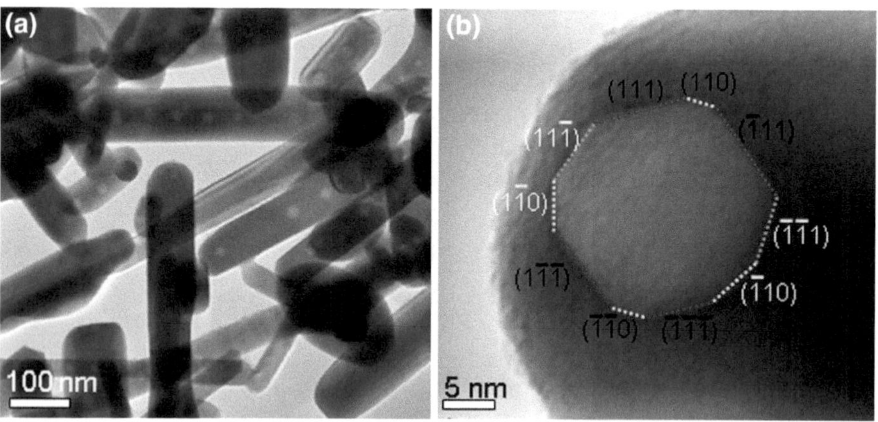

Fig. 5.8 **a** TEM images of IO with nanocavities. **b** HTTEM image of single nanocavity [50]

of light contrast regions along the length of the nanowire. The size of nanocavity was found in range of 10–30 nm. Figure 5.8b shows the TEM image of a typical region in IO nanowire. This micrograph clearly revealed that the light contrast region have the hexagonal shape.

The growth of flute-like MgO nanotubes, hematite nanorods and branched nanostructures with nanocavities was also reported [51]. Han and co-workers have reported dense nanocavities on titania nanorods confirmed by the electron energy loss spectroscopy in TEM study [51]. In our case, the light contrast region on the nanowire may be due to the decrease in thickness at that region along the electron beam direction. This could be a nanocavity either on the surface or inside the nanowires. Figure 5.9 schematically shown the different stages for the growth of cavities growth. The formation of nanocavity in the solid IO nanowires was attributed to the formation of indium rich reaction species during the growth and followed by the formation of nanovoids [43, 52].

During the nucleation stage, the indium rich reaction species condense on the silicon substrate and forms the nuclei. As the adsorption of reaction species continue, the growth of thermodynamically allowed crystallographic planes take place and different planes grow according to their surface energy. Since the reaction species was richer in indium vapors, therefore, the grown nanowires were expected to be rich in indium. The extra indium diffuses and leaves the vacancies behind, which is shown in Fig. 5.9. The out-diffusion of the metal has already been discussed for the formation of nanocavity in iron oxide nanoparticles and cobalt nanoparticles [53]. Without diffusion of In more and more vacancies created that finally coalesce and forms nanovoids. The energy minimization of nanovoids results in hexagonal shaped nanocavity.

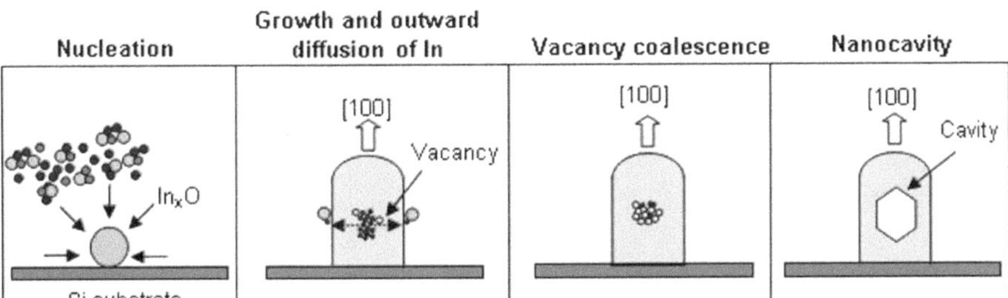

Fig. 5.9 The growth mechanism of hexagonal nanocavity decorated IO nanowires. The nucleation stage, out-diffusion, vacancy coalescence and nanocavity stages are illustrated. The different filled spheres are the symbolic representation of different reaction species. [•] Indium; [o] oxygen; [oo] oxygen deficient IO; [ooo] IO. The open circle shows the vacancy [50]

5.3.4 *Effect of Gas Flow Dynamics: Horizontal to Vertically Aligned Nanostructures*

IO is cubic crystalline materials and it is difficult to get anisotropic growth specially 1-D nanostructures vertically aligned on the substrate. Recently Kavita and co-workers have reported the array of vertically aligned crystalline indium oxide (In_2O_3) nanotubes on large area substrate via tube in tube horizontal CVD system using vapour phase deposition method on silicon (100) substrate with the assistance of Au catalysts [54]. Interestingly, they observed that in a tube in tube system the sequentially switching the argon gas flow with 200 and 50 sccm for different time intervals like 3, 4, 6, 7 and 8 min were helpful in controlling the dimensions and alignment of vertically standing nanotubes. Figure 5.10 shows scanning electron microscope images of as grown IO tubular nanostructures array under same deposition parameters like precursor amount, ethanol ambient, catalyst thickness and temperature in tube in tube CVD system, the only variable parameter was interruption time interval for gas flow.

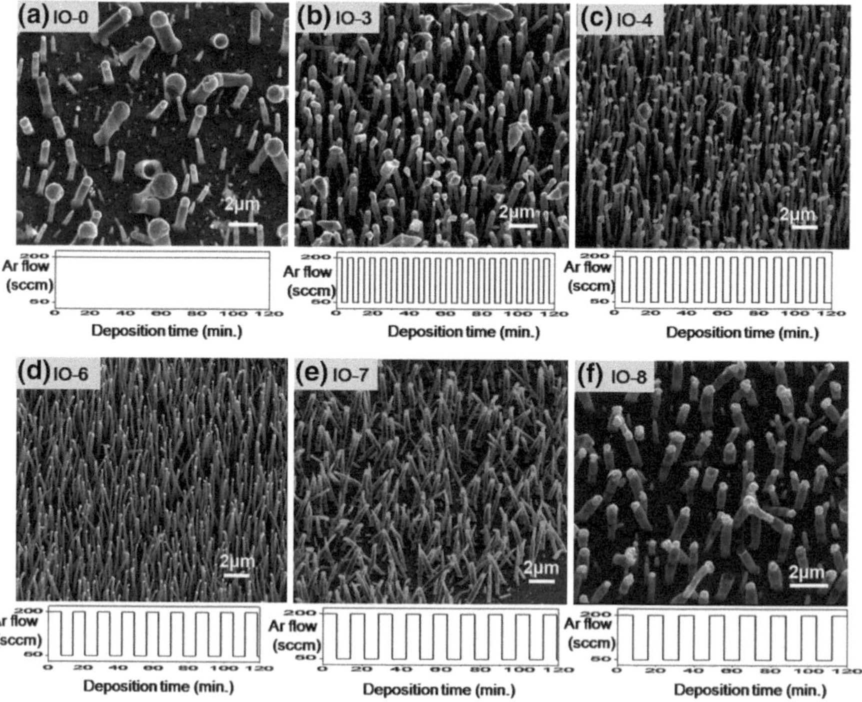

Fig. 5.10 SEM images of indium oxide nanostructures grown under different gas flow dynamics. IO-0 reveals the growth without any gas flow disturbance while IO-3 and so on showed the growth of indium oxide nanostructures with mentioned gas flow dynamics [54]

The uninterrupted/continuous flow of Ar gas at a flow rate of 200 sccm results in the growth of nanotubes with large variation in their diameter and were not well aligned as shown in Fig. 5.10a. On introducing sequential interruption in the gas flow rate means flow rate was sequentially interchanges with 200 and 50 sccm after particular time interval, dimensions and distribution of nanotubes affected by increasing the time interval from 3 to 4 and 6 min, it was observed that average diameter of nanotubes decreases and average length increases but further increase in interval time with 7 and 8 min average length decreases and average diameter increase. The outer diameter of nanotubes was 200–600 nm and wall thickness from 50–150 nm. Tubes diameter was not uniform but decreased along the height, cavity inside the tubes were also not straight and smooth incomplete portion of nanotubes.

5.4 Growth of Zinc Oxide Nanostructures (ZNT)

Zinc oxide is hexagonal crystalline material and it is easy for ZnO to follow the anisotropic growth of nanostructures without any catalyst. Lu et al. [55] reported the growth of ZnO nanotubes by simply thermal oxidation of zinc metal nanowires at 400 and 500 °C, Fig. 5.11.

The surface topography of the ZnO nanotubes is shown in Fig. 5.12. The TEM and HRTEM of ZNT are shown in Fig. 5.13a–c. The HRTEM and corresponding Fourier transform is shown in Fig. 5.13c.

Both the HRTEM image and the FFT pattern indicate a single-crystal structure of the nanocrystal. There are many others many other techniques were reported for the growth of ZnO nanostructures. The ZnO thin film was first deposited on the Si substrate via thermal evaporation and then ZnO nanowires were deposited using thermal CVD technique with ZnO and activated carbon as a source material. The nanowires grown on the Si (100) substrate was vertically aligned and about 150–200 nm in diameter.

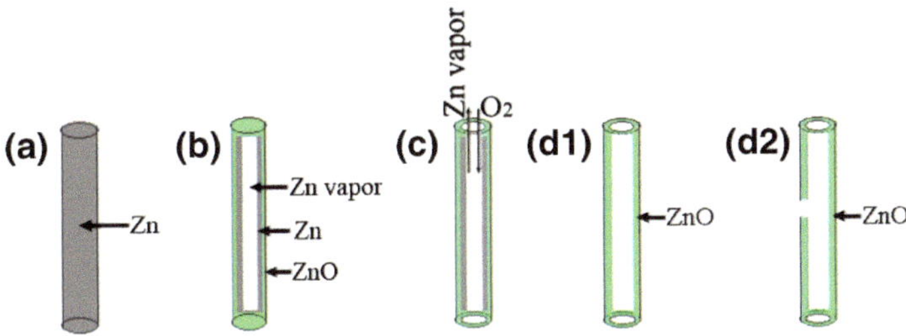

Fig. 5.11 **a** Schematic of Zn nanowires. **b–c** Shows the formation process from Zn to ZnO during the thermal treatment. **d_1–d_2** Two different stages of ZNT growth st two different temperatures [55]

Fig. 5.12 SEM image of ZnO nanotubes grown by thermal oxidation of Zn nanowires [55]

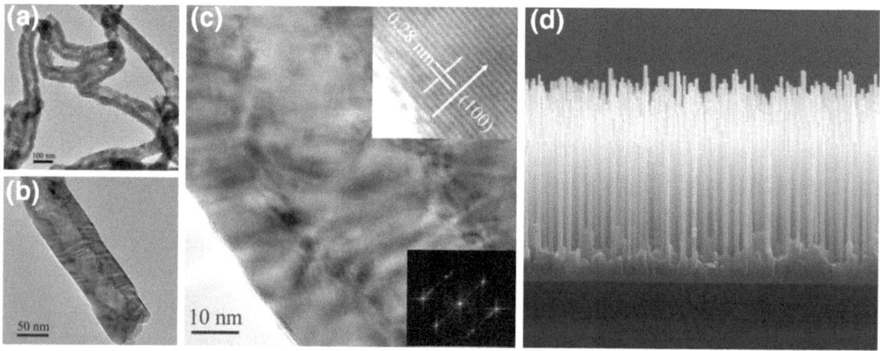

Fig. 5.13 **a–c** TEM and HRTEM of zinc oxide nanotubes, **d** is the vertically aligned zinc oxide nanostructures. [55]

5.5 Growth of 3-D Indium-Zinc Oxide Nanostructures

Karn and coworker have recently reported the growth of indium-zinc oxide mesocontainers and 1-D nanostructures by simply controlling the growth ambient from reducing to oxidizing. [56]. The 3-D nanostructures have impending applications as containers of nanograms of materials for drug delivery application. The morphology of mesocontainers is systematically investigated with time by scanning electron microscopy (SEM), as shown in Fig. 5.14.

The spherical particle of about 200 nm in size (Fig. 5.14a) was observed for 10 min deposition time which continued to grow bigger for higher deposition time. For 45 min of deposition time a different morphology of nanocable was observed.

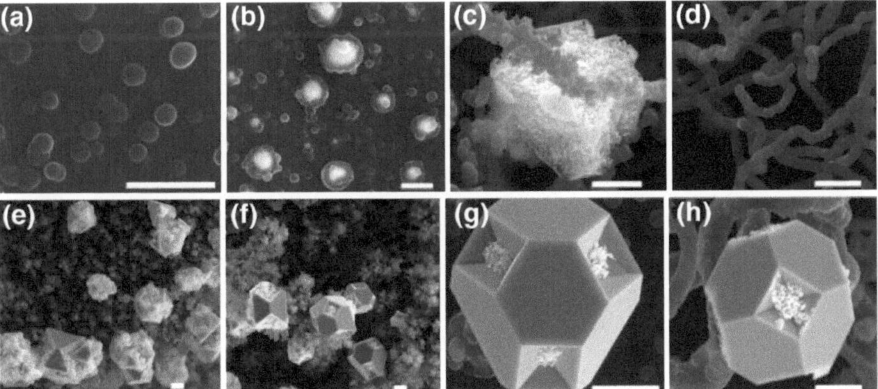

Fig. 5.14 Morphological evolution of indium zinc oxide structures at different time intervals **a** 10 min. **b** 20 min. **c** 30 min. **d** 45 min. **e** 60 min. **f** 75 min. **g**, **h** 90 min. The scale is 2 μm [56]

For deposition time longer than 45 min, the formation of nanocontainers was observed.

For higher deposition time, the nanocontainers did not change its morphology. Figure 5.14g–h revealed that the nanocontainers were filled with indium-zinc oxide materials [56]. The growth mechanism of nanocontainers formation is clearly understood from Fig. 5.15a–d. The nanocontainers evolved from the crystallization at different locations and finally leaded to the growth of nanocontainers. The sizes of the different nanocavities were measured and observed to have the material in nanogram to microgram.

5.6 Growth of Gallium Oxide Nanostructures

Ga_2O_3 is considered as potentially interesting deep ultraviolet (DUV) transparent oxide semiconductor. The bulk gallium oxide exhibits an original luminescence due to the defects. Monoclinic gallium oxide (β-Ga_2O_3) has optical band gap of E_g = 4.9 eV which makes this material suitable for DUV sensing application. Various growth techniques were reported for the growth of high quality Ga_2O_3 thin films and nanostructures [57]. Li and co-workers have reported the growth of Ga_2O_3 nanobelts by chemical vapor deposition technique with gallium and oxygen as the stating materials. The SEM images showing smooth nature of nanobelts, Fig. 5.16 [57].

In another study, Kuo and Huang [58] have reported the growth of ultra-long Ga_2O_3 nanowires and nanobelts using CVD technique. The length of nanostructures was in sub-millimeter or millimeter range. The constant supply for Ga vapours for extended hours and residual oxygen present in the growth chamber played a crucial role in the growth of extra-long Ga_2O_3 nanostructures. Two silicon

Fig. 5.15 Morphological evolution of indium zinc oxide structures at different time intervals **a** 10 min (ET10), **b** 20 min (ET20), **c** 30 min (ET30), **d** 45 min (ET45), **e** 60 min (ET60), **f** 75 min (ET75), **g**, **h** 90 min (ET90). The sample name is indicated in the parentheses [56]

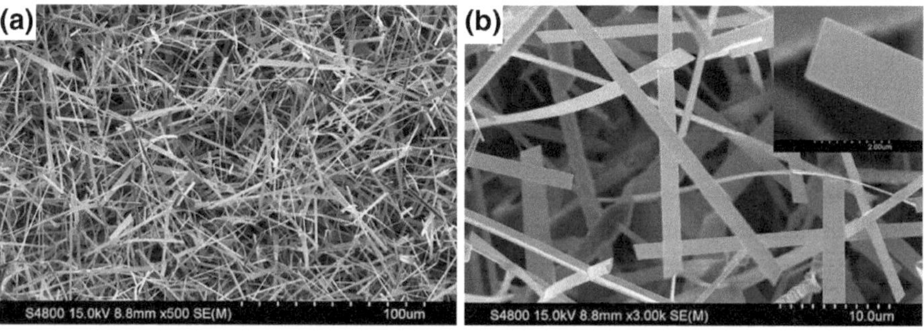

Fig. 5.16 **a** SEM image of Ga_2O_3 nanobelts. **b** SEM image at high resolution showing the smooth nature of Ga_2O_3 nanobrlts [57, 58]

substrates coated with gold catalyst were placed in the growth chamber at two different places. One substrate (position-1) was directly placed on the Ga source and second a substrate (position-2, 3–5 mm from the Ga source) was placed downstream direction near to the Ga source. The surface morphology of Ga_2O_3 nanostructures grown at position-1 and position-2 was drastically different, Fig. 5.17.

At position-1, nanowires and nanobelts were observed to grow with extended length. The nanowires were found to have diameters of around 100 nm. The majority of the 1D nanostructures are nanobelts. These nanowires and nanobelts can grow to exceptionally long lengths of hundreds of microns and even over 2 mm. The Ga_2O_3 nanostructures grown on substrate II vary significantly, depending on the distance of a region on the substrate from the gallium metal source, as shown in Fig. 5.18.

When the substrate is placed 3–5 mm away from the Ga source mostly nanowires and some nanobelts were formed. The nanowires generally have diameters on

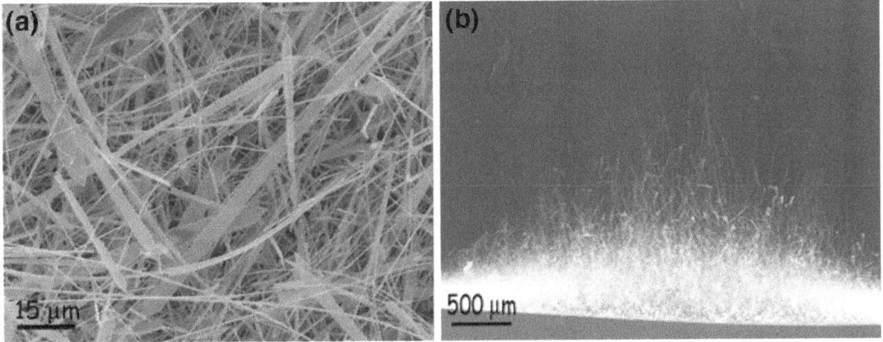

Fig. 5.17 **a** FE-SEM image of the as-synthesized Ga_2O_3 nanobelts and nanowires grown at position-1. **b** FE-SEM image of the side view of these nanobelts and nanowires [58]

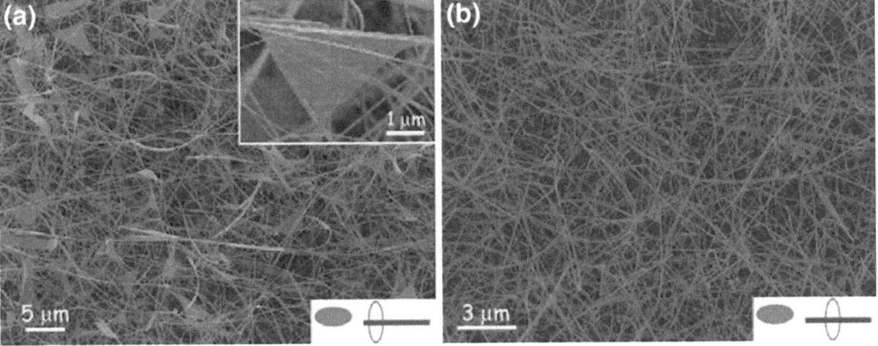

Fig. 5.18 SEM image of Ga_2O_3 nanostructures when the substrate placed at nearest to the Ga source **a** and when the substrate is placed at 6–8 mm away from the Ga source **b** [58]

the order of tens of nanometers and can reach lengths of hundreds of microns. While the nanowires grown at 6–8 mm away from the source were shorter in length, but can still reach lengths of tens of microns. The diameters of the nanowires were largely on the order of tens of nanometers.

5.7 Application: Environmental Sensors

5.7.1 *Hazardous Gas Sensors*

Metal oxide nanostructures are playing major role in environmental control by monitoring the potential hazardous gasses and toxic chemicals. Furthermore, the metal oxide based sensors are small dimensions, low power consumption, high sensor response, low detection limit, and high compatibility with microelectronic processing.

The gas sensing mechanism of metal oxide semiconductors is dependent on the interaction of oxygen and target gas at the metal-oxide surface. A molecule is considered to be chemisorbed if there is an electronic charge transfer between the gas and solid surface.

$$O_2(gas) + 2e^- \longrightarrow 2O^- \text{ (adsorption)} \tag{5.9}$$

Oxygen can be adsorbed in the form of O_2^-, O^- or O^{2-}. Near room temperature, O^{2-} species dominate the surface coverage whereas at higher temperature (423–923 K), O^- species dominate [59]. Figure 5.19a–b shows the gas sensing mechanism of metal-oxide materials.

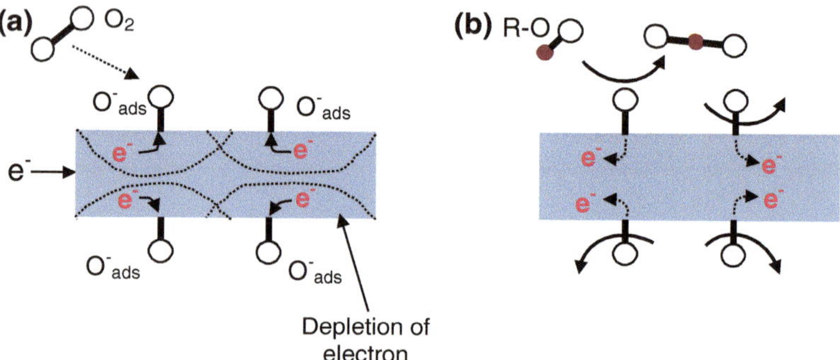

Fig. 5.19 Simplified schematic representation of chemical reaction occurring at the surface of IO. **a** Chemisorptions of oxygen from air take away electrons and creates depletion region which results in restricted flow of current. **b** After exposure to the reducing gas (R-O). The reducing gas extracts surface-bound oxygen, thereby releasing electrons back into the crystal and resulting free flow of current

Therefore, in ambient conditions, the near surface region of each grain is depleted of free electrons relative to the interior of the grain and hence a cause increases in the electrical resistance. There are number of the factor like grain size, porosity, thickness, voids and stoichiometry etc. that affect the gas sensing in bulk and thin films of IO oxide [60]. For improving the gas sensing characteristic, Gopel [61] emphasized the use of oxide nanostructures with diameter of the order of Debye's length (L_D) which is given by:

$$L_D = \left(\frac{\varepsilon \varepsilon_0 kT}{e^2 n} \right)^{1/2} \tag{5.10}$$

where ε, ε_0, T, e and n are the dielectric constant of free space and material, temperature, electronic charge and density of electrons, respectively. Nanostructures, in particular, are of great interest for gas sensing applications owing to their high surface/volume ratio, therefore providing more sites for oxygen molecules adsorption. In addition to the enhanced surface effect, the presence of different native defects on the surface of nanostructures has been proposed to strongly affect their gas sensing behavior [62, 63]. It is possible to grow the nanostructures with controlled crystallinity and stoichiometry which are the crucial parameters to control the gas sensing properties.

Kumar et al. [64] have reported that not only the high surface-to-volume ratio in nanomaterials was important to decide the high directivity of gas sensors but the stoichiometry of metal/oxygen ion also played a critical role. The nanowires and nanotubes with similar diameter distribution were grown by thermal CVD techniques in the presence of water, oxidizing agent, and ethanol, reducing agent, respectively.

The typical gas sensing response of IO nanowires and nanotubes for 1000 ppm ethanol at 300 °C is shown in Fig. 5.20a, where "ON" refers to the inclusion of ethanol in the artificial air and "OFF" refers to the suspension of the ethanol supply. The sensor response of IO nanowires and nanotubes for ethanol gas are 281 and 3.7

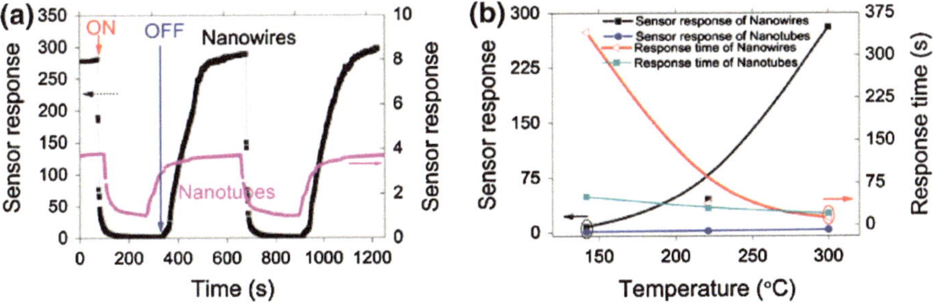

Fig. 5.20 **a** The gas sensor response of IO nanowires and nanotubes. **b** The temperature dependence of the sensor response and the response time for IO nanowires and nanotubes in the presence of 1000 ppm of ethanol [64]

with response time of 12 and 20 s, respectively. The sensor response and the response time of IO nanowires and nanotubes were measured as a function of the sample temperature (shown in Fig. 5.20b).

Although, the IO nanotubes (diameter: 40–250 nm, length: few tens of micrometer) have about similar dimension in the range of IO nanowires (60 ± 15 nm and length: few tens of micrometer), but the sensor response was found to be nearly two orders of magnitude smaller than the nanowires. Gas sensing is a surface property and the electron exchange between the surface states and bulk takes place at the surface layers of a thickness value of the order of Debye length [65]. Kumar and co-authors calculated the value of Debye length for nanowires and nanotubes and observed one order of magnitude difference, Debye length for nanowires and nanotubes was 48 and 0.4 nm. It is to be noted that double of Debye's length (48 nm) in case of IO nanowires is larger than its average diameter (65 nm), whereas for IO nanotubes Debye's length is only 0.4 nm. Therefore, at sample temperature of 300 °C the chemisorbed oxygen on IO nanowires surface fully depletes the conduction channel in nanowires and increases its electrical resistance. A schematic diagram for IO nanowires and nanotubes is shown in Fig. 5.21a–b. When the IO nanowires were exposed to ethanol gas, the chemisorbed oxygen gets removed. This restores the width of the conduction channel of IO nanowires and results in a decrease in its electrical resistance.

5.7.2 Deep Ultraviolet (DUV) Photodetector

The wide band gap of metal oxide nanostructures has been realized as the potential candidate for DUV photodetector. The UV detection reveals it's applications for numerous fields like in environmental control and bio applications. UV spectrum is the broad radiation which can be further divided into following regions: UV-A (400–320 nm), UV-B (320–280 nm), UV-C (280–200 nm), and far-UV (200–10 nm). Solar-blind photodetector based on metal-oxides or nitrides is only

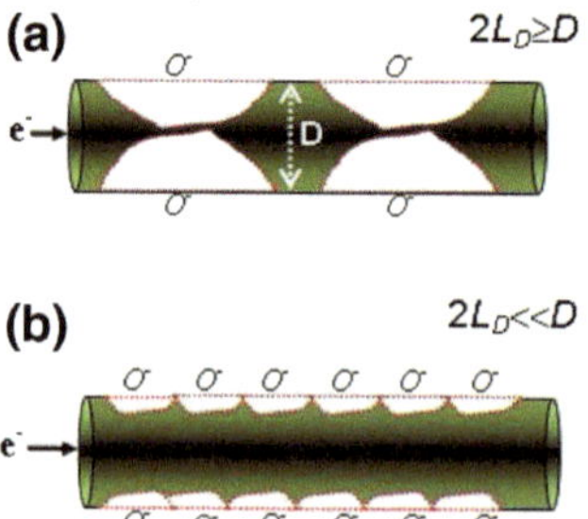

Fig. 5.21 Schematic diagram of depletion charge region due to chemisorbed oxygen on IO nanowires surface having correct stoichiometry (**a**) and IO nanotubes with oxygen vacancies (**b**). The L_D and D are Debye length and diameter of nanostructure, respectively [64]

sensitive to the radiation of wavelength lower than 280 nm. The photodetection mechanism of ZnO based UV detector is shown in Fig. 5.22 [66, 67].

In ZnO nanostructures, the surface adsorbed oxygen captured the electron and increased the resistivity of the material and create space charge region. If the created space charge region is equal or greater than the diameter of nanostructures then it seems like the OFF state and no conduction takes place through nanostructures. When a photon of energy more than the band gap of ZnO shine then it create electron-hole pair in the material. The photo generated holes can migrate to the surface of ZnO along the potential slope produced by band bending and release negatively charged absorbed oxygen ions. This leads to increase the photoconductivity of metal-oxide nanostructures. With respect to a traditional metal-oxide thin film, 1D metal-oxide nanostructures have huge dangling bonds on the surface and hence more sites for oxygen adsorption and more sensitivity upon light absorption. Therefore, 1-D nanostructures based photodetector have several advantages like high sensitivity, superior stability owing to high crystallinity, possible surface functionalization with target-specific receptor species, and field-effect transistor configurations that allow the use of gate potentials controlling the sensitivity and selectivity.

Recently, Li et al. [57] have reported solar blind deep UV photodetector based on the individual Ga_2O_3 nanobelt. The SEM image of the UV photodetector is shown in Fig. 5.23a. The irradiation single wavelength was illuminated on the

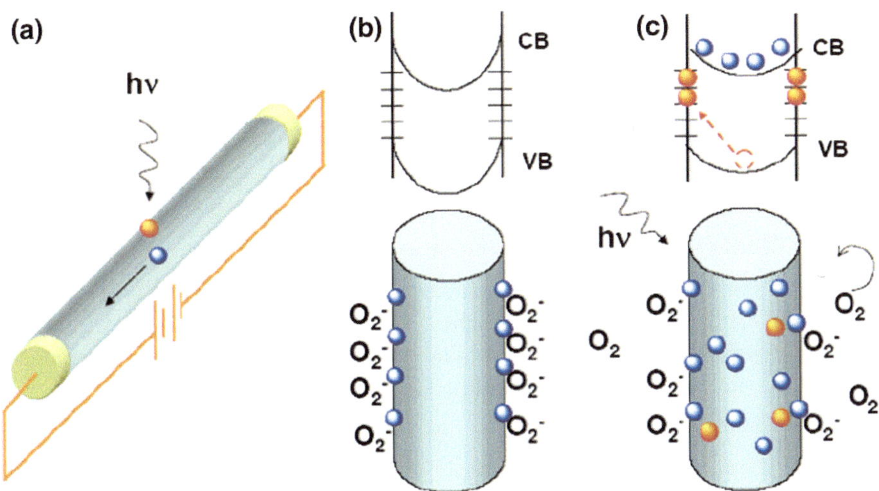

Fig. 5.22 Photoconduction mechanism in metal oxide nanowire: **a** Schematic of a metal oxide nanowire photodetector. **b**, **c** Trapping and photoconduction mechanism in ZnO nanowires. The top drawing in (**b**) shows the schematic of the energy band diagrams of a ZnO nanowire in dark, indicating band-bending and surface trap states. VB and CB are the valence and conduction bands, respectively. **c** Under light illumination, photogenerated holes migrate to the surface and become trapped, leaving behind unpaired electrons in the nanowire that contribute to the photocurrent [67]

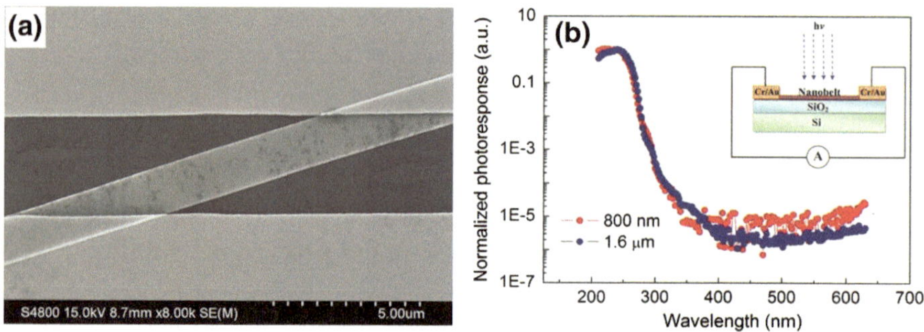

Fig. 5.23 SEM image of **a** Ga_2O_3 nanobelt lying across the two electrodes. **b** Normalized photoresponse of Ga_2O_3 different width nanobelt when shined with different photon energy [57]

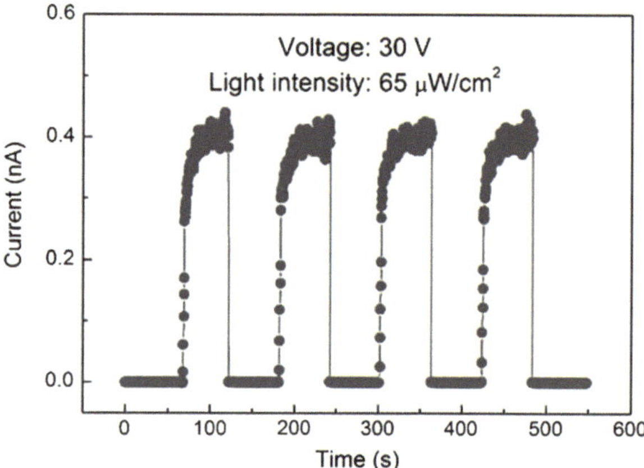

Fig. 5.24 Time response of a Ga_2O_3 nanobelt when shining with deep UV irradiation, 250 nm, with light intensity of 65 mW cm^{-2} and at a bias of 30 V across the photodetector [57]

nanobelt, and the current-voltage behavior was measured by simply the two-probe technique. The responsivity of gallium oxide Ga_2O_3 nanobelts with widths from 800 to 1.6 μm is shown in Fig. 5.23b. The results revealed that the photoresponse of nanobelt is independent of their width. The photoresponse with wavelength up to 300 nm has given no change while at shorter wavelength the sensitivity increases up to maximum valve and then drop off.

Figure 5.24 showed the ON-OFF characteristics of UV sensor irradiated at 250 nm with a bias of 30 V across the nanobelt. The response time and recovery time was measured which was observed to be 11.8 and <0.3 s. The fast response of DUV detector showed it's potential applications in futuristic applications.

5.8 Summary

This article has introduced the basics of transparent conducting oxide nanomaterials, their synthesis, characterization and application for environmental sensors. The effect of ambient, time, pressure, gas flow dynamics and other parameters in the context of chemical vapor deposition technique have been discussed for the growth of nanowires, metal-filled nanotubes, nanoflutes, octahedrons and mesocontainers. Recent studies show that the growth ambient, oxidizing or reducing ambient and gas flow dynamics critically affect sustainability of basic growth mechanism that controls not only the morphology but also the elemental composition of grown nanostructures. In addition to the intrinsic metal like electrical conductivity and high optical transparency of metal oxide nanostructures that are suitable for transparent conducting electrode applications, their chemical and physical properties make them suitable for a large number of other applications. The wide band gap and high temperature stability of these nanostructures demonstrate their applications in deep UV photodetector and futuristic single photon detection.

References

1. D.S. Ginley, H. Hosono, D.C. Paine (eds.), *Hand book of transparent conductors* (Springer, Berlin, 2010). ISBN 978-1-4419-1637-2
2. K.L. Chopra, S. Major, D.K. Pandya, Transparent conductors: a status review. Thin Film **102**, 1 (1983)
3. T. Kamiya, H. Hosono, Material characteristics and applications of transparent amorphous oxide semiconductors. NPG Asia Mater **2**, 15 (2010)
4. G.Z. Xing, J.B. Yi, D.D. Wang, L. Liao, T. Yu, Z.X. Shen, C.H.A. Huan, T.C. Sum, J. Ding, T. Wu, Strong correlation between ferromagnetism and oxygen deficiency in Cr-doped $In_2O_{3-\delta}$ nanostructures. Phys. Rev. B **79**, 174406 (2009)
5. H.T. Chen, S.J. Xiong, X.L. Wu, J. Zhu, J.C. Shen, Tin oxide nanoribbons with vacancy structures in luminescence-sensitive oxygen sensing. Nano Lett. **09**, 1926 (2009)
6. R.M. Pasquarelli, D.S. Ginley, R. O'Hayrea, Solution processing of transparent conductors: from flask to film. Chem. Soc. Rev. **40**, 5406 (2011)
7. M.P. Taylor, D.W. Readey, C.W. Teplin, M.F.A.M. van Hest, J.L. Alleman, M.S. Dabney, L.M. Gedvilas, B.M. Keyes, B. To, J.D. Perkins, D.S. Ginley, The electrical, optical and structural properties of $In_xZn_{1-x}O_y$ ($0 \le x \le 1$) thin films by combinatorial techniques. Meas. Sci. Technol. **16**, 90 (2005)
8. S. Limpijumnong, P. Reunchan, A. Janotti, C.G.V. de Walle, Hydrogen doping in indium oxide: an ab initio study. Phys. Rev. B **80**, 193202 (2009)
9. P. Nguyen, H.T. Ng, T. Yamada, M.K. Smith, J. Li, J. Han, M. Meyyappan, Direct integration of metal oxide nanowire in vertical field-effect transistor. Nano Lett. **04**, 651 (2004)
10. H. Cao, X. Qiu, Y. Liang, Q. Zhu, M. Zhao, Room-temperature ultraviolet-emitting In_2O_3 nanowires. Appl. Phys. Lett. **83**, 761 (2003)
11. R.G. Gordon, Criteria for choosing transparent conductors. Mater. Res. Soc. Bull. **25**, 52 (2000)
12. H. Jia, Y. Zhang, X. Chen, J. Shu, X. Luo, Z. Zhang, D. Yu, Polycrystalline silicon/$CoSi_2$ Schottky diode with integrated SiO_2 antifuse: a nonvolatile memory cell. Appl. Phys. Lett. **82**, 4163 (2003)

13. C. Li, W. Fan, B. Lei, D. Zhang, S. Han, T. Tang, X. Liu, Z. Liu, S. Asano, M. Meyyappan, J. Han, C. Zhou, Multilevel memory based on molecular devices. Appl. Phys. Lett. **84**, 1949 (2004)
14. P.C. Chen, G. Shen, S. Sukcharoenchoke, C. Zhou, Flexible and transparent supercapacitor based on In_2O_3 nanowire/carbon nanotube heterogeneous films. Appl. Phys. Lett. **94**, 043113 (2009)
15. M.F. Yu, O. Lourie, M.J. Dyer, K. Moloni, T.F. Kelly, R.S. Ruoff, Strength and breaking mechanism of multiwalled carbon nanotubes under tensile load. Science **287**, 637 (2000)
16. A. Thiaville, J. Miltat, Small is beautiful. Science **284**, 1939 (1999)
17. A.P. Alivisatos, P.F. Barbara, A.W. Castleman, J. Chang, D.A. Dixon, M.L. Klein, G.L. McLendon, J.S. Miller, M.A. Ratner, P.J. Rossky, S.I. Stupp, M.E. Thompson, From molecules to materials: current trends and future directions. Adv. Mater. **10**, 1297 (1998)
18. H.S. Nalwa, *Handbook of nanostructure materials and nanotechnology* (Academic Press, New York, 2000)
19. V.V. Mitin, V.A. Kochelap, M.A. Strasico, *Quantum heterostructures-microelectrnics and optoelectronics* (Cambridge University Press, USA, 1996)
20. I. Aruna, B.R. Mehta, L.K. Malhotra, S.M. Shivaprasad, A color-neutral, Gd nanoparticle switchable mirror with improved optical contrast and response time. Adv. Mater. **16**, 169 (2004)
21. P.E. Lippens, M. Lannoo, Calculation of the band gap for small CdS and ZnS crystallites. Phys. Rev. B **39**, 10942 (1989)
22. W.L. Hughes, Z.L. Wang, Nanobelt as nanocantilever. Appl. Phys. Lett. **82**, 2886 (2003)
23. B. Zenga, G. Xiong, S. Chen, S.H. Jo, W.Z. Wang, D.Z. Wang, Z.F. Rena, Field emission of silicon nanowires. Appl. Phys. Lett. **88**, 213108 (2006)
24. P. Zijlstra, J.W.M. Chon, M. Gu, Five-dimensional optical recording mediated by surface plasmons in gold nanorods. Nature **459**, 410 (2009)
25. X.W. Sun, J.Z. Huang, J.X. Wang, Z. Xu, A ZnO nanorod inorganic/organic heterostructure light-emitting diode emitting at 342 nm. Nano Lett. **8**, 1219 (2008)
26. A. Modi, N. Koratkar, E. Lass, B. Wei, P.M. Ajayan, Miniaturized gas ionization sensors using carbon nanotubes. Nature **424**, 171 (2003)
27. M.H. Huang, S. Mao, H. Feick, H. Yan, Y. Wu, H. Kind, E. Weber, R. Russo, P. Yang, Room-temperature ultraviolet nanowire nanolasers. Science **292**, 1897 (2001)
28. Y.L.F. Qian, J. Xiang, C.M. Lieber, Nanowire electronic and optoelectronic devices. Mater. Today **9**, 18 (2006)
29. X.C. Wu, J.M. Hong, Z.J. Han, Y.R. Tao, Fabrication and photoluminescence characteristics of single crystalline In_2O_3 nanowires. Chem. Phys. Lett. **373**, 28 (2003)
30. Y.F. Hao, G.W. Meng, C.H. Ye, L.D. Zhang, Controlled synthesis of In_2O_3 octahedrons and nanowires. Cryst. Growth Design **5**, 1617 (2005)
31. Y. Li, Y. Bando, D. Goldberg, Single-crystalline In_2O_3 nanotubes filled with In. Adv. Mater. **15**, 581 (2003)
32. Y. Yan, Y. Zhang, H. Zeng, J. Zhang, X. Cao, L. Zhang, Tunable synthesis of In_2O_3 nanowires, nanoarrows and nanorods. Nanotechnology **18**, 175601 (2007)
33. M.J. Zheng, L.D. Zhang, G.H. Li, X.Y. Zhang, X.F. Wang, Ordered indium-oxide nanowire arrays and their photoluminescence properties. Appl. Phys. Lett. **79**, 839 (2001)
34. B. Cheng, E.T. Samulski, Fabrication and characterization of nanotubular semi conductor oxides In_2O_3 and Ga_2O_3. J. Mater. Chem. **11**, 2901 (2001)
35. G.Q. Ding, W.Z. Shen, M.J. Zheng, Z.B. Zhou, Indium oxide "rods in dots" nanostructures. Appl. Phys. Lett. **89**, 063113 (2006)
36. J.S. Jeong, J.Y. Lee, C.J. Lee, S.J. An, G.-C. Yi, Synthesis and characterization of high-quality In_2O_3 nanobelts via catalyst-free growth using a simple physical vapor deposition at low temperature. Chem. Phys. Lett. **384**, 246 (2004)
37. C. O'Dwyer, M. Szachowicz, G. Visimberga, V. Lavayen, S.B. Newcomb, C.M. Sotomayor Torres, Bottom-up growth of fully transparent contact layers of indium tin oxide nanowires for light-emitting devices. Nature Nanotech. **4**, 239 (2009)

38. C. Li, D. Zhang, S. Han, X. Liu, T. Tang, C. Zhou, Diameter-controlled growth of single-crystalline In_2O_3 nanowires and their electronic properties. Adv. Mater. **15**, 143 (2003)
39. H.W. Kim, N.H. Kim, C. Lee, An MOCVD route to In_2O_3 one-dimensional materials with novel morphologies. Appl. Phys. A **81**, 1135 (2005)
40. Q. Liu, R. Zou, Y. Bando, D. Golberg, J. Hu, Nanowires sheathed inside nanotubes: manipulation, properties and applications. Prog. Mater. Sci. **70**, 1 (2015)
41. Mukesh Kumar, V.N. Singh, B.R. Mehta, J.P. Singh, Tunable synthesis of indium oxide octahedra, nanowires and tubular nanoarrow structures under oxidizing and reducing ambients. Nanotechnology **20**, 235608 (2009)
42. Y. Gao, Y. Bando, D. Goldberg, Melting and expansion behavior of indium in carbon nanotubes. Appl. Phys. Lett. **81**, 4133 (2002)
43. Z.R. Dai, Z.W. Pan, Z.L. Wang, Novel nanostructures of functional oxides synthesized by thermal evaporation. Adv. Funct. Mater. **13**, 9 (2003)
44. H.W. Kim, N.H. Kim, Growth of β-Ga_2O_3 nanobelts on Ir-coated substrates. Appl. Phys. A **80**, 537 (2004)
45. K. Hata, D.N. Futaba, K. Mizuno, T. Namai, M. Yumura, S. Iijima, Water-assisted highly efficient synthesis of impurity-free single-walled carbon nanotubes. Science **306**, 1362 (2004)
46. L.E. Jensen, M.T. Björk, S. Jeppesen, A.I. Persson, B.J. Ohlsson, L. Samuelson, Role of surface diffusion in chemical beam epitaxy of inas nanowires. Nano Lett. **4**, 1961 (2004)
47. P.M. Ajayan, S. Iijima, Capillarity-induced filling of carbon nanotubes. Nature **361**, 333 (1993)
48. Y. Gao, Y. Bando, Nanothermodynamic analysis of surface effect on expansion characteristics of Ga in carbon nanotubes. Appl. Phys. Lett. **81**, 3966 (2002)
49. Mukesh Kumar, V.N. Singh, B.R. Mehta, J.P. Singh, On the origin of photoluminescence in indium oxide octahedron structures. Appl. Phys. Lett. **92**, 171907 (2008)
50. Mukesh Kumar, V.N. Singh, B.R. Mehta, J.P. Singh, Tunable growth of indium oxide from nanoflute to metal-filled nanotubes. J. Phys. Chem. C **116**, 5450 (2012)
51. X. Gou, G. Wang, X. Kong, D. Wexler, J. Horvat, J. Yang, J. Park, Flutelike porous hematite nanorods and branched nanostructures: synthesis, characterisation and application for gas-sensing. Chem. Eur. J. **14**, 5996 (2008)
52. Y. Yin, R.M. Rioux, C.K. Erdonmez, S. Hughes, G.A. Somorjai, A.P. Alivisatos, Formation of Hollow nanocrystals through the nanoscale kirkendall effect. Science **304**, 711 (2004)
53. A. Cabot, V.F. Puntes, E. Shevchenko, Y. Yin, L. Balcells, A. Matthew, M.A. Marcus, S.M. Hughes, A.P. Alivisatos, Vacancy coalescence during oxidation of iron nanoparticles. J. Am. Chem. Soc. **129**, 10358 (2007)
54. K. Yadav, B.R. Mehta, J.P. Singh, Template-free synthesis of vertically aligned crystalline indium oxide nanotube arrays by pulsed argon flow in a tube-in-tube chemical vapor deposition system. J. Mater. Chem. C **2**, 6362 (2014)
55. H.B. Lu, H. Li, L. Liao, Y. Tian, M. Shuai, J.C. Li, M.F. Hu, Q. Fu, B.P. Zhu, Low-temperature synthesis and photocatalytic properties of ZnO nanotubes by thermal oxidation of Zn nanowires. Nanotechnology **19**, 045605 (2008)
56. A. Karn, M. Kumar, V.N. Singh, B.R. Mehta, S. Aravindan, J.P. Singh, Growth of indium oxide and zinc-doped indium oxide nanostructures. Chem. Vap. Deposition **18**, 295 (2012)
57. L. Li, E. Auer, M. Liao, X. Fang, T. Zhai, U.K. Gautam, A. Lugstein, Y. Koide, Y. Bando, D. Golberg, Deep-ultraviolet solar-blind photoconductivity of individual gallium oxide nanobelts. NanoScale **3**, 1120 (2011)
58. C.L. Kuo, M.H. Huang, The growth of ultralong and highly blue luminescent gallium oxide nanowires and nanobelts, and direct horizontal nanowire growth on substrates. Nanotechnology **19**, 155604 (2008)
59. S.C. Chang, Oxygen chemisorption on tin oxide: correlation between electrical conductivity and EPR measurements. J. Vac. Sci. Technol. **17**, 366 (1980)
60. G. Korotcenkov, V. Brinzari, J.R. Stetter, I. Blinov, V. Blaja, The nature of processes controlling the kinetics of indium oxide-based thin film gas sensor response. Sens. Actuators B **128**, 51 (2007)

61. W. Gopel, New materials and transducers for chemical sensors. Sens. Actuators **18**, 1 (1994)
62. Y. Zhang, J. Xu, Q. Xiang, H. Li, Q. Pan, P. Xu, Brush-like hierarchical ZnO nanostructures: synthesis, photoluminescence and gas sensor properties. J. Phys. Chem. C **113**, 3430 (2009)
63. H.T. Chen, S.J. Xiong, X.L. Wu, J.C. Shen, P.K. Chu, Tin oxide nanoribbons with vacancy structures in luminescence-sensitive oxygen sensing. Nano Lett. **09**, 1926 (2009)
64. M. Kumar, V.N. Singh, J.P. Singh, B.R. Mehta, The role of stoichiometry of indium and oxygen on gas sensing properties of indium oxide nanostructures. Appl. Phys. Lett. **96**, 123114 (2010)
65. H. Ogawa, M. Nishikawa, A. Abe, Hall measurement studies and an electrical conduction model of tin oxide ultrafine particle films. J. Appl. Phys. **53**, 4448 (1982)
66. T. Zhai, X. Fang, XXu Meiyong Liao, H. Zeng, B. Yoshio, D. Golberg, A comprehensive review of one-dimensional metal-oxide nanostructure photodetectors. Sensors **9**, 6504 (2009)
67. C. Soci, A. Zhang, B. Xiang, S.A. Dayeh, D.P.R. Aplin, J. Park, X.Y. Bao, Y.H. Lo, D. Wang, ZnO nanowire UV photodetectors with high internal gain. Nano Lett. **7**, 1003 (2007)

Chapter 6
Metal Matrix Nanocomposites and Their Application in Corrosion Control

Pallav Gupta, Devendra Kumar, M.A. Quraishi and Om Parkash

Abstract The present chapter gives an overview of nanocomposites which are novel materials for corrosion control. Nanocomposites comprise of more than one phase where size of each phase is less than 100 nm respectively. There are basically three types of nanocomposites: Ceramic-Matrix Nanocomposites, Metal-Matrix Nanocomposites and Polymer Matrix Nanocomposites. Several synthesis routes have been proposed for the fabrication of MMNCs such as Stir Casting, Powder Metallurgy, CVD, PVD etc. Major applications of metal matrix nanocomposites are in automobile and aerospace industries. Among various properties corrosion is an important property for determining the life expectancy of any nanocomposite material. In the present chapter a brief account of corrosion and its control using nanocomposites has been discussed. It is expected that the present chapter will help the readers to get a glimpse of nanocomposite materials for its wider use in industrial applications.

Keywords Nanocomposites and its types · Powder Metallurgy · Applications

P. Gupta (✉)
Department of Mechanical and Automation Engineering, A.S.E.T, Amity University, Noida 201313, U.P., India
e-mail: pgupta7@amity.edu

D. Kumar · O. Parkash
Department of Ceramic Engineering, Indian Institute of Technology (Banaras Hindu University), Varanasi 221005, U.P., India

M.A. Quraishi
Department of Chemistry, Indian Institute of Technology (Banaras Hindu University), Varanasi 221005, U.P., India

M. Husain and Z.H. Khan (eds.), *Advances in Nanomaterials*, Advanced Structured Materials 79, DOI 10.1007/978-81-322-2668-0_6

6.1 Nanocomposites

Multiphase solid material having more than two or three dimensions where size of each dimension is less than 100 nm are known as Nanocomposites. Present definition validates for porous media, colloids, gels and copolymers, but is more usually taken into consideration for bulk matrix and nano-dimensional structure which differentiate in various properties [1]. Structural, Mechanical, Electrical and Electrochemical properties of the nanocomposite will differ in comparison to the base material.

Common examples of nanocomposites are in the structure of the abalone shell and bone. Use of nanoparticle-rich materials increases the physical and chemical life of the component. Nanocomposites are better material in terms of mechanical strength. Reinforcement generally consists of sheets, fibres etc. [2]. Interfacial surface between the matrix and dispersant phase(s) is generally an order of magnitude greater than conventional composite materials. Property of matrix phase is significantly affected in the vicinity of the reinforcement. Different types of nano particulates results in different type of enhanced optical properties, dielectric properties, heat resistance or mechanical properties such as hardness, wear and deformation. During processing nano size reinforcements are dispersed in the matrix [3].

6.2 Types of Nanocomposites

6.2.1 Ceramic-Matrix Nanocomposites (CMNCs)

In ceramic matrix nanocomposites both matrix and reinforcement phase consists of a ceramic such as oxides, nitrides, borides, silicides etc. In some cases, ceramic-matrix nanocomposites encompass a metal as the second component. In ideal condition, ceramic is dispersed in each other so as to improve the nanoscopic properties [4]. Nanocomposites of such combination were found to improve the optical, electrical and magnetic properties along with tribological and corrosion properties respectively. Binary phase diagram is an important consideration while designing ceramic matrix nanocomposites and important measures should be taken to avoid a chemical reaction between both parts [5]. The safest measure thus is to carefully choose immiscible ceramic phases.

The major application of ceramic-matrix nanocomposites is in thin films. Thin films are solid layers of few nm to some tens of μm thickness which is deposited on an underlying substrate. It plays an important role in the functionalization of technical surfaces. An important technique for the preparation of nanocomposite layers is gas flow sputtering using hollow cathode technique. Current technique operates as a vacuum-based deposition technique providing high deposition rates up to some μm/s. Nanocomposite layers in the ceramics were fabricated by TiO_2 and Cu using hollow cathode technique. It showed a high mechanical hardness, small coefficients of friction and a high resistance to corrosion [6].

6.2.2 Metal-Matrix Nanocomposites (MMNCs)

Metal matrix nanocomposites (MMNCs) comprises of metal as the matrix and ceramic as the reinforcement. This type of composites can be classified as continuous and non-continuous reinforced materials. Another emerging class of nanocomposites is Carbon nanotube metal matrix nanocomposites (CNT-MMNC). CNT-MMNC provides high tensile strength and electrical conductivity as compared to carbon nanotube materials. Important considerations in the development of processing techniques of CNT-MMNCs are (a) economically producible, (b) provide a homogeneous dispersion of nanotubes in the metallic matrix, and (c) lead to strong interfacial adhesion between the metallic matrix and the carbon nanotubes. Apart from CNT-MMNC another important research area of MMNC are boron nitride reinforced metal matrix composites and carbon nitride metal matrix composites [7].

Nowadays, most metals and alloys could be used as matrix materials and they require dispersent material which needs to be stable over a range of temperature. The guiding aspect for the choice depends essentially on the matrix material. Only light metals are responsive with their low density proving an advantage. Titanium, aluminium and magnesium are the popular matrix metals which are particularly useful for aircraft applications. High modulus reinforcements are required if metallic matrix materials have to offer high strength. The strength-to-weight ratio of resulting composites will be higher than most of the alloys [8]. The other types of nanocomposite are known as super thermite nanocomposite such as hybrid sol–gel having silica base.

6.2.3 Polymer-Matrix Nanocomposites (PMNCs)

Class of nanocomposites in which the addition of ceramic nano particles to a polymer matrix leads to the improvement in the various mechanical and electrical properties is known as Polymer matrix nanocomposites (PMNCs). PMNCs are also termed as high performance composites. In PMNCs when homogeneity in the filler is achieved, the corresponding properties of the nanoscale filler are substantially different in comparison to the starting matrix. Polymer matrix nanocomposites are nowadays synthesized by using much stiffer nanoparticles of ceramics such as clays or carbon nanotubes. Improvement in mechanical properties is not limited to stiffness or strength [9]. High aspect ratio or high surface area of the fillers improves the physical properties of high performance nanocomposites. Future scope of multi-functional small scale devices i.e. sensor, actuator, medical equipment basically depends on polymer nanocomposites [10].

In the recent past polymer nanocomposites were fabricated using various one-dimensional carbon nanostructures. In order to improve the mechanical properties of polypropylene fumarate nanocomposites, for bone tissue engineering applications, reinforcing agents such as carbon nanotubes, graphene platelets, molybdenum disulfide nanoplatelets were used. For lower addition of nanomaterials, various mechanical properties which were improved include compressive yield strength, young's modulus, flexural modulus and flexural yield strength. It was observed that the property of reinforcement is found to be dependent on grain morphology, defects structure etc. Inorganic nanomaterials served as better reinforcements in comparison to carbon based nanomaterials. Electrical properties were found to improve for polymer nanocomposites reinforced with multi-walled carbon nanotubes.

6.3 Synthesis Routes for Fabricating Metal Matrix Nanocomposites

Metal matrix Nanocomposites can be processed in the liquid state or in the solid state. The easiest and cheapest method for processing nanocomposites is discussed in detail below.

6.3.1 Solid State Methods

1. ***Powder Metallurgy (P/M)***: Powdered metal and reinforcement are mixed and then bonded through a process of compaction, degassing, and thermo-mechanical treatment (possibly via hot isostatic pressing (HIP) or extrusion). Figure 6.1 shows different steps of powder metallurgy route [11].

Powder metallurgy (P/M) is the process of mixing fine powder particles, pressing the powder mixture into a required shape in a die, followed by heat treatment of the compact in a controlled atmosphere furnace to increase the bonding strength (Sintering). Powder metallurgy process involves four steps: (1) powder manufacture, (2) powder mixing and blending, (3) compacting, (4) sintering. Compaction is performed at room temperature, as well as at elevated-temperature. Sintering is conducted at atmospheric pressure. Secondary processing often follows to obtain special properties or enhanced precision.

Sintering and metal injection molding are the two major routes which are used to manufacture and consolidate the powder. In the recent past new techniques are developed which make use of the metal powders for the fabrication of new products. Mechanical strength can be improved by the use of this technique since the powder mixture is sintered and not melted.

Sintering process completes in three stages. In the initial stage, neck growth takes place rapidly but the powder particles remain discrete. During the second, densifying mechanism enacts and the structure recrystallizes by particles diffusion into each other. In the third stage, pores become spheroidal in shape and densification continues at a much lower rate. Solid state in solid state sintering refers to

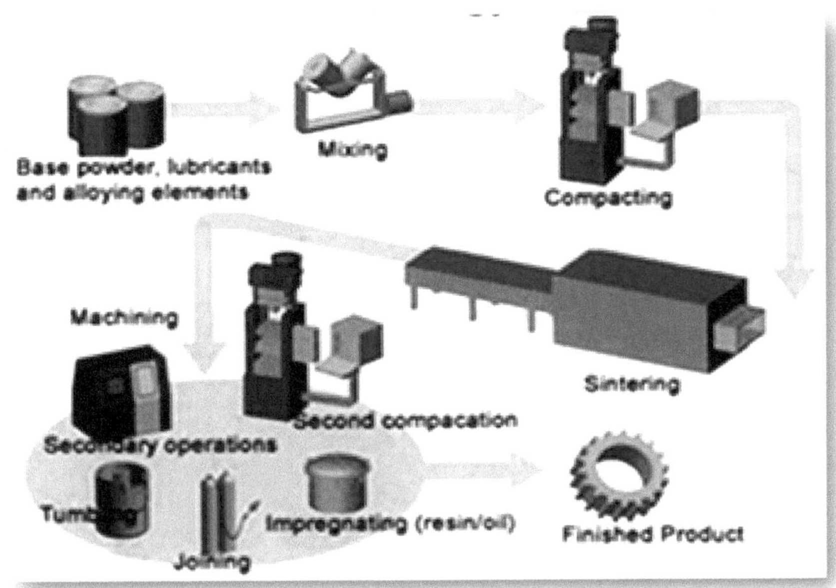

Fig. 6.1 Conventional powder metallurgy process

state of material when there is a perpetual bonding between particles, solid confirms that the material was not melted and it remains in its discrete form.

6.3.2 *Liquid State Methods*

1. ***Electroplating/Electroforming***: Electroplating/Electroforming is a technique in which a layer of composite material is deposited on a metallic or on a non-metallic electrode using electrolysis in an electrolytic cell.

Plating means coating over a metallic item, it is a spray formed thin layer and essentially becomes part of the object itself. Electroforming is metal built-up over a non-metallic surface, which is thick in size and can be removed from work piece as a standalone object. Some of the common uses for electroforming are mold making and reproduction of parts. Figure 6.2 shows electroplating/electroforming process.

2. ***Stir casting***: In Stir Casting technique discontinuous reinforcement is stirred into molten metal, followed by solidification. Figure 6.3 shows stir casting process. Stir Casting is a liquid process for the fabrication of the nanocomposite material, in which ceramic reinforcement is mixed with a molten metal using mechanical stirring. It is the simplest and cheapest method for the fabrication of the composite material [12].

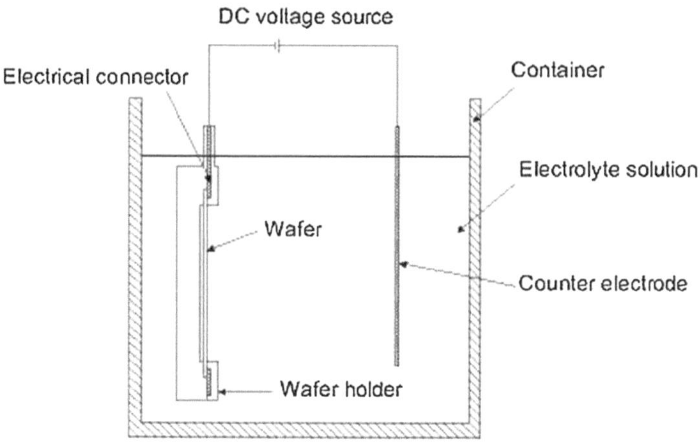

Fig. 6.2 Electroplating/electroforming process

Liquid composites are cast by conventional stir casting and metal forming techniques. Stir Casting is characterized by addition of low percentage of reinforcement phase. Reinforced phase throughout the matrix is not perfectly homogeneous due to:

1. Presence of local clusters of the dispersant phase.
2. Gravity segregation of the reinforcement because of the difference in the density of dispersant and matrix phase respectively.

Distribution of dispersed phase may be improved if the matrix is in semi-solid condition. The method using stirring metal composite materials in semi-solid state is called **Rheocasting**. High viscosity of the semi-solid matrix material enables better mixing of the dispersed phase.

3. ***Squeeze casting***: It is an important technique used for the fabrication of MMNCs. In this technique, after a slow mold filling the melt solidifies under very high pressure, leading to the formation of fine-grained structure. As compared with die-casted parts the squeeze-casted parts do not contain a gas inclusion, which helps in thermal treatment of the produced parts. One can differentiate between direct and indirect squeeze casting. With direct squeeze casting the pressure for the infiltration of the prefabricated preforms is applied directly to the melt [13]. The die is thereby part of the mold, which simplifies the structure of the tools substantially.

Disadvantage of the direct process include the following:

- Volume of the melt must be determined exactly, since no gate is present and thus the quantity of the melt determines the size of the cast construction unit.
- Appearance of oxidation products, formed in the cast part during dosage.

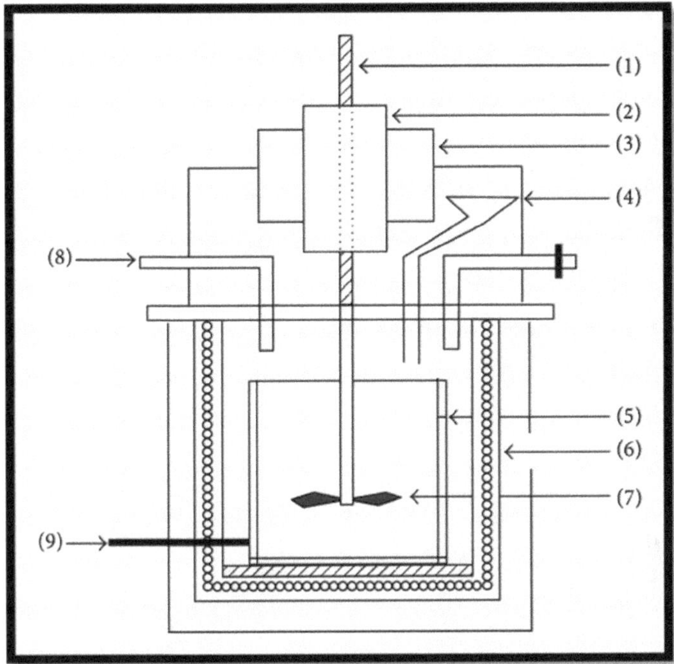

Fig. 6.3 Schematic of stir casting process (*1*) Stirrer spindle, (*2*) Sliding mechanism with impeller position control unit, (*3*) Electric motor, (*4*) Sprue, (*5*) Crucible, (*6*) Electric furnace, (*7*) Impeller, (*8*) Argon gas inlet, and (*9*) Thermocouple

In indirect squeeze casting technique molten material is forced to flow through a gate system. Flow rate of the molten material depends upon the diameter of the gate which results in a less turbulent mold filling.

In pressure casting processes used for the production of composite materials a prefabricated fiber or particle preforms is infiltrated with melt and the whole material is solidified using pressure system. For this technique a two-stage process is generally used. In the first stage the molten material is pressed into the form at low pressure followed by solidification phase formation at high pressure. This prevents damage to the preform by too fast infiltration. Due to low infiltration and response time, squeeze casting makes use of relatively reactive materials. Advantage of this technique is the possibility to produce intricate shaped components which are subjected to a higher stress during service. Figure 6.4 shows direct and indirect squeeze casting technique.

4. ***Spray deposition***: In spray deposition technique molten metal is gas atomized in the normal way and the spray is caused to impinge while still in the liquid or semi-solid state on a solid former where a layer of dense solid metal of a pre-determined shape is produced. The final product thus formed has a structure similar to that of powder-based material with all the attendant advantages of fine

grain, freedom from macro-segregation, etc. In common with the P/M process, spray deposition facilitates the production of alloy compositions that are difficult if not impossible to produce conventionally, and in certain cases the benefits that rapid solidification offers can also be obtained. Figure 6.5 shows Spray Deposition process.

Several deposition techniques which are being used and available are:

- Immersion plating.
- Spray deposition.
- Chemical Vapor Deposition (CVD).
- Physical Vapor Deposition (PVD).
- Spray forming techniques.

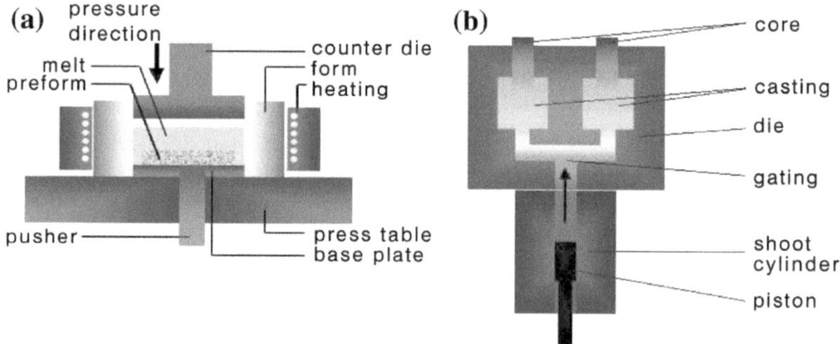

Fig. 6.4 Direct and indirect squeeze casting

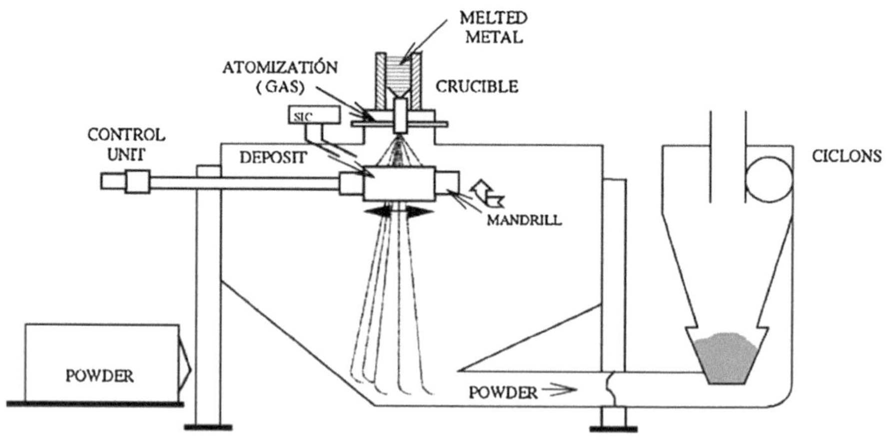

Fig. 6.5 Spray deposition process

6.4 Major Applications of Metal Matrix Nanocomposites

Metal Matrix Nanocomposites have large number of applications. A list of a few important applications is presented here:

1. The major applications served by MMNCs is in the automotive sector which include selectively reinforced pistons for diesel engines, selectively reinforced cylinder bores in Al engine blocks, intake and exhaust valves, driveshafts and propshafts, brake components (discs, rotors and calipers) and power module components for hybrid and electric cars.
2. Aeronautical MMC applications have been established in the aerostructural, aeropropulsion and subsystem categories. Aerostructural components include ventral fins and fuel access door covers on aircraft and rotor blade sleeves and swash plates on the helicopters. These components are all produced from P/M billet produced by Al Composites. The components are fracture-sensitive, and the helicopter rotor blade sleeve is fracture-critical. Aero propulsion components include fan exit guide vanes for the various engines used on aircraft and continuously reinforced Ti/SiC TMCs produced by FMW Composites for nozzle actuator links in the General Electric engine [14].
3. MMNCs are used for industrial, recreational and infrastructure applications. Industrial applications include cemented carbide and cermet materials, electroplated and impregnated diamond tools, Cu and Ag MMNCs for electrical contacts, erosion-resistant cladding for the petrochemical industry, Cu-infiltrated steel components, and TiC-reinforced Fe and Ni alloys.

TiC-reinforced Fe and Ni MMNCs are widely used for operations like cutting, rolling, pelletizing, stamping, piercing, warm metalworking, drawing, forming and

Fig. 6.6 Ultra-hard and wear-resistant component manufactured from Fe/TiC MMNCs

punching. Other applications of MMNCs are in hammers, impact dies, canning tools, crimp rollers, check valves, extruder nipples, bending dies, extrusion dies and hot forging die inserts. Superior performance is provided relative to tool steels and hardened Ni alloys that are conventionally used in these situations. Figure 6.6 shows the Ultra-hard and wear-resistant component manufactured from Fe/TiC MMNCs.

6.5 Corrosion

Corrosion is the deterioration of metals by electrochemical process with its environment. It involves oxidation of metals at anode and reduction at cathode. Rusting of iron (hydrated iron oxides) is a well known example of corrosion. Corrosion degrades the useful properties of materials and structures including strength, appearance and permeability to liquids and gases [15].

Many alloys corrode in presence of moisture and alloys. Corrosion can be confined locally in the form of a pit or crack. Since corrosion is a diffusion-controlled process, it occurs on exposed surfaces. Corrosion can be controlled by passivation and chromate conversion and by increasing corrosion resistance of the material. Corrosion can be classified on the basis of the appearance of the corroded material in following forms [16]:

1. **Uniform Corrosion**: It occurs uniformly all over the metal surface.
2. **Localized Corrosion**: It is confined to small area. It can be classified into following three types:

   - **Pitting Corrosion**: It is confined to small areas and appears in the form of small cavities and holes.
   - **Crevice Corrosion**: It occurs in shielded areas for example under gaskets.
   - **Filliform Corrosion**: It occurs under painted surface and appears in the form of filaments.

3. **Intergranular Corrosion**: It occurs along the grain boundaries.
4. **Galvanic Corrosion**: It occurs when two different metals are coupled together in contact with corrosive environment. More positive metal act as anode while less positive metal act as cathode.
5. **Stress Corrosion Cracking**: It occurs due to the combined action of stress and environmental conditions.
6. **Fretting Corrosion**: It occurs due to friction between moving surfaces.
7. **Dealloying**: Selective removal of one metal from an alloy under specific environment for example, dezincification of brass in ammonical environment.
8. **Microbial Corrosion**: Microbial corrosion is a corrosion caused by the activity of microorganism in the presence or absence of oxygen. Sulfate-reducing bacteria are active in the absence of oxygen (anaerobic); they produce hydrogen sulfide, causing sulfide stress cracking. In the presence of oxygen (aerobic), some bacteria directly oxidize iron to iron oxides and hydroxides, other bacteria like thiobacillus oxidize sulfur and produce sulfuric acid causing sulfide corrosion.

6.5.1 Mechanism of Corrosion

According to the mixed potential theory when a metal specimen is immersed in corrosive medium metal acquires a steady state potential which is known as E_{corr}. At E_{corr} rate of oxidation is equal to the rate of reduction.

It is important to note that when the specimen is polarized slightly more positive than E_{corr}, then the anodic current predominates over the cathodic current. As the specimen potential is driven further positive, the cathodic current component becomes negligible with respect to the anodic component. A mathematical relationship exists which relates anodic and cathodic currents to the magnitude of the polarization. Specimen can be polarized in +ve and −ve direction with respect to E_{corr} [17]. Metal specimen can be polarized in positive as well as in negative direction with respect to the E_{corr}.

The polarization characteristics can be measured experimentally by plotting the current versus applied potential. This plot is termed as potentiodynamic polarization plot. Potentials negative of E_{corr} give rise to cathodic currents, while potentials positive of E_{corr} give rise to anodic currents.

The potentiodynamic anodic polarization plot of a sample of 430 stainless steel is shown in Fig. 6.7. In this Fig. region A represents active region. When potential is applied the metal begins to corrode. At point B, further increase in potential the rate of corrosion decreases and the onset of passivation begins. The loss of chemical reactivity at this point is due to the formation of a passive film on the metal surface. This point is characterized by two coordinate values, the primary passive potential (EPP) and the critical current density (I_c). In region C, the current decreases rapidly as the passivating film forms on the specimen. At point D the passivating film begins to breakdown in region E. This region is known as transpassive region.

A potentiodynamic anodic polarization plot Fig. 6.7 gives following information:

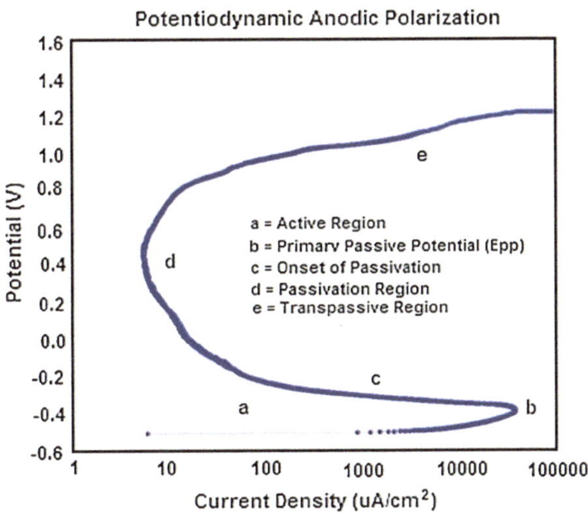

Fig. 6.7 Standard potentiodynamic anodic polarization plot of 430 stainless steel

1. The ability of the material to spontaneously passivate in the particular medium.
2. The corrosion rate in the passive region.
3. The potential region over which the specimen remains passive.

6.5.2 *Tafel Plots*

A Tafel plot shown in Fig. 6.8 is carried out on a metal by polarizing the specimen about −250 mV anodically and cathodically from the corrosion potential, E_{corr}.

The anodic or cathodic Tafel plots are described by the Tafel equation:

$$\eta = \beta \log \frac{I}{I_{corr}}$$

where,

η overvoltage, the difference between the potential of the specimen and the corrosion potential.

β Tafel constant.

I_{corr} current at overvoltage η, μA.

6.5.3 *Formulas Used in Corrosion Measurements*

$$C.R.(\text{mpy}) = \frac{0.13 I_{corr}(E.W.)}{d}$$

where,

mpy milli-inches per year.

I_{corr} corrosion current density (μAcm^2).

E.W. equivalent weight of the corroding species, (g).

d density of the corroding species, (g/cm^3).

$$\mu_p\% = (I^o_{corr} - I^i_{corr})/I^o_{corr} * 100$$

where,

I^o_{corr} values of corrosion current density in absence of reinforcement particles.

I^i_{corr} values of corrosion current density in presence of reinforcement particles.

μ_p Corrosion Protection Efficiency.

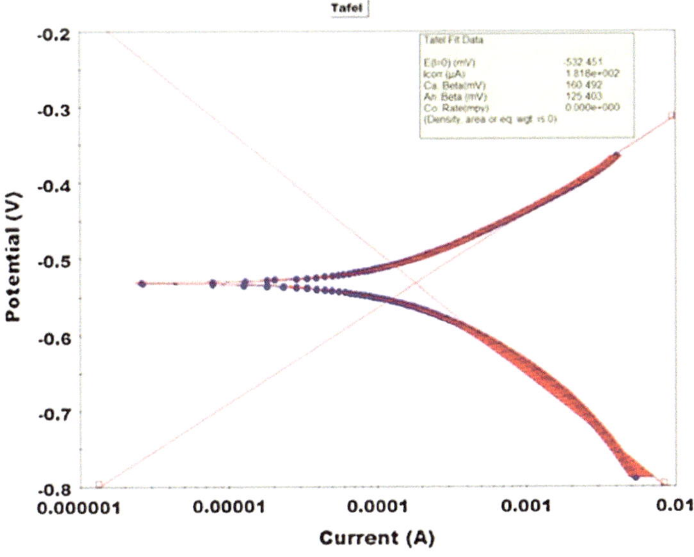

Fig. 6.8 Experimentally measured tafel plot [Princeton Applied Research]

6.6 Corrosion Control by Metal Matrix Nanocomposites

Following methods are used for corrosion control:

1. By using corrosion resistant materials
2. Design improvement
3. Coatings
4. Inhibitors
5. Cathodic protection.

Corrosion control using MMNCs plays an important role in preventing corrosion. In the present chapter synthesis of Fe-Al_2O_3 metal matrix nanocomposites has been described using powder metallurgy technique and corrosion inhibiting properties has been discussed using electrochemical methods.

Specimens were synthesized by compaction followed by sintering in argon atmosphere controlled furnace in the range of 900–1100 °C for 1–3 h respectively. Few specimens were also synthesized by adding CoO and CeO_2 as dopant in Fe-10 % Al_2O_3 metal matrix nanocomposites. CoO and CeO_2 doped specimens were synthesized by compaction and sintering at 1100 °C for 1 h. Corrosion behavior of all the specimens was carried out in 1 N HCl solution and it was found that:

1. In (5/10 %) Al_2O_3 reinforced specimen the corrosion inhibition efficiency of most of the specimens was found to be above 90 %. The improvement in the corrosion characteristics can be attributed to the formation of aluminium chlorate ($AlCl_3O_{12}$) phase. Aluminium chlorate phase forms a film on the surface of

the specimen due to the chemical reaction of alumina with hydrochloric acid. Formation of aluminium chlorate film is in nano size [18, 19]. Figures 6.9 and 6.10 shows the tafel plots of Fe-(5/10 %) Al_2O_3 reinforced metal matrix nanocomposite.

2. Corrosion inhibition efficiency of 1.0 % CoO doped nanocomposite specimen was found to be highest. 0.5 % CoO doping in pure Fe also showed improved corrosion inhibition efficiency. CoO doping in Fe-Al_2O_3 metal matrix nanocomposite system leads to formation of nano size film of aluminium chlorate ($AlCl_3O_{12}$) and cobalt chlorate ($CoCl_2$) takes place. Due to the chemical reaction between cobalt oxide and hydrochloric acid a cobalt chlorate phase forms whereas aluminium chlorate forms due to the chemical reaction between alumina and hydrochloric acid. Finer nano size particles of aluminium chlorate are found after the corrosion [20]. Figure 6.11 shows the tafel polarization plots of all the specimens doped with CoO.

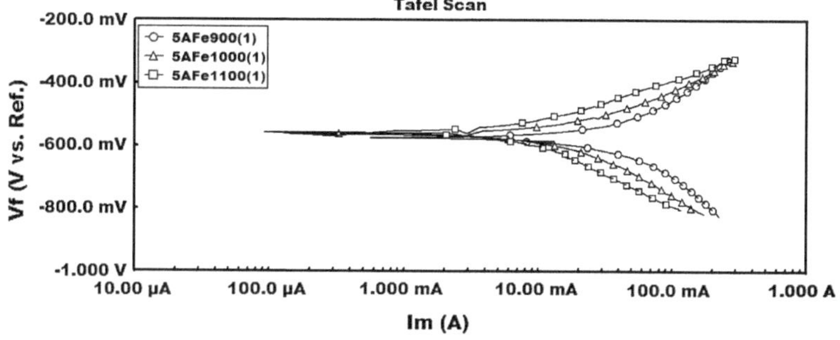

Fig. 6.9 Tafel plots of Fe-5 % Al_2O_3 reinforced metal matrix nanocomposite [18]

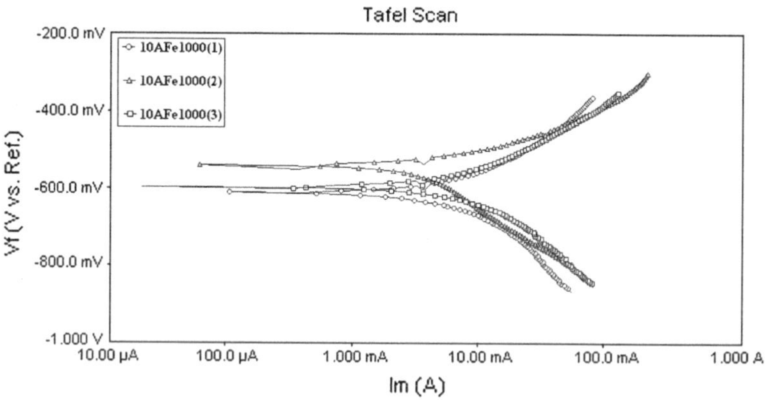

Fig. 6.10 Tafel plots of Fe-10 % Al_2O_3 reinforced metal matrix nanocomposite [19]

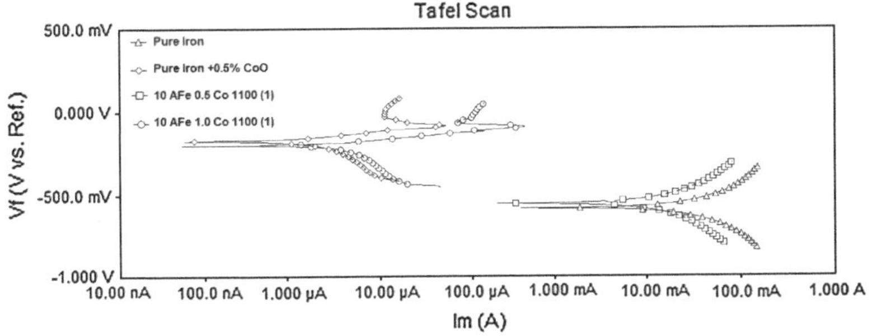

Fig. 6.11 Tafel polarization plots of all the specimens doped with CoO [20]

3) CeO_2 doped Fe-Al_2O_3 metal matrix nanocomposite system showed the formation of nano amorphous layer on the specimen surface. 0.5 % doped CeO_2 specimen showed the highest corrosion inhibition efficiency.

Corrosion is an electrochemical reaction which involves oxidation at anode releasing electrons. These electrons go to cathode side and reduce H^+ into H_2. CeO_2 inhibit/prevent corrosion in two ways:

(a) By passivating the metal.
(b) By plugging pore present on metal matrix nanocomposite by CeO_2 particles itself.

References

1. N. Chawla, K.K. Chawla, *Metal Matrix Composites* (Kluwer Academic Publishers, Boston, 2004)
2. Z. Zhang, D.L. Chen, Contribution of Orowan strengthening effect in particulate-reinforced metal matrix nanocomposites. Mater. Sci. Eng. A **483–484**, 148–152 (2008)
3. K.U. Kainer, *Basics of Metal Matrix Composites* (Wiley-VCH Verlag Gmbh & Co. KGaA, Weinheim, 2006). ISBN 3-527-31360-5
4. P. Istomin, A. Nadutkin, V. Grass, Fabrication of Ti_3SiC_2-based ceramic matrix composites by a powder-free SHS technique. Ceram. Int. **39**, 3663–3667 (2013)
5. D. Sarkara, S. Adak, N.K. Mitra, Preparation and characterization of an Al_2O_3–ZrO_2 nanocomposite, Part I: powder synthesis and transformation behavior during fracture. Compos. A Appl. Sci. Manuf. **38**, 124–131 (2007)
6. T. Hernandez, M.C. Bautista, The role of the synthesis route to obtain densified TiO_2-doped alumina ceramics. J. Eur. Ceram. Soc. **25**, 663–672 (2005)
7. J.M. Torralba, C.E. da Costa, F. Velasco, P/M aluminum matrix composites: an overview. J. Mater. Process. Technol. **133**, 203–206 (2003)
8. S.C. Tjong, Z.Y. Ma, Microstructural and mechanical characteristics of in situ metal matrix matrix composites. Mater. Sci. Eng. **29**, 49–113 (2000)
9. C. Bathias, An engineering point of view about fatigue of polymer matrix composite materials. Int. J. Fatigue **28**, 1094–1099 (2006)

10. M. Yashima, T. Kaṭo, M. Kakihana, M.A. Gulgun, Y. Matsuo, M. Yoshimura, Crystallization of hafnia and zirconia during the pyrolysis of acetate gels. J. Mater. Res. **12**, 2575–2583 (1997)
11. J.M. Torralba, C.E. da Costa, F. Velasco, P/M aluminum matrix composites: an overview. J. Mater. Process. Technol. **133**, 203–206 (2003)
12. T.S. Srivatasan, I.A. Ibrahim, F.A. Mohamed, E.J. Lavernia, Processing techniques for particulate-reinforced metal aluminium matrix composites. J. Mater. Sci. **26**, 5965–5978 (1991)
13. Y.H. Seo, C.G. Kang, The effect of applied pressure on particle dispersion characteristics and mechanical properties in melt-stirring squeeze-cast SiC/Al composites. J. Mater. Process. Technol. **55**, 370–379 (1995)
14. S.P. Rawal, Metal-Matrix Composites for Space Applications. J. Miner. Met. Mater. Soc. **53** (4), 14–17 (2001)
15. P.P. Trzaskoma, E. McCafferty, C.R. Crowe, Corrosion Behavior of SiC/Al Metal Matrix Composites. J. Electrochem. Soc. **130**(9), 1804–1809 (1983)
16. J. Xu, J. Tao, S. Jiang, Z. Xu, Investigation on corrosion and wear behaviors of nanoparticles reinforced Ni-based composite alloying layer. Appl. Surf. Sci. **254**, 4036–4043 (2008)
17. R. Escalera-Lozano, C.A. Gutiérrez, M.A. Pech-Canul, M.I. Pech-Canul, Corrosion characteristics of hybrid Al/SiCp/$MgAl_2O_4$ composites fabricated with fly ash and recycled aluminium. Mater. Charact. **58**, 953–960 (2007)
18. P. Gupta, D. Kumar, M.A. Quraishi, O. Parkash, Corrosion behavior of Al_2O_3 reinforced Fe metal matrix nanocomposites produced by powder metallurgy technique. Adv. Sci. Eng. Med. (American Scientific Publishers) **5**(4), 366–370 (2013)
19. P. Gupta, D. Kumar, M.A. Quraishi, O. Parkash, Effect of sintering parameters on the corrosion characteristics of iron-alumina metal matrix nanocomposites. J. Mater. Environ. Sci. **6**(1), 155–167 (2015)
20. P. Gupta, D. Kumar, M.A. Quraishi, O. Parkash, Effect of cobalt oxide doping on the corrosion behavior of iron-alumina metal matrix nanocomposites. Adv. Sci. Eng. Med. (American Scientific Publishers) **5**(12), 1279–1291 (2013)

Chapter 7
Diamond Nanogrinding

Mark J. Jackson, Michael D. Whitfield, Grant M. Robinson, Jonathan S. Morrell and Waqar Ahmed

Abstract Nanoscale fabrication requires a substrate made from an engineering material to be truly flat so that "bottom-up" nanofabrication techniques such as lithographically induced self-assembly and soft lithography can be used to deposit nanofeatures. The coating of piezoelectric materials with sub-micron size diamond particles has enabled the production of truly flat substrates so that nanofeatures can be created on engineering materials using a new manufacturing process known as "piezoelectric nanogrinding". The principle of the process relies on applying an electric current to the diamond coated piezoelectric material that causes the material to strain. When the diamond-coated piezoelectric material is placed in close proximity to the substrate, the diamonds remove extremely small fragments of the substrate when the electric current is applied to the material. The magnitude of the applied current controls the material removal rate. The process can be used to process biomedical materials especially in the production of nanoscale ducts and channels in micro- and nanofluidic devices. To achieve the generation of truly flat surfaces, the process must be executed within a specially constructed vibration dampening space frame. The chapter describes the principle of the process of nanogrinding using coated piezoelectric materials, and correlates the wear of diamonds with stresses induced into the diamonds when an electric current is applied to the piezoelectric in order to remove very small amounts of material. The removal of material can also be performed using a porous tool with abrasive materials embedded in them such as diamonds that increases the material removal rate as long as the porous tool is engineered in such a way that the loss of abrasive fragments is eliminated. This is achieved by laser assisted dressing, by engineering the bond of

M.J. Jackson (✉) · M.D. Whitfield · G.M. Robinson
Micro Machinists Corporation, Massachusetts Avenue, Cambridge, MA 02139, USA
e-mail: bondedabrasive@gmail.com

J.S. Morrell
Technology Development, Y12 National Security Complex, Oak Ridge, TN 37381, USA
e-mail: morrelljs@y12.doe.gov

W. Ahmed
School of Medicine and Dentistry, University of Central Lancashire, Preston PR1 2HE, UK
e-mail: wahmed4@uclan.ac.uk

M. Husain and Z.H. Khan (eds.), *Advances in Nanomaterials*,
Advanced Structured Materials 79, DOI 10.1007/978-81-322-2668-0_7

the porous tool to resist wear, and by laser assisted microstructural modification of the surface of the porous tool. The chapter describes how the bonds in porous tools are engineered to minimize abrasive grain loss and how vitrified bonding bridges can be processed using a laser to form extremely sharp nanoscale cutting wedges. The porous nanogrinding tool can be bonded to a piezoelectric material so that it can be used in the piezoelectric nanogrinding process.

Keywords Machining · Diamond · Nanogrinding · Nanomachining · Nanomaterials

7.1 Introduction

Nanoscale fabrication requires a substrate made from an engineering material to be truly flat so that "bottom-up" nanofabrication techniques such as lithographically induced self-assembly and soft lithography can be used to deposit nanofeatures. The coating of piezoelectric materials with sub-micron size diamond particles has enabled the production of truly flat substrates so that nanofeatures can be created on engineering materials using a new manufacturing process known as "piezoelectric nanogrinding". The principle of the process relies on applying an electric current to the diamond coated piezoelectric material that causes the material to strain. When the diamond-coated piezoelectric material is placed in close proximity to the substrate, the diamonds remove extremely small fragments of the substrate when the electric current is applied to the material. The magnitude of the applied current controls the material removal rate. The process can be used to process biomedical materials especially in the production of nanoscale ducts and channels in micro- and nanofluidic devices. To achieve the generation of truly flat surfaces, the process must be executed within a specially constructed vibration dampening space frame. The chapter describes the principle of the process of nanogrinding using coated piezoelectric materials, and correlates the wear of diamonds with stresses induced into the diamonds when an electric current is applied to the piezoelectric in order to remove very small amounts of material. The removal of material can also be performed using a porous tool with abrasive materials embedded in them such as diamonds that increases the material removal rate as long as the porous tool is engineered in such a way that the loss of abrasive fragments is eliminated. This is achieved by laser assisted dressing, by engineering the bond of the porous tool to resist wear, and by laser assisted microstructural modification of the surface of the porous tool. The chapter describes how the bonds in porous tools are engineered to minimize abrasive grain loss and how vitrified bonding bridges can be processed using a laser to form extremely sharp nanoscale cutting wedges. The porous nanogrinding tool can be bonded to a piezoelectric material so that it can be used in the piezoelectric nanogrinding process.

7.2 Piezoelectric Nanogrinding

The piezoelectric nanogrinding process is a process that relies on using a nickel-coated ceramic material with microscale diamond particles bonded to it that are cubo-octahedral in shape to machine nanoscale features in a variety of work-piece materials. The diamonds are bonded to the piezoelectric material by gaseous deposition, laser cladding, or directly bonding a porous tool to the material via an adhesive paste. The process is executed by applying a known sinusoidal frequency to the piezoelectric crystal in order to achieve a desired oscillatory displacement. Rapid vibration of the crystal will allow material removal rates to be increased, thus making it a nanomanufacturing process. The nanogrinding process is accompanied by wear of the diamond grains, and the rate of this wear plays an important role in determining the efficiency of the nanogrinding process and the quality of the nanomachined surface. Wear mechanisms in nanogrinding processes appear to be similar to that of single-point cutting tools, the only difference being the size of swarf particles generated. Figure 7.1 shows the arrangement of grinding swarf and abrasive grains that have been lost during the nanogrinding process.

Figure 7.1 shows abrasive grains with blunted cutting edges (wear flats), and abrasive grains with sharp cutting edges that are released from the surface of the piezoelectric crystal before they have chance to grind nanoscale chips from the

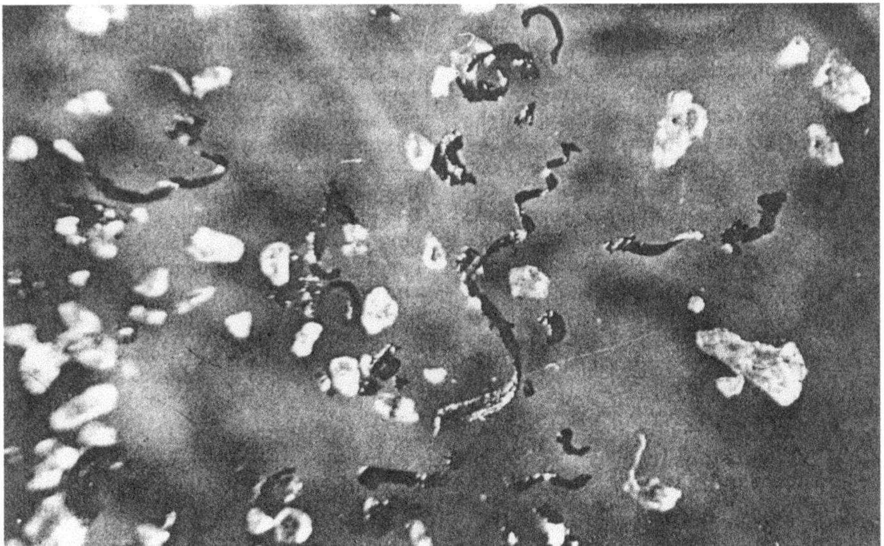

Fig. 7.1 Nanoscale grinding detritus showing blunt and sharp abrasive grains of diamond and metal chips. Re-printed with kind permission from Springer Science+Business Media B.V.

surface of the workpiece. The process suffers with a loss of diamond grains even when the interfacial adhesion between diamond and piezoelectric material is very good. A more closely related process that has been reported widely is that of the wear of probes used in atomic force microscopy [1, 2]. However, these observations were purely experimental with no explanation of how to design probes that inhibit, or retard, wear. A performance index used to characterize diamond wear resistance is the grinding ratio, or G-ratio, and is expressed as the ratio of the change in volume of the workpiece removed, Δv_W, to the change in the volume of the diamond abrasive grain removed, Δv_S, and is shown in Eq. (7.1),

$$G = \Delta v_w / \Delta v_s \tag{7.1}$$

Grinding ratios for processes at the nanoscale have not yet been characterized. However, the complexities of wear of abrasive materials at any scale lead us to believe that the variety of different and interacting wear mechanisms involved, namely, plastic flow of abrasive, crumbling of the abrasive, chemical wear etc., makes the wear of diamond at the nanoscale too complicated to be explained using a single theoretical model [3]. The following analysis of diamond grains represented by loaded wedges assumes that grain fracture is the dominant wear mechanism when grinding at the nanoscale using the piezoelectric nanogrinding process.

7.3 Stress Analysis in a Nanogrinding Grain

7.3.1 Analysis of Loaded Nanogrinding Grains

Diamond grains are blocky in nature and possess sharp cutting points prior to nanogrinding workpiece materials. Figure 7.2 shows a collection of diamond abrasive grains that have well defined cutting points that form a wedge at their apex. When bonded into a strong matrix, these grains can be considered to be representative infinite wedges.

An infinite wedge represents the cutting point of an abrasive grain in contact with the workpiece material (Fig. 7.3).

The wedge is loaded at the apex by a load P in an arbitrary direction at angle ω to the axis of symmetry of the wedge. Resolving the force into components P.cos ω in the direction of the axis, and P.sin ω perpendicular to that the stresses due to each of these forces can be evaluated from two-dimensional elastic theory [4]. The state of stress in the wedge, due to force P.cos ω, can be obtained from the stress function,

$$\phi = C.r.\theta.\sin\theta \tag{7.2}$$

Fig. 7.2 A collection of diamond grains showing cutting points located at their apex, and locations of {100} and {110} planes. Diamond grains are approximately 60 μm in diameter. Re-printed with kind permission from Springer Science+Business Media B.V.

Fig. 7.3 The single point, loaded infinite wedge. Re-printed with kind permission from Springer Science+Business Media B.V.

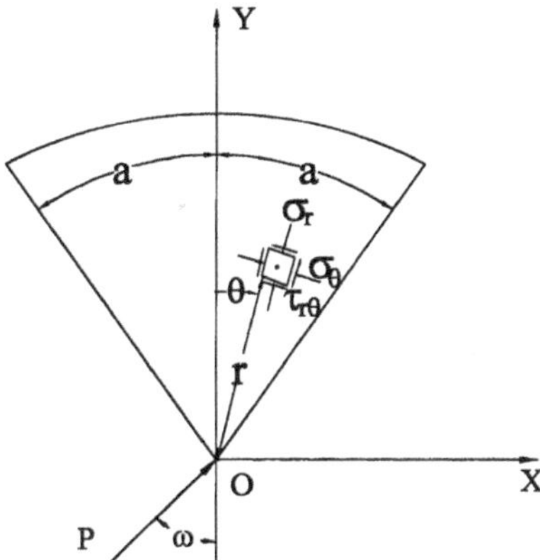

where r and θ are polar coordinates, or the point N in Fig. 7.4, and C is a constant.

The stress function yields the following radial, tangential, and shear stress components,

$$\sigma_r = -2C\frac{\cos\theta}{r} \tag{7.3}$$

$$\sigma_\theta = 0 \tag{7.4}$$

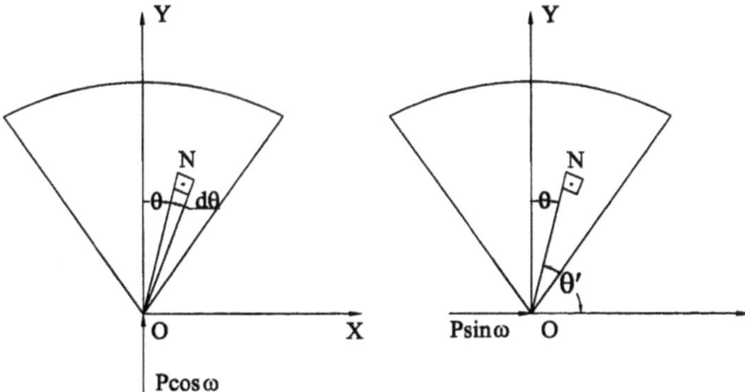

Fig. 7.4 The single-point, loaded infinite wedge showing force components, and the point N within the wedge at polar co-ordinates, r and θ. Re-printed with kind permission from Springer Science+Business Media B.V.

$$\tau_{r\theta} = 0 \tag{7.5}$$

To determine the constant, C, the equilibrium of forces along the y-axis is:

$$P\cos\omega - \int_{-a}^{a} \sigma_{r.} \cos\theta dA = 0 \tag{7.6}$$

Where dA is an element of cross sectional area within the wedge. If, t, is the thickness of wedge, then,

$$\begin{aligned} \cos\omega P &= \int_{-a}^{a} 2C\frac{\cos\theta}{r}.t.r.\cos\theta d\theta \\ &= 2Ct\int_{-a}^{a} \cos^2\theta d\theta \\ &= \mathrm{Ct}[2\mathrm{a} + \sin 2\mathrm{a}] \end{aligned} \tag{7.7}$$

Therefore,

$$C = \frac{P\cos\omega}{t(2a + \sin 2a)} \tag{7.8}$$

And,

$$\sigma_r = -\frac{2P\cos\theta.\cos\omega}{r.t(2a + \sin 2a)} \tag{7.9}$$

Note that the negative sign denotes that the stress is compressive. The state of stress in the wedge, due to force P sin ω, can be obtained from the stress function,

$$\phi = C'.r.\theta'.\sin\theta' \tag{7.10}$$

Therefore,

$$\sigma_r = -2C\frac{\cos\theta'}{r} \tag{7.11}$$

$$\sigma_\theta = 0 \tag{7.12}$$

$$\tau_{r\theta} = 0 \tag{7.13}$$

Equilibrium of forces along the x-axis (Fig. 7.4) yields the following solution for the constant, C,

$$P\sin\omega - \int_{\pi/2-a}^{\pi/2+a} \sigma_r.t.r.\cos\theta' d\theta' = 0 \tag{7.14}$$

$$P\sin\omega = -\int_{\pi/2-a}^{\pi/2+a} 2C\frac{\cos\theta'}{r}.t.r.\cos\theta' d\theta'$$

$$= 2Ct.\int_{\pi/2-a}^{\pi/2+a} \cos^2\theta' d\theta'$$

$$= -\mathrm{C.t}(2\mathrm{a} - \sin 2\mathrm{a}) \tag{7.15}$$

$$C = \frac{P\sin\omega}{t(2a - \sin 2a)} \tag{7.16}$$

Thus,

$$\sigma_r = -\frac{2P\cos\theta'.\sin\omega}{r.t(2a - \sin 2a)} \tag{7.17}$$

Expressing in terms of the angle θ (where θ′ is negative), yields,

$$\sigma_r = -\frac{2P\cos\theta.\sin\omega}{r.t(2a - \sin 2a)} \tag{7.18}$$

Therefore, the combined stresses are:

$$\sigma_r = -\frac{2P}{r.t}\left[\frac{\cos\omega\cos\theta}{2a + \sin 2a} + \frac{\sin\omega\cos\theta}{2a - \sin 2a}\right] \tag{7.19}$$

It follows that σ_r vanishes for angle θ_o defined using the expression:

$$\tan\theta_o = \frac{1}{\tan\omega}\cdot\frac{2a - \sin 2a}{2a + \sin 2a} \tag{7.20}$$

This equation corresponds to a straight line through the apex as shown in Fig. 7.5. This natural axis separates the regions of compressive and tensile stresses. It can be seen that for values of angle ω which gives, $|\theta_o| > |a|$, the neutral axis lies outside the included angle of the wedge. This means that the whole area of the wedge will be under stresses of uniform sign. Expressing Eq. (7.19) in terms of the rake angle of the abrasive grain, β, and force components F_t and nF_t (Fig. 7.6), yields,

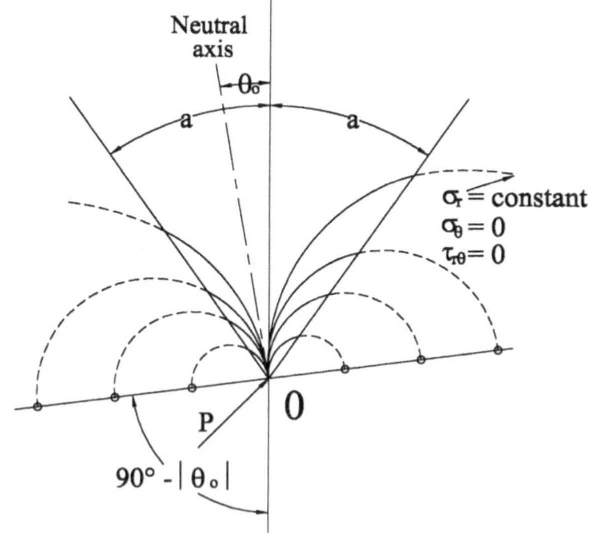

Fig. 7.5 Stress analysis of a single-point loaded wedge. Re-printed with kind permission from Springer Science+Business Media B.V.

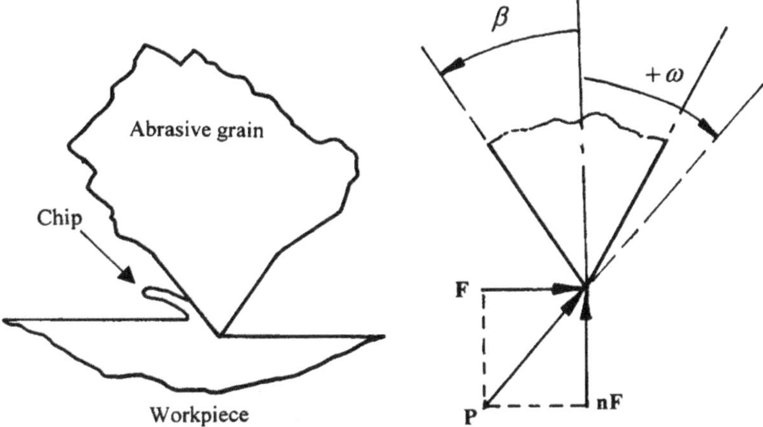

Fig. 7.6 Ideal wedge-shaped cutting point and grinding force diagram. Re-printed with kind permission from Springer Science+Business Media B.V.

$$\sigma_r = \left\{ -\frac{2F_t}{r.t} \frac{[n.\cos(a-\beta) + \sin(a-\beta)]\cos\theta}{2a + \sin 2a} + \frac{[\{\cos(a-\beta) - n.\sin(a-\beta)\}\cos\theta]}{2a - \sin 2a} \right\} \tag{7.21}$$

It can be observed that:

$$\tan\omega = \frac{\cos(a-\beta) - n.\sin(a-\beta)}{n.\cos(a-\beta) + \sin(a-\beta)} \tag{7.22}$$

In the simple case of a wedge with the normal force nF_t along the wedge axis, a is equal to β, hence,

$$\tan\omega = 1/\mathrm{n} \tag{7.23}$$

It is interesting to examine the radial stresses on the left-hand face of the wedge, which corresponds to the leading face of idealized wedge. Thus, for the left-hand face, θ is equal to −a, and from Eq. (7.19),

$$\sigma_r = -\frac{2P}{r.t}\left[\frac{\cos\omega\cos a}{2a + \sin 2a} + \frac{-\cos a.\sin\omega}{2a - \sin 2a}\right] \tag{7.24}$$

This stress is zero, i.e., the neutral axis coincides with the left-hand limit of the wedge, when,

$$\frac{1}{\tan\omega}=\frac{\sin a(2a+\sin 2a)}{\cos a(2a-\sin 2a)} \tag{7.25}$$

Thus if,

(a) a = β, then,

$$n=\frac{\sin a}{\cos a}\cdot\frac{2a+\sin 2a}{2a-\sin 2a} \tag{7.26}$$

(b) $a-\beta=\frac{\pi}{2}-a$

(as is the case when F_t is parallel to the right-hand face of the wedge). From Eq. (7.23),

$$\tan\omega=\frac{\sin a-n.\cos a}{n.\sin a+\cos a}=\frac{\sin a-n\cos a}{n.\sin a+\cos a} \tag{7.27}$$

And substituting in (7.25), yields,

$$\frac{n\sin a+\cos a}{\sin a-n\cos a}=\frac{\sin a}{\cos a}\cdot\frac{2a+\sin 2a}{2a-\sin 2a} \tag{7.28}$$

$$\frac{\frac{1}{2}.n\sin 2a+\cos^2 a}{\sin^2 a-\frac{1}{2}.n.\sin 2a}=\frac{2a+\sin 2a}{2a-\sin 2a} \tag{7.29}$$

$$\therefore\quad n=\frac{1}{2a}-\cot 2a \tag{7.30}$$

Equation (7.25) expresses the condition for the whole of the wedge's cross-sectional area to be under compressive stress. It can be seen that this depends not only upon the rake angle, β, but also upon the ratio, n. In general the relative size of the region of compressive to the region of tensile stresses depends upon β and n as Eqs. (7.20) and (7.22) indicate. Also, from Eq. (7.21), the magnitude of the stress on the left-hand face of the wedge is found to be dependent upon the tangential force component, F_t, and the force component ratio, n. Referring to Eq. (7.19), it can be seen that for constant stress, σ_r = constant,

$$r.C_1=C_2.\cos\theta+C_3.\sin\theta \tag{7.31}$$

where C_1, C_2, C_3 are constants. Equation (7.31) represents, in polar co-ordinates, the circumference of a circle tangent to the line,

$$0 = C_2 . \cos\theta + C_3 . \sin\theta \tag{7.32}$$

that is to the neutral axis at the point r = 0. However, the point r = 0 must be considered separately because the stress at that point approaches infinity, since by definition P is a point load. The central point of these circles are of constant radial stress, and so the point of constant maximum shear stress must lie on a line perpendicular to the neutral axis at the point where r is equal to zero. The radius of each of those circles depends upon the magnitude of the radial stress, σ_r.

7.4 Fracture Dominated Wear Model

Brittle materials exhibit high strength properties when loaded in compression than in tension. The ratio of rupture strengths is usually between 3:1 and 10:1. The existence of relatively low tensile stresses in the abrasive grains may cause failure by fracture to occur. To model the action of diamonds bonded to piezoelectric ceramics, one must consider a single active cutting grain to be classed as a wedge of constant width loaded at its inverted apex with point loads, F, and, nF, which represent the radial and tangential force components with reference to the grinding wheel in which the grain is supported, and P is the resultant force (Fig. 7.6). The stress distributions within point-loaded wedges can be determined analytically, and the results of such an analysis indicate that if tensile stresses exist within the wedge then it will occur at its maximum along the rake face. The existence of a tensile stress depends on the magnitude of the force ratio, n. If the ratio is especially small that a tensile stress exists in the wedge, then for a specific force ratio the tensile stress is proportional to the tangential grinding force, F. Stresses of this nature would extend to and beyond the abrasive grain-bonding interface. The fracture of abrasive grain, bonding phase, or the interface between the two, depends on the particular type of piezoelectric ceramic material used and the magnitude of the tensile stress induced during nanogrinding.

Grains of diamond are 10 times stronger in compression than in tension. The probability of grain fracture is likely to increase with an increase in tensile stress exerted in the grain although the magnitude of the stress may be slightly higher than one-fifth the magnitude of the maximum compressive stress in the grain. A significant barrier to the acceptance of stress patterns evaluated for such situations arises because point loads applied to perfectly sharp wedges produce infinitely high stresses at, and about, the point of contact. Loads must be applied over a finite area. It seems likely that higher tensile stresses are associated with higher grain fracture probability resulting in rapid loss of diamond grains and, consequently, lower grinding ratios. The wear model should incorporate the fact that the loads are spread over a finite area. This implies that point loads are applied along the rake face. The model should allow the examination between the wear rate of a diamond coated piezoelectric ceramic material and the general nature of stresses established in active cutting grains subjected to nanogrinding forces. This means that it is necessary to estimate the force components of grinding on each active cutting grain.

7.5 Nanogrinding

7.5.1 Nanogrinding Apparatus

The experimental apparatus consisted of holding a polished specimen between the jaws of a vice so that a piezoelectric crystal oscillator traverses back-and-forth the specimen thus machining the specimen by creating a depth of cut between the diamonds adhered on the piezoelectric crystal and the workpiece material. Workpiece materials were polished with a 100 nm sized polishing compound. All samples were divided into four sections and each section was analyzed prior to machining and after machining occurred. The workpieces were mounted in a vice that was attached to a x-y-z linear slide in order to achieve accurate positioning of the workpiece. The piezoelectric crystal was mounted on a steel framework that was orthogonal to the workpiece. The whole unit was located within a tetrahedral space frame to dampen excess vibrations (Fig. 7.7).

7.5.2 Nanogrinding Procedure

When the crystal and workpiece were aligned, the depth of cut was incremented in stages of 10 nm. The motion of the diamonds attached to the piezoelectric crystal generates a machining effect that is caused by the action of diamonds grinding into the workpiece material. Tracks or trenches are created by the diamond grain gouging the surface of the material when an electric current is applied. The material is removed until the end of the oscillating motion creates the material to plow.

The mechanism of oscillation can be described as a restricted bending mode that simulates a shear displacement of the crystal. At this point it is normal procedure to know how to estimate the number of grains contacting the surface of the workpiece. The estimation of the number of active cutting grains is estimated quite simply by driving the diamond coated piezoelectric ceramic at the prescribed specific metal removal rate into a piece of lead. The impression that the grinding wheel produces in the length of lead is equal to the number of cutting points that are active during the grinding stroke at that particular depth of cut. The effect of oscillating the diamond-coated crystal is shown in Fig. 7.8.

The motion of the diamonds imparted by oscillating the crystal in the bending mode that causes a shear displacement to occur, which contributes to plowing of the material at the end of the nanogrinding stroke, is shown in Fig. 7.9.

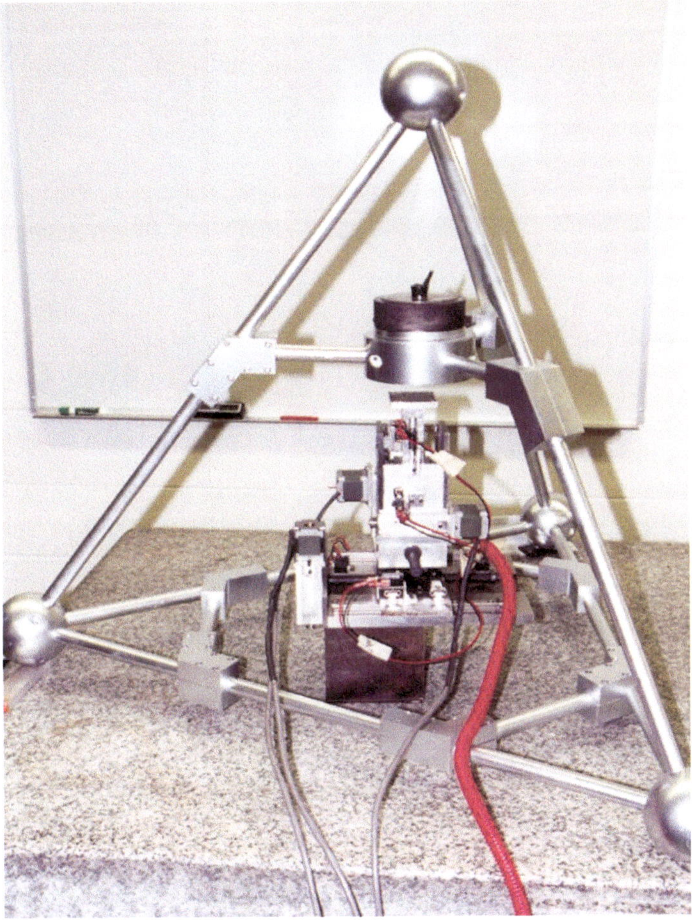

Fig. 7.7 Piezoelectric machining center capable of nanogrinding. Re-printed with kind permission from Springer Science+Business Media B.V.

The effect of using the piezoelectric to machine tracks, or trenches, in engineering materials opens up the prospect of nanomanufacturing products that require geometric features such as channels so that fluids and mixed phase flows can be manipulated in devices such as micro- and nanofluidic "lab-on-a-chip" products. Figure 7.10 clearly shows such a channel produced using the piezoelectric nanogrinding process.

The measured force components of the nanogrinding operation are measured using a dynamometer. These components of force are then applied to a model abrasive grain by dividing the grinding force data into the number of active cutting grains over an area that simulates the abrasive grain-workpiece contact area. Stresses established in this area are calculated using finite elements. The wear of the

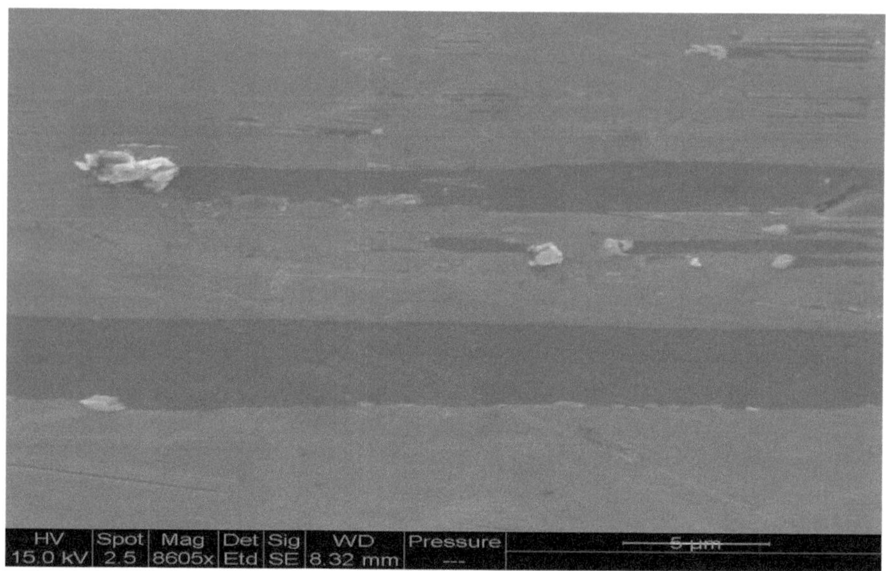

Fig. 7.8 Material removal by oscillating a diamond-coated piezoelectric ceramic crystal. Re-printed with kind permission from Springer Science+Business Media B.V.

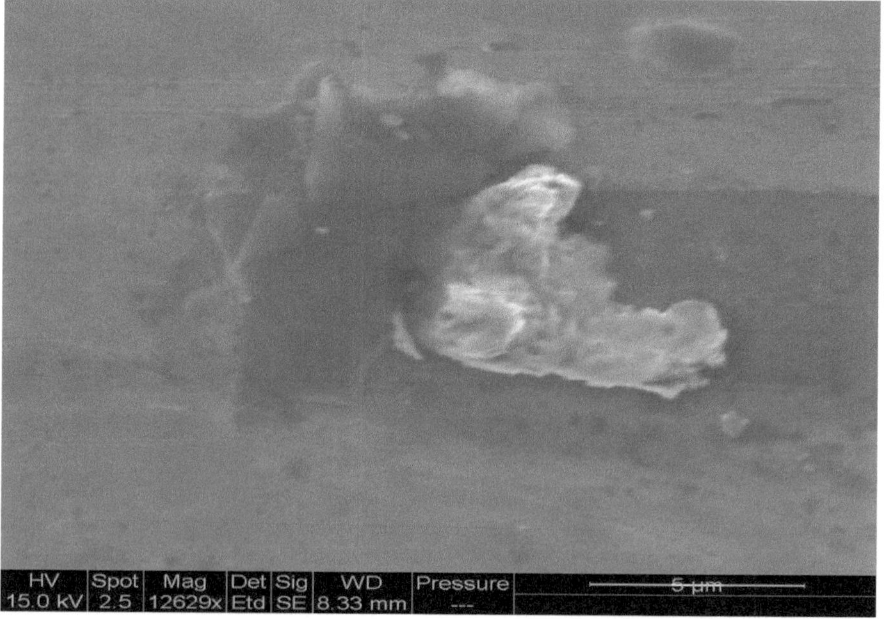

Fig. 7.9 Plowed material at the end of a nanomachined track. Re-printed with kind permission from Springer Science+Business Media B.V.

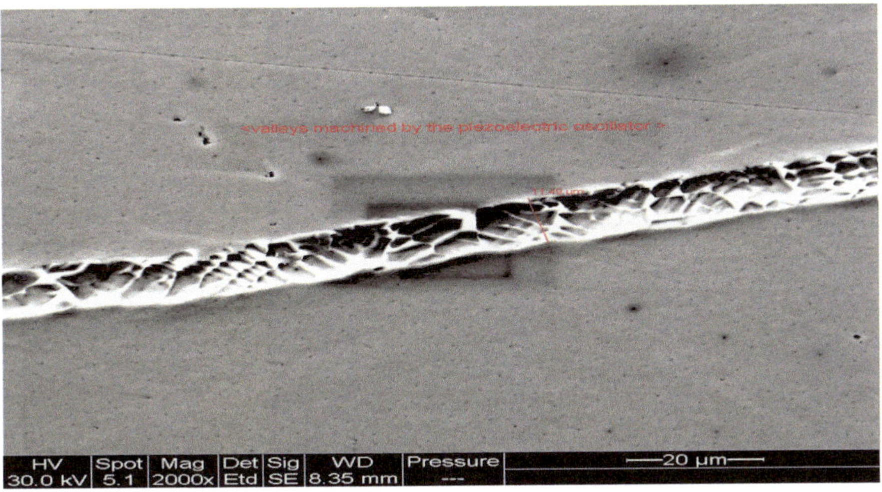

Fig. 7.10 Nanomachined tracks in steel. Re-printed with kind permission from Springer Science +Business Media B.V.

piezoelectric material by diamond loss, expressed in terms of a grinding ratio, and its relationship to the stress levels set up in the model grain is investigated using a stress analysis method.

7.5.3 Stress Analysis

The assumed geometry of an ideal grain in the vicinity of its cutting edge is a simple symmetrical wedge of constant width with an included angle of 70° that results in a rake angle of −35°. There is no wear flat on the model cutting grain. In order that a finite element method is used to evaluate stresses in the wedge, the wedge was subdivided into 210 diamond-shaped elements with a total of 251 nodes. Forty-one nodes were constrained at the boundary of the wedge and the leading five nodes on the rake face were loaded. The tangential and normal grinding forces were replaced by equivalent forces acting perpendicular to (normal load) and along (shear load) the rake face of the wedge.

The concentrated loads at the five nodes are representative of the distributed and normal loads acting on the rake face over the abrasive grain-chip contact length. The normal force distribution on the rake face was taken to be maximum at the cutting edge and decreasing linearly to zero at the end of the abrasive grain-chip contact length. The shear force was taken to be constant over the first half of the contact length, decreasing linearly to zero over contact length. Grinding loads were also applied directly to the rake face and at the tip of the grain without calculating equivalent forces. This was performed in order to compare and contrast the effect of different force distributions on the stresses generated within the wedge. To measure the value of using the maximum tensile stress as a way to estimate grain fracture tendency, the correlation between the two sets of data were calculated for each set of data. The region of fracture initiation was also located using Griffith's criterion of fracture. For,

$$\frac{\sigma_c}{\sigma_t}.\sigma_1 + \sigma_3 > 0 \tag{7.33}$$

Then,

$$\sigma_1 = \sigma_t \tag{7.34}$$

But for,

$$\frac{\sigma_c}{\sigma_t}.\sigma_1 + \sigma_3 < 0 \tag{7.35}$$

Then,

$$(|\sigma_1| - |\sigma_3|)^2 + 8\sigma_t(|\sigma_1| - |\sigma_3|) = 0 \tag{7.36}$$

where σ_1 and σ_3 are the principal stresses, assuming that $\sigma_1 > \sigma_3$, σ_t is the ultimate tensile strength of the abrasive grain, and σ_c is the ultimate compressive strength. For diamond grain material, the ratio of σ_t and σ_c is 0.1. The results of the two-dimensional stress analyses were consistent with the experimentally determined stress distribution obtained by Loladze [5] when cutting soft metal with photoelastic tools. The maximum tensile stress always occurs at the rake face at a distance from the cutting edge ranging from 1.5 to 4 times the abrasive grain-chip contact length, the exact magnitude of the coefficient depends on the loading conditions for a particular machining event. For a given value of the tangential force component, F, the higher the force ratio, F/nF, the greater the distance the maximum tensile stress is away from the cutting edge.

These results indicate that mechanically induced fracture occurs at a finite distance away from the cutting edge. When using Griffith's criterion, the influence of mechanically induced stresses indicate that fracture initiation zones are established. Figure 7.11 shows the occurrence of such zones in an idealized wedge. The first zone is located around the point of maximum tensile stress and is always at the rake face. Failure in this zone is tensile in nature and would initiate fracture at a point on the rake face of the order of two-to-three times the abrasive grain-chip contact length away from the cutting edge. This type of fracture is consistent with fracture on scale comparable with the chip thickness. The second much smaller zone is located at the immediate vicinity of the cutting edge. Failure is compressive in this region and results in small-scale crumbling of the cutting edge leading to the formation of a wear flat on the abrasive grain.

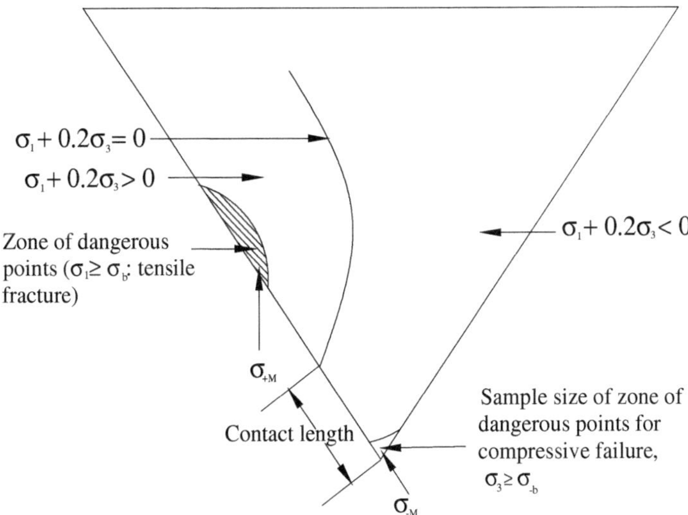

Fig. 7.11 Griffith's criterion applied to the idealized wedge showing tensile and compressive fracture initiation zones. Re-printed with kind permission from Springer Science+Business Media B.V.

The correlation between the magnitude of the maximum tensile stress in the model abrasive grains and the appropriate grinding ratio (Table 7.1) is high and is dependent on the way the forces are applied to the grains. It would be expected that the higher the tensile stress, the greater is the rate of diamond wear and consequently the corresponding grinding ratio. Perfect linear correlation in accordance with this would result in a correlation coefficient of −1. The correlation coefficient between the maximum tensile stress and the grinding ratio is significant. This is to be expected as the force ratio may vary slightly. However, if the tangential component of the grinding force changes significantly without a change in force ratio, then it is expected that the maximum tensile stress will change significantly and reduce the grinding ratio. The calculation and application of equivalent grinding loads produce a lower correlation coefficient compared to directly applied grinding loads. This implies that grinding loads are simply not point loads acting act the tip of the inverted apex and along the abrasive grain-chip contact length of the diamond grain. In fact, directly applied grinding forces produce better correlation coefficients. This means that for perfectly sharp diamond grains, one must apply the component grinding loads directly to the rake face.

It can be seen from Table 7.1 that induced tensile stresses account for the loss of grain material from the diamond coated piezoelectric ceramic material. Therefore, the maximum tensile stress is the best indicator of diamond performance, in terms of grinding ratio, during a nanogrinding operation. The analysis performed on perfectly sharp diamond grains has provided a strong correlation between

Table 7.1 Correlation coefficient between maximum tensile stress and grinding ratio for an idealized wedge using experimental data

Workpiece material	Exact wedge model with point loads applied to apex of wedge	Approximate finite element model: equivalent grinding forces applied to rake face of wedge	Approximate finite element model: grinding forces applied directly to the rake face of the wedge
Diamond on single crystal LiF	−0.7	−0.8	−0.9
Diamond on single crystal MgO	−0.6	−0.7	−0.8
Diamond on GaAs	−0.67	−0.68	−0.74
Diamond on sapphire	−0.72	−0.89	−0.97
Diamond on silicon	−0.85	−0.87	−0.9

Comparison is also made between the methods of applying loads to the idealized wedge models. Re-printed with kind permission from Springer Science+Business Media B.V.

maximum tensile stress induced in the grain material and the wear parameter, grinding ratio, for the experimental data used in this chapter.

Correlations with other data sets have not proved so fruitful. From this, we can safely assume that the mechanism of grain fracture is not the dominant mechanism, which implies that other mechanisms are operating. The correlation coefficient demonstrates that a tougher grain material must be used in order to limit the effects of abrasive wear and the formation of wear flats, or a stronger bond, and possibly a higher volume of bond between diamond and piezoelectric crystal, is required to nanogrind under the current experimental conditions. Therefore, the present method of calculating the correlation coefficient between the maximum tensile stress and the grinding ratio demonstrates its potential application to the wider problem of selecting abrasive grains based on specific metal removal rates and the nature of the nanogrinding operation. When porous tools are used to embed diamonds or any other abrasive material, the same analysis can be used but account of the properties of the bonding bridge must be made. The bonding bridge can be made of a variety of different materials but the most common one used for dressable applications is the vitrified type, which is made from a mixture of clays, glasses, and minerals. The emphasis on using dressable types for nanogrinding is based on their ability to be re-sharpened by dislodging worn grains and by microstructural phase transformations by focusing optical energy on the bonding bridges that hold the grains in place.

7.6 Porous Nanogrinding Tools

Porous nanogrinding tools are composed of abrasive particles (sub micron size) embedded in a vitrified bond with porosity interspersed between grinding grains and bonding bridges. The porosity level is approximately 15–21 %. Figure 7.12

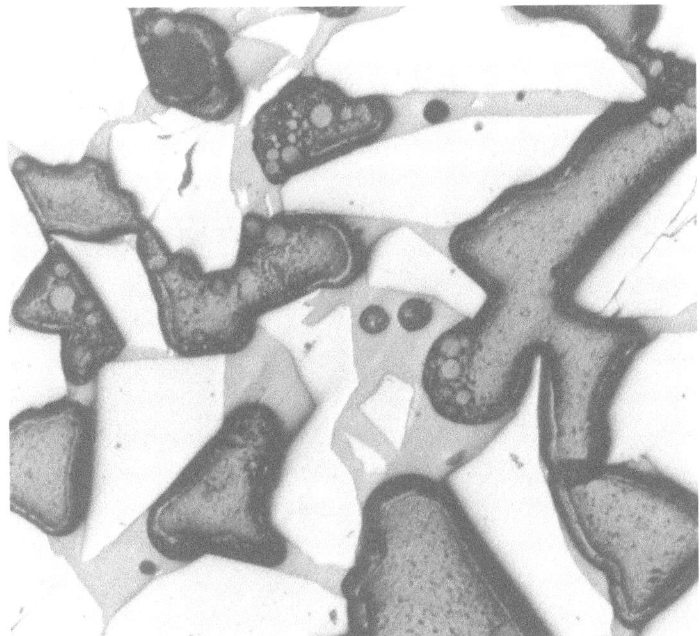

Fig. 7.12 Structure of the porous tool used for nanogrinding. Re-printed with kind permission from Springer Science+Business Media B.V.

shows the image of a nanogrinding tool prior to laser modification. The vitrified bonds are specially engineered to promote the formation of texture that creates ridges of cutting planes and nanogrinding "peaks" of α-Al_2O_3 in the preferred (012),(104), and (110) planes. The peaks created due to laser modification of the surface aid the nanogrinding process. Vitrified bonds are composed of glasses that are formed when clays, ground glass frits, mineral fluxes such as feldspars, and chemical fluxes such as borax melt when the grinding wheel is fired at temperatures in the range, 1000–1200 °C. With reference to raw material nomenclature, a "frit" is a pre-ground glass with a pre-determined oxide content, a "flux" is a low melting point siliceous clay that reduces surface tension at the bond bridge-abrasive grain interface, a "pre-fritted" bond is a bond that contains no clay minerals (i.e., clays and fluxes), and "firing" refers to vitrification heat treatment that consolidates the individual bond constituents together [6]. Considering individual bond constituents, mineral fluxes and ground glass frits have little direct effect on the ability to manufacture grinding wheels. However, most clays develop some plasticity in the presence of water (from the binder), which improves the ability to mould the mixture so that the wheel, in its green state, can be mechanically handled.

Clays and clay-based fluxes contain an amount of free quartz that has a detrimental effect on the development of strength during vitrification heat treatment. Clays are used to provide vitrified grinding wheels with green strength during the

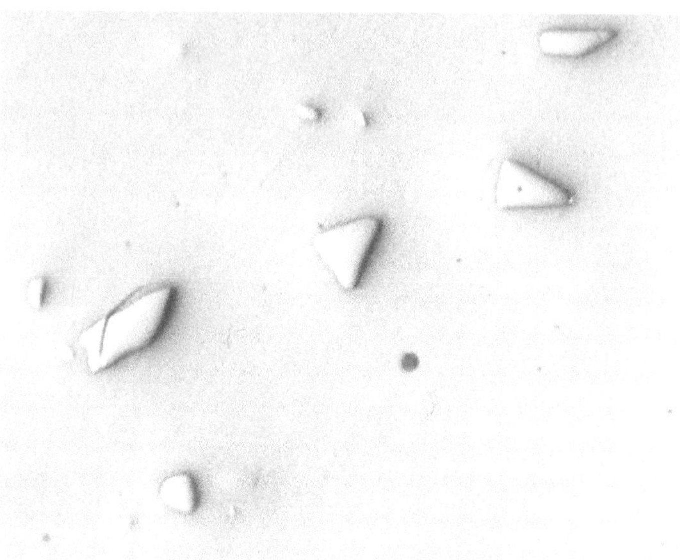

Fig. 7.13 A collection of quartz particles in a vitrified bonding system. The quartz particle on the left has a circumferential crack extending into the dissolution rim. Re-printed with kind permission from Springer Science+Business Media B.V.

heat treatment process. However, when the glass material solidifies around the particles of clay and quartz, the displacive transformation of quartz during the cooling stage of vitrification leads to the formation of cracks in the glass around the quartz particle (Fig. 7.13). The strength of the bonding bridge is impaired and leads to the early release of the abrasive particle during the cutting of metal. The basic wear mechanisms that affect vitrified grinding wheels are concerned with grain fracture during metal cutting, fracture of bond bridges, mechanical fracture of abrasive grains due to spalling, and fracture at the interface between abrasive grain and bond bridge. Failure in vitrified silicon carbide grinding wheels is more probable due to the lack of a well-developed bonding layer between abrasive grain and glass bond-bridge. The bonding layer is approximately a few micrometers in thickness, and is caused by the use of a high clay content bonding system. High glass content bonding systems tend to aggressively decompose the surface of silicon carbide abrasive grains. In vitrified corundum grinding wheels, high glass content bonding systems are used extensively and lead to bonding layers in excess of one hundred micrometers in thickness.

In addition to the formation of very thin bonding layers in vitrified silicon carbide grinding wheels, the use of high clay content bonding systems implies that there is an increase in the amount of quartz in the bond bridges between abrasive grains. Although the likelihood of decomposition of silicon carbide surfaces is reduced, the probability of bond bridge failure is increased due to the increased quartz content. Therefore, the dissolution of quartz is of paramount importance in order to compensate for thinner interfacial bonding layers. The dissolution of quartz in a liquid phase does not require a nucleation step. One process that determines the rate of the overall reaction is the phase-boundary reaction rate that is fixed by the movement of ions across the interface. However, reaction at the phase boundary leads to an increased concentration at the interface. Ions must diffuse away from the reaction interface so that the reaction can continue. The rate of material transfer and the diffusion rate are controlled by molecular diffusion in the presence of a high-viscosity liquid phase. For a stationary solid in an unstirred liquid, or in a liquid with no fluid flow produced by hydrodynamic instabilities, the rate of dissolution is governed by molecular diffusion.

The effective diffusion length over which mass is transported is proportional to $\sqrt{Dt}$, where D is the diffusion coefficient and t is time, and therefore the change in thickness of the solid, which is proportional to the mass dissolved, varies with $\sqrt{t}$. Natural, or free, convection occurs because of hydrodynamic instabilities in the liquid which gives rise to fluid flow over the solid. This enhances the kinetics of dissolution. Generally, a partially submerged solid undergoes more dissolution near to the solid-liquid interface. Below this interface the kinetics of dissolution of the solid can be analyzed using the principles of free convection. The boundary layer thickness is determined by the hydrodynamic conditions of fluid flow. Viscous liquids form much thicker boundary layers that tend to impede material transfer. Higher liquid velocities promote the formation of thinner boundary layers and permit more rapid material transfer. Considering the dissolution of quartz in glass materials, the high viscosity and slow fluid flows combine to give thick boundary layers. Also, the diffusion rate is much slower in viscous silicate liquids than in aqueous solutions, thus giving a tendency for the reaction process to be controlled by material-transfer phenomena rather than by interface reactions.

Difficulties encountered when developing a dissolution model arise from the fact that the phase boundary between quartz particle and molten glass moves during the diffusion process. The problem of a fixed boundary can be solved without difficulty although this is not equivalent to the conditions associated with a moving boundary between quartz particle and a highly viscous glass melt. The development of dissolution models is required to determine the magnitude of quartz remaining in the bonding system after a period of heat treatment. The models are then compared with experimentally determined quartz content of the bonding systems using X-ray diffraction techniques.

7.6.1 Dissolution Models for Quartz in Bonding Bridges

When densification occurs in a vitrified grinding wheel, the cooling rate is reduced to prevent thermal stress cracking in the bonding layer between abrasive particles. Cooling rates are reduced when crystalline inversions occur that involve volume changes. The inversion range for quartz and cristobalite are 550–580 °C and

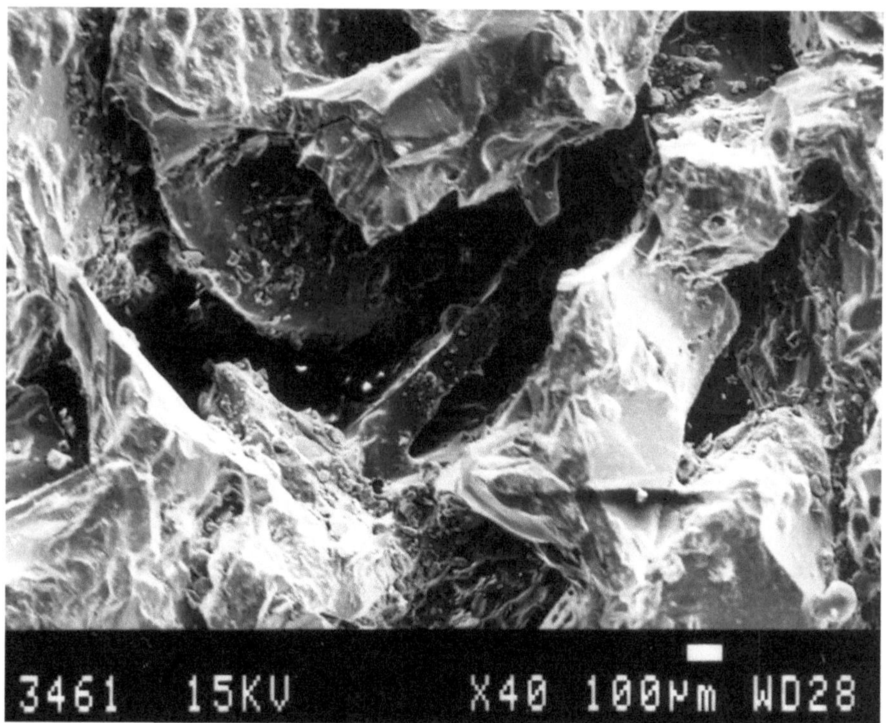

Fig. 7.14 Bonding bridge failure in a vitrified grinding wheel caused by the displacive transformation of quartz at high temperature during heat treatment. Re-printed with kind permission from Springer Science+Business Media B.V.

200–300 °C, respectively. Since the formation of cristobalite is rare in most vitrified bonding systems used for grinding wheels, the rapid displacive transformation of quartz tends to promote the formation of cracks in bonding bridges (Fig. 7.14). Once the grinding grain is lost the remaining bonding bridges can be modified using a high power laser to create an oriented texture that forms "peaks" of α-Al_2O_3 in the preferred (012), (104), and (110) planes (Fig. 7.15).

When quartz-containing bonds begin to cool form the soaking, or vitrification, temperature it is thought that the liquid phase relieves stresses resulting from thermal expansion mismatch between itself and the phases, β-quartz, β-cristobalite, and mullite, to at least 800 °C. At 800 °C, stresses will develop in quartz particles and the matrix that causes micro-cracking to occur. The shrinkage behaviour of quartz and the glass phase has been described by Storch et al. [7]. Between the temperature range, 573 and 800 °C, the glass phase shrinks more than the quartz phase that causes tangential tensile stresses to form cracks in the matrix. At 573 °C, β-quartz transforms to α-quartz that causes residual stresses to produce circumferential cracking around quartz particles (Fig. 7.14). Some of these cracks have been seen to propagate into the glass phase [8]. Similar observations occur in the

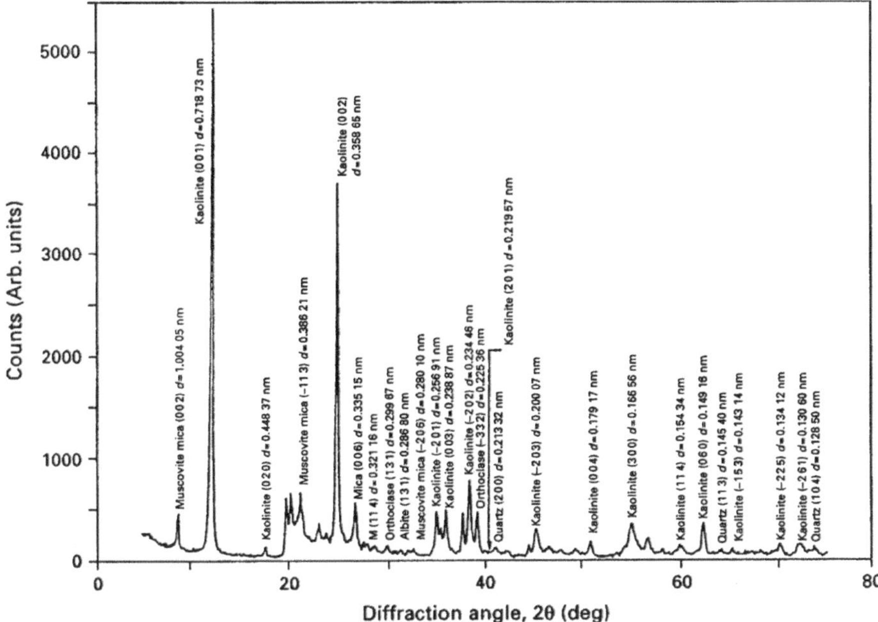

Fig. 7.15 X-ray diffraction spectrum of China clay showing crystallographic planes and interplanar distances of various mineral phases in the clay. Re-printed with kind permission from Springer Science+Business Media B.V.

cristobalite phase. Spontaneous cracking of quartz has been found to occur over a temperature range that depends on the size of the quartz particles [9]. Particles larger than 600 μm diameter cracked spontaneously at 640 °C, whereas smaller particles of less than 40 μm diameter cracked at 573 °C. This observation agrees with temperature-dependent micro-cracking reported by Kirchhoff et al. [10]. To maintain the integrity of the bond bridges containing coarse quartz particles, the grinding wheel must remain at the vitrification temperature until the quartz particles have dissolved.

The dissolution model assumes that at a constant absolute temperature, T, a particle of quartz melts in the surrounding viscous glass melt, and that the rate of change of the volume of quartz present in the melt at a particular instant in time is proportional to the residual volume of quartz. The above assumption is based on the fact that alkali ions diffuse from the viscous glass melt to the boundary of the quartz particle, thus producing a dissolution rim around each quartz particle. A high reaction rate will initially occur which continuously decreases as the quartz particle is converted to a viscous melt. Jackson and Mills [11] derived a mathematical relationship that accounts for the change in density when β-quartz transforms to α-quartz on cooling from the vitrification temperature, thus,

$$m_{T,t} = M\gamma \exp\left(-At^{1/2}\exp\left[\frac{-B}{T}\right]\right) \tag{7.37}$$

where, $m_{T,t}$, is the residual mass fraction of quartz at a constant time and temperature couple, M is the original mass fraction of quartz prior to heat treatment, γ is the ratio of densities of β-quartz and α-quartz, A and B are constants, t is time, and T is absolute temperature. The model was compared with experimental data determined using the powder X-ray diffraction method. The experimental work was divided into two parts. The first part concentrates on comparing the dissolution model with X-ray diffraction data using "sintering" bond compositions that are used in vitrified silicon carbide nanogrinding tools, while the second part focuses on comparing the model with "fusible" bond compositions that are used in high-performance vitrified corundum nanogrinding tools.

7.6.2 Preparation of Bonding Bridges for Nanogrinding Wheels

The raw materials used in the experimental study were Hymod Prima ball clay, standard porcelain China clay, potash feldspar, and synthetic quartz (supplied as silica flour). The chemical analysis of the raw materials is shown in Table 7.2. Rational analysis of the raw materials was performed to reveal the mineralogical composition of the raw materials. The rational analysis appears in Table 7.3. The characteristic X-ray diffraction spectra for ball clay and China clay are shown in Figs. 7.1 and 7.16. The bond mixture described is one typically used in vitrified silicon carbide grinding wheels where the erosion of the abrasive grain is reduced by using high clay content bonding systems.

Fusible bonding systems using a mixture of ball clay and potassium-rich feldspar were made to test the model developed by Jackson and Mills [11]. The ball clay

Table 7.2 Chemical analyses of raw materials

Oxide (wt%)	China clay	Ball clay	Potash feldspar	Quartz
Al_2O_3	37	31	18.01	0.65
SiO_2	48	52	66.6	98.4
K_2O	1.65	1.8	11.01	0.35
Na_2O	0.1	0.2	3.2	0.04
CaO	0.07	0.2	0.09	0.00
MgO	0.03	0.3	0.09	0.00
TiO_2	0.02	0.9	0.00	0.07
Fe_2O_3	0.68	1.1	0.11	0.03
Loss on ignition	12.5	16.5	0.89	0.20

Table 7.3 Mineralogical analyses of raw materials

Compound (wt%)	China clay	Ball clay	Potash feldspar	Quartz
Quartz	4.05	12.77	4.93	98.40
Orthoclase	0.00	15.23	64.96	0.00
Kaolinite	79.70	62.71	2.17	0.00
Mica	13.94	0.00	0.00	0.00
Soda feldspar	0.8	1.69	27.07	0.00
Miscellaneous oxides/losses	1.51	7.60	0.87	1.60

Re-printed with kind permission from Springer Science+Business Media B.V.

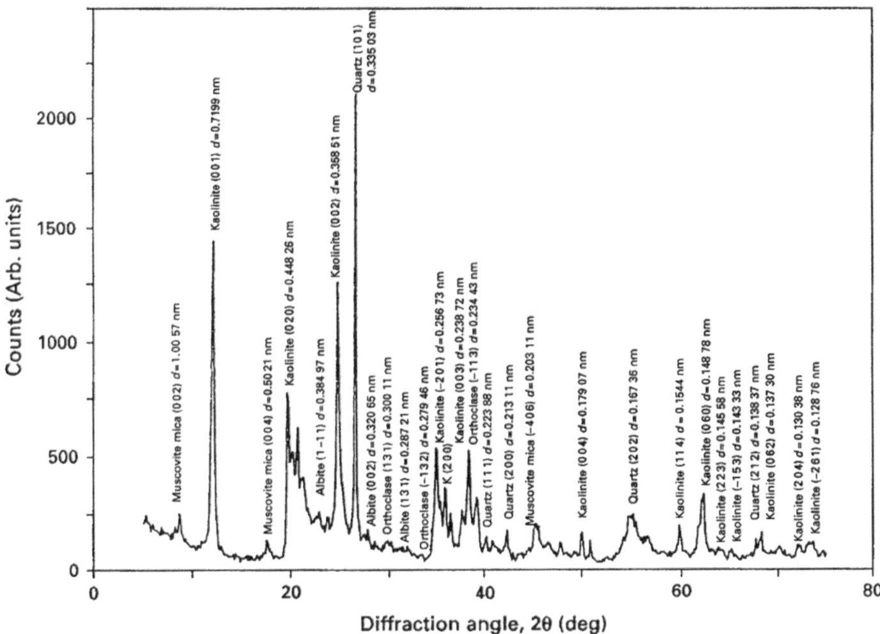

Fig. 7.16 X-ray diffraction spectrum of ball clay showing crystallographic planes and interplanar distances of various mineral phases in the clay. Re-printed with kind permission from Springer Science+Business Media B.V.

used contained 12.77 wt% quartz, and the feldspar contained 4.93 wt% quartz. The bonding system was composed of 66 wt% ball clay, and 34 % feldspar. The initial quartz content, M, of the bond mixture was 10.1 wt%. The bond mixture described is one typically used in high-performance vitrified corundum grinding wheels.

The raw materials were mixed in a mortar, pressed in a mold, and fired at various temperatures. A heating rate of 2.9 °C min^{-1} was employed until the vitrification temperature was reached.

The typical soaking temperature was varied between 1200 and 1400 °C for "sintering" bond compositions, and 950 and 1050 °C for "fusible" bond compositions in order to simulate industrial firing conditions. The samples were cooled at a rate of 1.8 °C min^{-1} to avoid thermal stress fracture. The fired samples were crushed to form a fine powder in preparation for X-ray diffraction.

7.6.3 X-ray Diffraction of Bonding Systems

The dissolution model was compared with experimental data using the X-ray powder diffraction method. X-ray diffraction of the raw materials was performed on a Phillips 1710 X-ray generator with a 40 kV tube voltage and a 30 mA current. Monochromatic Cukα radiation, λ = 0.154060 nm, was employed. A scanning speed of 2° per minute for diffraction angles of 2θ was used between 2θ angles of 10° and 80°, and the X-ray intensity was recorded using a computer. The spectrum was then analyzed and compared with known spectra. Powder specimens were prepared by crushing in a mortar and pestle in preparation for quantitative X-ray diffraction. To eliminate the requirement of knowing mass absorption coefficients of ceramic samples for quantitative X-ray diffraction, Alexander and Klug [12] introduced the use of an internal standard. First, the ceramic sample is crushed to form a powder—the sizes of particles should be small enough to make extinction and micro-absorption effects negligible. Second, the internal standard to be added should have a mass absorption coefficient at a radiation wavelength such that intensity peaks from the phase(s) being measured are not diminished or amplified. It should be noted that the powder diffraction mixture should be homogeneous on a scale of size smaller than the amount of material exposed to the X-ray beam, and should be free from preferred orientation. The powder bed that is subjected to "X-rays" should be deep enough to give maximum diffracted intensity.

The expected equilibrium phases from the fired mixtures are quartz (unreacted and partially dissolved), mullite, cristobalite and glass. However, from the samples tested, the compounds quartz, mullite and glass were successfully detected. A calibration curve was constructed using a suitable internal standard (calcium fluoride), a diluent (glass made by melting potash feldspar), and a synthetic form of the phase(s) to be measured. Synthetic mullite had a purity greater than 99.8 %, whilst powdered quartz had a purity greater than 99.84 % SiO_2. The method used for quantitative analysis of ceramic powders was developed by Khandelwal and Cook [13]. The internal standard gave a fairly intense (111) reflection (d = 0.1354 nm) lying between the (100) reflection for quartz (d = 0.4257 nm) and the (200) reflection for mullite (d = 0.3773 nm). Using copper kα radiation (λ = 0.15405 nm), the corresponding values of diffraction angle 2θ are: (100) quartz = 20.82°; (111) calcium fluoride = 28.3°; and (200) mullite = 32.26°. Figure 7.17 shows the calibration curve generated by varying proportions of calcium fluoride, synthetic quartz and mullite. Mass fractions of the crystalline phases in the mixture can be read from the calibration lines by measuring the intensity ratio of the phase(s) to the internal standard.

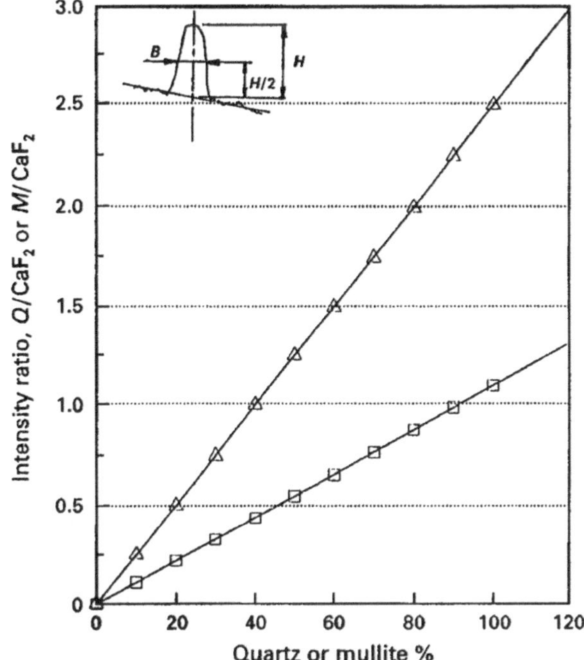

Fig. 7.17 Calibration curve for quantitative analysis of X-ray determined quartz and mullite using the CaF_2 (111) plane generated by the internal standard. Re-printed with kind permission from Springer Science+Business Media B.V.

Figure 7.18 shows the diffraction peaks of interest for quantitative analysis lying between 15° and 40° of the diffraction angle 2θ. The figure shows the reflections of the (111) plane of calcium fluoride, (200) plane of mullite, and the (100) plane of quartz. In order to calculate the mass fractions of quartz and mullite in the mixture, the height of the chosen diffraction peak and its width at half-height were measured from the diffraction spectrum. The product of these two measures were then compared with that of the internal standard, and the resultant intensity ratio was used to find the exact mass fraction of the phase(s) measured in the glass that has been subjected to X-rays.

7.6.4 Refractory Bonding Systems—Verification and Comparison of Dissolution Models for Quartz

In addition to comparing the experimental results to the dissolution model, results published in the literature were also used to test the accuracy of the model. The composition of the experimental mixtures was matched to those specified by Lundin [6]. Lundin's experimental mixtures were composed of 25 wt% quartz (13.2 μm particle size), 50 wt% clay (kaolin), and 25 wt% flux (potassium feldspar—25 μm particle size). The constants A and B for the sintering bonding system were calculated,

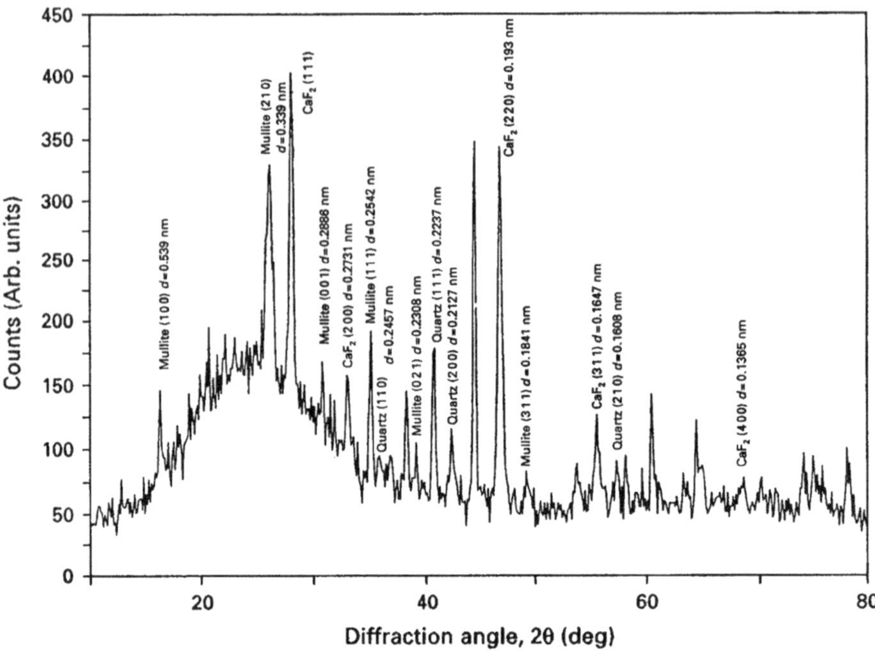

Fig. 7.18 X-ray diffraction spectrum of a vitrified bonding system showing the interplanar distances of crystallographic planes of mullite, quartz and calcium fluoride. Scan rate was 2° per minute. Re-printed with kind permission from Springer Science+Business Media B.V.

$$A = 5.62 \times 10^8 \tag{7.38}$$

$$B = 33374 \tag{7.39}$$

From which the experimental activation energy, Q, is 132.65 kcal/mole. The residual quartz content for the sintering bonding system is,

$$m_{T,t} = 26.25.\exp\left[-5.62 \times 10^8.t^{1/2}.e^{\frac{-33374}{T}}\right] \tag{7.40}$$

The data comparing Lundin's experimental results, the author's experimental results, and the dissolution model are shown in Table 7.4. When the data are plotted as the logarithm of $(-\ln[m/M]/t^{1/2})$ versus the reciprocal of absolute temperature, 1/T, then all data fits a straight-line relationship. The gradient was calculated to be 33,374, the constant B, using two data points. Lundin's experimental gradient gave a value of 32,962 using the least squares method, and 34,000 for the present work. The corresponding activation energies for both systems are 131 kcal/mole for Lundin's work [6], and 135 kcal/mole for the present work, respectively. Figures 7.19 and 7.20 show the effects of time on residual quartz content at different temperatures according to Eq. (7.40) together with comparative experimental data.

A comparison was made with dissolution models published in the literature. One of the earliest models was derived by Jander [14]. The equation can be expressed:

$$\left(1 - \sqrt[3]{1 - Z}\right)^2 = \left\{\frac{C_1.D}{r^2}\right\}.t \tag{7.41}$$

where Z is the volume of quartz that has been dissolved, r is the original particle radius, and D is the diffusion coefficient for the diffusing species. This equation can be transformed into mass fractions using Archimedes' law, thus,

$$\left(1 - \sqrt[3]{\frac{m}{M}}\right)^2 = C_2.t \tag{7.42}$$

where C is a constant dependent on soaking temperature and initial particle size of quartz. Krause and Keetman [15] expressed the dissolution of quartz as a function of isothermal firing time, viz,

$$M - m = C_3.\ln t \tag{7.43}$$

where M is the initial quartz content, and m is the residual quartz content after time, t. The unit of time here is seconds such that after one second of firing the residual quartz content is equal to the initial quartz content. Monshi's dissolution model [16] can be transformed into the following equation assuming isothermal firing conditions:

$$\ln\left\{\frac{m}{M}\right\} = -C_6\sqrt{t} \tag{7.44}$$

Jackson and Mills' model [11] for isothermal firing conditions is transformed into:

$$\ln\left\{\frac{m}{\gamma.M}\right\} = -C_7\sqrt{t} \tag{7.45}$$

where γ is the ratio of densities of $\beta-$ and α-quartz. Constants for all the equations presented here are calculated using quartz mass fraction data after 18 h firing. The constants are dimensioned in seconds. The equations shown were compared with experimental data generated by Lundin [6] for a clay-based material containing 40 wt% kaolin, 40 wt% quartz, and 20 wt% feldspar. According to the transformed equations, the mass fraction of quartz can be calculated as follow

Table 7.4 Residual quartz content of a sintering bonding system at various vitrification temperatures

Temp. (°C)	Time (h)	Lundin's exp. result (wt%)	Exp. result (wt%)	Jackson and Mills' [11] result (wt%)
1200 (1473 K)	1	24.1	24.2	24.2
1200	1	24.7	24.3	24.2
1200	1	26.1	24.8	24.2
1200	2	23.7	23.8	23.4
1200	2	23.6	23.9	23.4
1200*	2	23.4	23.4	23.4
1200	4	21.3	22.2	22.3
1200	8	20.3	20.9	20.8
1200	18	19.0	18.5	18.6
1200	18	18.9	18.6	18.6
1200	48	15.2	15.1	14.9
1250 (1523 K)	1	22.7	22	22.1
1250*	2	20.6	20.6	20.6
1250	4	18	18.5	18.6
1250	8	15.5	16	16.2
1250	18	12.6	12.5	12.6
1250	48	8.3	7.8	8.0
1300 (1573 K)	0.5	22.6	20.4	20.6
1300	0.5	21	20.9	20.6
1300	1	20	18.3	18.6
1300	2	16.1	15.9	16.2
1300	4	13.4	12.8	13.2
1300	8	10	9.7	9.9
1300	18	5.9	5.8	6.1
1300	50	1.6	1.8	2.3
1300	120	0.3	0.2	0.6

Lundin's [6] experimental data are compared with the author's experimental data and the model [11]. The *asterisk* indicates values used for deriving the constants used in the theoretical model. Re-printed with kind permission from Springer Science+Business Media B.V.

- *Jander's model* [14]

$$m = 41.9.\left(1 - \left\{1.55 \times 10^{-6}.t\right\}\right)^{3/2} \tag{7.46}$$

- *Krause and Keetman's model* [15]

$$m = 41.9 - (2.58.\ln t) \tag{7.47}$$

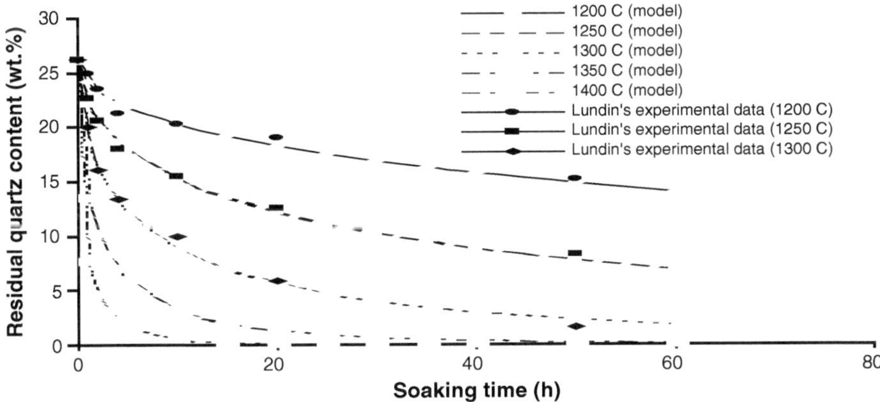

Fig. 7.19 Effect of time on residual quartz content of a sintering bonding system according to Jackson and Mills's model [11], and compared with Lundin's experimental data [6]. Re-printed with kind permission from Springer Science+Business Media B.V.

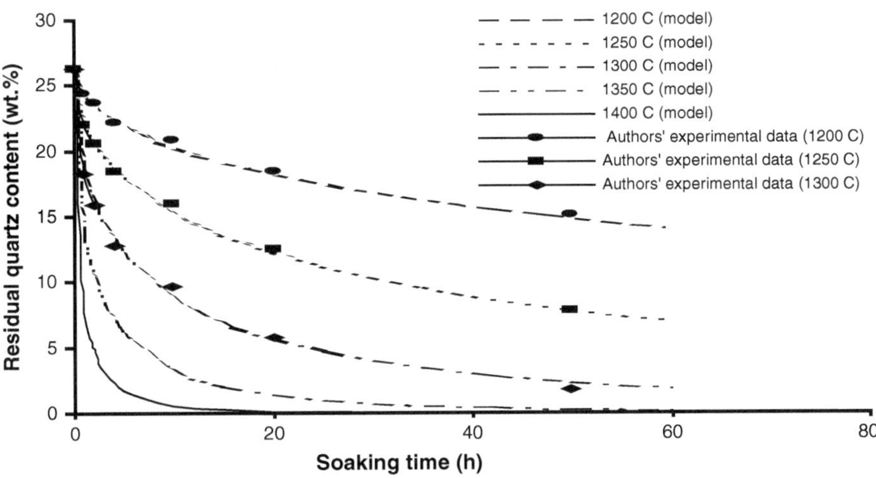

Fig. 7.20 Effect of time on residual quartz content of a sintering bonding system according to Jackson and Mills's model [11] and compared with the authors' experimental data. Re-printed with kind permission from Springer Science+Business Media B.V.

Table 7.5 Residual quartz content for different soaking times at 1300 °C for a sintering bonding system composed of 40 wt% kaolin, 40 wt% quartz, and 20 wt% feldspar (Lundin's mixture number M21 [6]) compared with other dissolution models

Time (h)	Lundin's exp. data [6]	Jander [14]	Krause and Keetman [15]	Monshi [16]	Jackson and Mills [11]
0	41.9	41.9	0.00	41.9	41.9
0.5	35.9	41.72	22.55	34.61	34.76
1	32.8	41.54	20.76	31.97	32.12
2	29.2	41.19	18.97	28.58	28.72
4	23.2	40.49	17.18	24.39	24.51
8	19.5	39.11	15.39	19.49	19.59
18	13.3	35.72	13.30	13.30	13.36
24	10.7	33.74	12.56	11.13	11.19
48	6.9	26.18	10.77	6.43	6.51
120	3.6	7.85	8.96	2.17	2.17
190	2.7	0.00	7.22	1.00	1.01
258	2.0	0.00	6.43	0.54	0.55

- *Monshi's model* [16]

$$m = 41.9.e^{-4.5\times10^{-3}\sqrt{t}} \tag{7.48}$$

- *Jackson and Mills' model* [11]

$$m = 41.73.e^{-4.5\times10^{-3}\sqrt{t}} \tag{7.49}$$

The transformed equations are then tested using data provided by Lundin [6]. Referring to Table 7.5, it can be shown that the mass fraction of quartz obtained using the equations derived by Jander [14] and Krause and Keetman [15] did not agree with Lundin's experimental results [6].

The results obtained using Monshi's model [16] are in much better agreement compared to Lundin's data. However, the results obtained using Jackson and Mills' model [11] is more accurate at predicting the mass fraction of quartz remaining owing to the differences in the density of quartz. After long periods of heat treatment, the model predicts lower magnitudes of mass fractions of quartz when compared to Lundin's experimental results [6].

7.6.5 Fusible Bonding Systems—Verification and Comparison of Dissolution Models for Quartz

The constants, A and B, for the fusible bonding system were determined using time and temperature couples at 2 and 10 h and were calculated to be, -5.2×10^8 and $-33{,}205$, respectively. The dissolution equation then becomes,

$$m_{T,t} = 10.06\exp\left[-5.2 \times 10^8 t^{1/2}.e^{\frac{-33205}{T}}\right] \tag{7.50}$$

Equation (7.50) is used to compare the experimentally determined mass fraction of quartz remaining after heat treatment with the predicted values. The calculated mass fraction of quartz remaining after a period of heat treatment is calculated using the equation derived by Jackson and Mills [11]. The results of the dissolution model compare well with the experimental data over short periods of time. However, over longer periods of heat treatment the model tends to become less accurate (Fig. 7.21). A comparison was made with published dissolution models. The equations shown were compared with experimental data at 1050 °C. The equations shown were compared with experimental data at 1050 °C. According to the transformed equations, the mass fraction of quartz can be calculated as follows:

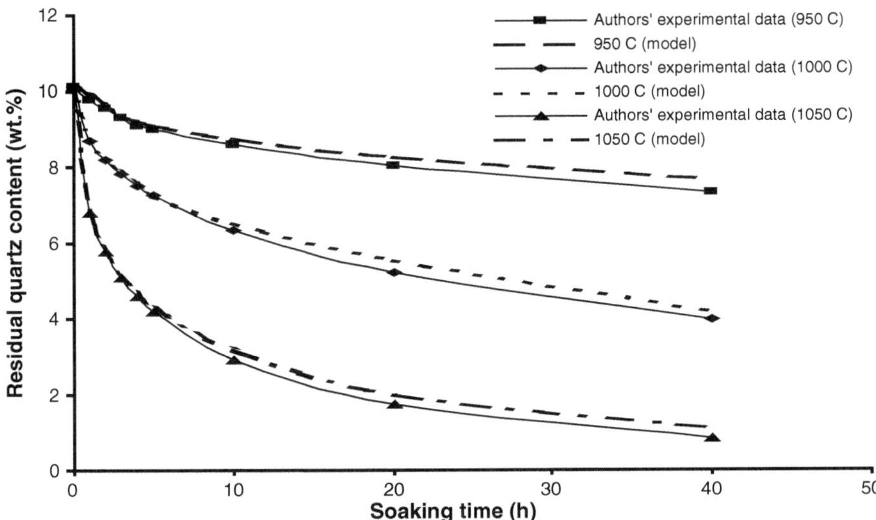

Fig. 7.21 Effect of time on residual quartz content of a fusible bonding system according to Jackson and Mills's model [11] and compared with the author's experimental data. Re-printed with kind permission from Springer Science+Business Media B.V.

Jander's model [14]

$$m = 10.1.\left(1 - \left\{3.44 \times 10^{-6}.t\right\}\right)^{3/2} \tag{7.51}$$

Krause and Keetman's model [15]

$$m = 10.1 - (0.59.\ln t) \tag{7.52}$$

Monshi's model [16]

$$m = 10.1.e^{-6.4\times 10^{-3}\sqrt{t}} \tag{7.53}$$

Jackson and Mills' model [11]

$$m = 10.06.e^{-6.37\times 10^{-3}\sqrt{t}} \tag{7.54}$$

With reference to Table 7.6, it can be shown that the mass fraction of quartz obtained using the equations derived by Jander [14] and Krause and Keetman [15], did not agree with the experimental results at 1050 °C. The results obtained using Monshi's model [16] are in much better agreement compared to the experimental data. However, the results obtained from Jackson and Mills's model are more accurate at predicting the mass fraction of quartz remaining owing to the differences in the density of quartz. After long periods of heat treatment, the model predicts slightly lower magnitudes of mass fractions of quartz when compared to the

Table 7.6 Residual quartz content for different soaking times at 1050 °C for a fusible bonding system compared with other dissolution models

Time (h)	Exp. data	Jander model [14]	Krause and Keetman model [15]	Monshi's model [16]	Jackson and Mills' model [11]
0	10.1	10.1	0	10.1	10.1
1	6.84	9.91	5.23	6.88	6.86
2	5.79	9.72	4.82	5.87	5.86
3	5.13	9.54	4.58	5.21	5.19
4	4.7	9.36	4.41	4.7	4.68
5	4.28	9.18	4.28	4.28	4.28
10	3.2	8.28	3.87	2.99	3
20	2	6.6	3.46	1.81	1.82
40	1.1	3.62	3.04	0.89	0.89

Re-printed with kind permission from Springer Science+Business Media B.V.

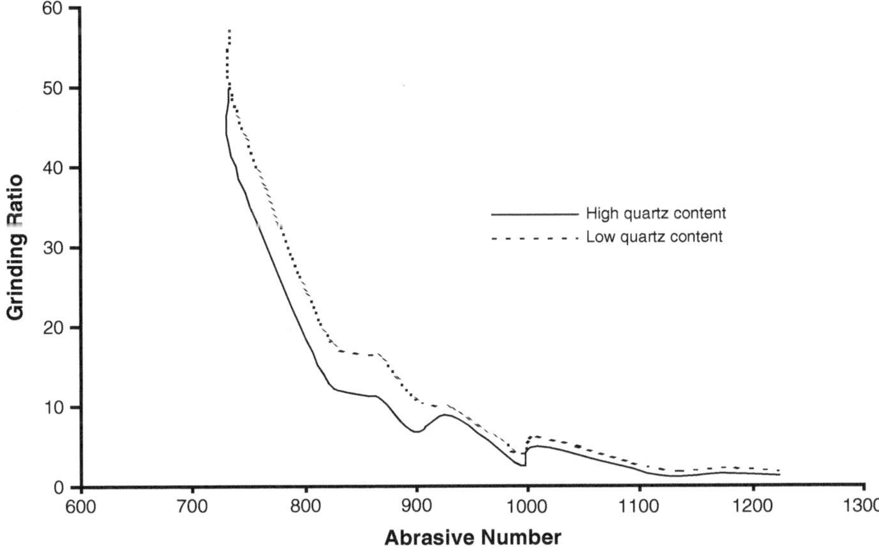

Fig. 7.22 Effect of the abrasive number on the grinding ratio for a high-quartz content and a low quartz-content bonding system. Re-printed with kind permission from Springer Science+Business Media B.V.

experimental results. The dissolution of quartz during heat treatment has a significant effect on the wear of vitrified grinding wheels. Figure 7.22 shows the effect of using a high and a low quartz content bonding system on the wear of vitrified corundum grinding wheels grinding a large number of tool steel materials [17]. The classification of tool steels is in the form of an abrasive hardness number, which is a weighted average of the number of carbides contained within the tool material. As shown in Fig. 7.22, the grinding ratio, or G-ratio, is a measure of the efficiency of the grinding wheel. It is the quotient of the volume of workpiece material removed and the volume of the wheel material removed. The figure demonstrates the effectiveness of reducing the quartz content of the bonding system of porous nanogrinding tools.

7.7 Laser Dressing of Nanogrinding Tools

The processing of abrasive materials by locally restructuring their surface using a high power laser allows one to provide a localized heat source at high speed that produces surface features that are able to remove material at the nanoscale. The depth of thermal processing of the surface that allows the material to melt, solidify and cool is controlled by parameters such as fluence (energy per unit area), laser beam scanning speed, and input power. Previous studies on laser dressing of grinding wheels have resulted in the localized modification of the morphology of corundum abrasive materials [18–20]. The grinding wheel is initially made by sintering corundum grains with a low melting point binding material, which is subsequently melted, vaporized, and selectively removed using a concentrated beam of optical energy that changes surface topography and composition. The refinement of grain size, densification of grinding layer, and evolution of competitive grains with sharp vertices is shown to occur during laser dressing [18]. These changes result in an increase in grinding ratios, especially when grinding tool steels [18].

X-ray diffraction (XRD) of the laser dressed surface indicated that crystallographic texture occurs on the surface of the grinding wheel, and the identification of planar texture is considered important as it controls the formation of multifaceted grains that take part in removing workpiece material. Pole figure analysis of laser dressed corundum composite materials indicates that crystallographic texture is preferred along the (110) plane [18–20]. Grinding properties are particularly sensitive to preferred orientation and as such variations in grinding are apparent. To predict the performance of these materials, it is important to identify crystallographic texture using XRD techniques. Orientation imaging microscopy (OIM) is used to characterize micro-texture [21] as it gives phase specific crystallographic texture. The OIM technique uses Kikuchi patterns to predict crystallographic texture at various points on the surface of the laser dressed material. Microstructural variations are observed on the laser dressed surface owing to temperature gradients and variations in cooling rates. Corundum, or alumina, has poor thermal conductivity compared to metals that allows thermal energy to be retained in the surface for a longer period of time compared to the workpiece material. The occurrence of porosity between grains also reduces thermal conductivity, which results in the retardation of heat lost from the surface of the abrasive material [22]. In addition to the effects of porosity, consolidation of the corundum abrasive material with the bonding material changes the thermal properties during laser dressing [23–25]. Therefore, the combined thermophysical properties of the composite material must be used. Table 7.7 shows the thermophysical properties of the abrasive material. It is important to quantify the cooling rate in order to understand how the microstructure evolves into preferred orientation during laser dressing. Many

Table 7.7 Thermophysical properties of vitrified corundum abrasive material

Thermophysical property	Numerical value of property[a]
Density, ρ	2.7×10^3 kg/m^3
Melting temperature, T_m	2045 K
Latent heat of fusion, ΔH_m	1.03 × 106 J/mol
Specific heat, Cp	1137 J/kg K
Thermal diffusivity, D	1.08×10^{-6} m^2/s

[a]*Source* Vitrified Technologies Inc., Central Research and Development Laboratory, P.O. Box 3724, West Lafayette, IN 47996

models have been generated that predict how microstructures form under rapid solidification conditions [26–28]. The chapter presents an understanding of how cooling rates affect the formation of multiple cutting points on the surface of the abrasive material and how it affects grindability of a hypereutectoid steel such as AISI 52100 bearing steel. This is achieved by observing cooling rates using a two-color pyrometer, by characterizing the microstructural texture using OIM, and by determining grinding ratios when grinding AISI 52100 bearing steel.

7.7.1 Experimental Procedures

Vitrified corundum abrasive products were supplied by Vitrified Technologies Inc., and were composed of corundum grains doped with 0.25 % wt. chromia that were bonded together using a bonding system composed of borosilicate glass and orthoclase feldspar. The chemical composition of the abrasive grain material is 99.52 % Al_2O_3, 0.18 % Na_2O, 0.05 % Fe_2O_3, and 0.25 % Cr_2O_3. A 2.5 kW Hobart continuous-wave Nd:YAG laser equipped with a fiber optic beam delivery system was used for dressing the surface of the grinding wheel. Laser power was varied from 400 to 1000 W, whilst the overlap between dressed layers was less than 20 %. The laser beam was scanned at a linear speed of 250 cm/min on the surface of the grinding wheel. Full experimental details of the dressing procedure are provided in Refs. [18, 20]. Laser dressed grinding wheels were finally compared to conventionally dressed grinding wheels grinding a hypereutectoid steel typically used for bearing applications that is capable of being super finished.

7.7.1.1 Measurement of Dressing Temperatures

A two-color optical pyrometer was used to measure the temperature of the abrasive surface during dressing. The temperature range of the pyrometer is 600–1500 °C with a response time of 0.5 ms. The pyrometer is calibrated against National Institute standards to within 10 °C over the range 800–3000 °C. A specially constructed telescope with filter was constructed in order to measure the laser beam temperature at the surface, and a program was written to directly record temperature against processing time. When the laser beam dresses the surface, a large plume of light is emitted that interferes with the intensity of the light collected by the telescope. To overcome the large emission of light, a delay was programmed in the software program to prevent the erroneous measurement of temperature. As the plume of light fades away, the pyrometer records the temperature. The temperature is measured at 400 and 500 W of laser power, but at very high laser power such as 1000 W the temperature is almost impossible to record because of the intensity of light. Therefore, data are presented at the lower values of laser power because the plume lasts for less than 1 s.

7.7.1.2 Orientation Imaging Microscopy

To gain high-quality Kikuchi patterns from laser dressed grinding wheels, samples were cross-sectioned by polishing with a water solution on SiC abrasive papers then followed by polishing with a diamond paste to achieve a fine surface. The samples were then carbon coated with a uniform thickness of 3 nm. Electron backscattered diffraction patterns (EBSD), or backscattered Kikuchi patterns, were generated using a Philips FEG SEM equipped with an orientation-imaging microscope [29–31]. OIM analysis was conducted at Oak Ridge National Laboratory. To produce crystallographic orientation maps, the electron beam is scanned over a selected surface area and the resulting EBSD patterns are automatically indexed and the local orientation is determined [32, 33]. For the present study, a sample processed at 1000 W was used in order to provide a crystallographic map of the whole re-solidified layer. From XRD and energy dispersive analysis (EDS), it was found that the primary phase created was α-alumina [18, 19] and so the hexagonal crystal structure of alumina was selected to index the EBSD Kikuchi patterns obtained during the analysis. For OIM analysis, the 1000 W laser dressed surface was used whereas the lower power processed samples were used to measure cooling rates. This does not affect solidification behavior as all samples follow the same morphological features during solidification, the only variable being the depth of the re-solidified layer.

7.7.1.3 Nanogrinding Experiments

To characterize the differences between conventional abrasives, sol-gel abrasives, monocrystalline alumina abrasives and laser dressed chromium-doped alumina, a series of grinding experiments were conducted in order to confirm results gained in field trials. In order to duplicate the field trials, a series of controlled experiments were conducted under increasing feed rates. The experiments were terminated when a condition of severe burn, chatter, or grinding wheel breakdown was reached. The initial trial wheels used were: angular white alumina wheel with low temperature bonding system (Wheel specification: 77A60JV); sol-gel and angular white abrasive mixture wheel with low temperature bonding system (Wheel specification: 73A60JV); monocrystalline alumina grinding wheel with a low temperature bonding system (Wheel specification: SA60JV); and laser dressed chromium-doped alumina bonded with a high temperature bonding system. All grinding wheels were manufactured with a vitrified bond, 60 mesh-size abrasive (approximately 220 μm in diameter), J-mechanical grade, and a fairly open structure. Experiments were performed on a Jones and Shipman series 10 cylindrical grinding machine using a 450 mm diameter grinding wheel rotating at 33 m/s surface speed. The mechanically-dressed wheels were dressed using a D25 blade tool supplied by Unicorn-Van Moppes, and the dressing conditions were: 3 passes at 0.15 mm/rev at a depth of 0.03 mm plus one pass at 0.015 mm depth at the start of each batch. Stock removal was 0.25 mm on diameter grinding AISI 52100 ball bearing steel (59 HRC) using 2 % concentration soluble oil at 0.5 bar pressure and a flow rate of

15 l/min. Forty experimental test pieces were ground per experiment. The laser dressed samples were mounted in a modified grinding wheel that holds segments of samples that have been laser dressed. The segmented grinding wheel acts in the same way as a conventional grinding wheel but has the advantages of quick and easy replacement of laser-processed samples. The most important measurements made during the grinding experiments are grinding power, surface roughness of the workpiece, and the wheel wear parameter, grinding ratio, which is the ratio of workpiece machined to the amount of grinding wheel lost during grinding.

7.7.2 *Experimental Results and Discussion*

7.7.2.1 Laser Dressing Temperatures

The rapid solidification of the laser dressed samples show that a refined microstructure and reduced segregation of phases is possible. Very high cooling rates are possible over a small region of the workpiece where non-equilibrium structures are produced owing to rapid solidification of the material. High cooling rates can be achieved using a laser beam that tends to diminish as the laser beam moves along the track. Therefore, the microstructure will change at the edges of the track toward the unaffected part of the grinding wheel. Laser irradiation is considered to be a point source with heat energy flowing radially away from the point. The system is assumed to attain quasi-steady-state conditions when the melt velocity at the melt/solid interface becomes constant [34]. Incident radiation is partly reflected and absorbed by the abrasive material, whilst some absorbed energy is lost by re-radiation and convection and some is conducted into the material. The melt zone is heated from the surface at the center of the track, which established convection currents within the melt pool and transports heat away very quickly from the melt/solid interface (Fig. 7.23). Experimentally determined cooling rates during laser processing for processing at two different power levels (400 and 500 W) are shown in Table 7.8. The data are calculated from temperature and time data collected using the two-color pyrometer. The data is obtained by curve fitting the time dependent equation for the processing temperature so that cooling rates can be estimated close to the melting point of the abrasive material. Experimental data is provided for two tracks and cooling rates are calculated at the melting temperature of the abrasive material (2045 K), which is assumed to be the maximum cooling rate (Fig. 7.24). The cooling rates in samples processed at higher laser power are lower compared to that at lower laser power. This is explained by the fact that the abrasive material is a poor conductor of heat compared to metals. Therefore, an increase in laser beam power will provide a greater input of heat to the material that will provide a larger melt zone and an increase in the interaction volume. Thus, it takes more time to dissipate the heat built-up and causes a drop in cooling rates. Therefore, for high laser power the cooling rate is expected to be lower compared to those calculated for 400 and 500 W (Fig. 7.25).

Fig. 7.23 Scanning electron micrograph showing center of laser processed track, resolidified layer, and original structure [18, 19, 45]. Re-printed with kind permission from Springer Science +Business Media B.V. *Source* Vitrified Technologies Inc., Central Research and Development Laboratory, P.O. Box 3724, West Lafayette, IN 47996

Table 7.8 Cooling rates for laser processed grinding wheel [18, 19, 45]

Laser power (W)	Maximum cooling rates, dT/dt, (°C/s), for track 1 at the melting temperature	Cooling rates, dT/dt, (°C/s), for track 1 at the start of solidification	Cooling rates, dT/dt, (°C/s), for track 1 at the start of solidification	Cooling rates, dT/dt, (°C/s), for track 2 at the start of solidification
400	−770.8	−437	−702.8	−382.2
500	−701.5	−401	−580.6	−299.7

Re-printed with kind permission from Springer Science+Business Media B.V.

The cooling rate in the second track is slightly lower than track 1 for both 400 W and 500 W samples. As the laser passes the over the first track it pre-heats the material adjacent to it that results in a reduced cooling rate for the second track owing to the decrease in the thermal gradient.

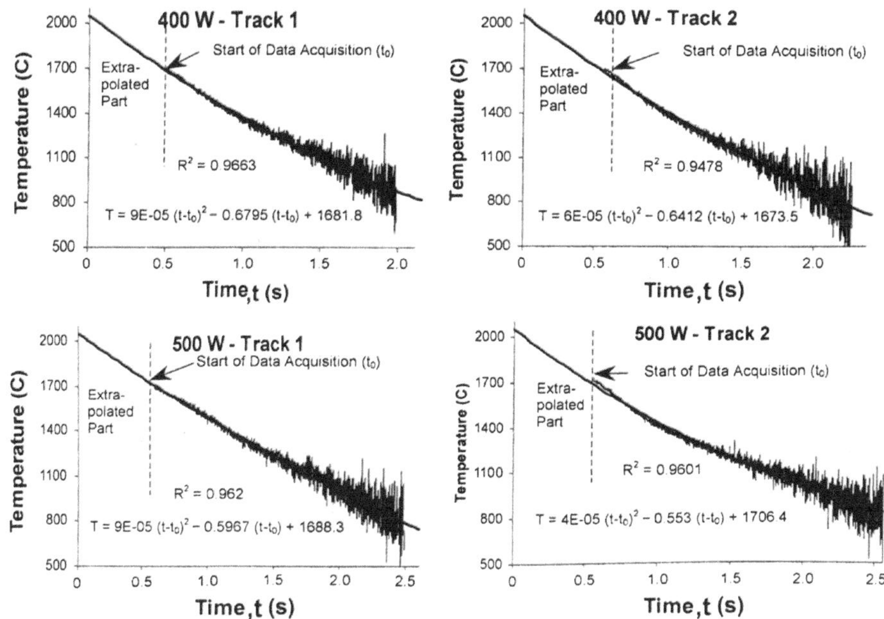

Fig. 7.24 Temperature and time profiles of laser processed tracks [18, 19, 45]. Re-printed with kind permission from Springer Science+Business Media B.V.

Table 7.8 shows the cooling rate at 2.2 s after solidification, which corresponds to the temperature of the sample near to the lower calibrated temperature range of the pyrometer. The trend observed shows that the cooling rate is similar to that seen at the solidification temperature, but cooling rates are much lower due to thermal gradients and heat loss from the sample. Solidification occurs in directions that are opposite to the direction of heat extraction. It is also assumed that grain growth is equal in all directions owing to the fact that the directed laser energy is a point source of heat dissipation. Microstructural texture is assumed to form in accordance with the model of heat extraction. A full range of microstructures such as planar, cellular, and dendritic, are observed in laser processed samples [35, 36].

A planar solidification front is observed when the interface is at the equilibrium liquidus temperature and when every point at the front of the interface is above the liquidus temperature. Any instability dissolves back in the melt and if it is supercooled, then the perturbation formed on the interface will grow fast into the liquid melt. Under the influence of a positive temperature gradient, constitutional supercooling is solute driven. The type of solidification microstructure formed during constitutional supercooling depends on the G/R ratio, where G is the thermal gradient and R is the solidification rate [35, 36]. The solidification structure observed in the laser-processed tracks of the abrsive material changes the surface morphology. Referring to Fig. 7.26, grains are multifaceted at the center of the laser track and are equiaxed possessing a dendritic structure. The edges of the tracks

Fig. 7.25 SEM image showing equiaxed grains with dendritic structure in the center of the track (**a**) and columnar growth of grains near the edge of the track (**b**) [18, 19, 45]. Re-printed with kind permission from Springer Science+Business Media B.V. *Source* Vitrified Technologies Inc., Central Research and Development Laboratory, P.O. Box 3724, West Lafayette, IN 47996

show elongated columnar grains that grow away from the point source in a direction normal to the laser beam. The columnar structures are similar to those reported by other authors [37–39] who have conducted laser based processing of alumina ceramics.

The heat transfer equation for a moving heat source is given by the following equation, where the substantial derivative of temperature, i.e., T = T(y, t), and is given a partial derivative [40–43],

$$\frac{d}{dt}[t(y,t)] = \frac{\partial T}{\partial t} + \frac{\partial y}{\partial t} \cdot \frac{\partial T}{\partial y} = D\frac{\partial^2 T}{\partial y^2} + \frac{\dot{q}}{C_p \rho} \tag{7.55}$$

For the time-independent situation, i.e., quasi-steady state heat transfer for a moving boundary problem, the time derivative tends to zero in Eq. (7.55). Hence,

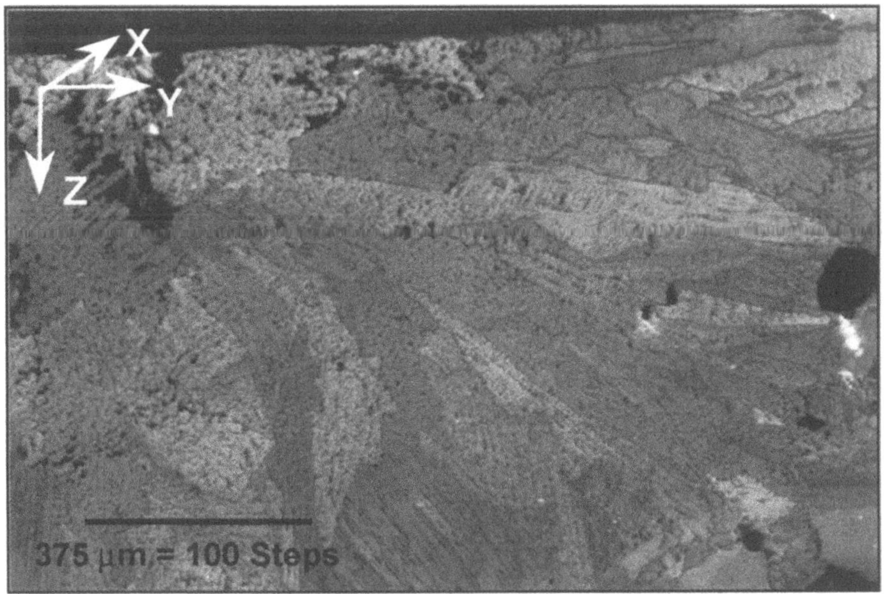

Fig. 7.26 Image quality cross section of the laser processed track showing the re-solidified region laser dressed at 1000 W using OIM [18, 19, 45]. Re-printed with kind permission from Springer Science+Business Media B.V.

the surface cooling rate dT/dt in the direction opposite to the laser under the quasi-steady state (T/t = 0) can be estimated as [42],

$$\frac{dT}{dt} = \frac{dT}{dy} \cdot \frac{dy}{dt} \quad \text{or,} \quad G = \frac{1}{v} \cdot \frac{dT}{dt} \tag{7.56}$$

where v is the velocity of the laser beam.

Thus, the thermal gradients (G) are calculated from the temperature and time diagrams assuming quasi-steady state conditions using Eq. (7.56). During solidification, the velocity of the liquid/solid interface or growth rate, R, is provided by the Neumann solutions to the Stefan problem [43]. Assumptions made include: strong convective currents exist in the melt pool and the total energy absorbed at the free surface of the liquid is transported to the interface by convection. The solution to the Stefan problem is,

$$V_i \rho \Delta H_m - \kappa_s \cdot \frac{\partial T}{\partial y} = V_i \rho \left[\Delta H_m + C_p (T_m - T_o) \right] = I_a \tag{7.57}$$

where Vi is the liquid/solid interface velocity, and y is the distance in the melt pool. This equation gives the average velocity of the interface during the time that the treated zone solidifies. The solidification front velocity at the start of solidification

Table 7.9 G/R ratios for laser processing parameters [18, 19, 45]

Track position	G/R ratio for laser power of 400 W: track 1	G/R ratio for laser power of 400 W: track 2	G/R ratio for laser power of 500 W: track 1	G/R ratio for laser power of 500 W: track 2
Center	4.43×10^6	4.04×10^6	4.03×10^6	3.34×10^6
Edge[a]	1.74×10^9	1.74×10^9	1.74×10^9	1.74×10^9

[a]G/R ratios are the same for all conditions at the edge of the track because Eq. (7.58) is used for calculating the growth rate, R, which is independent of G

is greater than Eq. (7.57) estimates. Therefore, at the start of solidification the local temperature gradient is very large and this leads to a maximum interface solidification velocity that is approximated by [44]:

$$V_i^{\max} \approx \varsigma \left(\frac{D}{\tau_1} \right)^{0.5} \tag{7.58}$$

where τ_1 is the dwell time of the laser and ζ is provided by the solution of,

$$\varsigma[1 + erf(\varsigma)] \exp(\varsigma^2) = \frac{C_p(T_m - T_o)}{\sqrt{\pi}.\Delta H_m} \tag{7.59}$$

It can be shown from Eq. (7.59) that the average and maximum interface solidification rates are dependent on laser beam intensity and velocity. By using Eqs. (7.57) and (7.58), the velocity of the solid/liquid interface, or growth rate (R), is calculated at the center of the track. G/R ratios are calculated and are given in Table 7.9. The ratios are high near to the edge of the track where solidification starts and decreases toward the center of the track.

The observed solidification structure varied from equiaxed grains with dendritic structure within the track to columnar grains at the edges of the track. Constitutional supercooling theory predicts that increasing the G/R ratio causes a progressive change in solidification structure from fully dendritic to cellular dendritic, then to cellular and then finally to a planar morphology. A similar trend is observed in the laser track. Therefore, the observed microstructure agrees with the predictions made by theory.

7.7.2.2 Orientation Imaging Microscopy of Laser-Dressed Materials

Figure 7.29 shows the image quality (IQ) micrograph obtained from the OIM. The area selected for analysis is of a cross-section of the re-solidified layer of one of the tracks. IQ represents the sharp character of the EBSD pattern and thus local crystal perfection. Hence, the IQ represents the real microstructure as demonstrated in an optical microscope.

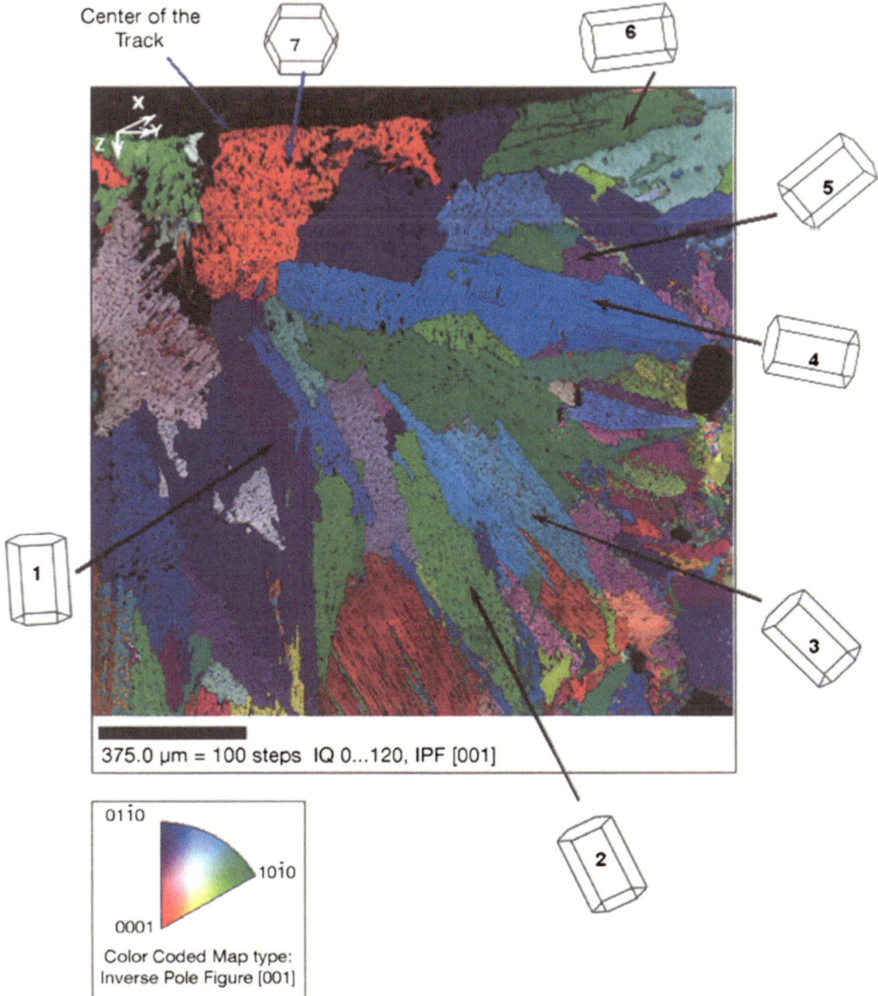

Fig. 7.27 Inverse pole figure (IPF) color-coded map of the resolidified region shown in Fig. 7.26 generated using a laser power level of 1000 W [18, 19, 45]. Re-printed with kind permission from Springer Science+Business Media B.V.

The dark regions of the image are pores where the IQ intensity is zero. Grain boundaries are easily detected, which is in sharp contrast to a scanning electron microscopic image that does not resolve grain boundaries at that particular magnification. The growth direction is radial from the surface of the center of the track where the beam is focused. Figure 7.28 shows the inverse pole figure (IPF) map of the area shown in Fig. 7.27.

The grains are color-coded that represents the crystallographic orientation of the normal surface. The key used for color-coding is provided in the stereographic

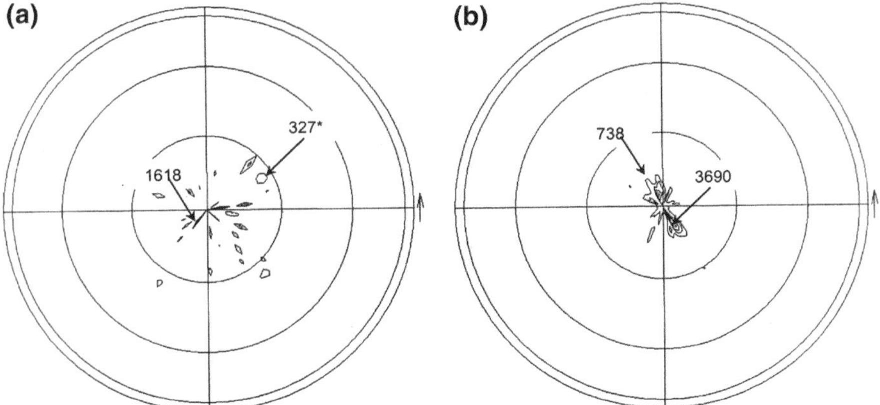

Fig. 7.28 Pole figure for (110) plane in (**a**) undressed sample and (**b**) laser dressed sample (1000 W). *Note Asterisk* denotes intensity [18, 19, 45]. Re-printed with kind permission from Springer Science+Business Media B.V. *Source* Vitrified Technologies Inc., Central Research and Development Laboratory, P.O. Box 3724, West Lafayette, IN 47996

triangle at the base of the figure. The orientation of some of the grains are in the form of the hexagonal cell for α-alumina. It is observed from the figure that growth takes place along the c-axis in the hexagonal α-alumina lattice during solidification of the molten abrasive bonding material. Grains numbered 1–4 possess an orientation that extends radially away from the center of the track as shown by the heat extraction model.

These grains lie within the re-solidified layer, whilst the outer grains lie parallel to the surface of the track. There are smaller grains that have their c-axes oriented differently than other smaller grains and not along the direction radially outward from the center of the laser-processed track.

These grains do not grow in size compared to other grains. Thus, a competitive growth mechanism is observed in which the growth is very fast for grains whose c-axis is parallel to the growth direction, and very slow for all other grains. Pole figure analysis and XRD during previous studies [18, 45] indicated that a preferred orientation occurs on the (110) plane, as shown in Fig. 7.28.

The (110) planes are prismatic planes that are parallel to the c-axis in the hexagonal α-alumina cell. OIM maps suggest that growth takes place along the c-axis direction, which tends to favor (110) planes.

7.7.2.3 Nanogrinding Experiments

Figures 7.29, 7.30 and 7.31 show the grinding results of mechanically-dressed and laser-dressed grinding wheels. None of the grinding wheels tested produced chatter or burn when grinding the experimental test pieces. Each experiment was terminated as a result of wheel breakdown, which showed itself as changes in grinding

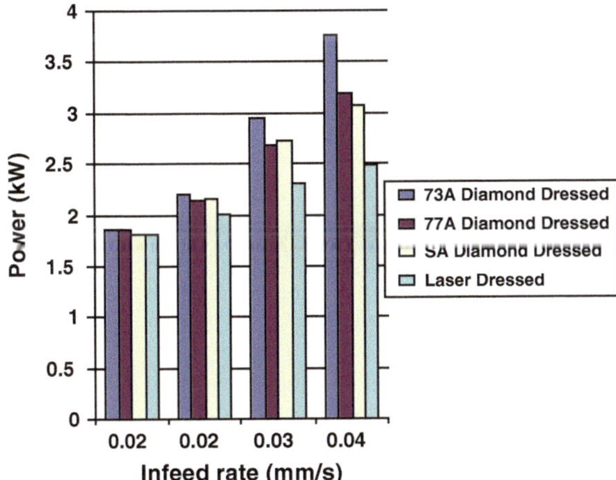

Fig. 7.29 Grinding power as function of feed rate for vitrified corundum grinding wheels with different types of abrasive and dressing conditions. Re-printed with kind permission from Springer Science+Business Media B.V. *Source* Vitrified Technologies Inc., Central Research and Development Laboratory, P.O. Box 3724, West Lafayette, IN 47996

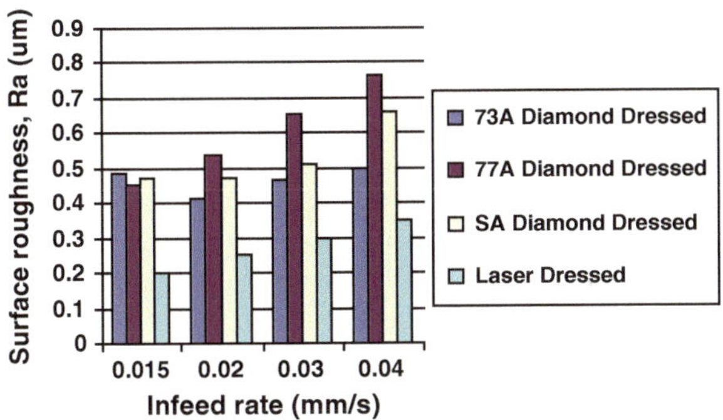

Fig. 7.30 Surface roughness as a function of feed rate for vitrified corundum grinding wheels with different types of abrasive and dressing conditions. Re-printed with kind permission from Springer Science+Business Media B.V. *Source* Vitrified Technologies Inc., Central Research and Development Laboratory, P.O. Box 3724, West Lafayette, IN 47996

power, such as a rapid decrease in power or in the form of erratic fluctuations in power. At the lowest feed rate of 0.015 mm/s, all wheels gave a similar performance regarding power level, surface finish, and grinding ratio. At the intermediate feed rate of 0.02 mm/s, it was possible to differentiate between all three grinding wheels. The sol-gel abrasive wheel gave the best surface roughness and the highest grinding ratio (G-ratio) while the angular white alumina wheel gave the worst performance. At a feed rate of 0.03 mm/s, the white alumina (77A) wheel ground between 20 and 25 components satisfactorily and then the power became erratic and the surface

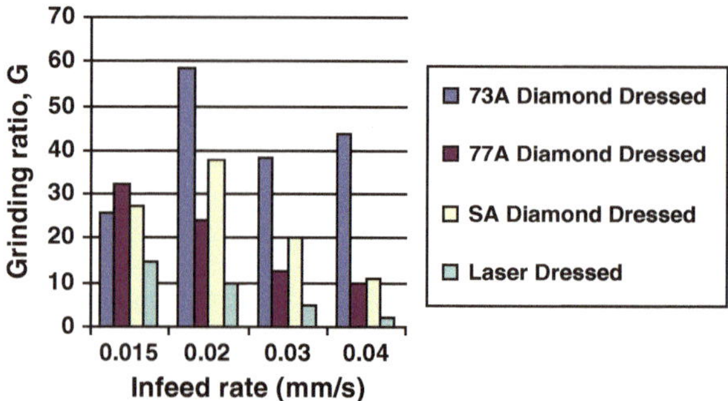

Fig. 7.31 Grinding ratio as a function of feed rate for vitrified corundum grinding wheels with different types of abrasive and dressing conditions. Re-printed with kind permission from Springer Science+Business Media B.V. *Source* Vitrified Technologies Inc., Central Research and Development Laboratory, P.O. Box 3724, West Lafayette, IN 47996

roughness deteriorated, indicating wheel breakdown. The monocrystalline alumina wheel ground around 30 test pieces before surface roughness deteriorated, although grinding power showed a slight fluctuation. The sol-gel abrasive wheel showed no sign of breakdown. Subsequent tests were conducted up to 0.08 mm/s. However, only the sol-gel wheel could grind twenty test pieces after which it slowly broke down. In conclusion, although the sol-gel wheel operates at a higher power level, it is capable of working at much higher stock removal rates than either the white alumina or monocrystalline alumina grinding wheels. At moderately high stock removal rates, white alumina and monocrystalline alumina abrasives can remove metal at the same rate, but the monocrystalline alumina wheel tends to give a better surface finish and works longer before it requires dressing. At low stock removal rates there is no significant difference between sol-gel, angular white alumina, and monocrystalline alumina grinding wheels.

The effects of dressing conditions show interesting results in that laser-dressed grinding wheels require less grinding power at increasing feed rates (Fig. 7.29).

This implies that laser-dressed wheels are breaking down at a faster rate than mechanically-dressed grinding wheels as the feed rate increases, which is confirmed by the reduction in grinding ratio as shown in Figs. 7.30 and 7.31. However, even though more of the grinding wheel is lost through grinding, a significant improvement in surface roughness is gained at the expense of wheel loss.

Figures 7.29, 7.30 and 7.31 show that laser-dressed grinding wheels possess a free-cutting ability compared to the mechanically-dressed grinding wheels, but in contrast to those wheels they are able to provide a finer surface roughness to the workpiece due to the predominance of oriented vertices of α-alumina created by laser dressing that remove very small amounts of workpiece material. This process is known as "nanogrinding".

7.8 Future Directions

For perfectly sharp diamond coated piezoelectric ceramic materials, grain fracture appears to be the dominant cause of abrasive material loss during a grinding operation. Grain fracture is much more likely to be caused by mechanically induced tensile stresses within abrasive grains than by mechanically induced compressive stresses. The best indicator of diamond grain performance during a nanogrinding operation under different operating conditions is the level of tensile stress established in abrasive grains. High tensile stresses are associated with grain fracture and low grinding ratios in perfectly sharp diamond-coated piezoelectric ceramic materials. Finite element models of perfectly sharp grinding grains can be applied to the piezoelectric nanogrinding process where the dominant wear mechanism is grain fracture. Piezoelectric nanogrinding is a process that has demonstrated its capability of being developed into a nanomanufacturing process of the future. The dissolution model derived by Jackson and Mills has been compared with experimental data using sintering and fusible vitrified bonding systems that are used extensively with high performance nanogrinding tools. The results predicted by the model compare well with the experimental results presented in this chapter. However, over longer periods of isothermal vitrification, the model becomes less accurate due to the assumptions made in the dissolution model. The model may be of use when predicting the mass fraction of quartz using high temperature firing cycles that are characterized by short soaking periods. The development of porous nanogrinding tools lies in their ability to be dressed using a directed photon beam that sharpens worn grains or promotes the formation of textured peaks from the vitreous bonding system. The development of quartz-free bonding systems that has high corundum content is of paramount importance if porous nanogrinding tools are to be effective when machining engineering materials at the nanoscale. Cooling rates of the order of approximately 100 °C/s are observed during laser dressing of vitrified corundum grinding wheels. High laser power tends to generate slower cooling rates because of the thermal properties of these complex ceramic materials. The observed solidification structures tend to be dendritic at the center of the laser-processed track, and columnar near to the edges of the track. OIM analyses of the laser processed tracks showed a preferred growth of grains along the c-axis and in a direction opposing heat flow. In association with XRD and pole figure analysis, preferred orientation favors growth on the (110) planes that are parallel to the c-axis in the hexagonal alumina cell. Competitive growth of grains takes place in the re-solidified layer during which time the grains with their c-axes not orientated from the center of the track are prevented from growing in their favored directions. Grinding experiments suggest that laser-dressed grinding wheels cut freely but produce lower surface roughness on AISI 52100 workpiece material owing to the predominance of oriented vertices of α-alumina that remove very small amounts of material even at increasing rates of feed.

Acknowledgements The authors thank Springer publishers for allowing the authors permission to re-print and update this chapter that was originally published in, 'Micro and Nanomachining', originally published by Springer in 2007 (ISBN 978-0387-25874-4). Re-printed with kind permission from Springer Science+Business Media B.V.

References

1. Q. Zhao, S. Dong, T. Sun, Research on diamond tip wear mechanism in atomic force microscope-based micro-and nanomachining. High Technol. Lett. **7**(3), 84–89 (2001)
2. Q.L. Ahao, S. Dong, T. Sun, Investigation of an atomic force microscope diamond tip wear in micro- and nanomachining. Key Eng. Mater. **2**(13), 315–350 (2001)
3. M.J. Jackson, Vitrification heat treatment during the manufacture of corundum grinding wheels. J. Manuf. Proc. **3**, 17–28 (2001)
4. S.P. Timoshenko, J.N. Goodier, in *Theory of Elasticity*, 3rd edn—International student edn (McGraw-Hill Kogakusha Ltd., NY, 1970), pp. 109–113, 139–144
5. T.N. Loladze, Requirements of tool materials. in *Proceedings of the 8th International Machine Tool Design and Research Conference* (Pergamon Press, Oxfoard, 1967), pp. 821–842
6. S.T. Lundin, *Studies on Triaxial Whiteware Bodies (Almqvist and Wiksell)* (Stockholm, Sweden, 1959)
7. W. Storch, H. Ruf, H. Scholze, Berichte Deut. Keram. Ges. **61**, 325 (1984)
8. E. Binns, Sci. Ceram. **1**, 315 (1962)
9. W.F. Ford, J. White, Trans. J. Brit. Ceram. Soc. **50**, 461 (1951)
10. G. Kirchoff, W. Pompe, H.A. Bahr, J. Mater. Sci. **17**, 2809 (1982)
11. M.J. Jackson, B. Mills, Dissolution of quartz in vitrified ceramic materials. J. Mater. Sci. **32**, 5295–5304 (1997)
12. I.E. Alexander, H.P. Klug, X-ray diffraction procedures. Anal. Chem. **20**, 886 (1948)
13. S.K. Khandelwal, R.L. Cook, Effect of alumina additions on crystalline constituents and fired properties of electrical porcelain. Am. Ceram. Soc. Bull. **49**, 522–526 (1970)
14. W. Jander, Reaktion im festen zustande bei hoheren temperaturen (Reactions in solids at high temperature). Z. Anorg. U. Allgem. Chem. **163**, 1–30 (1927)
15. P. Krause, E. Keetman, Zur kenntnis der keramischen brennvorgange (On combustion processes in ceramics). Sprechsaal **69**, 45–47 (1936)
16. A. Monshi, Investigation into the strength of whiteware bodies, Ph.D. Thesis, University of Sheffield, United Kingdom, 1990
17. M.J. Jackson, A study of vitreous-bonded abrasive materials, Ph.D. Thesis, Liverpool John Moores University, United Kingdom, Dec 1995
18. A. Khangar, N. Dahotre, Morphological modification in laser-dressed grinding wheel material for microscale grinding. J. Mater. Process. Technol. **170**, 1–10 (2005)
19. A. Khangar, N.B. Dahotre, M.J. Jackson, G.M. Robinson, Laser dressing of alumina grinding wheels, in *Proceedings of the International Surface Engineering Congress and Exposition.* ASM Int. pp. 423–426 (2003)
20. D.J. Dingley, V. Randle, J. Mater. Sci. **27**, 4585 (1992)
21. M.J. Jackson, G.M. Robinson, Femtosecond laser micromachining of aluminum surfaces under controlled gas environments. J. Mater. Eng. Perform. **15**, 155–160 (2006)
22. D. Triantafyllidis, L. Li, F.H. Stott, Surface treatment of alumina-based ceramic using combined laser sources. Appl. Surf. Sci. **186**, 140–144 (2002)
23. J. Lawrence, L. Li, J.T. Spencer, The effects of high power laser diode laser radiation on the wettability adhesion, and bonding characteristics of an alumina–silica based oxide and vitreous enamel. Surf. Coat. Technol. **115**, 273–281 (1999)
24. Y. Yuanzheng, Z. Youlan, L. Zhengyi, C. Yuzhi, Laser remelting of plasma sprayed alumina ceramic coatings and subsequent wear resistance. Mater. Sci. Eng. **A291**, 168–172 (2000)

25. L. Bradley, L. Li, F.H. Stott, Flame assisted laser surface treatment of refractory materials for crack free densification. Mater. Sci. Eng. **A278**, 204–212 (2000)
26. W. Kurz, R. Trivedi, Trans. ASME **114**, 450 (1992)
27. A. Kar, J. Mazumdar, J. Appl. Phys. **61**, 2645 (1987)
28. M. Rappaz, S. David, J.M. Vitek, L.A. Boatner, Metall. Trans. A **20A**, 1125 (1989)
29. D.J. Dingley, D.P. Field, Mater. Sci. Technol. **12**, 1–9 (1996)
30. D.P. Field, Ultramicroscopy **67**, 1–9 (1997)
31. B.K. Kim, J.A. Szpunar, Scripta. Met. **44**, 2605 (2001)
32. J. Farrer, J. Michael, C. Carter, *Electron Backscatter Diffraction in Materials Science* (Kluwer Academic Publishers, New York, 2000), p. 299
33. K.Z. Baba-Kishi, J. Mater. Sci. **37**, 1715 (2002)
34. J. Mazumdar, W.M. Steen, J. Appl. Phys. **51**, 941–947 (1980)
35. M.C. Fleming, *Solidification Processing* (McGraw-Hill, New York, 1974)
36. O. Esquivel, J. Mazumdar, M. Bass, S.M. Copeley, Microstructural formation according to the theory of constitutional supercooling. in *Rapid Solidification Processing, Processing and Technologies II* (Claitors, Baton Rouge, 1980), pp. 150–173
37. L. Bradley, L. Li, F. Stott, Appl. Surf. Sci. **138–139**, 522–528 (1998)
38. S.Z. Lee, K.H. Ghar, Mat. Wiss. Wekstofftech. **23**, 117–123 (1992)
39. J. Shieh, S. Wu, Appl. Phys. Lett. **59**, 1512–1514 (1991)
40. D. Rosenthal, J. Welding, Supplement **20**, 220–234 (1941)
41. D. Rosenthal, Trans. ASME **68**, 849 (1946)
42. H. Carslaw, J.C. Jaeger, *Conduction of Heat in Solids* (Clarendon Press, Oxford, 1962)
43. S. Kou, R. Mehrabian, *Modelling of Casting and Welding Processes III* (The Metals and Minerals Society, TMS, Warrendale, 1986)
44. D. Bauerle, *Laser Processing and Chemistry*, 3rd edn. (Springer-Verlag, Berlin, Germany, 2000), pp. 168–176
45. M.J. Jackson, G.M. Robinson, N. Dahotre, A. Khangar, R. Moss, Laser dressing of vitrified aluminium oxide grinding wheels. Br. Ceram. Trans. **102**, 237–245 (2003)

Chapter 8
Epitaxial GaN Layers: Low Temperature Growth Using Laser Molecular Beam Epitaxy Technique and Characterizations

Sunil S. Kushvaha and M. Senthil Kumar

Abstract Generally, the GaN growth by conventional techniques like metal organic chemical vapor deposition (MOCVD) and plasma assisted- molecular beam epitaxy (PA-MBE) techniques employ a higher growth temperature (900–1000 °C in MOCVD and >720 °C in PA-MBE), in which, the probability of forming unwanted alloys or compounds with the substrate at interface will be quite high. To minimize the formation of undesirable interfacial compounds, a low temperature growth is favorable. Here, we have explored the possibility of low temperature growth of GaN layers on sapphire (0001) substrates using an ultra-high vacuum laser assisted molecular beam epitaxy (LMBE) system under different growth conditions. GaN epitaxial layers have been grown by laser ablating liquid Ga metal and polycrystalline solid GaN targets in the presence of active nitrogen environment supplied by radio frequency nitrogen plasma source. The structural and optical properties of the epitaxial GaN layers were characterized using reflection high energy electron diffraction, high resolution x-ray diffraction, atomic force microscopy, Raman spectroscopy, Rutherford Backscattering Spectroscopy, secondary ion mass spectroscopy and photoluminescence spectroscopy. The low temperature LMBE grown GaN layers showed high crystalline structures with a screw dislocation density in the range of 10^7 cm^{-2} as calculated from the x-ray rocking measurements along (0002) plane, which is the lowest value obtained so far using LMBE technique. A strong near band-edge photoluminescence emission has been obtained for the grown GaN layers at room temperature with a relatively weak yellow band emission. Our results indicate that the LMBE technique is capable of growing high quality III-nitride crystalline films at a relatively lower growth temperature compared to the conventional growth techniques.

S.S. Kushvaha (✉) · M. Senthil Kumar (✉)
Physics of Energy Harvesting Division, CSIR-National Physical Laboratory, Dr. K.S. Krishnan Marg, New Delhi 110012, India
e-mail: kushvahas@nplindia.org

M. Senthil Kumar
e-mail: senthilmk@nplindia.org

M. Husain and Z.H. Khan (eds.), *Advances in Nanomaterials*,
Advanced Structured Materials 79, DOI 10.1007/978-81-322-2668-0_8

Keywords GaN · Laser molecular beam epitaxy · Sapphire · High resolution x-ray diffraction · Atomic force microscopy · Raman spectroscopy · Rutherford backscattering spectroscopy · Secondary ion mass spectroscopy · Photoluminescence spectroscopy

8.1 Introduction

The III-nitride based compounds (GaN, AlN and InN) and alloys have been demonstrated to be promising materials for applications in the field of optoelectronic devices such as laser diodes (LDs), light emitting diodes (LEDs), ultra-violet detectors and solar cells [1–8]. The most important property of III-nitrides is their wide, direct bandgap ranging from 0.7 eV for InN, to 3.4 eV for GaN, to 6.2 for AlN, therefore covering a large part of electromagnetic spectrum that is not covered by other semiconductor technology, as shown in Fig. 8.1. Also, there is a small lattice mismatch among III-nitrides with broad range of energy band gap in comparison to the other III-V compound semiconductors. Among these III-nitrides, GaN is most important since its bandgap can be engineered between 0.7 and 6.2 eV by proper alloying of In and Al metals with a minimum lattice-mismatch. In addition, GaN has excellent material properties such as high saturation velocity, large electric critical field (3 MV/cm) and excellent thermal and chemical stability.

At present, after III-nitride technology, semiconductor optical devices emitting all the three primary colors of the visible spectrum are established and thus the generation of pure white light has become possible, which have a great impact on the solid state lighting and display industry. High power and high temperature electronic devices have also been fabricated using nitrides due to their wide band gap, high breakdown field and good thermal conductivity (1.3 W/cm-K) [9, 10]. GaN devices

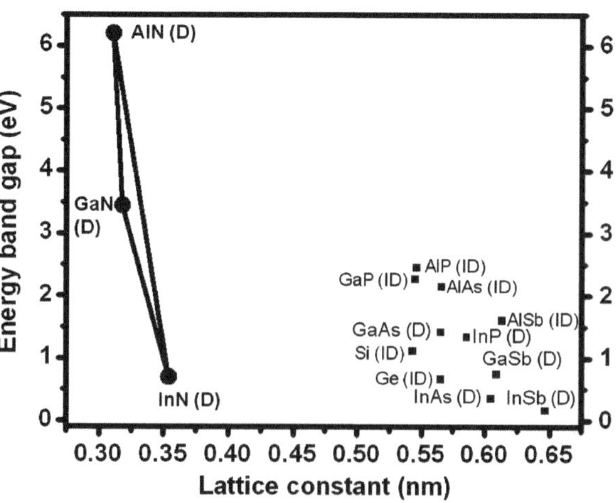

Fig. 8.1 Energy band gap versus lattice constant plot of various technologically important semiconductors. D—Direct band gap, ID—Indirect band gap

are also emerging as a replacement for vacuum electron devices in radar system due to high power and high linearity which is required at high frequencies. Although several researchers reported the applications of GaN based materials in developing monolithic microwave integrated circuits for use in wireless broadband communication and satellite, still better performance are desirable for applications [11].

Generally, epitaxial GaN films are grown using metal organic chemical vapor deposition (MOCVD) [12–17], metalorganic vapour phase epitaxy (MOVPE) [18–20], plasma assisted molecular beam epitaxy (PA-MBE) [21–27] and hydride vapor phase epitaxy (HVPE) [28–30]. In MOCVD, the film is deposited through a chemical reaction which occurs between metal organic precursors and hydrides on a hot substrate surface maintained at atmospheric or low pressure regime. The Ga precursors used for GaN growth are mainly trimethylgallium and triethylgallium. The vapor pressure of the source metal organic is important as it greatly influences the reaction and deposition rates. In HVPE growth technique, Ga and NH_3 are used for GaN growth. Commonly, HCl gas is reacted with liquid Ga at 820–850 °C to form GaCl, which is transported to the growth zone of the reactor and then reacted with precursor NH_3 at 1000–1100 °C forming GaN molecules. The reactor pressure in HVPE is kept at around atmospheric pressure [29]. The growth rate is very high in case of MOCVD, MOVPE and HVPE techniques. However, these techniques involve very high growth temperature with involvement of various chemical and poisonous gases. In case of MBE, the growth of GaN can be controlled in a very precise way, to the extent, that a monolayer of thin film could be accomplished. However, the MBE also involves moderately higher growth temperature in the range of 750–800 °C.

The conventional techniques employ a high growth temperature for the GaN layer, which leads to the formation of undesirable interfacial compounds or alloys. This affects the functionalities of the film due to the surface-interface imperfections, especially in case of ultra-thin layer and multi-layer structures. Since the reproducibility and reliability of the multi-layer device structures depend on the nature of interface and a sharp or abrupt interface is very much desirable. To get an abrupt interface, one of the easy ways is to use low temperature growth. In this respect, the laser molecular beam epitaxy (LMBE) (also referred as ultra high vacuum pulsed laser deposition or UHV-PLD) has the advantage that GaN growth could be achieved at a relatively low temperature due to high-kinetic energy of the film precursors produced by laser ablation of the targets [31–37]. Kim et al. [37] reported that epitaxial GaN films were grown on nearly lattice-matched SiC substrate even at room temperature using LMBE technique. The MBE consists several Knudsen cells and plasma sources to get the high quality film in UHV environment with several in situ characterization tools. The difference in conventional MBE and LMBE is that the material is laser ablated in case of LMBE instead of Knudsen cells in presence of a radio-frequency (r.f.) N_2 plasma. Previous reports describe the growth of GaN using LMBE or PLD system either using the compressed GaN powder as a polycrystalline solid GaN target or liquid Ga target in an atomic nitrogen source or NH_3 gas environments [31, 34]. The sintered GaN targets normally contain trapped oxygen and the grown GaN layers are characterized with substantial level of oxygen contamination [34]. Here, we employed two different targets such as liquid Ga metal and HVPE grown solid GaN target to ensure the high purity source without any trapped oxygen.

For GaN growth, sapphire is the most commonly used substrate because of its low cost, crystal symmetry, high thermal stability and availability of large-sized wafers [3]. For example, GaN based LED devices composed of multi-layers of p- and n-type doped GaN and their alloys are fabricated on sapphire substrates which has large lattice-mismatch (~ 16 %). In this chapter, we discuss on the LMBE growth of good quality epitaxial GaN layers on sapphire (0001) substrates using liquid Ga metal and HVPE grown solid GaN targets in presence of r.f. N_2 plasma at different growth temperatures and deposition rates. The effect of growth temperature on the crystalline properties of GaN film has been studied extensively. We obtained good structural quality GaN epitaxial layers employing both liquid Ga and solid GaN targets in the temperature range of 600–700 °C, which is lower than the conventional MOCVD and MBE techniques. The obtained results suggest that the LMBE growth technique could be an alternative for developing efficient GaN based devices.

8.2 Experimental Section

The growth of epitaxial GaN films on sapphire (0001) substrates were carried out in a LMBE system equipped with reflection high energy electron diffraction (RHEED) for in-situ monitoring, a r.f. N_2 plasma source to supply nitrogen radicals to the growth surface and various laser ablation targets (liquid Ga and solid GaN). The schematic of LMBE system is shown in Fig. 8.2. The base pressure of the

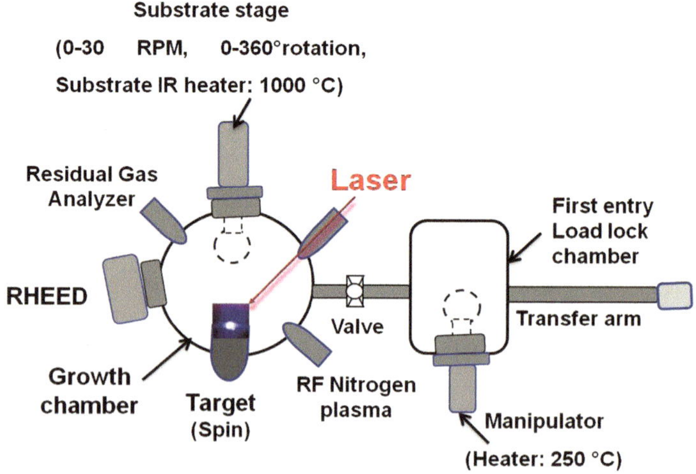

Fig. 8.2 Schematic diagram of laser molecular beam epitaxy (LMBE) system

LMBE growth chamber is ~2×10^{-10} Torr that can provide a clean environment to minimize the residual gas contamination. Liquid Ga metal (99.99999 % purity), HVPE grown polycrystalline solid GaN target (99.9999 % purity) and semiconductor grade N_2 gas were used as the growth precursors. The resistive heater has the capability to raise the substrate temperature maximum up to 1000 °C via infra red (IR) radiation. The back side of the sapphire substrate was coated with a 1 μm thick layer of molybdenum by magnetron sputtering to increase the absorption of IR heat radiation and uniform heat distribution. The substrates were cleaned using standard organic solvents and de-ionized water. The substrates were then degassed for several hours at 250 °C in the entry load lock chamber before introducing them into the growth chamber. The sapphire substrates were thermally cleaned at 850 °C in the growth chamber followed by nitridation at 700 °C using r.f. generated nitrogen radicals with RF plasma power of 400 W and N_2 gas flow of 1.2 sccm. A KrF excimer laser (248 nm wavelength, 25 ns pulse) was used to ablate the targets with an energy density of ~2–8 J/cm^2 and a laser repetition rate ranging 5–45 Hz, depending on the nature of targets and the requirement of the growth rate. Nitrogen radicals are also supplied via r.f. N_2 plasma source during the ablation of Ga or GaN targets to avoid any N_2 deficiency [31]. All GaN samples were grown on sapphire (0001) for 2 h with 2 rpm clock wise rotation of substrate and anti-clock wise spinning of target. Sapphire cleaning, nitridation and GaN growth were monitored in-situ using RHEED. The liquid N_2 was circulated to the cryo- shroud of the growth chamber to minimize any cross contamination from the chamber during heating of the sapphire at high temperature.

The structural properties of the grown epitaxial GaN films were characterized using high resolution x-ray diffraction (HR-XRD), atomic force microscopy (AFM), Raman spectroscopy, secondary ion mass spectroscopy (SIMS) and Rutherford backscattering spectroscopy (RBS). A PANalytical HR-XRD system was employed to characterize the quality of crystalline GaN films grown on sapphire substrate using CuKα1 radiation. The 2θ scan was performed to check the orientation of the film on the substrates while the rocking curve measurement was used to characterize the crystalline quality. The broadening of rocking curve is generally used to estimate the defects such as dislocation density and mosaic spread in the epitaxial layer. Here, the rocking curves of the GaN layers for (0002) and (10-12) diffraction planes were recorded in incident beam configuration using a multilayer graded mirror with channel-cut 2-bounce Ge (220) monochromator. A scintillation detector was used to record the diffracted beam from the GaN layers for measurements of rocking curve and ϕ-scan. An AFM was employed in the tapping mode to characterize the surface morphology of the GaN films using Si tips of curvature radii less than 10 nm. The micro-Raman spectra were collected in backscattering geometry by triple monochromator spectrometer (T-64000, Jobin-Yvon/Horiba group) using an excitation source of 514.5 nm wavelength with resolution 0.1 cm^{-1}. The quantitative analysis of crystalline quality of LMBE grown GaN was also analyzed using RBS technique. A well-collimated He^{++} ion beam of energy 3 MeV from 1.7 MV Pelletron accelerator source was used for the channeling measurements. A Si based detector at an angle of 165° with respect to the incident beam was used to detect the backscattered particles correspond to Ga, N, O and Al. SIMS data was acquired by using time of flight (TOF)-SIMS 5 (ION-TOF GmbH Germany). The pulsed primary ion beam of Bi^+ with 25 keV energy was incident on the sample to produce secondary ions. The Bi current was kept at 1.1 pA with a dose density of 8.72×10^{13} ions/cm^2. On the other hand, the depth profile data was obtained by sputtering the sample by 1 keV Cs^+ ions. The Cs current was 86.1 nA with a dose density of 4.81×10^{17} ions/cm^2.

For photoconductivity measurements, ohmic contacts were prepared by depositing aluminum dots with a separation of $\sim$1 mm. The radiation from xenon lamp (250–700 nm) was passed through a monochromator to select monochromatic light beam. A conventional optical setup was used to focus the light on the samples from the front side of the monochromator. Photoconductivity data were collected in DC mode with continuous illumination of different monochromatic light beam of desired wavelengths in ambient conditions. An Agilent B2900A source meter was used for the DC photoconductivity measurements. Two different photon energies of 365 and 490 nm were used. The beam power was measured by a calibrated DH-Si detector.

8.3 Result and Discussion

Several researchers have used LMBE/UHV-PLD technique to grow heteroepitaxial GaN films [31–36]. The key factor to grow high quality GaN film depends on several parameters such as substrate cleaning, substrate nitridation and buffer layers [38], as well as the type and purity of the target materials, background pressure and growth environment during ablation of targets. In our study, we have grown GaN on pre-nitridated sapphire substrates using liquid Ga and solid GaN targets in the presence of r.f. N_2 plasma with varying laser parameters and substrate temperatures. The process involved has been discussed in more details in below sections.

8.3.1 Cleaning and Nitridation of Sapphire

GaN has been heteroepitaxially grown on various substrates such as SiC [37, 39], Si (111) [40, 41], sapphire (0001) [31, 35, 36, 42, 43], ZrB_2 [44, 45], and other substrates [46–49]. Despite poor structural mismatch, sapphire is preferred over other substrates as it is chemically and thermally stable which is important for epitaxial growth. Different orientation of the sapphire has been used as substrate, some of which are the (0001), (10-10), (11-20) and (1-102) orientations commonly known as c-plane, m-plane, a-plane and r-plane, respectively. The best orientation for GaN growth is the m-plane while considering crystal symmetry and lattice-match. However, undesirable formation of GaN crystalline twins was obtained due to the non alignment of c-axis GaN on m-plane sapphire. Hence out of these orientations, c-plane is commonly used for growth of smooth and flat epitaxial GaN films [31, 33].

The degree of in-plane alignment of GaN relative to the sapphire (0001) substrate, determined by ϕ-scan of HR-XRD, is shown in Fig. 8.3. The scanning planes used in ϕ-scan were (10-1-10) for sapphire and (10-12) for GaN. The three peaks at 120° intervals are seen in the ϕ-scan for sapphire substrates (Fig. 8.3a). The diffraction peaks from the (10-12) plane of GaN film were observed at intervals of 60° as seen in Fig. 8.3b, which confirms the hexagonal structure of the epitaxial GaN film. From Fig. 8.3, it is observed that there is a 30° rotation of the GaN unit cell with respect to the sapphire. This indicates the in-plane epitaxial relationship of GaN grown on sapphire as $(10\text{-}10)_{GaN}/(11\text{-}20)_{sapphire}$. This is also observed during in-situ RHEED analysis.

In-situ RHEED analysis and ex-situ AFM characterization were performed to check the surface modification of the sapphire substrates as a result of thermal cleaning and nitridation processes. The RHEED provides information about the surface reconstruction as it is sensitive only to the first few atomic layers. The RHEED pattern of thermally cleaned c-plane sapphire substrate was obtained as shown in Fig. 8.4a. The streaky RHEED pattern of sapphire indicates that the

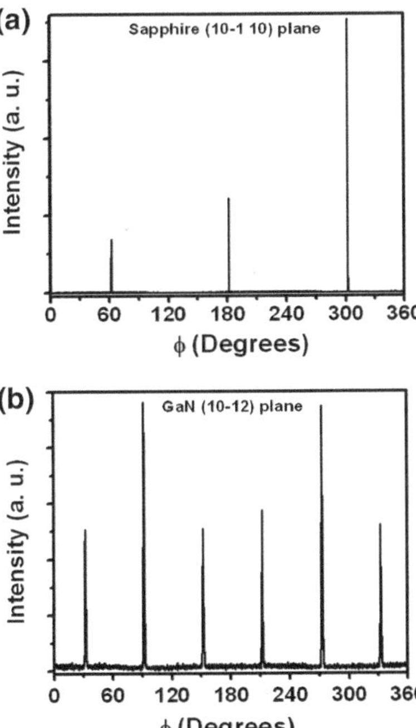

Fig. 8.3 XRD in-plane ϕ-scan of: **a** sapphire (10-1-10) plane; **b** (10-12) plane of GaN layer grown by LMBE system

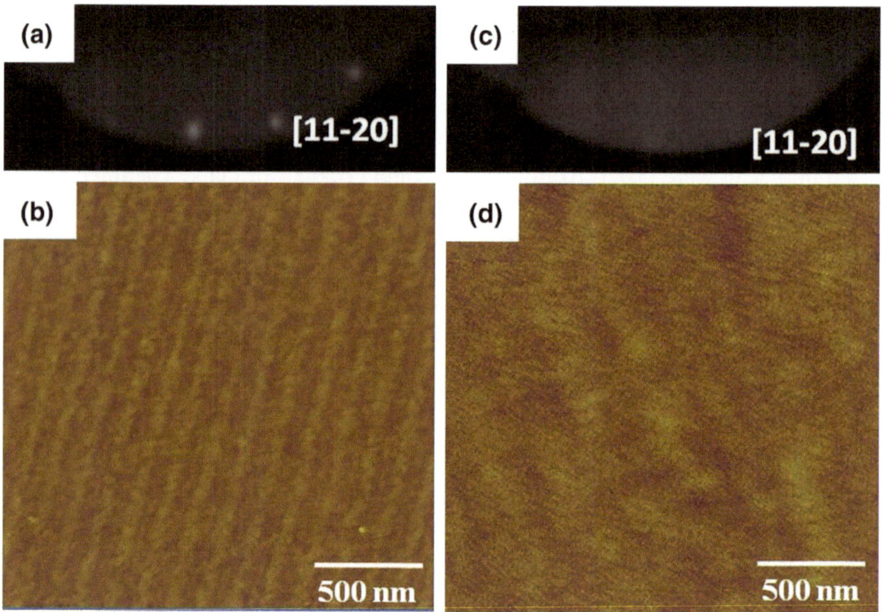

Fig. 8.4 A typical **a** RHEED pattern along [11-20] direction and **b** AFM image of sapphire (0001) substrate after cleaning at 850 °C for 10 min in UHV. **c** RHEED pattern and **d** AFM image of sapphire (0001) substrate after nitridation at 700 °C for 35 min; AFM scan area: 2 μm × 2 μm

surface is flat and clean. In addition, AFM image as shown in Fig. 8.4b reveals that the UHV cleaned sapphire surface is atomically flat root-mean-square (rms) surface roughness: 0.16 nm) with steps and terraces. The rms roughness of the surfaces was determined from AFM images of 2 μm × 2 μm area.

It has been reported that the buffer layer play a critical role on the growth of high quality GaN film on sapphire substrates. For example, the presence of AlN on sapphire substrate will lead to growth of Ga-face GaN whereas without any buffer layer, N-face GaN growth occurs. Generally, the Ga-face GaN has smooth surface and is preferred for device applications [38]. The pre-nitridation of sapphire substrates using nitrogen radicals forms a thin AlN which acts as an intermediate (or buffer) layer for growing a high quality GaN film. Figure 8.4c shows the diffused RHEED pattern of the sapphire substrate after nitridation at 700 °C. The AFM image as shown in Fig. 8.4d also shows that the surface of sapphire has been modified with a surface roughness of 0.1 nm. Thus, the RHEED and AFM analyses show the signature of AlN formation on the substrate surfaces by the nitridation process. Here, all the GaN films have been grown on pre-nitridated sapphire substrates.

8.3.2 Growth of GaN Films on Sapphire Using Liquid Ga Target

Firstly, the laser energy density and repetition rates were varied between 5 and 10 J/cm^2 and 10 and 50 Hz, respectively, to optimize the Ga flux, keeping the r.f. N_2 plasma condition constant. A well defined laser plume pattern was obtained for a energy density of 5 J/cm^2 and a laser repetition rate of 45 Hz [33, 43]. The laser ablated Ga plume contains a variety of Ga species such as neutral Ga, Ga^+ and Ga^{2+} ions. Mckiernan and Mosnier [50] also studied the dynamics of laser ablated plumes from Ga and GaN targets using stigmatic emission spectroscopy and found that the laser plume contains only neutral Ga and the plume ionization is comparatively small if the distance between target and substrate kept more than 4 cm. In our studies, the substrate-to-target distance was kept ≥6 cm and hence, it is expected that the neutral Ga atoms will mainly participate in the GaN growth.

For growth of GaN, the laser parameters were fixed as follows: energy density ~5 J/cm^2, repetition rate at 45 Hz, and the substrate-to-target distance as ~10 cm to obtain a uniform Ga flux. The growth of the GaN films on sapphire were monitored using in-situ RHEED and the RHEED patterns of GaN layer after GaN growth along [11–20] direction are shown in Fig. 8.5a, b. The GaN layers grown at <500 °C exhibited streaky pattern after growth while slightly spotty RHEED pattern was observed for the films grown at >500 °C. It has been clearly observed that there is 30° in-plane rotation of GaN (11-20) to sapphire (10-10) as confirmed from in-situ RHEED and HR-XRD ϕ-scan [51]. XRC of GaN (0002) plane was also measured for the samples grown at growth temperature ranging from 300 to 700 °C. The (0002) plane XRC of GaN epilayer grown on sapphire (0001) at 500, 600 and 700 °C is presented in Fig. 8.5c. The crystalline quality of GaN layer shows a strong dependence on the growth temperature. The XRC GaN (0002) FWHM values dramatically decreased from 2030 arc sec at 300 °C to 245 arc sec at 500–600 °C. The good crystalline quality GaN layers have been achieved at the growth temperature of 500–600 °C, which is about 150 °C lower than the conventional MBE growth [21]. The low temperature growth can be attributed to the high kinetic energy of ablated species in the laser MBE process. The obtained (0002) rocking curve width of 245 arc sec is one of the lowest values reported for GaN layer grown on sapphire (0001) by LMBE technique and is comparable to that of GaN layers grown on sapphire by other conventional techniques. Ohta et al. (2003) have obtained 1260 arc sec for the GaN grown on sapphire with AlN buffer by LMBE [52]. The growth rate was decreased from 300 to 160 nm/h when the growth temperature increased from 300 to 700 °C possibly due to increase in desorption of Ga from the surface at higher growth temperatures [33].

The XRCs of different orientation planes have been used to indirectly estimate the presence of dislocation density in the GaN epitaxial layers [53]. The FWHM value of the GaN (0002) plane reflects lattice distortion from screw dislocations and mixed dislocations whereas the FWHM of asymmetric GaN (10-12) plane is related with lattice distortion from edge dislocations and mixed dislocations [53–57].

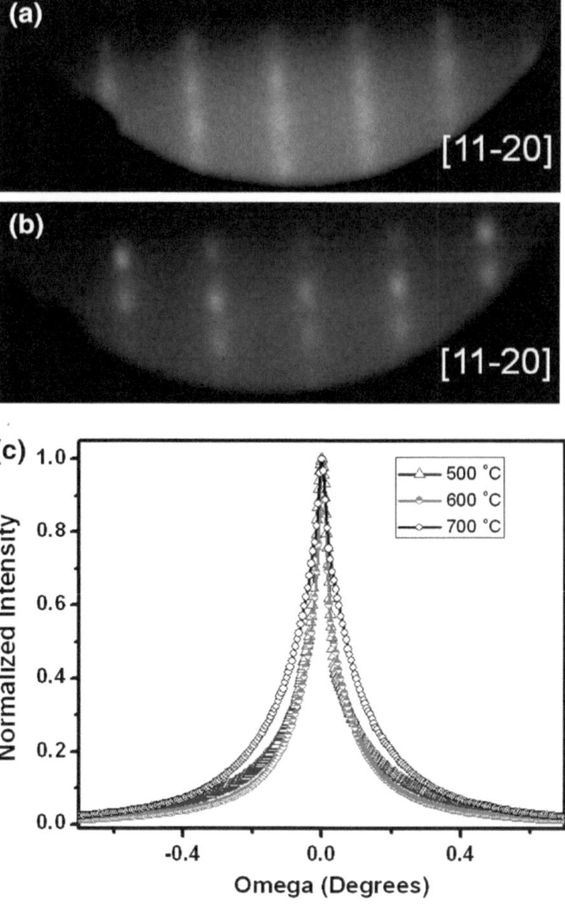

Fig. 8.5 A typical RHEED pattern along [11–20] direction after growth of GaN on pre-nitritated sapphire (0001) by laser MBE at **a** below and **b** above 500 °C growth temperature. **c** Normalized x-ray rocking curves of GaN (0002) plane for GaN growth on sapphire (0001) using liquid Ga target at substrate temperature 500, 600 and 700 °C

We have estimated the dislocation density in laser MBE grown GaN (0001) film on sapphire at 600 °C using the following equations [42, 54–57]:

$$D_{screw} = \frac{\beta^2_{(0002)}}{9b^2_{screw}}, \quad D_{edge} = \frac{\beta^2_{(10-12)}}{9b^2_{edge}} \tag{8.1}$$

$$\mathrm{D_{dis}} = \mathrm{D_{screw}} + \mathrm{D_{edge}} \tag{8.2}$$

where $\mathrm{D_{screw}}$ is the screw dislocation density, $\mathrm{D_{edge}}$ is the edge dislocation density, β is the FWHM values measured for HR-XRD rocking curves of (0002) and (10-12) planes in the degree and b is the Burgers vector length ($\mathrm{b_{screw}}$ = 0.5185 nm, $\mathrm{b_{edge}}$ = 0.3189 nm). The values for $\beta_{(0002)}$ and $\beta_{(10\text{-}12)}$ obtained from HR-XRD rocking curves for GaN grown at 600 °C are 0.068° and 0.62°, respectively. The calculated screw and edge threading dislocation densities are 5.82×10^7 and

8.3×10^9 cm^{-2}, respectively. It indicates that a high density of edge dislocation is present in these GaN layers compared to that of screw dislocation.

Figure 8.6 shows the AFM images of the GaN films grown at various temperatures. For 300 °C growth, we obtained un-coalesced flat-surface GaN islands with a rms surface roughness of ~5.6 nm (Fig. 8.6a). With increasing substrate temperature to 400 °C, a smooth surface GaN with few large pits is obtained as shown in Fig. 8.6b. Mostly large, flat hexagonal GaN islands are found for the GaN film grown at 500 °C as shown in Fig. 8.6c. The morphology of the GaN film abruptly changed when growth of GaN film was done at 600 °C. Mostly uniform particles of size ~60–80 nm is observed from the AFM image (see Fig. 8.6d). The rms surface roughness of the GaN layers grown between 400 and 600 °C is obtained to be same ~4.5 nm. The rms value increased to 5.2 nm for the film grown at 700 °C (Fig. 8.6e), with random-sized island morphology. The conversion of GaN growth from flat island to rough particles with increasing temperature implies that the growth mode is converted from 2D to 3D due to enhanced desorption of Ga flux at the higher growth temperatures which alters the flux condition from Ga-rich to N-rich as observed by in-situ RHEED [33].

In order to identify the structural quality of the GaN films, we also performed micro-Raman spectroscopy using an excitation source of 514.5 nm wavelength at room temperature. Figure 8.7a shows the typical micro-Raman spectrum of the LMBE grown GaN film. For hexagonal GaN films grown on sapphire, there are several Raman scattering phonon modes such as A_1(LO), A_1(TO), E_1(LO), E_1(TO) and E_2(high). The E_1(LO) and E1(TO) modes are forbidden in the back scattering geometry measurements. There exists few reports in which E_1(TO) peak is detected in the backscattered Raman measurements. This is attributed to the high density of dislocations or defects in the heteroepitaxially grown GaN film [58]. According to the selection rules, the A_1(LO) and E_2(high) phonon modes are expected to be more pronounced. The line shape, FWHM value and shift in E_2(high) peak disclose the crystalline quality and the stress present in the GaN films, whereas A_1(LO) peak determines the free carrier concentration (or doping level) and electron mobility [59, 60]. The two active Raman modes E_2(high) and A_1(LO) have been observed at 568 and 736 cm^{-1}, respectively. These peak values are consistent with the previously reported values for hexagonal GaN films [61, 62]. The shoulder peaks at 576 and 750 cm^{-1} arise from the sapphire (0001) substrate [63]. The quantitative analysis of crystalline quality near interface and film region of the grown GaN layer has also been examined using RBS in channeling condition as shown in Fig. 8.7b. The RBS signals from Ga, Al, O and N were clearly observed. The channeling is poor with large yield (χ) (the ratio of the backscattering yield of aligned direction to that from the random direction in the near surface region) at the GaN/sapphire interface region due to the formation of high dislocation density and low angle grain boundaries [64]. The ion channeling yield in the film region is ~6 %, which is comparable to the previous reported value [31] and it indicates the good crystalline property in bulk region.

We have also performed SIMS depth profile analysis to know the impurity levels present in the LMBE grown GaN film as well as the nature of the interface. Figure 8.8 shows a typical SIMS depth profile of GaN film grown on sapphire using

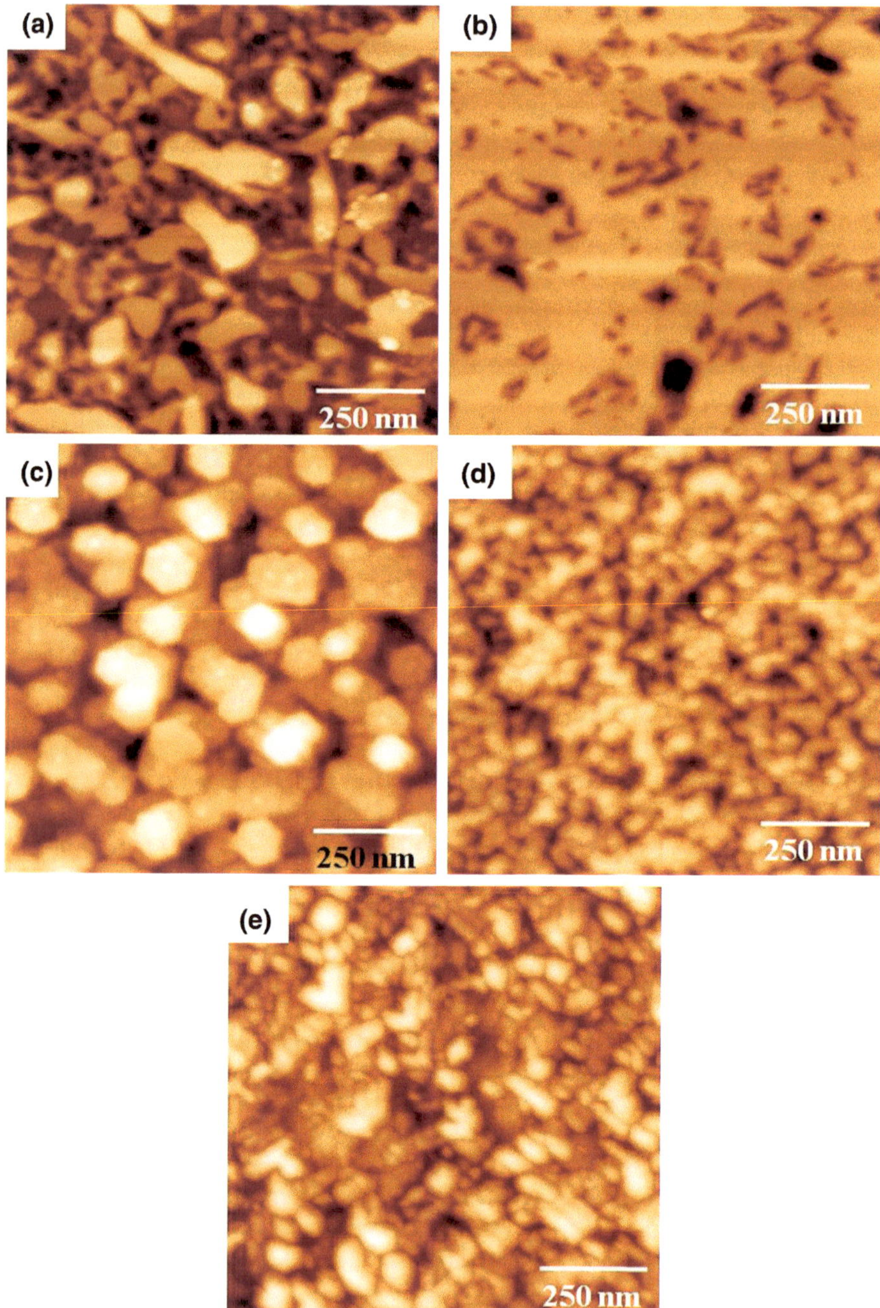

Fig. 8.6 AFM images of GaN films grown on sapphire (0001) using liquid Ga target in presence of r.f. nitrogen plasma at different growth temperatures: **a** 300 °C, **b** 400 °C, **c** 500 °C, **d** 600 °C and **e** 700 °C. Scan area: 1 μm × 1 μm

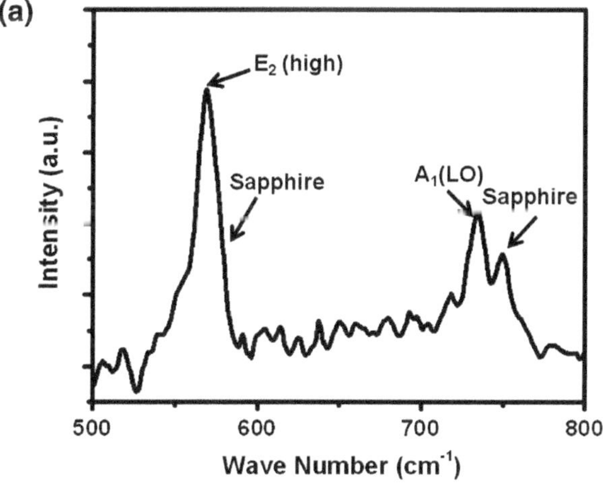

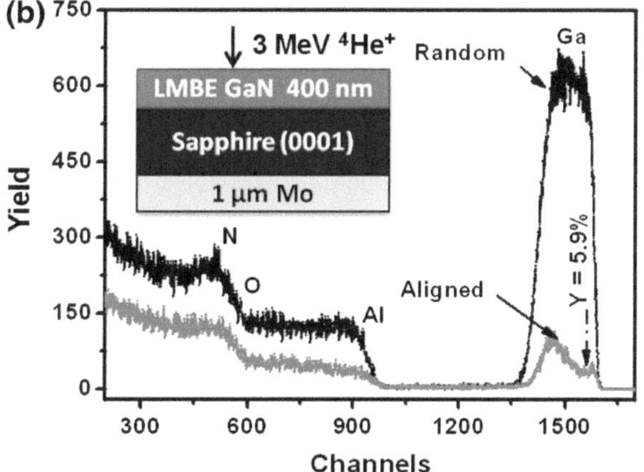

Fig. 8.7 Room temperature **a** Raman and **b** RBS spectra of GaN films grown on sapphire (0001) using liquid Ga target in presence of r.f. N_2 plasma at 600 °C

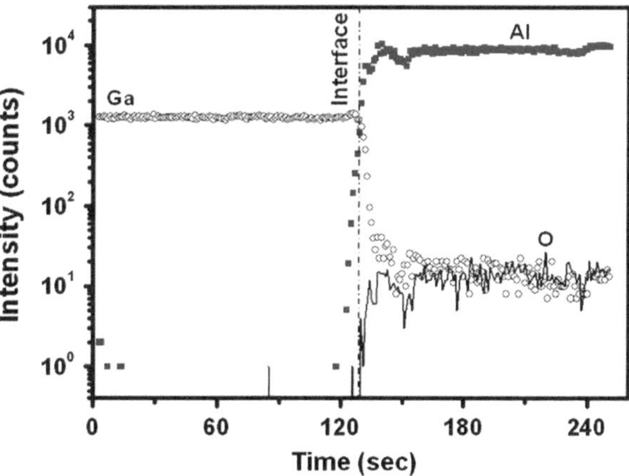

Fig. 8.8 SIMS depth profile on GaN film grown on sapphire (0001) substrate using liquid Ga target in presence of r.f. N_2 plasma at 600 °C

liquid Ga target. A steady signal of Ga species is observed in the film region, which falls sharply at the interface indicating an abrupt interface between the GaN film and the substrate. The substrate region shows the signals corresponding to Al and O species sharply rising from the interface. The low intensity of O is because of its low sensitivity factor for detection by Bi^+ ion. The presence of carbon and oxygen impurity is below the detectable limit in the GaN film. Thus, the LMBE technique yields GaN films with uniform composition and a sharp interface with negligible impurity levels [33].

8.3.3 Growth of GaN Layers on Sapphire Using Solid HVPE GaN Target

We have also grown GaN layers on pre-nitrated sapphire substrates by laser ablating a HVPE grown polycrystalline solid GaN target. During ablation, nitrogen radicals were additionally supplied using the r.f. plasma cell to compensate any nitrogen deficiency. In this section, we discuss the effect of growth temperature, laser repetition rate and laser energy density on the structural properties of GaN layers grown using GaN solid target.

8.3.3.1 Effect of Growth Temperature

The structural quality of GaN film was studied in the growth temperature ranging 500–750 °C keeping the laser energy density of ~5 J/cm^2 and the laser repetition rate of 10 Hz constant in the presence of r.f. N_2 plasma. From in situ RHEED monitoring, it was observed that the GaN grew mostly in 3D mode within the experimental conditions adopted in this study as shown in Fig. 8.9. The XRD 2θ scans of all GaN films grown on pre-nitridated sapphire at different growth

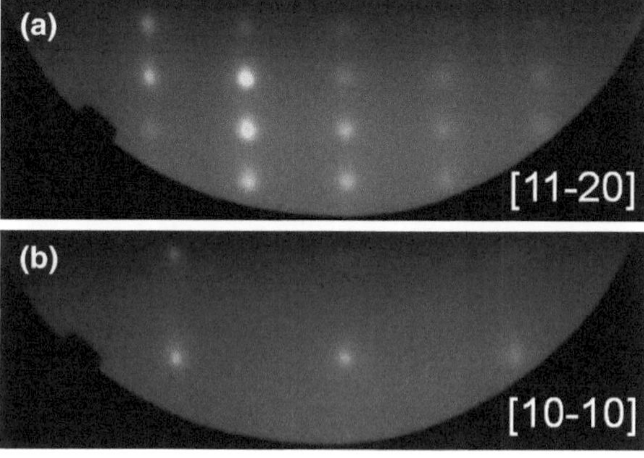

Fig. 8.9 A typical RHEED pattern along **a** [11–20] and **b** [10-10] directions after growth of GaN on pre-nitritated sapphire (0001) by LMBE using solid GaN target

temperature showed a high degree of texturing of {000 1} wurtzite GaN indicating the heteroepitaxial relationship between $(0001)_{GaN}$ and $(0001)_{sapphire}$. Figure 8.10a shows a typical XRD 2θ scan of the LMBE grown GaN film on sapphire (0001) substrate using HVPE grown solid GaN target. In addition, ϕ-scan for GaN film grown on sapphire using solid GaN target show a 30° in-plane rotation between GaN and sapphire as shown in Fig. 8.3. Figure 8.10b shows the normalized XRCs for GaN (0002) plane on sapphire grown at different temperatures. The GaN layers grown at 500 °C revealed a poor crystalline property with a (0002) plane XRC FWHM of 1582 arc sec. However, the value dramatically decreased to 153 arc sec when the growth temperature was increased from 500 to 600 °C indicating the high crystalline quality. The FWHM value further decreased with the increase of growth temperature and a lowest value of 110 arc sec has been achieved for the GaN layers grown at 700 °C [42, 64]. The FWHM for the GaN film grown at 750 °C was as large as ~610 arc sec, where the layer thickness was ~40 nm. The GaN growth

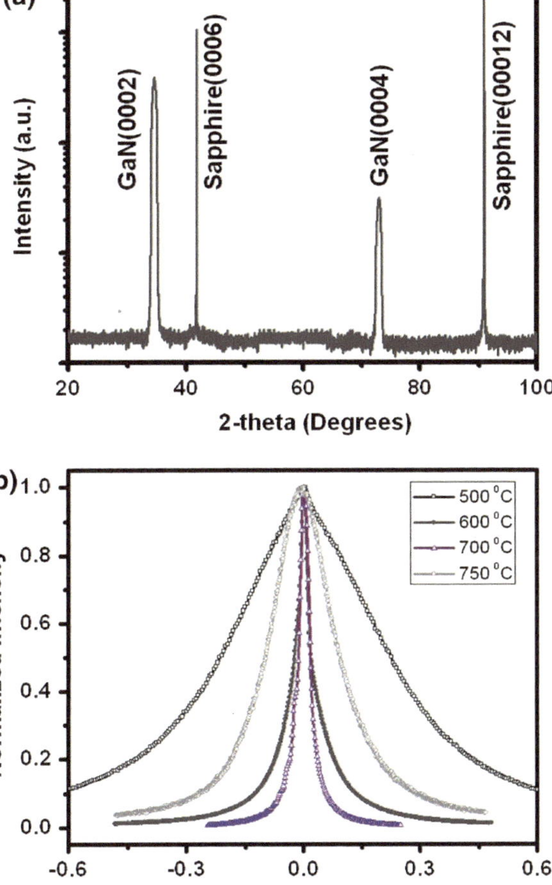

Fig. 8.10 **a** Typical XRD 2θ-scan of GaN layer grown on sapphire (0001); **b** Normalized XRCs of GaN (0002) plane for GaN grown on sapphire using HVPE GaN solid target with laser energy density of ~5 J/cm^2 and laser repetition rate of 10 Hz at substrate temperature between 500 and 750 °C

rate decreased with increasing growth temperature and a practically low growth rate was noticed for the layers grown above 700 °C. The XRC measurements revealed that the GaN film grown at 700 °C has better crystalline property with a low spread of mosaicity. The screw dislocation density has been calculated for the film grown at 700 °C taking the XRC FWHM (0002) value of 0.03° into account. Using Eq. (8.1), we obtained a lower screw dislocation density of 1.13×10^7 cm^{-2} in comparison to that of GaN grown using the liquid Ga target (5.82×10^7 cm^{-2}). This indicates that the GaN layers grown using high purity HVPE grown polycrystalline GaN solid target in presence of r.f. N_2 plasma have less crystalline defects.

Surface morphology of the GaN films grown on sapphire at different growth temperatures were investigated using AFM measured in tapping mode. The AFM image of the GaN film grown at 500 °C shows mostly hazy features with grain size of 75–85 nm with rms roughness of ~10 nm as presented in Fig. 8.11a. For the GaN

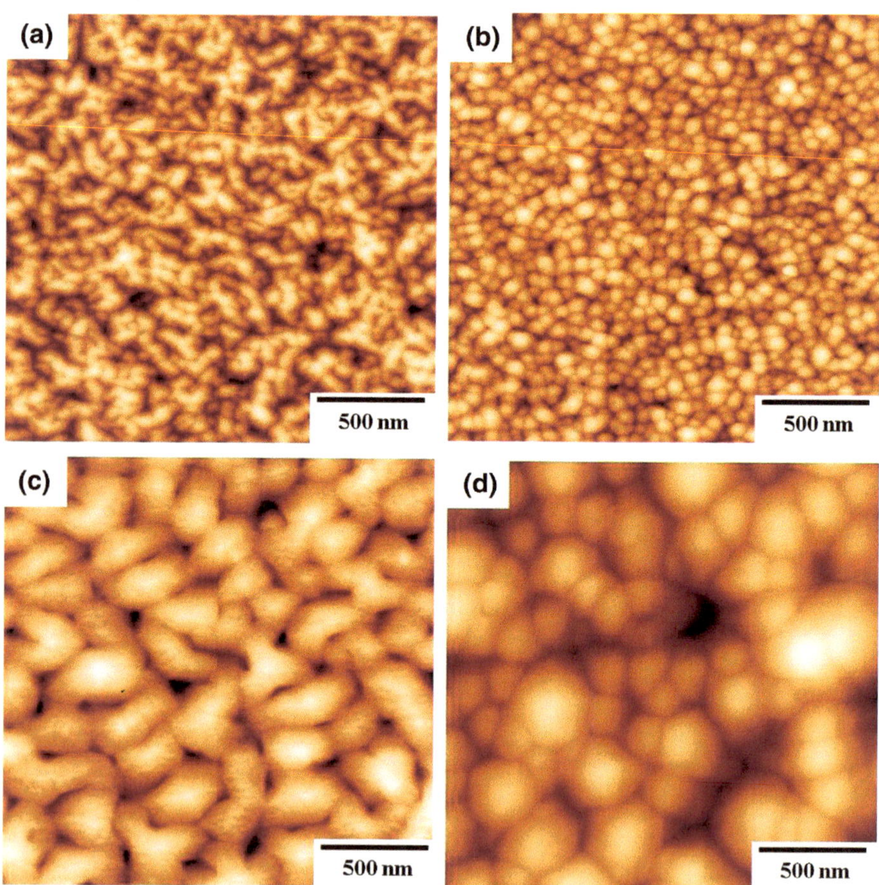

Fig. 8.11 AFM images of GaN films grown on sapphire using solid GaN target with laser energy density of ~5 J/cm^2 at different growth temperatures: **a** 500 °C, **b** 600 °C, **c** 700 °C, and **d** 750 °C. AFM scan area: 2 μm × 2 μm

film grown on sapphire at 600 °C in Fig. 8.11b, very uniformly grown grains of size 80–85 nm with a rms surface roughness ~4.9 nm have been obtained [4]. The lateral size of the grains is large (~200–280 nm) as shown in Fig. 8.11c for GaN film grown at 700 °C with a rms surface roughness of ~6.2 nm. Whereas, the GaN layer grown at 750 °C shows random grown grains of size varying between 180 and 300 nm (rms roughness of ~9.8 nm) as shown in Fig. 8.11d. The high temperature growth generally promotes the coalescence of crystallite islands or particles due to enhanced migration of surface adatoms or clusters. However, non-uniform GaN growth obtained at growth temperature 750 °C might be due to increased desorption rate of Ga flux at higher growth temperatures [31]. It has been observed from HR-XRD and AFM characterizations that good crystalline film quality is only obtained at growth temperature higher than 500 °C but below 750 °C [42].

The differences in the surface morphologies of GaN on sapphire at 600 and 700 °C are related with the diffusivity and characteristic coalescence time of GaN adatoms on sapphire during the growth. The surface diffusion coefficient mainly depends on the growth temperature as [65]:

$$D(T) = D_0 \exp\left[-{}^{E_a}\!/_{kT}\right] \tag{8.3}$$

where D(T) is the surface diffusion coefficient, D_0 the attempt frequency, E_a the activation energy, T the growth temperature in Kelvin and k the Boltzmann constant. The surface diffusion coefficient increases with growth temperature. At growth temperature of 700 °C, the diffusion coefficient will be about one order higher than that at 600 °C. For a given island or crystallites, the characteristic coalescence time is inversely proportional to the surface diffusion coefficient [66]. The shorter coalescence time is the reason why bigger GaN grains were obtained on sapphire at 700 °C compare to low growth temperature 600 °C in the same growth conditions.

8.3.3.2 Effect of Laser Repetition Rates

After optimization of the growth temperature, the effect of laser repetition rate on the properties of the GaN layers has been studied. Keeping the laser energy density and the substrate temperature constant, respectively, at ~2.5 J/cm^2 and 700 °C, we grew the GaN film on pre-nitridated sapphire substrates at different laser repetition rates ranging from 10 to 40 Hz. The variation of FWHM values of (0002) plane XRC as a function of laser repetition rate is presented in Fig. 8.12a. The (0002) plane XRC FWHM of GaN layer grown at 10 Hz is 540 arc sec. However, the value dramatically decreased to 189 arc sec when the laser repetition rate was increased from 10 to 20 Hz indicating the high crystalline quality. The FWHM value further decreased with the increase in laser repetition rate. A lowest value of 121 arc sec was achieved for the GaN layers grown at 30 Hz [67]. However, with further increase to 40 Hz, the FWHM (0002) XRC value increased as ~1580 arc sec.

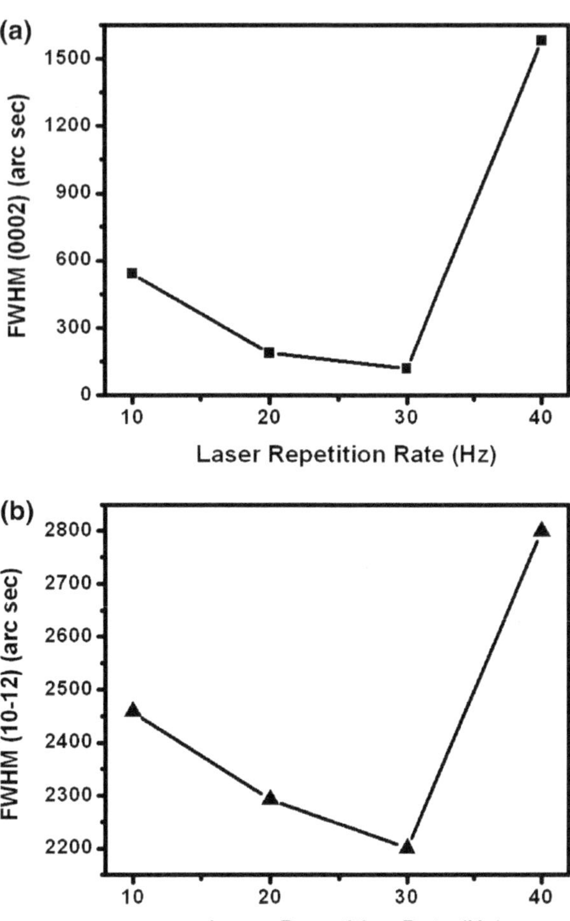

Fig. 8.12 XRC FWHM of **a** (0002) and **b** (10-12) GaN planes as a function of laser repetition rate. All the samples were grown with a laser energy density of ~2.5 J/cm^2 at 700 °C

The increase is attributed to the low surface migration of the adatoms or clusters due to high flux rate at 40 Hz. The previously reported FWHM XRC value for LMBE GaN film on sapphire was 420 arc sec grown by laser ablating a polycrystalline GaN solid target under NH_3 ambient [31]. A similar trend is also observed in the XRC FWHM value for the GaN (10-12) plane (Fig. 8.12b). The screw dislocation density present in the film grown at 30 Hz has been calculated using FWHM XRC (0002) value of 0.0336°. From Eq. (8.1), the screw dislocation density present in the GaN film is 1.42×10^7 cm^{-2}. This screw dislocation density is almost similar with the GaN film grown at 10 Hz with the laser energy density of ~5 J/cm^2 at 700 °C. At high growth temperature, the desorption of surface adatoms is quite high due to the lower growth rate [31] and therefore, a relatively higher growth rate should be maintained to achieve good quality layers with a considerable growth rate either using high laser energy density or increasing laser repetition rate at low energy density.

Surface morphologies of the GaN films grown at temperature 700 °C with the laser energy density $\sim$2.5 J/cm^2 for various repetition rates from 10 to 40 Hz were also characterized using AFM. Figure 8.13a represents the AFM image of the GaN film grown using the laser repetition rate of 10 Hz. Mostly granular film with grain sizes of 100–220 nm have been observed. With further increase in laser repetition rates to 20 and 30 Hz, the lateral grain size increased as shown in Fig. 8.13b, c. However, the film grown at 40 Hz shows the un-coalesced islands as in Fig. 8.13d. This observation indicates that the GaN film grown at the laser energy density of $\sim$2.5 J/cm^2 with laser repetition rate 20–30 Hz has better surface properties.

The GaN layers of better crystalline and surface properties grown at 700 °C with high laser energy density $\sim$5 J/cm^2 at laser repetition rate 10 Hz were further characterized using Raman and RBS channeling. Figure 8.14a shows the typical Raman spectroscopy of the GaN layer grown using HVPE solid target. Two active

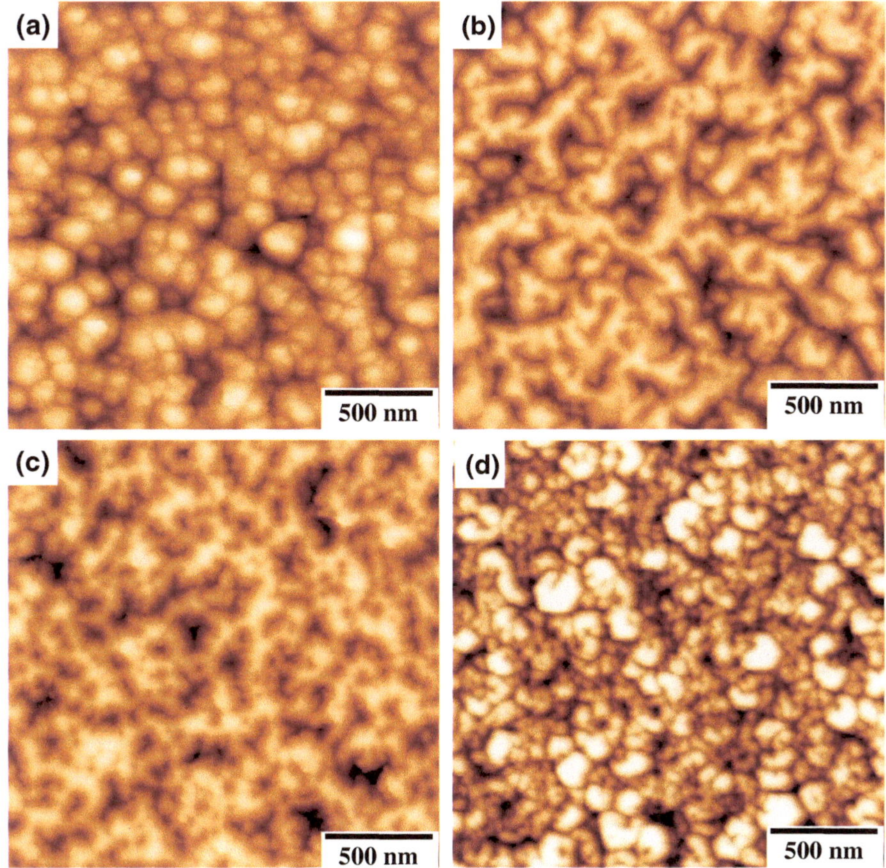

Fig. 8.13 AFM images of GaN films grown on sapphire using solid GaN target at growth temperature of 700 °C with various laser repetition rates: **a** 10 Hz, **b** 20 Hz, **c** 30 Hz, and **d** 40 Hz. AFM scan area: 2 μm × 2 μm

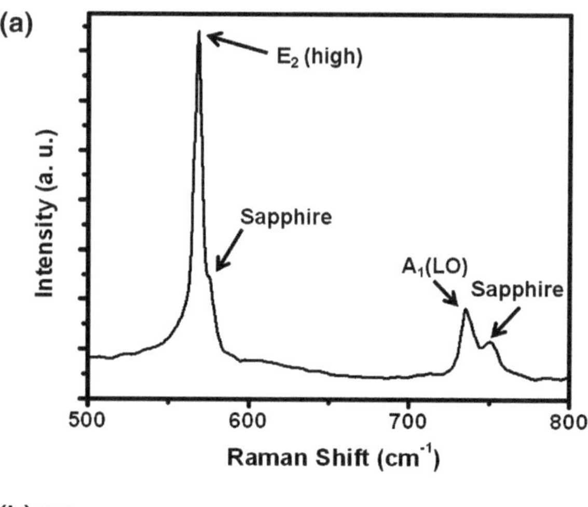

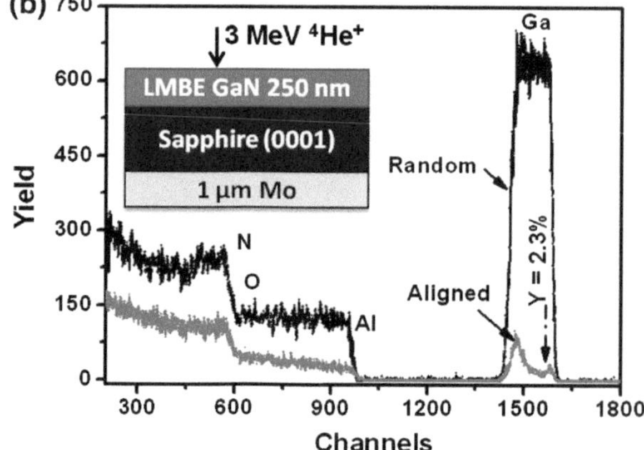

Fig. 8.14 Room temperature **a** Raman and **b** Typical RBS spectra of GaN films grown on sapphire (0001) using solid GaN target at substrate temperature of 700 °C with laser energy density of ~5 J/cm^2

Raman modes A_1 (LO) and E_2 (high) have been observed at 568 and 736 cm^{-1} and these are the Raman allowed peaks for wurtzite GaN in backscattered geometry [61]. The shoulder peaks obtained in the spectrum at 576 and 750 cm^{-1} is from the sapphire substrate itself [63]. The RBS channeling spectrum was also measured for the GaN films grown using solid target with high laser energy density ~5 J/cm^2 and it has been shown in Fig. 8.14b. The ion channeling yield (χ) is very low as ~2.3 %. Vispute et al. [31] previously reported 3 % yield for the GaN layers grown on sapphire (0001) by laser ablating a polycrystalline, stoichiometric GaN target in the presence of NH_3 ambient. The low RBS yield value in the aligned direction is an indication of the high crystalline quality of the grown GaN layer. The layers are also found to be stoichiometric. It should be noted that the defects created near the interface is confined within a short distance in the GaN layer grown using HVPE GaN target compared to the liquid Ga one (χ ~ 6 %) [64]. Further, the presence of oxygen at the surface of the film has also been observed in the sample that may be due to the oxide formation after exposure to the atmosphere.

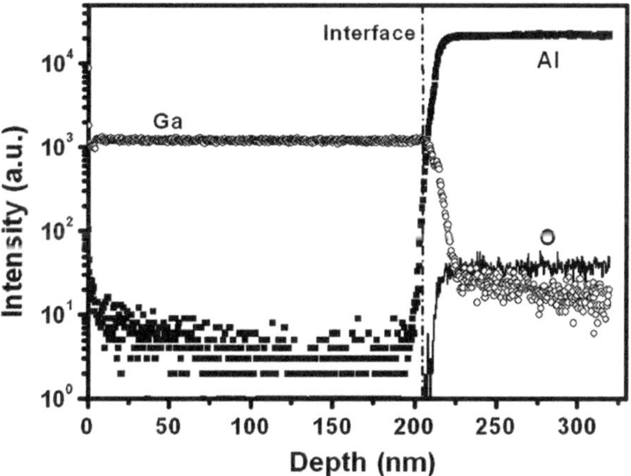

Fig. 8.15 Typical SIMS depth profile on GaN film grown on sapphire substrate using HVPE grown solid GaN target at 700 °C at high laser energy density ~5 J/cm^2.

We have performed TOF-SIMS depth profile analysis to find the impurity levels present in GaN film grown using HVPE GaN solid target. Figure 8.15 shows the SIMS depth profile of a typical GaN film grown on sapphire using LMBE. In the GaN film region, we observed uniform distribution of Ga species throughout the layer and a sharp GaN/sapphire interface as well. The oxygen impurity level was obtained to be below detection limit. Whereas, GaN layers grown using PLD technique are generally found with substantial levels of oxygen contamination as the sintered GaN targets contain trapped oxygen [34]. The lack of oxygen impurity or other residual gas contamination is achieved due to the low background pressure in the growth chamber as well as the use of high purity HVPE grown GaN solid target [42].

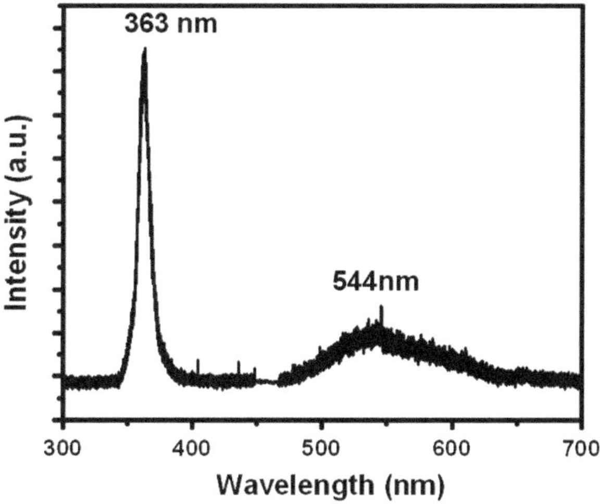

Fig. 8.16 A typical PL spectrum of GaN film grown on sapphire (0001) using solid GaN target at high laser energy density ~5 J/cm^2

The typical PL spectrum recorded for the LMBE grown GaN on sapphire using solid GaN target with laser energy density ~ 5 J/cm^2 is shown in Fig. 8.16. The GaN film was excited with a laser line of 266 nm at room temperature. The emitted signal from the sample was passed through an emission monochromator and recorded using a CCD detector. Wurtzite bulk GaN has the near band edge (NBE) emission at ~3.4 eV and defects related peaks in the range of 1.4–2.8 eV [21, 68–73]. Here, we obtained a broad yellow luminescence (YL) band at wavelength ~544 nm (2.28 eV) and the NBE related luminescence feature at ~363 nm (3.41 eV). The origin of the NBE transition peak is known from earlier experiments [74, 75]. Cazzanelli et al. [76] also observed the band edge emission at 3.36 eV for a PLD grown GaN films on sapphire. The NBE region is mainly dominated by free exciton transition, donor-to-valence band transition and donor-bound exciton transitions among others. Thermal broadening and defect related disorder in the film also increase the broadening of the NBE peak. Generally, YL peak in GaN films has been associated to various kinds of defect related transitions such as transition from shallow donors (O and C complexes, O_N: oxygen is substituted for nitrogen, C_N: carbon is substituted for nitrogen) to deep acceptors (Ga vacancies, V_{Ga}) [69–71]. The FWHM of YL is quite large as compared to the NBE transitions peak, which is a signature of the contribution of multiple defect bands in the mid-gap region. The high intensity NBE to YL band ratio normally indicates the high quality nature of the LMBE grown GaN film.

In general, unintentionally doped GaN epitaxial thin films are mostly n-type [77, 78] and a high carrier concentration is seen to reduce the performance of opto-electronic devices like, LEDs, photodetectors and diode lasers. The photoconductivity (PC) effect has been studied using PC measurement to know the nature of the unintentional doping in our LMBE grown GaN. The PC can persist for a long time after the excitation light source is turned off. This persistent photoconductivity (PPC) has been reported in GaN thin films grown by MOCVD on c-plane sapphire substrates [79, 80]. Figure 8.17 shows the time scan of the photocurrent behavior

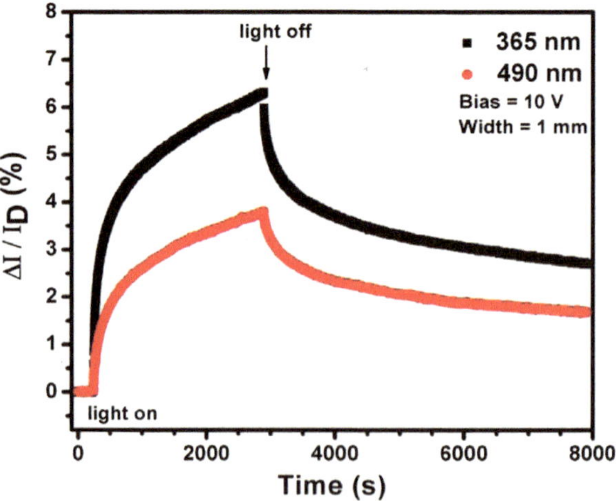

Fig. 8.17 Room temperature photocurrent response of GaN film grown on sapphire substrate using HVPE grown solid GaN target with laser energy density ~5 J/cm^2

using 365 and 490 nm excitation wavelengths. The sample was stored in dark condition for up to 24 h prior to each illumination. For both the wavelengths, the photocurrent initially increased rapidly, and then settled to a slower increase. The photocurrent saturation was slow for the total of 50 min of its exposure. The rapid rise in photocurrent can be explained by the initial excitation of electrons to higher energy states. The slower increase in photocurrent is attributed to photo-excited states that exhibit a short lifetime, and it may be due to the surface lattice relaxation. As suggested in Ref. [81], such a mechanism could convert shallow donor to deep donor states, allowing electron to remain in a metastable state. As time progresses, the surface state relaxation decreases and the relaxation time increases. The photocurrent decay follows a similar pattern as the rise, but in reverse way i.e. with an exponential decay. We have defined normalized photocurrent by $\Delta I/I_D$ (%), where $\Delta I = I(t) - I_D$, I_D is the current at dark condition and I(t) is the current under light illumination. There is a significant difference in photocurrent for two excitation wavelengths, as shown in Fig. 8.17. The photocurrent increases to 6.3 and 3.8 % with 365 and 490 nm radiations, respectively. When illuminated by 365 nm radiation, the rise in photoconductivity is observed to be higher than that by 490 nm excitation. Observed photo-induced current using below-band gap visible light (490 nm) can be attributed to defect states like Ga vacancies (V_{Ga}) [70, 71]. A decrease in photoconductivity observed with 490 nm radiation is consistent with our observation of decreased yellow emission in PL spectrum (Fig. 8.16).

To describe persistent behavior of photoconductivity, we used the stretched exponential function, which is a convenient and relatively flexible fitting function [82]. The function consists of a "stretching" parameter (γ) that distinguishes it from a classical exponential decay function. Mathematically,

$$I(t) = A\exp[-(t/\tau)^{\gamma}] \tag{8.4}$$

where A is the saturation photocurrent, and τ is the relaxation time constant. The parameter γ falls in the range 0 to 1. As γ approaches 1, the function approaches to a classical single-exponential behavior without stretching. One so called, Kohlrausch analysis is a convenient fitting function for processes in which slight deviations from single-exponential behavior occur due to the combination of multiple energy transfer mechanisms. The fit yields the γ values as 0.31 for the 365 nm and 0.37 for the 490 nm consistent with the previous reports [83, 84]. The estimated τ values are 8522 and 7987 s for 365 and 490 nm excitations, respectively. There is no significant difference in either τ or in γ for the two radiations suggesting that the 365 and 490 nm radiation do not play a significant role in the photocurrent relaxation process [70].

The photocurrent relaxation involves two physical processes. The control decay, which is rapid, refers to a recombination of conduction band electrons with photo-generated holes at the surface. Thereafter, the decay slows which may be attributed to the lattice relaxation process of surface states or defect states. The 365 nm illumination creates more photo-generated carriers compared to 490 nm excitation. Therefore, surface band bending is large at 365 nm illumination. The surface band bending can increase the surface energy barrier, which slows transport of photo-generated carriers. Therefore, the small differences in decay time (τ) between two excitations could be due to more surface band bending caused by the large number of photo-generated carriers to the surface at 365 nm illumination.

8.4 Conclusion and Future Remarks

The epitaxial GaN films were grown on sapphire (0001) substrates by LMBE system under various growth conditions such as using different targets, laser energy density, laser repetition rates and growth temperatures. The grown GaN layers were characterized by HRXRD, RHEED, AFM, Raman, SIMS and PL spectroscopy for their structural and optical properties. In case of liquid Ga metal target in presence of r.f. N_2 plasma, the III/V flux ratio is found critical in determining the growth mode of GaN and has been successfully optimized for GaN epitaxial layer growth with a smooth surface. The influence of growth temperature in the range of 300–700 °C on the structural properties of GaN layer has been also studied and it is found that the structural quality of GaN layers improves with increase in growth temperature. The GaN layers grown on sapphire exhibited a narrow (0002) rocking curve width of 245 arc sec at the temperature range of 500–600 °C, which is 150 °C lower when compared to conventional PA-MBE growth.

The GaN film growth on sapphire using HVPE grown GaN polycrystalline target at different growth temperature, laser energy density and laser repetition rate has also been studied. AFM images showed that the surface roughness of the GaN epilayer is largely dependent on the growth temperature and laser repetition rate. For the GaN layers grown with the laser parameters of ~5 J/cm^2 and 10 Hz, the average grain size of the GaN film increases with growth temperature and flat hexagonal islands of size ~200–280 nm are observed for the GaN layer grown 700 °C whereas small grains of uniform size 80–85 nm has been obtained for GaN layer grown at 600 °C. The FWHM value of XRC (0002) plane is obtained to be 110 arc sec for the film grown at 700 °C, indicating highly c-axis oriented growth. Raman spectroscopy studies also complimented the growth of high quality wurtzite GaN films at 700 °C. Besides, the PL measurement also confirmed the high quality of the GaN layers with minimum yellow defect levels. Similar results have also been observed for the GaN layers grown with a lower energy density of ~2.5 J/cm^2 and a high laser repetition rate of 30 Hz.

The use of liquid Ga metal and solid polycrystalline GaN targets in LMBE system for the growth of epitaxial GaN films in presence of the r.f. nitrogen plasma was studied systematically. It is found that the crystalline quality of GaN layers grown using bulk solid GaN target show high crystallinity (110 arc sec) compared to that grown using liquid Ga target (245 arc sec), as characterized by the narrow XRC FWHM value and the low backscattering yield obtained from RBS measurements. The SIMS analysis revealed that there is non-traceable amount of oxygen and carbon present in the LMBE grown epitaxial GaN films. These results ensure the capability of LMBE system to produce high quality GaN epitaxial film on sapphire at a relatively low growth temperature compared to other conventional epitaxial growth techniques. Use of LMBE techniques opens up a unique possibility to explore a variety of substrate materials that are chemically vulnerable and softening at higher growth temperatures but well-suited for growth of III-nitrides and their alloys for III-nitrides based device applications.

Acknowledgments Prof. R.C. Budhani is thanked for constant encouragement and support. The authors would like to thank Dr. K.K. Maurya, Dr. N.D. Sharma, Dr. Prabir Pal, Dr. M.K. Dalai, Mr. Sandeep Singh (CSIR-NPL), Dr. P.K. Tyagi (DTU, Delhi), Dr. B.S. Yadav (SSPL, Delhi) and Dr. S. Ojha (IUAC, New Delhi) for their help in different characterization used in this chapter. This work has been supported by Council of Scientific and Industrial Research (CSIR) under networking projects PSC-0109 and NWP-55.

References

1. S. Nakamura, The roles of structural imperfections in InGaN-based blue light-emitting diodes and laser diodes. Science **281**, 956 (1998)
2. D. Li, X. Sun, H. Song, Z. Li, Y. Chen, H. Jiang, G. Miao, Realization of a high-performance GaN UV detector by nanoplasmonic enhancement. Adv. Mater. **24**, 845 (2012)
3. T. Wang, Y.H. Liu, Y.B. Lee, Y. Izumi, J.P. Ao, J. Bai, H.D. Li, S. Sakai, Fabrication of high performance of AlGaN/GaN-based UV light-emitting diodes. J. Cryst. Growth **235**, 177 (2002)
4. J.-K. Sheu, F.-W. Huang, C.-H. Lee, M.-L. Lee, Y.-H. Yeh, P.-C. Chen, W.C. Lai, Improved conversion efficiency of GaN-based solar cells with Mn-doped absorption layer. Appl. Phys. Lett. **103**, 063906 (2013)
5. M.X. Feng, J.P. Liu, S.M. Zhang, D.S. Jiang, Z.C. Li, K. Zhou, D.Y. Li, L.Q. Zhang, F. Wang, H. Wang, P. Chen, Z.S. Liu, D.G. Zhao, Q. Sun, H. Yang, High efficient GaN-based laser diodes with tunnel junction. Appl. Phys. Lett. **103**, 043508 (2013)
6. H.Y. Ryu, W.J. Choi, K.S. Jeon, M.G. Kang, Y. Choi, J.S. Lee, Analysis of below-threshold efficiency characteristics of InGaN-based blue laser diodes. J. Appl. Phys. **112**, 083109 (2012)
7. P. Chen, M.X. Feng, D.S. Jiang, D.G. Zhao, Z.S. Liu, L. Li, L.L. Wu, L.C. Le, J.J. Zhu, H. Wang, S.M. Zhang, H. Yang, Improvement of characteristics of InGaN-based laser diodes with undoped InGaN upper waveguide layer. J. Appl. Phys. **112**, 113105 (2012)
8. D. Li, X. Sun, H. Song, Z. Li, H. Jiang, Y. Chen, G. Miao, B. Shen, Effect of asymmetric Schottky barrier on GaN-based metal-semiconductor-metal ultraviolet detector. Appl. Phys. Lett. **99**, 261102 (2011)
9. S.L. Rumyantsev, N. Pala, M.S. Shur, R. Gaska, M.E. Levinshtein, M.A. Khan, G. Simin, X. Hu, J. Yang, Effect of gate leakage current on noise properties of AlGaN/GaN field effect transistors. J. Appl. Phys. **88**, 6726 (2000)
10. Y.-J. Lee, Y.-C. Yao, C.-Y. Huang, T.-Y. Lin, L.-L. Cheng, C.-Y. Liu, M.-T. Wang, J.-M. Hwang, High breakdown voltage in AlGaN/GaN HEMTs using AlGaN/GaN/AlGaN quantum-well electron-blocking layers. Nanoscale Res. Lett. **9**, 434 (2014)
11. S.J. Pearton, C.R. Abernathy, M.E. Overberg, G.T. Thaler, A.H. Onstine, B.P. Gila, F. Ren, B. Lou, J. Kim, New applications advisable for gallium nitride. Mater. Today **5**, 24 (2002)
12. A. Guillén-Cervantes, Z. Rivera-Álvarez, M. López-López, A. Ponce-Pedraza, C. Guarneros, V.M. Sánchez-Reséndiz, Structural and optical properties of GaN thin films grown on Al_2O_3 substrates by MOCVD at different reactor pressures. Appl. Surf. Sci. **258**, 1267 (2011)

13. D. Kapolnek, X.H. Wu, B. Heying, S. Keller, B.P. Keller, U.K. Mishra, S.P. DenBaars, J.S. Speck, Structural evolution in epitaxial metalorganic chemical vapor deposition grown GaN films on sapphire. Appl. Phys. Lett. **67**, 1541 (1995)
14. B.L. Liu, M. Lachab, A. Jia, A. Yoshikawaa, K. Takahashi, MOCVD growth of device-quality GaN on sapphire using a three-step approach. J. Cryst. Growth **234**, 637 (2002)
15. S. Raghvan, X. Weng, E. Dickey, J.M. Redwing, Effect of AlN interlayers on growth stress in GaN layers deposited on (111) Si. Appl. Phys. Lett. **87**, 142101 (2005)
16. P.R. Hageman, J.J. Schermer, P.K. Larsen, GaN growth on single-crystal diamond substrates by metalorganic chemical vapour deposition and hydride vapour deposition. Thin Solid Films **443**, 9 (2003)
17. B.-R. Shim, H. Okita, K. Jeganathan, M. Shimizu, H. Okumura, Structural properties of GaN films with AlN buffer layers with varying growth temperatures by plasma-assisted molecular beam epitaxy. Jpn. J. Appl. Phys. **42**, 2265 (2003)
18. P. Gupta, A.A. Rahman, N. Hatui, M.R. Gokhale, M.M. Deshmukh, A. Bhattacharya, MOVPE growth of semipolar III-nitride semiconductors on CVD graphene. J. Cryst. Growth **372**, 105 (2013)
19. M. Katagiri, H. Fang, H. Miyake, K. Hiramatsu, H. Oku, H. Asamura, K. Kawamura, MOVPE growth of GaN on Si substrate with 3C-SiC buffer layer. Jpn. J. Appl. Phys. **53**, 05FL09 (2014)
20. C. Cobet, T. Schmidtling, M. Drago, N. Wollschlager, N. Esser, W. Richter, Surface termination during GaN growth by metalorganic vapor phase epitaxy determined by ellipsometry. J. Appl. Phys. **94**, 6997 (2003)
21. S.S. Kushvaha, P. Pal, A.K. Shukla, A.G. Joshi, G. Gupta, M. Kumar, S. Singh, B.K. Gupta, D. Haranath, Effect of growth temperature on defects in epitaxial GaN film grown by plasma assisted molecular beam epitaxy. AIP Adv. **4**, 027114 (2014)
22. G. Koblmuller, F. Reurings, F. Tuomisto, J.S. Speck, Influence of Ga/N ratio on morphology, vacancies, and electrical transport in GaN grown by molecular beam epitaxy at high temperature. Appl. Phys. Lett. **97**, 191915 (2010)
23. T. Schumann, T. Gotschke, F. Limbach, T. Stoica, R. Calarco, Selective-area catalyst-free MBE growth of GaN nanowires using a patterned oxide layer. Nanotechnology **22**, 095603 (2011)
24. G. Koblmüller, R.M. Chu, A. Raman, U.K. Mishra, J.S. Speck, High-temperature molecular beam epitaxial growth of AlGaN/GaN on GaN templates with reduced interface impurity levels. J. Appl. Phys. **107**, 043527 (2010)
25. M. Kumar, B. Roul, T.N. Bhat, M.K. Rajpalke, S.B. Krupanidhi, Structural characterization and ultraviolet photoresponse of GaN nanodots grown by molecular beam epitaxy. Appl. Phys. Exp. **5**, 085202 (2012)
26. K. Anand, M.P. Mamgain, A.K. Shukla, S.S. Kushvaha, P. Pal, S.P. Khanna, Growth and characterization of plasma assisted MBE grown GaN films at different plasma powers. Adv. Sci. Lett. **20**, 1401 (2014)
27. M. Kesaria, S. Shetty, S.M. Shivaprasad, Evidence for dislocation induced spontaneous formation of GaN nanowalls and nanocolumns on bare C-plane sapphire. Cryst. Growth Des. **11**, 4900 (2011)
28. K. Yamane, M. Ueno, H. Furuya, N. Okada, K. Tadatomo, Successful natural stress-induced separation of hydride vapor phase epitaxy-grown GaN layers on sapphire substrates. J. Cryst. Growth **358**, 1 (2012)
29. L. Zhang, J. Yu, X. Hao, Y. Wu, Y. Dai, Y. Shao, H. Zhang, Y. Tian, Influence of stress in GaN crystals grown by HVPE on MOCVD-GaN/6H-SiC substrate. Sci. Rep. **4**, 4179 (2014)
30. D. Gogova, E. Talik, I.G. Ivanov, B. Monemar, Large-area free-standing GaN substrate grown by hydride vapor phase epitaxy on epitaxial lateral overgrown GaN template. Phys. B Cond. Matter. **371**, 133 (2006)
31. R.D. Vispute, V. Talyansky, R.P. Sharma, S. Choopun, M. Downes, T. Venkatesan, K.A. Jones, A.A. Iliadis, M.A. Khan, J.W. Yang, Growth of epitaxial GaN films by pulsed laser deposition. Appl. Phys. Lett. **71**, 102 (1997)

32. G. Li, S.J. Shih, Z. Fu, A new system for synthesis of high quality nonpolar GaN thin films. ChemComm **46**, 1206 (2010)
33. M.S. Kumar, S.S. Kushvaha, K.K. Maurya, Low temperature growth of GaN epitaxial layers on sapphire (0001) by pulsed laser deposition using liquid gallium target. Sci. Adv. Mater. **6**, 1215 (2014)
34. P. Merel, M. Chaker, M. Tabbal, H. Pepin, Structural and electrical characteristics of epitaxial GaN thin films grown using pulsed laser deposition assisted by an atomic nitrogen source. Appl. Surf. Sci. **177**, 165 (2001)
35. L. Ying-Ying, Z. Jun, L. Won Bo, H. Lan-Zhong, Z. Ying, L. Yan-Rong, Effect of thickness on the microstructure of GaN films on Al_2O_3 (0001) by laser molecular beam epitaxy. Chin. Phys. B **20**, 108102 (2011)
36. P. Sanguino, M. Niehus, L.V. Melo, R. Schwarz, S. Koynov, T. Monteiro, J. Soares, H. Alves, B.K. Meyer, Characterisation of GaN films grown on sapphire by low-temperature cyclic pulsed laser deposition/nitrogen rf plasma. Solid-State Electron. **47**, 559 (2003)
37. M.H. Kim, M. Oshima, H. Kinoshita, Y. Shirakura, K. Miyamura, J. Ohta, A. Kobayashi, H. Fujioka, Investigation of the initial stage of GaN epitaxial growth on 6H-SiC (0001) at room temperature. Appl. Phys. Lett. **89**, 031916 (2006)
38. K. Uchida, A. Watanabe, F. Yano, M. Kouguchi, T. Tanaka, S. Minagawa, Nitridation process of sapphire substrate surface and its effect on the growth of GaN. J. Appl. Phys. **79**, 3487 (1996)
39. L. Neumann, J.W. Gerlach, B. Rauschenbach, Initial stages of the ion-beam assisted epitaxial GaN film growth on 6H-SiC(0001). Thin Solid Films **520**, 3936 (2012)
40. S. Dasgupta, Nidhi, F. Wu, J.S. Speck, U.K. Mishra, Growth and characterization of N-polar GaN films on Si(111). Jpn. J. Appl. Phys. **51**, 115503 (2012)
41. Z.J. Xu, L.X. Zhang, H.T. He, J.N. Wang, M.H. Xie, Growth of GaN on Si(111): surfaces and crystallinity of the epifilms and the transport behavior of GaN/Si heterojunctions. J. Appl. Phys. **110**, 093514 (2011)
42. S.S. Kushvaha, M. Senthil Kumar, K.K. Maurya, M.K. Dalai, N.D. Sharma, Highly c-axis oriented growth of GaN film on sapphire (0001) by laser molecular beam epitaxy using HVPE grown GaN bulk target. AIP Adv. **3**, 092109 (2013)
43. M. Senthil Kumar, S.S. Kushvaha, K.K. Maurya, Low temperature growth of GaN epitaxial layer on sapphire (0001) substrate by laser molecular beam epitaxy technique. Phys. Semicond. Devices **807** (2014)
44. Y. Tomida, S. Nitta, S. Kamiyama, H. Amano, I. Akasaki, S. Otani, H. Kinoshita, R. Liu, A. Bell, F.A. Ponce, Growth of GaN on ZrB_2 substrate by metal-organic vapor phase epitaxy. Appl. Surf. Sci. **216**, 502 (2003)
45. Y. Kawaguchi, J. Ohta, A. Kobayashi, H. Fujioka, Room-temperature epitaxial growth of GaN on lattice-matched ZrB_2 substrates by pulsed-laser deposition. Appl. Phys. Lett. **87**, 221907 (2005)
46. A. Kobayashi, H. Fujioka, J. Ohta, M. Oshima, Room temperature layer-by-layer growth of GaN on atomically-flat ZnO. Jpn. J. Appl. Phys. **43**, L53 (2004)
47. W. Wang, S. Zhou, Z. Liu, W. Yang, Y. Lin, H. Qian, F. Gao, G. Li, Investigation on the structural properties of GaN films grown on $La_{0.3}Sr_{1.7}AlTaO_6$ substrates. Mater. Res. Express **1**, 025903 (2014)
48. W. Yang, W. Wang, Z. Liu, Y. Lin, S. Zhou, H. Qian, G. Li, Epitaxial growth and its mechanism of GaN films on nitrided $LiGaO_2$ (001) substrates by pulsed laser deposition. CrystEngComm **17**, 1073 (2015)
49. H. Takahashi, J. Ohta, H. Fujioka, M. Oshima, M. Kimura, G-GIXD characterization of GaN grown by laser MBE. J. Cryst. Growth **237–239**, 1158 (2002)
50. A.P. McKiernan, J.-P. Mosnier, Study of expansion of laser ablation plumes of Ga and GaN in various N_2 atmospheres using stigmatic emission spectroscopy. Appl. Surf. Sci. **197–198**, 325 (2002)

51. S.S. Kushvaha, M. Senthil Kumar, M. Maheshwari, A.K. Shukla, P. Pal, K.K. Maurya, Structural and electronic properties of epitaxial GaN layer grown on sapphire (0001) using laser molecular beam epitaxy. Mater. Res. Express **1**, 035903 (2014)
52. J. Ohta, H. Fujioka, M. Oshima, K. Fujiwara, A. Ishii, Experimental and theoretical investigation on the structural properties of GaN grown on sapphire. Appl. Phys. Lett. **83**, 3075 (2003)
53. M.A. Moram, M.E. Vickers, X-ray diffraction of III-nitrides. Rep. Prog. Phys. **72**, 036502 (2009)
54. E. Arslan, M.K. Ozturk, A. Teke, S. Ozcelik, E. Ozbay, Buffer optimization for crack-free GaN epitaxial layers grown on Si(1 1 1) substrate by MOCVD. J. Phys. D Appl. Phys. **41**, 155317 (2008)
55. X.H. Zheng, H. Chen, Z.B. Yan, Y.J. Han, H.B. Yu, D.S. Li, Q. Huang, J.M. Zhou, Determination of twist angle of in-plane mosaic spread of GaN films by high-resolution X-ray diffraction. J. Cryst. Growth **255**, 63 (2003)
56. P. Gay, P.B. Hirsch, A. Kelly, The estimation of dislocation densities in metals from X-ray data. Acta Metall. **1**, 315 (1953)
57. M.A. Moram, C.C. Ghedia, D.V.S. Rao, J.S. Barnard, Y. Zhang, M.J. Kappers, C.J. Humphreys, On the origin of threading dislocations in GaN films. J. Appl. Phys. **106**, 073513 (2009)
58. T.B. Wei, R.F. Duan, J.X. Wang, J.M. Li, Z.Q. Huo, Y.P. Zeng, Hillocks and hexagonal pits in a thick film grown by HVPE. Microelectron. J. **39**, 1556 (2008)
59. Z.C. Feng, W. Wang, S.J. Chua, P.X. Zhang, K.P.J. Williams, G.D. Pitt, Raman scattering properties of GaN thin films grown on sapphire under visible and ultraviolet excitation. J. Raman Spectrosc. **32**, 840 (2001)
60. A. Kasic, D. Gogova, H. Larsson, C. Hemmingsson, I. Ivanov, B. Monemar, C. Bundesmann, M. Schubert, Micro-Raman scattering profiling studies on HVPE-grown free-standing GaN. Phys. Stat. sol. (a) **201**, 2773 (2004)
61. VYu. Davydov, YuE Kitaev, I.N. Goncharuk, A.N. Smirnov, J. Graul, O. Semchinova, D. Uffmann, M.B. Smirnov, A.P. Mirgorodsky, R.A. Evarestov, Phonon dispersion and Raman scattering in hexagonal GaN and AlN. Phys. Rev. B **58**, 12899 (1998)
62. H. Harima, Properties of GaN and related compounds studied by means of Raman scattering. J. Phys. Condens. Matter. **14**, R967 (2002)
63. X. Hai-Ying, N. Ping-Juan, X. Yu-Xin, The properties of GaMnN lms grown by metalorganic chemical vapour deposition using Raman spectroscopy. Chin. Phys. B **21**, 077801 (2012)
64. M.S. Kumar, S.S. Kushvaha, K.K. Maurya, K. Saravanan, S. Ojha, High resolution X-ray diffraction and Rutherford backscattering spectroscopy studies on laser molecular beam epitaxy grown GaN layers on sapphire (0001). Adv. Sci. Lett. **20**, 1406 (2014)
65. P. Jensen, Growth of nanostructures by cluster deposition: Experiments and simple models. Rev. Mod. Phys. **71**, 1695 (1999)
66. R. Alayan, L. Arnaud, M. Broyer, E. Cottancin, J. Lermé, J.L. Vialle, M. Pellarin, Morphology and growth of metal clusters in the gas phase: a transition from spherical to ramified structures. Phys. Rev. B **73**, 125444 (2006)
67. S.S. Kushvaha, M. Senthil Kumar, B.S. Yadav, P.K. Tyagi, S. Ojha, K.K. Maurya, B.P. Singh, Influence of laser repetition rate on the structural and optical properties of GaN layers grown on sapphire (0001) by Laser Molecular Beam Epitaxy. CrystEngComm **18**, 744 (2016)
68. M.A. Reshchikov, H. Morkoc, Luminescence properties of defects in GaN. J. Appl. Phys. **97**, 061301 (2005)
69. C. Kisielowski, J. Krüger, S. Ruvimov, T. Suski, J.W. Ager III, E. Jones, Z. Liliental-Weber, M. Rubin, E.R. Weber, M.D. Bremser, R.F. Davis, Strain-related phenomena in GaN thin films. Phys. Rev. B **54**, 17745 (1996)
70. L. Macht, J.L. Weyher, A. Grzegorczyk, P.K. Larsen, Statistical photoluminescence of dislocations and associated defects in heteroepitaxial GaN grown by metal organic chemical vapor deposition. Phys. Rev. B **71**, 073309 (2005)

71. H.C. Yang, T.Y. Lin, Y.F. Chen, Nature of the 2.8-eV photoluminescence band in Si-doped GaN. Phys. Rev. B **62**, 12593 (2000)
72. M.A. Reshchikov, R.Y. Korotkov, Analysis of the temperature and excitation intensity dependencies of photoluminescence in undoped GaN films. Phys. Rev. B **64**, 115205 (2001)
73. D.O. Demchenko, I.C. Diallo, M.A. Reshchikov, Yellow Luminescence of gallium nitride generated by carbon defect complexes. Phys. Rev. Lett. **110**, 087404 (2013)
74. M.H. Kim, Y.C. Bang, N.M. Park, C.J. Choi, T.Y. Seong, S.J. Park, Growth of high-quality GaN on Si(111) substrate by ultrahigh vacuum chemical vapor deposition. Appl. Phys. Lett. **78**, 2858 (2001)
75. J. Cao, D. Pavlidis, A. Eisenbach, A. Philippe, C.B. Chevallier, G. Guillot, Photoluminescence properties of GaN grown on compliant silicon-on-insulator substrates. Appl. Phys. Lett. **71**, 3880 (1997)
76. M. Cazzanelli, D. Cole, J.F. Donegan, J.G. Lunney, P.G. Middleton, K.P. O'Donnell, C. Vinegoni, L. Pavesi, Photoluminescence of localized excitons in pulsed-laser-deposited GaN. Appl. Phys. Lett. **73**, 3390 (1998)
77. T. Kuykendall, P. Pauzauskie, S. Lee, Y. Zhang, J. Goldberger, P. Yang, Metalorganic chemical vapor deposition route to GaN nanowires with triangular cross sections. Nano Lett. **3**, 1063 (2003)
78. Y. Huang, X. Duan, Y. Cui, C.M. Lieber, Gallium nitride nanowire nanodevices. Nano Lett. **2**, 101 (2002)
79. S.J. Chung, M.S. Jeong, O.H. Cha, C.H. Hong, E.K. Suh, H.J. Lee, Y.S. Kim, B.H. Kim, Optical absorption and anomalous photoconductivity in undoped n-type GaN. Appl. Phys. Lett. **76**, 1021 (2000)
80. H.M. Chen, Y.F. Chen, M.C. Lee, M.S. Fang, Persistent photoconductivity in n-type GaN. J. Appl. Phys. **82**, 899 (1997)
81. J.Z. Li, J.Y. Lin, H.X. Jianga, M.A. Khan, Q. Chen, Persistent photoconductivity in a two-dimensional electron gas system formed by an AlGaN/GaN heterostructure. J. Appl. Phys. **82**, 1227 (1997)
82. V.V. Ursaki, I.M. Tiginyanu, P.C. Riccia, A. Anedda, S. Hubbard, D. Pavlidis, Persistent photoconductivity and optical quenching of photocurrent in GaN layers under dual excitation. J. Appl. Phys. **94**, 3875 (2003)
83. J.Z. Li, J.Y. Lin, H.X. Jiang, A. Salvador, A. Votchkarev, H. Morkoc, Nature of Mg impurities in GaN. Appl. Phys. Lett. **69**, 1474 (1996)
84. H.M. Chen, Y.F. Chen, M.C. Lee, M.S. Feng, Yellow luminescence in n-type GaN epitaxial films. Phys. Rev. B **56**, 6942 (1997)

Chapter 9
Aperiodic Silicon Nanowire Arrays: Fabrication, Light Trapping Properties and Solar Cell Applications

Sanjay K. Srivastava, C.M.S. Rauthan, Vikram Kumar and P.K. Singh

Abstract Solar photovoltaic (SPV) is capable of providing the most feasible carbon-free route to the worldwide traditional power consumption. During the last decade, there has been tremendous development in silicon wafer based photovoltaic (PV) cells technology and today commercial silicon PV cells over 20 % efficiencies have been achieved. However, large-scale implementation of silicon wafer PV is currently not economical because of their high cost as compared to traditional power sources. One of the primary cost components for silicon PV cells is the starting silicon wafer, which requires extensive purification to maintain reasonable performance. Therefore, development of efficient and low cost PV devices is extremely important. Silicon nanowires (SiNWs) are a very promising candidate for next generation PV. The SiNW arrays exhibit low reflection, strong broadband light absorption and may be used as antireflection surface in solar cells. In addition to enhanced optical properties, nanowire arrays also have the potential for efficient charge carrier collection across the nanowire diameter for radial junction (homo/hetro p-n junctions) solar cells and therefore may relax high quality material requirement, enabling lower-cost PV cells. In the chapter, a short review of aperiodic SiNW arrays fabrication by silver assisted wet chemical etching method, their light trapping properties and PV applications with emphasis on SiNW arrays based solar cells would be presented. Finally, challenges in effective use of SiNW arrays in PV devices and future perspective would also be briefly discussed.

Keywords Metal assisted chemical etching · Silicon nanowires · Light trapping · Antireflection · Solar cell · Thin solar cell

---

S.K. Srivastava (✉) · C.M.S. Rauthan · P.K. Singh
CSIR-National Physical Laboratory, Dr. K.S. Krishnan Marg, New Delhi 110012, India
e-mail: srivassk@nplindia.org; sksrivastava78@gmail.com

V. Kumar
Department of Physics, Indian Institute of Technology Delhi, New Delhi 110016, India

M. Husain and Z.H. Khan (eds.), *Advances in Nanomaterials*,
Advanced Structured Materials 79, DOI 10.1007/978-81-322-2668-0_9

9.1 Introduction

During the last decade, solar energy production, in particular solar photovoltaic (PV) has increased rapidly. The solar PV is capable of providing the most feasible carbon-free route to displacing tera-Watt (TW) of worldwide non-renewable power consumptions. However, large-scale implementation of silicon wafer PV is currently not economical because their cost is still high as compared to the traditional power sources. There has been tremendous development in silicon wafer based photovoltaic (PV) cells technology. Silicon based PV contribution has increased steadily and presently it is ~90 % of the global production. Moreover, commercial silicon PV cells over 20 % efficiencies have been achieved. However, present R&D efforts are focused on improvements of device efficiency and optimal uses of the material in terms of gm/W. One of the primary cost components for silicon PV cells is the starting silicon wafer, which requires relatively large quantity (~10–11 gm for ~200 μm thick 156 × 156 mm^2 silicon wafer) material. Moreover, It requires high quality silicon material to realize reasonable performance of the solar cells. Therefore, reducing the required quality and quantity of Si materials and simultaneously development of efficient device will help in making low cost PV devices which may consequently lead to large-scale implementation of silicon PV. In order to reach this goal, in the recent past unique optical and physical properties of complex nanostructures as well as novel materials and compounds are being exploited.

Silicon based PV devices offer several inherent advantages, including the natural silicon abundance, lack of toxicity, and compatibility with mature micro-electronics fabrication techniques. Silicon nanowires (SiNWs) are a very promising candidate for next generation PV [1–13]. The SiNW arrays exhibit ultra-low reflection, strong broadband light absorption and may be used as antireflection surface in solar cells. In addition to enhanced optical properties, nanowire arrays also have the potential for efficient charge carrier collection across the nanowire diameter for radial junction (homo/hetro p-n junctions) solar cells. Therefore may relax the requirement of high quality Si material, enabling lower-cost PV cells. The SiNWs solar cells with radial homo/hetero-junctions offer unique advantages over conventional solar cells due to their orthogonal photon absorption and improved carrier collection [1, 6, 14, 15]. This device structure open doors for the use of thin and low grade silicon materials on flexible/glass substrates hence cheap PV devices [16, 17]. Prime focus has been to achieve simple ways of synthesis/fabrication of nanowire (NW) arrays, to increase the light absorption and generation of charge carriers and designing of SiNWs based PV device for efficient harvesting of generated charge carriers and therefore electrical energy generation.

In this chapter, we focus on fabrication of aperiodic SiNW arrays by a simple metal assisted electroless wet chemical etching (MA-EWCE) in particular, Ag-assisted EWCE method, their light trapping/antireflection properties and recent developments and challenges in SiNW arrays based p-n junction solar cells.

9.2 Large Area Fabrication of Silicon Nanowire Arrays

Fabrication of SiNWs in a controlled manner is extremely important for their solar photovoltaic and other opto-electronic applications. However, a simple and efficient way to fabricate highly oriented, tunable length SiNWs with desired electronic characteristics at low temperature has been a challenge for the scientific community. In general, there are two main approaches for the SiNWs fabrication viz 'bottom-up' and 'top-down'. Several bottom–up methods such as the vapor–liquid–solid or VLS [18–20] method by chemical vapor deposition, molecular beam epitaxy [21], laser ablation [22, 23], thermal evaporation [24, 25], and few more [26, 27] have been successfully developed to prepare SiNWs. However, these processes generally require complex and expensive equipments, high temperatures, templates and hazardous silicon precursors which make them time consuming and expensive and therefore limits their suitability for cost effective solar PV. Moreover, SiNWs fabrication over large area is also difficult to achieve by bottom-up approaches. In top-down approach, there are mainly three methods, namely lithographic methods [28], reactive ion etching (RIE) [29] and the MA-EWCE [30–32]. The first two approaches are also expensive where large area and homogeneous SiNWs fabrication is not easy. The MA-EWCE method, on the other hand, is free from all these limitations. It is a very simple, room temperature, solution based beaker process that enables production of vertically aligned SiNW arrays over wafer scale [33]. In addition, MA-EWCE offers production of SiNWs with identical structural and electronic properties such as doping type and conductivity as that of parent wafers which, in general, is very difficult to achieve in bottom-up approaches. Both aperiodic (randomized) and periodic arrays of SiNWs can be fabricated by this process. However, for periodic arrays, an additional lithographic process employing colloidal or nanospheres lithography etc. is essential prior to the MA-EWCE process [34, 35]. On the other hand, conventional MA-EWCE process results aperiodic array of SiNWs. Herein, the scope of this chapter is limited to aperiodic SiNW arrays only.

In MA-EWCE process, there are two known approaches; (a) single-step MA-EWCE, and (b) two-step MA-EWCE. The former is a self-controlled process and simply involves placing of clean silicon wafer in an aqueous solution of HF containing metal ions such as Ag, Au, Fe etc. [30, 33, 35–37] whereas in the later one metal (Ag, Au) particles are first deposited either by a wet chemical process (electroless plating) or by a physical vapor deposition followed by placing of metal coated silicon wafer in the etching solution (aqueous HF and H_2O_2) to fabricate SiNW arrays [38–41]. The schematic of the experimental set-up commonly used for SiNWs fabrication by single- or two-step MA-EWCE process is presented in Fig. 9.1.

The fabrication of SiNW arrays by MA-EWCE has been extensively reported wherein most of the studies are based on Ag assisted EWCE process [30–35, 38–43]. In addition to Ag, metals such as Au, Fe, Cu, Pt etc. have also been used (as catalytic agent) for the fabrication of SiNW arrays by MA-EWCE method [36–39], however, the investigations are limited. One possible reason for the preference of Ag as

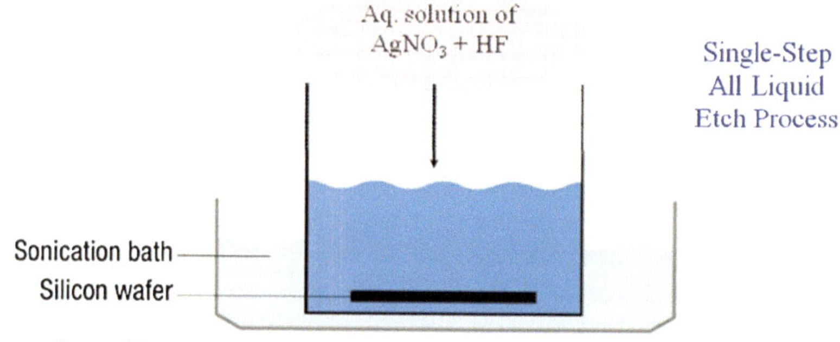

Fig. 9.1 Schematic of experimental set-up commonly used for SiNW arrays fabrication by single-step MA-EWCE method. Etching bath is mainly comprised of aqueous HF solution containing either of metal (Ag, Au, Cu etc.) ions

catalyst over others could be due to its advantages such as redox potential of Ag^+/Ag system is overlapping with Si valence band and therefore ease the silicon etching by the self-sustained EWCE process [38, 39, 42–44]; Other possibilities may be related to avoid the side effects of Au, Fe, Cu, etc. on the electronic properties of SiNWs formed since these metals are known deep level impurities and have adverse effect on the electronic properties (particularly life time of minority charge carriers) of silicon based devices. The minority carrier lifetime is a very critical parameter for solar cells applications [45–47]. The etch bath in the Ag-assisted EWCE process of SiNW arrays fabrication consists of aqueous HF solution containing Ag^+ ions (aq. $AgNO_3$ solution). This bath is also known as oxidizing HF solution.

In general, fabrication of SiNW arrays is carried out on both p and n-type conductive silicon wafers with (1–10 Ω cm) doping density in a beaker using aqueous HF (~5.0 mol/L) solution containing ~0.02 mol/L $AgNO_3$ at room temperature. Most of the reports are focused on silicon wafers with (100) crystal orientation [30–35, 38–43], since for PV applications silicon wafers with these electronic properties and crystal orientations are preferred. However, effect of crystal orientation on the nanowire formation have also been investigated using (110), (111) oriented silicon wafers [48–50]. In general, the MA-EWCE method consists of the following sequential steps for SiNW arrays fabrication. (i) Prior to the etching, the wafers are sequentially cleaned with acetone, ethanol, de-ionized (DI) water and boiling in piranha solution (H_2SO_4:H_2O_2 = 3:1 by volume, for 30–60 min). (ii) The wafers are then rinsed thoroughly with de-ionized water followed by dipping in 5–10 % HF solution to remove silicon oxides either native or grown during the process on the surface. (iii) Cleaned silicon wafers are then immersed in the etching solution for the desired time duration for etching. (iv) Finally residual/deposited Ag particles are removed. This can be achieved by treating the samples in NH_4OH:H_2O_2::3:1 or dilute HNO_3 solution [30–35, 38–43] followed by rinsing in DI water copiously and drying. Recently, Srivastava et al. [43] have reported that this method (single step MA-EWCE) provides a wide range of flexibility in the processing parameters for fabrication of aligned SiNW arrays. In this article, they have reported extensive investigations on the influence of several etch parameters such as etch time, etch bath temperature, etch bath components like HF and $AgNO_3$ concentrations and also the surface area of the silicon samples under a given etch conditions on the formation of SiNW arrays. Another very important feature of this method is that it is a room temperature beaker process and there is no limitation on the size of the samples. Further, Kumar et al. [33] have reported fabrication of SiNW arrays on 50 mm diameter silicon wafers. With suitable design of the set-up it can easily be scaled up for large scale fabrication of SiNW arrays [51, 52].

9.3 Morphology and Structure of SiNWs

Typical cross-sectional and planer view SEM images of the as-prepared SiNW arrays are shown in Fig. 9.2. In this method a thick film composed of tree-like dendritic structures (a cluster of Ag particles/dendrites) is deposited on and around the silicon samples. Underneath of this layer, vertically aligned SiNWs are formed as can be seen in Fig. 9.2a. Energy dispersive x-ray spectroscopy analysis has revealed that thick film of dendrites is made of pure Ag. The Ag particles are also found at the bottom of the pores as can be seen from Fig. 9.2b wherein a highly magnified view is also shown in the figure inset. The presence of Ag particle in the nanowires valley is, indeed, the confirmation of Ag assisted etching of Si. A typical cross-sectional view of the vertically aligned SiNW arrays formed on Si samples after etching out Ag particles/dendrites (in NH_4OH and H_2O_2 solution) is shown in Fig. 9.2c. The SiNWs are, in general, bundled together (after Ag etching in $NH_4OH + H_2O_2$ or HNO_3 solution, cleaning with DI water followed by drying) and are firmly attached with the substrate. The top view of bunching of SiNW arrays is shown in Fig. 9.2d.

The etched silicon wafers consist of dense and vertically aligned one dimensional nanostructures. The lateral widths or diameters of SiNWs are randomly distributed within 50–300 nm range (as estimated by SEM investigations). However, little percentage of the SiNWs could also be of ≥500 nm lateral width. These nano-structures are similar to surface relief structures with dimensions less than the wavelength of visible light. An important aspect of the single-step MA-EWCE method is that the length of the SiNW arrays can be tailored in a controlled manner by judiciously controlling the etch parameters. For example, Srivastava et al. [43] reported that the length of the SiNW arrays varies linearly with etch time (up to 120 min.) at ambient temperature (10–50 °C), and for given set of etch bath composition (HF and $AgNO_3$). The SiNW arrays length versus etching time curve is shown in Fig. 9.3c. The dimensions of the nanostructures, particularly the diameter, remain practically the same throughout the length as can be seen from

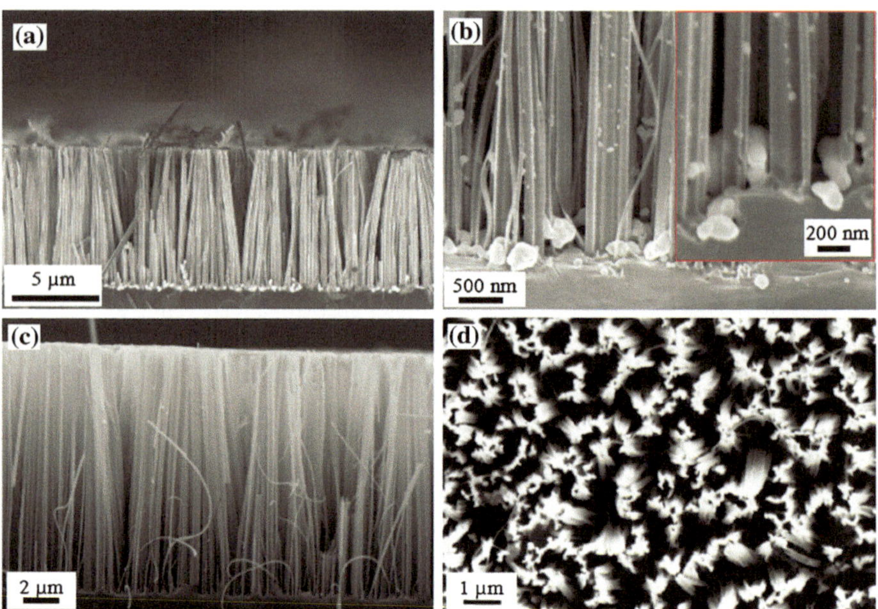

Fig. 9.2 SEM images showing the (**a**) cross sectional view (45° tilted) of the as prepared SiNW array samples covered with Ag dendritic film, (**b**) cross sectional (45° tilted) view of the *bottom* section showing Ag particles at the *bottom* of the pores (a magnified view is shown in the inset), (**c**) cross sectional view of the SiNW arrays after removal of the Ag dendrites and (**d**) planar *top* view of the SiNW arrays shown in (**c**). Ref. [43] © IOP Publishing. Reproduced by permission of IOP Publishing. All rights reserved

Fig. 9.3a, b for etching time of 45 and 90 min where the NWs length are ~13 and ~ 25 μm respectively. The estimated average etching rate is ~0.25 μm min^{-1} for HF and $AgNO_3$ concentration of 5 and 0.02 mol L^{-1} respectively at ~25 °C. Similar observations have been made by others also [52–55]. For longer etching durations (>120 min) in a given set of etch parameters, the SiNW arrays formation rate slows down (to ~0.18 μm min^{-1}) as can be seen from Fig. 9.3c [43]. The observed slowdown in the etch rate for prolonged etching (>120 min) has been attributed to the followings: (i) continuous consumption of Ag^+ content in a given etching solution with time; (ii) The thick Ag film deposited over the sample may decrease the efficiency of charge transfer between Si and Ag and hence etch rate beneath this film; (iii) Decrease in HF reactivity, due to ambient exposure (as the experiment is normally performed in an open vessel); (iv) Continuous consumption of HF (as H_2SiF_6 during etching); and (v) etching of already formed SiNWs from the top also cannot be ruled out. As shown in Fig. 9.3, It is evident that longer SiNWs are tangled with each other near the top, and are held together in bundles. This could be attributed to the flexibility of SiNWs and surface tensional forces which are effective during the drying process. Subsequently the van der Waals forces hold SiNWs together and these forces become dominant with the progress of

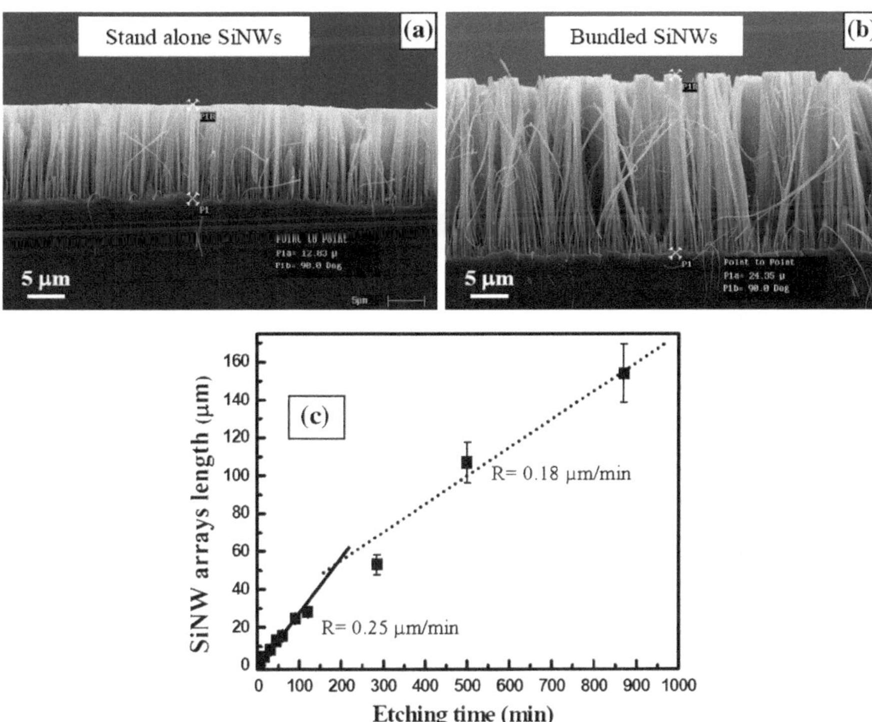

Fig. 9.3 Cross sectional SEM images of SiNW arrays prepared on p-type (boron doped, (100) Si wafer at room temperature for etch time of 45 min (**a**) and 90 min (**b**). Longer SiNWs are bundled. Length of the NW arrays is almost doubled on increasing the etch time two times under given etch conditions like etch bath concentration, temperature, Si wafer property, sample size etc. (**c**) Length of SiNW arrays as a function of etch time prepared by single step MA-EWCE method using HF and $AgNO_3$ solution at room temperature using boron doped Si (100). Length increases linearly with etch time up to 120 min after which it starts slowing down if etched for longer etch time

etch time (for longer SiNW arrays). At times, SiNWs collapse as their length becomes too large to stand alone (as can be seen in Fig. 9.3b).

Another important feature of this top-down approach is that it enables to produce SiNWs of almost identical structural (in terms of crystal structure, orientation) and electronic properties (such as doping conductivity/type etc.) as that of the silicon wafer used for etching (i.e.; identical to the starting silicon wafer property) [35, 47] without any metal impurities as normally observed in VLS approaches of SiNWs preparation [56, 57]. For example, the structural and elemental analyses (as shown in Fig. 9.4) have indicated that the SiNWs produced from (100) oriented Si wafer were single crystalline with axial orientation along <100> direction, the orientation of the starting silicon wafer used [33].

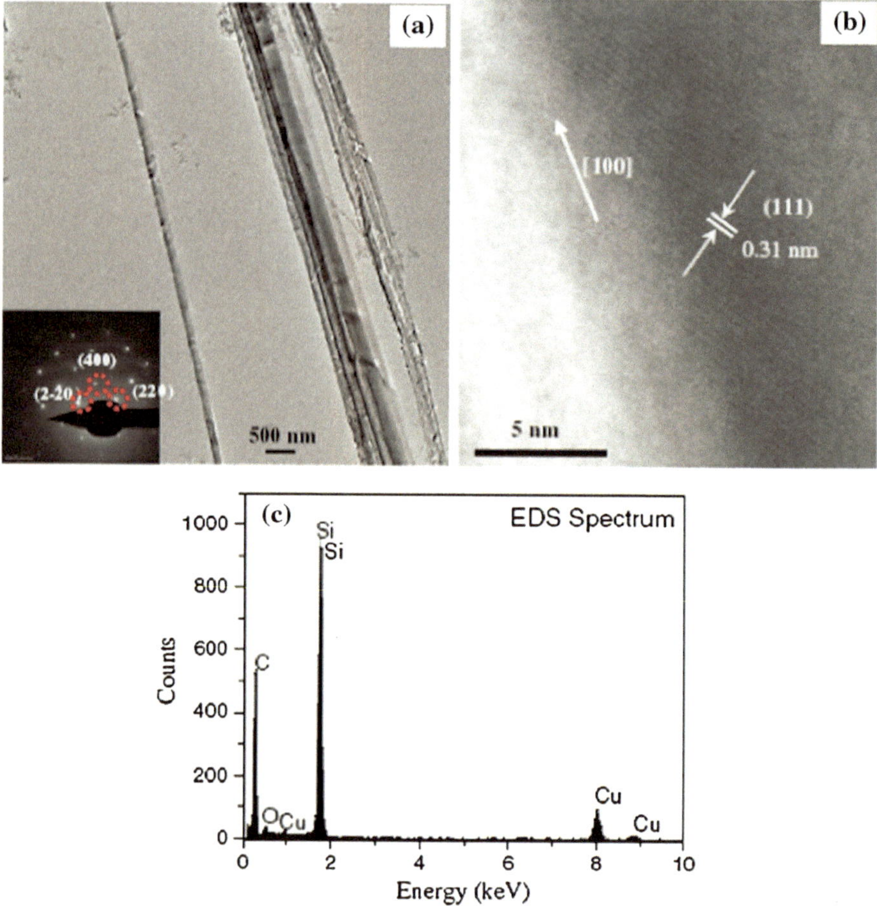

Fig. 9.4 **a** Bright field TEM image of SiNWs, corresponding selected area electron diffraction (SAED) pattern is shown in the inset. **b** High resolution transmission electron microscope (HRTEM) image of a nanowire showing single crystalline. **c** Representative energy dispersive specroscopic (EDS) spectrum of a single nanowire. Reprinted from Ref. [33] © Springer Science +Business Media B.V. 2009; with kind permission from Springer Science and Business Media

9.4 Fabrication Mechanism

Catalytic activity of Ag nanoparticles is well known in electrochemical etching of silicon in HF solution where deep cylindrical nano holes were produced in (100) silicon using Ag nanoparticles as catalyst. This process is called as metal assisted chemical etching of silicon [44]. The mechanism of formation of vertically aligned SiNW arrays can be understood as being a self assembled Ag induced selective etching process based on localized microscopic electrochemical cell model as presented schematically in Fig. 9.5. The process is based on the continuous

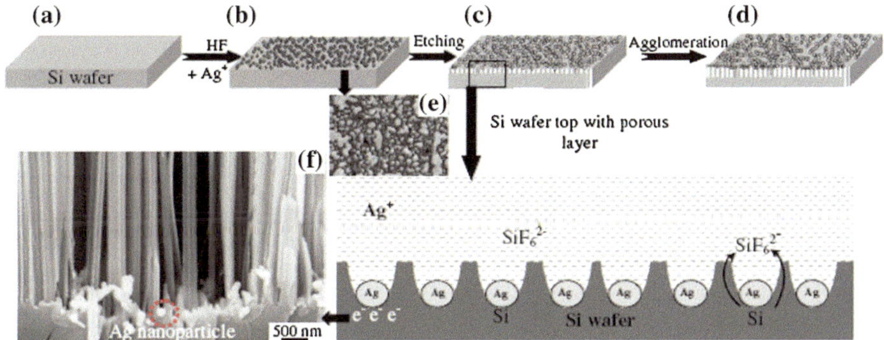

Fig. 9.5 Schematics of the SiNWs growth mechanism by Ag- induced selective chemical etching of silicon surface in HF-$AgNO_3$ solution via self assembled nano-electrochemical cell model. **a** Virgin cleaned Si sample, **b** Deposition of Ag-nanclusters as a result of Ag^+ reduction on Si (as shown by SEM image in **e**), **c** Subsequent Ag-assisted localized etching of Si and trapping of Ag particles in the pores formed (as shown in the magnified view), **d** Agglomeration of Ag nanclusters forming non-compact Ag dendrites and evolution of SiNWs on larger etching duration, **f** A magnified SEM image showing the Ag nanopartilces in the bottom of the SiNW arrays. Reprinted from Ref. [33] © Springer Science+Business Media B.V. 2009; with kind permission from Springer Science and Business Media

galvanic displacement of Si by Ag^+ via $Ag^+ \rightarrow Ag$ reduction on the silicon surface and hence forming many local nano-electrochemical cells on the silicon surface. The deposited Ag nanoparticles act as cathode and silicon surface in contact with the Ag nanoparticles as anode. In brief, Ag^+ ions get reduced to Ag onto the silicon surface by injecting holes (h^+) into the Si valence band (VB) and oxidizing the silicon surface locally in contact with Ag nanoparticles. The oxidized Si surface is subsequently etched away by HF as illustrated in Fig. 9.5. The complete self-controlled redox reaction occurring in aq. HF + $AgNO_3$ etch bath on Si surface can be divided into two parts: cathode and anode reactions which are described below as [33, 39, 43].

At cathode site:

$$Ag^+ + e^-(VB) \rightarrow Ag(s)$$

At anode site:

$$Si(s) + 2H_2O \rightarrow SiO_2 + 4H^+ + 4e^-(VB)$$
$$SiO_2(s) + 6HF \rightarrow H_2SiF_6 + 2H_2O$$

The initial reduction of Ag^+ forms Ag nanoclusters which act as catalytic site for subsequent cathodic reaction (Fig. 9.5a). Subsequently Ag^+ ions get reduced to Ag preferably on the Ag nanoclusters (but not on the uncovered Si surface) by electron transfer from the underlying wafer (since Ag is relatively more electronegative than the silicon and hence provide easy injection path for holes) (Fig. 9.5b). The complete self-assembled Ag assisted redox reaction is also illustrated schematically

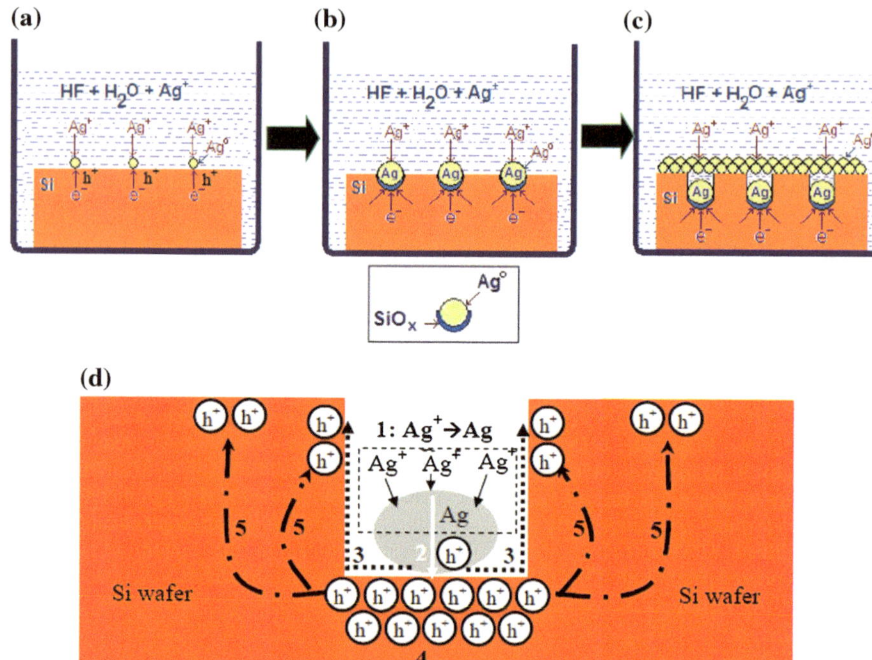

Fig. 9.6 Schematic presentation of the mechanism of etching of the Si wafer leading to SiNW array formation in an Ag assisted single step EWCE process. **a** Reduction of Ag^+ ions to elemental Ag via hole (h^+) injection into the VB of Si, leading to the formation of Ag nuclei at the Si surface, and simultaneously hole injection into the Si substrate, oxidizing the Si to SiO_x. **b** Growth of Ag nuclei into large particles with increasing etching time, oxidative dissolution of Si atoms (in the presence of HF) under the Ag particles and sinking of the particles into the Si substrate. **c** The charge transfer preferentially occurring at the etching front (Ag/Si interface) and vertical propagation for the Ag/Si interface. **d** Schematic representation of the complete etching mechanism of Si in aqueous HF/$AgNO_3$ solution: (1) reduction of (Ag^+) to elemental Ag at the silicon surface; (2) injection of holes (h^+) into the Si valence band at the Ag/Si interface, oxidation of Si at the Ag/Si interface and dissolution of oxides by HF; (3) diffusion of the etchants (HF) and byproducts (H_2SiF_6) along the Ag/Si interface; (4) preferential charge transfer at the Ag/Si interface leading to faster etching of the Ag/Si interface as compared to non-Ag regions and vertical propagation of the Ag/Si interface (i.e. the pore *bottom*); and (5) diffusion of excess holes at the Ag/Si interface to off Ag regions (*side walls*) leading to rough side walls (NW surface). Ref. [43] © IOP Publishing. Reproduced by permission of IOP Publishing. All rights reserved

in the magnified view of Fig. 9.5c. In the next step, agglomeration of Ag nanoclusters in the form of dendritic structures (a non-compact Ag layer) occurs as shown in Fig. 9.5d, which lead to selective oxidation and dissolution of Si in the aqueous HF + $AgNO_3$ solution. Once the Ag nanoclusters get trapped into the silicon pores they move deeper and deeper resulting into the 1-D nanostructures or SiNWs. The complete etching process of SiNW arrays fabrication by Ag-assisted EWCE method using HF and $AgNO_3$ solution can be represented in five basic steps and are presented schematically in Fig. 9.6.

The SiNW arrays of desired length can be produced by controlling the etch time and temperature for a given etch-solution composition and sample size. It has been shown that surface area of the Si samples exposed in the etch-solution has an strong influence on the formation kinetics of the NWs. The $AgNO_3$ concentrations (0.005–0.04 M) and HF concentrations (2–12 M) are ideal for uniform, controlled formation of SiNWs [43]. Short and well aligned NW arrays could be made at low $AgNO_3$ concentrations, which may be important for PV device applications [6]. Generally, the higher $AgNO_3$ and HF concentrations lead to side etching and tapering of the NWs.

It has been observed that the lateral dimension or diameter of the SiNWs prepared by the MA-EWCE process depends on the dimension and distribution of the Ag nanoparticles formed on the Si surface. It is difficult to control the cluster size in such a self-assembled MA-EWCE process. In practice, controlling the diameters of SiNWs in a single-step MA-EWCE process (without using any lithography process/templates) is rather difficult. Even though based on experimental observations, efforts have been made to develop a model which can estimate the distribution and diameter of the SiNWs by this method [58]. It is also important to note that the MA-EWCE method can also be applied to produce SiNW arrays on multi-crystalline (mc-) silicon substrates. Though the process is anisotropic in nature and SiNWs orientation and morphology are found to be strictly grain orientation dependent [43].

9.5 Light Trapping and Enhanced Optical Absorption in SiNW Arrays

9.5.1 SiNW Arrays on Bulk Silicon Wafer

The optical absorption properties of solar cells in the solar spectrum are important in determining cell efficiency. For example, polished silicon surface has a high natural reflectivity (>35 %) with a strong spectral dependence and therefore, minimization of reflection losses is a prerequisite for efficient Si solar cells [59]. To achieve this, the top surface of the Si solar cell is typically textured or coated with an antireflection layer. Anisotropic etching in aqueous alkaline solution (such as KOH or NaOH) is a standard process for mono-crystalline Si and is widely applied in Si-wafer based PV industry. Such approach of surface-texturing of Si forms densely packed random micro-pyramids which results in the reduction of reflection losses in a broad spectral range [60, 61]. With this process reflection losses can be reduced to $\sim$10 %. For further reduction of reflection (to $\sim$2–3 %), antireflection coating (ARC) based on quarter wavelength transparent layers of dielectrics such as SiO_x, TiO_x, or Si_xN_y is used [59, 61]. Such ARCs have resonant structures and work effectively only in a limited spectral range and at an specific angles of light incidence. Recently, sub-wavelength structures (SWSs) based non-reflecting surfaces have also been investigated using dry etching methods such as fast atom beam

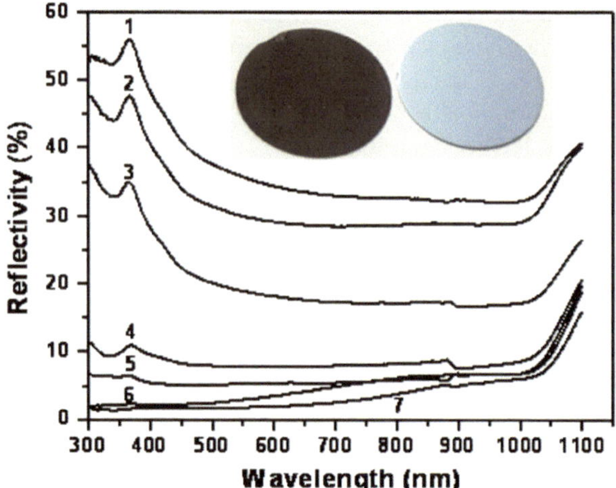

Fig. 9.7 Reflectivity (R_λ) as function of wavelength (λ) of silicon (or sample) surface with Si wires obtained after different etching times of (1): 0 min (polished wafer); (2): 1 min; (3): 1.5 min; (4): 2 min; (5): 5 min; (6): 15 min and (7): 45 min. Inset shows the optical photograph of a polished Si wafer and a silicon wafer with vertically aligned-SiNW arrays prepared for 45 min (*black appearance*). Reprinted from Ref. [72]. Copyright (2010), with permission from Elsevier

and reactive ion etching employing nano-imprint lithography, different masks and templates [62–69]. However, these methods are complex, expensive and therefore have limited applications to produce cost-effective and efficient Si solar cells at industrial scale. On the other hand, the MA-EWCE is a cost effective approach for large area fabrication of SWS, and can be upgraded to mass production required for Si solar cells with minor changes in the fabrication protocol.

The SiNW arrays possess several unique optical properties owing to their high surface areas and structural features smaller than the wavelength of light and therefore have potential applications in solar cells. Excellent anti-reflection or light trapping properties of the aperiodic SiNWs arrays prepared by MA-EWCE method has been reported by many groups in the recent past [3, 7, 43, 51–54, 70–83]. For example, Peng et al. [70] reported that SiNW arrays fabricated on single-crystal Si wafers drastically suppress light reflection (<1.4 %) over a wide spectral range (300–600 nm). Srivastava et al. [72] performed systematic investigations on the anti-reflectance property of the SiNWs fabricated by MA-EWCE method for different arrays length (etch time). The hemispherical reflectance of SiNW arrays for different length along with the reflectance of the polished planar silicon wafer is shown in Fig. 9.7 which clearly illustrates that SiNW arrays drastically suppress light reflection over a broad spectral range. They observed that reflectance decreases with increase in etching time (i.e., with increasing SiNW arrays length) and reaches a minimum of ~1.5 % in the 300–600 nm range and ~4 % in the spectral range 600–1000 nm in the samples etched for 15 min or more (corresponding to the NW arrays length of ~4 μm or more). Digital image of the SiNW arrays sample is shown in the inset of Fig. 9.7 along with that of a

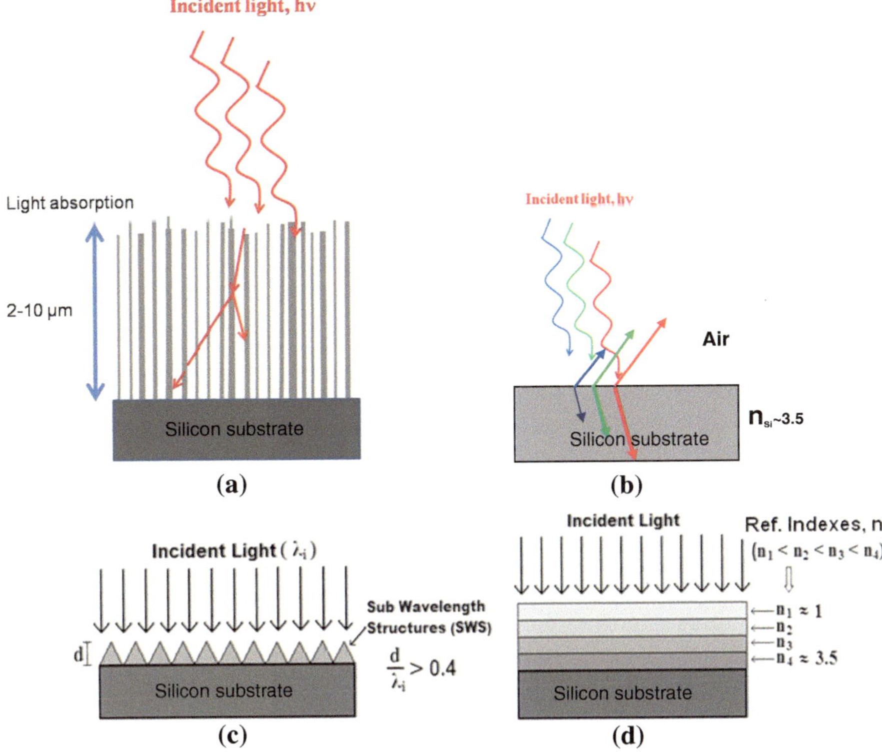

Fig. 9.8 Schematic presentation of (**a**) diffuse scattering of incident light from highly rough surfaces of SiNWs enhances the path length of incident light and therefore reduces vertical distance for light absorption as compared to the planar silicon surface, (**b**) light interaction in a planar silicon surface, (**c**) sub-wavelength structures (*SWSs*), (**d**) effective media with gradually varying effective refractive index, n, (from *top* to *bottom*). **c** and **d** are reprinted from Ref. [72]. Copyright (2010), with permission from Elsevier

chemically-mechanically polished silicon surface for comparison. The corresponding diffused reflectance plot for polished and SiNW arrays based silicon wafers are also shown. The black surface is due to extremely low reflectance and therefore such surfaces are also known as 'black' silicon [71]. The excellent anti-reflection (AR) property of the SiNW arrays confers them as a potential candidate for an effective antireflective layer in silicon solar cells. Till date a number of articles have been reported on the SiNW arrays based solar cells exploiting the efficient light trapping/anti-reflection characteristics of the aperiodic SiNW arrays prepared by MA-EWCE method and will be discussed later.

The excellent AR or light trapping property of the aperiodic SiNW arrays has been attributed to the followings: (i) highly rough surfaces and network assembly of the NWs may lead to strong light scattering interactions among the NWs making light to travel path many folds than the NWs array thickness. This eventually leads to almost complete light trapping and its absorption as shown in Fig. 9.8a in

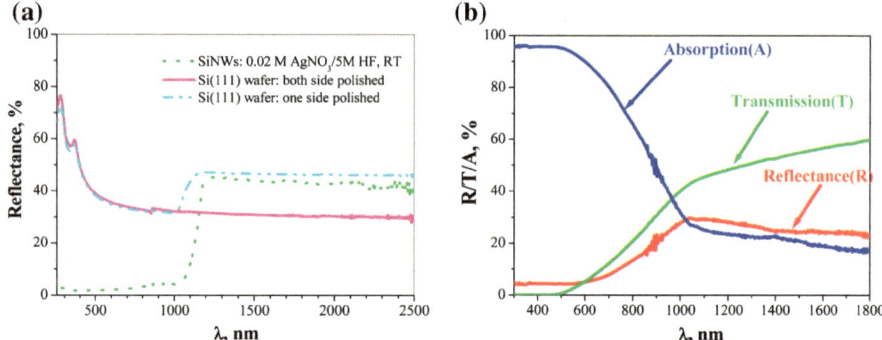

Fig. 9.9 **a** Reflectance spectra of SiNWs prepared via $AgNO_3$/HF etching of mc-p$^+$nn$^+$-Si layers on glass; as a reference, single or double side polished single crystalline silicon wafers were used. **b** Optical transmission (*T*), reflectance (*R*), and absorption ($A = 1 - T - R$) of SiNWs prepared by etching of ~2.7 μm thick mc-p$^+$nn$^+$-Si layers on glass. Reprinted with permission from Ref. [17], Copyright (2009) American Chemical Society

contrast to the light striking the planar silicon surface only once as shown in Fig. 9.8b [70–72]; (ii) the NWs arrays have structural morphology similar to sub-wavelength structures (SWSs) surface as shown schematically in Fig. 9.8c, which can suppress the reflection loss over a wide spectral range [62–69, 84, 85]; (iii) the NW arrays surface has varying porosity from top (air–SiNWs interface) to the bottom end of the NWs (NWs–Si substrate interface). The porosity gradient causes a graded refractive index layer from top to bottom (refractive index 'n' increases with decrease in porosity in the SiNW arrays layer) as shown schematically in Fig. 9.8d [52, 72, 80, 81, 83, 86–90] which closely resembles a multi layer antireflection coating [72, 88].

9.5.2 SiNW Arrays on Silicon Thin Film on Glass Substrates

The MA-EWCE method has also been used to fabricate the SiNW arrays on thin silicon films supported on glass substrates to examine the excellent light harvesting quality of SiNW arrays for thin solar cell concepts. The optical absorption properties of SiNW array films fabricated on glass substrates by MA-EWCE have been measured by Tsakalakos et al. [3] and Sivakov et al. [17]. The SiNWs array films on glass exhibited very low reflectance (<10 % at 300–800 nm), strong broadband optical absorption (>90 % at 500 nm), and an optical absorption much higher than Si films of the equivalent thickness (as shown in Fig. 9.9). The enhanced broadband absorption has been attributed to the strong resonance among the vertical SiNW arrays, while the observed absorption is partly due to the high-density surface states in the SiNW film also [3, 17].

9.5.3 *Theoretical Modelling/Simulation of ARC/Light Trapping for MA-EWCE SiNW Arrays*

As described above with the help of experimental observations, it is of no doubt that the optical properties of the SiNW arrays are strikingly different from those of Si bulk wafer and thin films. Recently, detailed simulation of the optical properties of SiNWs has also been performed by several research groups looking at their potential application in efficient and cost effective solar PV devices or SiNW-based solar cells [8, 91–93]. However, most of these simulations/analyses have been done for ordered SiNWs. For example, Hu et al. [8] simulated the effects of NWs diameter, length and filling ratio on the absorptance of NW arrays. They found that NW arrays with moderate filing ratio have much lower reflectance compared to the thin films of equivalent thicknesses. Lin et al. [92] used the transfer matrix method (TMM) to calculate the optical absorptance of vertically-aligned SiNW arrays. It was found that for a fixed filling ratio, significant optical absorption enhancement occurs when the lattice constant was increased from 100 to 600 nm. The enhancement was attributed to an increase in field concentration within the NW as well as excitation of guided resonance modes. It was also shown that an optimized SiNW array with lattice constant of 600 nm and wire diameter of 540 nm has a 72.4 % higher ultimate efficiency than a Si thin film of equal thickness.

Similarly, there are several efforts made to simulate the optical properties for aperiodic SiNW arrays. For example, Bao et al. [94] numerically investigated the optical properties for vertically aligned SiNW arrays with three types of structural randomness, random position, diameter, and length. It was shown that the NW arrays with random position exhibited slight absorption enhancement, while those with random diameter or length exhibited significant absorption enhancement, which was attributed to the stronger optical scattering in a random structure as depicted in Fig. 9.8a. Further, It was also shown that SiNW arrays structural randomness can further enhance optical absorption compared to ordered NW arrays [94]. Rigorous coupled-wave analysis (RCWA) has also been widely used to simulate the experimental observations of excellent light trapping of the vertical SiNW arrays [52, 82, 95]. For example, Xie et al. [95] have experimentally found that the reflectance can be significantly suppressed (< 1 %) over a wide solar spectrum (300–1000 nm) in the as-prepared SiNW arrays. Using the bundled model, they have used RCWA to simulate the reflectance in SiNW arrays, and found that the calculated results are in good agreement with the experimental data.

9.6 Omni-directional Light Trapping Properties

In addition to excellent AR property, the SiNW arrays are found to exhibit other favorable AR properties, including omni-directionality and polarization-insensitivity [52, 95]. Critical features of the AR coatings, such as broadband, angularly insensitive, and polarization- insensitive characteristics have been demonstrated in the SiNW arrays under both unpolarized and polarized light by Chang et al. [52]. The effective elimination of spectral and angular reflection was accounted to the fact that controlled geometrical configurations, such as density of SiNW arrays and interfaces of air/NW arrays layers and NW arrays layers/substrate, which resulted in an intermediate transition region, exhibit a favorable gradient of effective refractive index. The smooth refractive index gradient of SiNW arrays due to increased interface roughness and decreased filling fraction was shown to be critical to considerable reflection elimination, leading to the excellent broadband and omni-directional AR characteristics of the aperiodic SiNW arrays. They could achieve the averaged specular and diffuse reflectance of the NWs arrays as low as 0.03 and 1.46 % respectively within the wavelength ranges of 200–850 nm. TE and TM reflectance can be as low as 0.1 % for the angle of incidence up to 55° measured at wavelength of 528 nm [52].

9.7 Solar Cells Basics and Importance of Silicon Nanowires

The solar cell operation principle can be understood from Fig. 9.10. It is based on an unbiased p-n junction diode connected to a load (R_L) to generate power [59, 96]. There is a region of high (in-built) electric field, called the depletion region (labeled as J_d), at the junction of the p- and n-type semiconductors where photo-generated charge carriers (electrons and holes) are separated by the in-built electric field, producing a current. In addition, carriers generated in either the p- or n-type regions within a minority charge carrier diffusion length (l_d) of the depletion region can be collected at the junction and contribute to the total current. The key criteria for efficient solar cells are high absorption of the incident light in the active region of carrier collection and minimal loss of carriers due to recombination (bulk and surface defects). Conventional wafers based crystalline silicon solar cells rely primarily on thick diffusion regions (~200 μm), compared to the much thinner drift region (J_d), for carrier collection as schematically illustrated in Fig. 9.11a. This scheme is inevitable due to the poor absorption of light by silicon across the solar spectrum [59], and therefore a substantial thick silicon is used to absorb the incident light effectively in wide spectral range. However, as illustrated in the cross sectional view of typical high efficiency planar silicon solar cell geometry (Fig. 9.11a), the charge carriers generated away from the junction have to diffuse long distance to reach the junction (without recombination) for collection, therefore, requires large diffusion length. This requirement restricts the quality of the silicon materials to be reasonably high. Thus, high-efficiency silicon solar cells require defect-free material with long minority carrier lifetimes (or large diffusion lengths). The current record efficiency under terrestrial illumination conditions (AM 1.5 G, 100 mW/cm^2) for silicon solar cell is ~25 % which was achieved over two decades ago at Martin Green's laboratory in Australia (UNSW). This efficiency is quite close to the maximum theoretical

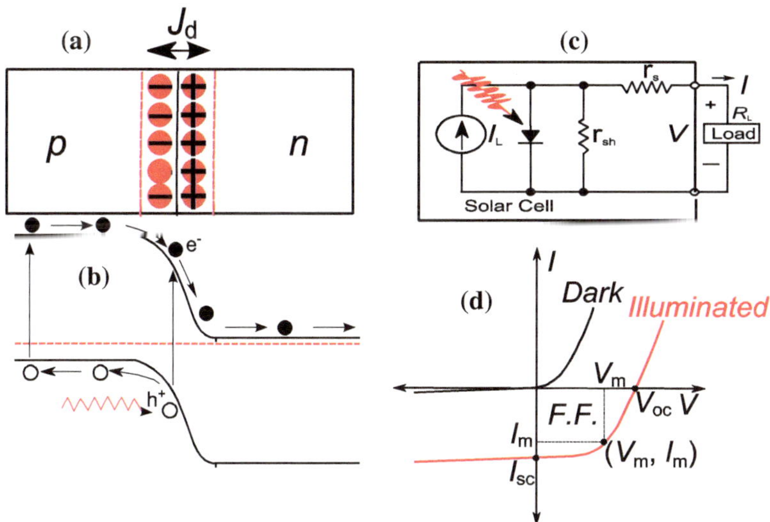

Fig. 9.10 **a** *Top* is schematic of a p-n junction showing depletion region (J_d) with ionized immobile dopant charges. **b** *Bottom* is the correspondent energy-band diagram. Incident light creates an electron–hole pair that is swept across the junction by the in-built drift electric field to the charge-neutral regions. **c** An equivalent circuit of a practical solar cell connected to an external load (R_L). In the circuit, parasitic resistances, series (r_s) and shunt (r_{sh}) are also shown. **d** Typical *I-V* characteristics of a p-n junction solar cell under dark and illumination conditions. The solar cell's basic performance parameters are also shown in the *I-V* characteristics

efficiency ~31 % predicted for a single junction solar cell by Shockley–Queisser, also known as Shockley–Queisser limit [97]. However, the fabrication costs are quite high for such record efficiency cells. Commercial silicon solar cells efficiency typically range from ~15 to 22 % depending on solar cell structure and design [98].

9.7.1 Solar Cells Basic Parameters

The expression for net current (I) flowing across the load can be expressed as,

$$I = I_L - I_0\{\exp(q(V + Ir_s)/nkT) - 1\} - (V + Ir_s)/r_{sh}$$

I_{sc}	Short circuit current; J_{sc}: Short circuit current density
I_L	light generated current,
I_0	dark saturation current,
$V_{oc} = kT/q\{\ln(I_L/I_o) + 1\}$	open circuit voltage {since at V_{oc}, $I = 0$)
V_m and I_m	Operating point yielding the maximum power output

Fill Factor ($F.F.$) = $V_m I_m / V_{oc} I_{sc}$
Power conversion efficiency (η):

$$\eta = P_{max}/P_{in} = V_m I_m / P_{in} = F.F.V_{oc} I_{sc}/P_{in}$$

In solar cell operation, generally, key aim is to generate power by (1) generating a large short circuit current, I_{sc}, (2) a large open-circuit voltage, V_{oc} and (3) minimise parasitic power loss mechanisms (particularly series and shunt resistances) and therefore, maximize the fill factor ($F.F.$) [59, 96].

9.7.2 Silicon Nanowire Arrays Based Solar Cell Concept

In crystalline silicon solar cells, the solar grade silicon materials growth (which involves extensive purification, ingot growth) and slicing account for up to ~50 % of the overall cost of wafer based Si solar cells [97]. Single junction solar cell geometries based on SiNWs with radial p-n junctions are attracting immense interest as they provide new opportunities for enhanced light absorption and improved device performance, as well as using significantly less silicon compared with wafer based cells [6, 16, 17, 99]. The radial geometry is illustrated in Fig. 9.11b wherein the direction of solar radiation is parallel to the SiNW axis such that light is absorbed parallel to the p-n junction. Therefore, light absorption and charge carriers separation are orthogonalised, with generation of photo carriers in the close vicinity of the junction leading to their efficient collection. This relaxes the stringent requirement on carrier diffusion lengths and therefore low quality wafers with much shorter diffusion length than that in wafer based Si solar cells (as illustrated in Fig. 9.11a) can be used to make cells. This may lead to lower device cost. Therefore, the potential for improved performance and cost reductions of SiNW arrays over their bulk PV counterparts is possible which may be primarily due to (i) increased absorption (due to diffuse light scattering in the NW arrays) relaxing the requirement of thick silicon materials, (ii) short collection lengths of minority carriers (if radial p-n junction) that are radially separated and collected normal to the light absorption direction relaxing the quality material requirement, and (iii) flexibility of cell integration on a variety of low-cost supporting substrates (such as glass for example). It is possible to achieve thin layers of NWs fabricated on or transferred to low-cost substrates [16, 17]. Therefore, a NW approach enables to use a less purity Si (almost 10 times lesser) along with lower cost material and device fabrication techniques. Although improvements in efficiency over conventional crystalline silicon solar cells are not anticipated at present but achieving reasonably good efficiencies (~15–18 %) in combination with low-cost thin-film processing could dramatically impact the application of solar cells. Measurements on single wire solar cells and optoelectronic simulations have projected that large area SiNWs array solar cells have potential to achieve >17 % energy conversion efficiency [100].

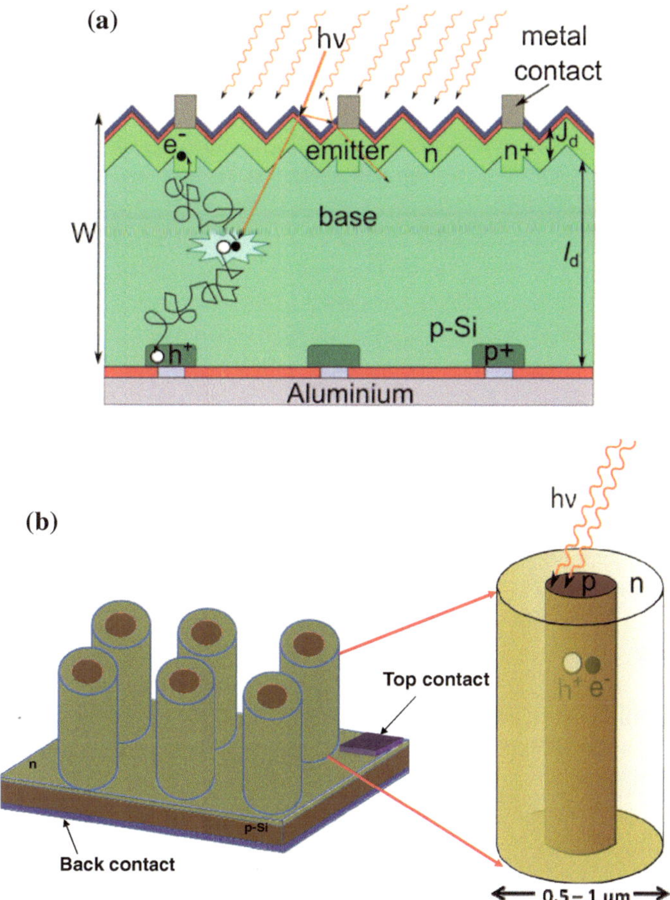

Fig. 9.11 **a** Schematic of a conventional planar Si solar cell with antireflection coating and surface (μ-pyramid) texturing. In this scheme, typical cell thickness, W ~ 200 μm, is used to absorb the light and the charge carriers collection is primarily from the diffusion region (l_d) where $l_d \gg J_d$, the junction drift region. Further, the light absorption and carriers collection are in the same direction (longitudinal to the solar cell surface). **b** Schematic of solar cell employing SiNWs in radial p-n junction geometry, the charge carriers are collected radially (have to travel ~0.5-1 μm only in the radial direction) compared to ~200 μm in planar cells and light is absorbed along the longitudinal (NW axis) direction normal to radial direction

9.8 Silicon Nanowire Arrays Based Photovoltaic Devices: Recent Developments

To exploit the excellent light trapping properties of vertical aperiodic SiNW arrays for solar cells, Peng et al. [70] first reported SiNW arrays based p-n junction solar cells following the conventional fabrication protocol using thermal diffusion process. They fabricated the SiNW based solar cells using the following processing sequence: (i) fabrication of SiNWs arrays on p-type Si wafers; (ii) RCA (the Radio Corporation of American) cleaning to remove the residual metal and organic contaminants followed by removal of SiO_x in a buffered HF solution; (iii) formation of thin emitter (n^{++} layer) on SiNWs via phosphorous (P) diffusion at >900 °C; (iv) evaporation of an Al layer onto the rear Si surface and removal of the rear parasitic p-n junction; and (v) deposition of Ti/Pd/Ag grid contacts onto the front sides of SiNW arrays. Such a solar cell on a mono-crystal Si substrate exhibited a V_{oc} of 548.5 mV and a *F.F.* of 65 %, giving $\eta = 9.31$ %, which was lower than that of a planar solar cell (without SiNWs) fabricated in the same batch. The low conversion efficiency was mainly attributed to the surface or interfacial recombination, high series resistance, and low-current-collection efficiency of the front-grid electrodes. Later on, Fang et al. [101] fabricated solar cells on slantingly aligned SiNWs on a Si (111) wafer, which yielded an improved efficiency of 11.37 % under AM 1.5 G illumination. Though, this efficiency was still lower than that of the planar solar cell fabricated under identical processing conditions. The improved SiNWs based cell performance was attributed to the combined effects of excellent antireflection properties and better electrical contacts as compared to the vertical SiNW arrays used in earlier approach [70]. However, the high surface recombination of minority carriers in SiNW arrays remained the main challenging issues to the solar cell performance.

As mentioned earlier in Sect. 9.1, the SiNWs prepared have tendency to be bunched/bundled particularly for longer lengths and also affect the optical as well as solar cell characteristics. To address this, Jung et al. [83] demonstrated that vertically aligned, tapered SiNWs prepared using MA-EWCE process followed by post-KOH treatment were very promising for solar cells because of their stronger light trapping ability. It was shown that compared to the bunched NW arrays, tapered SiNWs could further suppress the optical reflectance over a broad spectral range due to a gradual transition of the effective refractive indexes. The conversion efficiency of the tapered SiNW solar cell was reported higher than that of bunched NWs arrays. However, the cell efficiency was quite low (~6.56 %) as compared to that achieved earlier by Peng et al. [70] and Fang et al. [101].

It is to be noted that in most of the efforts made in the MA-EWCE prepared SiNW arrays based solar cells as discussed earlier used vertically aligned SiNWs directly on Si wafers [70, 83, 101]. In such cases, metal contact quality was rather poor due to non-compact contacts between electrodes and SiNW arrays and thus might lead to significant electrical loss. Fang et al. [101] tried to overcome the problem by fabricating solar cells on slantingly aligned SiNW arrays on p-type Si (111) substrate. However, highly compact contacts could not even be made on slantingly-aligned SiNW arrays as that on the planar silicon surface made in the conventional silicon solar cells. Kumar et al. [102] tried to address the problem of metal contacts on SiNW arrays surfaces by fabricating SiNW arrays selectively in the active area (i.e. emitter exposed to light) of the solar cell and keeping metal grid pattern on planar surface. The n^+-p-p^+ structure based solar cells were fabricated by conventional cell fabrication protocol (similar to that used by Peng et al. [70] except the selective SiNWs fabrication) only as presented schematically in Fig. 9.12. In this approach they improved the quality of front metal electrodes of the SiNW arrays based solar cells while retaining the excellent AR property. SEM images of different regions of such SiNW arrays based solar cell are also shown in Fig. 9.12. With this approach, the J_{sc} increased significantly to 37 mA/cm^2 but V_{oc} decreased marginally (544 mV) as compared with the cell made on planar Si substrates (without SiNWs). The efficiency (13.7 %) realized was improved by absolute 1 % compared with the cell made on planar Si substrates (without SiNWs) which was basically attributed to enhanced light absorption due to SiNW arrays. They also concluded that low efficiency of SiNW arrays cell was due to enhanced recombination losses caused by large surface area of the SiNWs.

Formation of a selective emitter in crystalline silicon solar cells improves photovoltaic conversion efficiency by decoupling emitter regions for light absorption (moderately doped) and metallization (degenerately doped). Recently, Um et al. [103] demonstrated a self-aligned selective emitter successfully integrated into an antireflective SiNW solar cell. Using the single-step MA-EWCE, NW arrays formed only at light absorbing areas between top-metal grids while selectively retaining Ohmic contact regions underneath the metal grids. They observed a remarkable ~40 % enhancement in blue responses of internal quantum efficiency (IQE), corresponding to a conversion efficiency of 12.8 % in comparison to the 8.05 % of a conventional NW solar cell. In a similar development, Chen et al. [104] demonstrated a simple and convenient double-step diffusion process to form a highly doped n^+ region at the tips of the SiNW arrays. With this approach, the electrode-contact enhancement was achieved reducing the series resistance effectively which eventually lead to an increase in the J_{sc} of the cell. They could achieve a 40.2 % increase in J_{sc} in two step diffused processed solar cells as compared to that fabricated by one-step diffusion process. Further, there has been number of reports investigating the influence of SiNW arrays length, their aspect ratio and density [105–111] including few on large area (e.g.; 125 × 125 mm^2 or 156 × 156 mm^2) SiNW arrays based solar cells with efficiencies as high as over 16 % depending on processing conditions. However, in almost all the cases it was observed that SiNWs reduces the reflectance of solar cells significantly but this optical gain does not improve the performance of the solar cells, such as power conversion efficiency, quantum efficiency etc. The solar cells performance parameters are strongly influenced by the length of the NWs, their aspect ratio, and density. Researchers have unanimously pointed out the concern of increased recombination losses (both bulk as well as surface recombination) of charge carriers, and the series resistance due to enhanced surface area. Indeed an increase in these parameters has a negative effect on the performance of the solar cells in contrast to the expected improvement due to the enhanced optical absorption. Therefore, the design principle of the SiNWs based solar cell is that the ultra-low reflectance is not the only pursuit but optimizing the surface passivation, improving the electrode-contact property must also be considered for high efficiency SiNW arrays based solar cells.

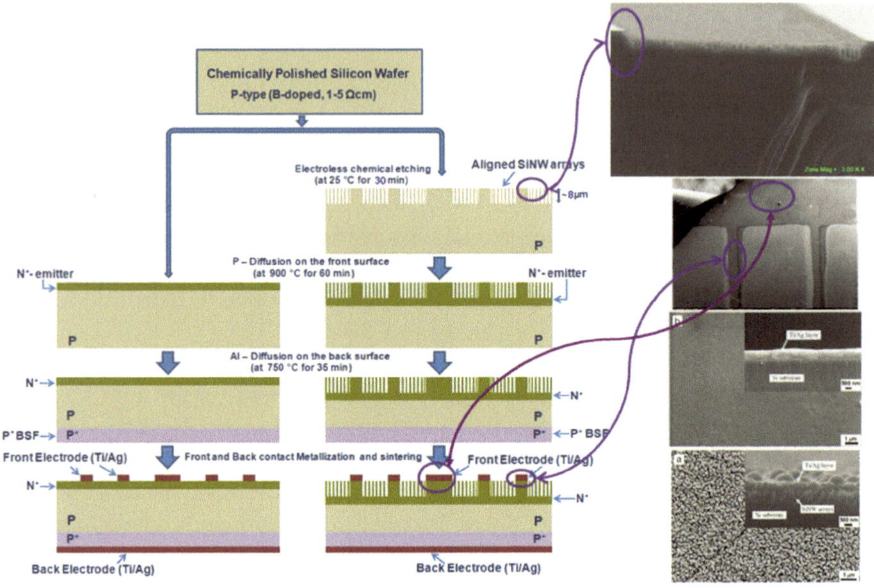

Fig. 9.12 Schematic illustration of the fabrication process for preparing a SiNWs based silicon solar cell (flow chart in the centre): (i) initial p-type silicon wafer, (ii) fabrication of vertically aligned SiNW arrays on exposed silicon wafer by selective chemical etching in an aqueous HF/$AgNO_3$ solution at room temperature, (iii) conversion of p-type SiNW arrays to n-type SiNW emitters by phosphorus diffusion using solid dopant source, (iv) coating of Al layer on the rear surface by thermal evaporation method and alloying it to make back surface field, (v) making of metal contacts of Ti/Ag on the pattern of un-etched silicon surface in the front and on entire area on the back surface followed by sintering to make Ohmic contacts. The solar cell fabrication on conventional planar silicon wafer is also shown in the *left* side. SEM images of the selective area SiNW arrays, *top* and *side* view of the metal contacts on planar area and NWs regions are also shown in the *right* side

9.9 Problems and Challenges of Aperiodic SiNW Arrays Based Solar Cells

It is of no doubt that there has been tremendous progress in the MA-EWCE based SiNW arrays based solar cells in the recent past. However, there are several obvious questions to be answered like it is not fully clear whether the p-n junction is also formed coaxially in such SiNWs or the SiNWs get converted fully into n-type after high temperature dopant diffusion process. As described in the previous section, the majority of the work reported on aperiodic SiNW arrays based solar cells is on bulk Si wafer substrates where photogeneration from the substrate most likely contributes to device performance but has rarely been quantified. In most cases the NWs are heavily doped and utilized purely for their antireflective property only in a silicon wafer solar cell (as indicated in Fig. 9.13b). Consequently, the as-prepared SiNW solar cells are similar to a conventional planar p-n junction solar cell with the

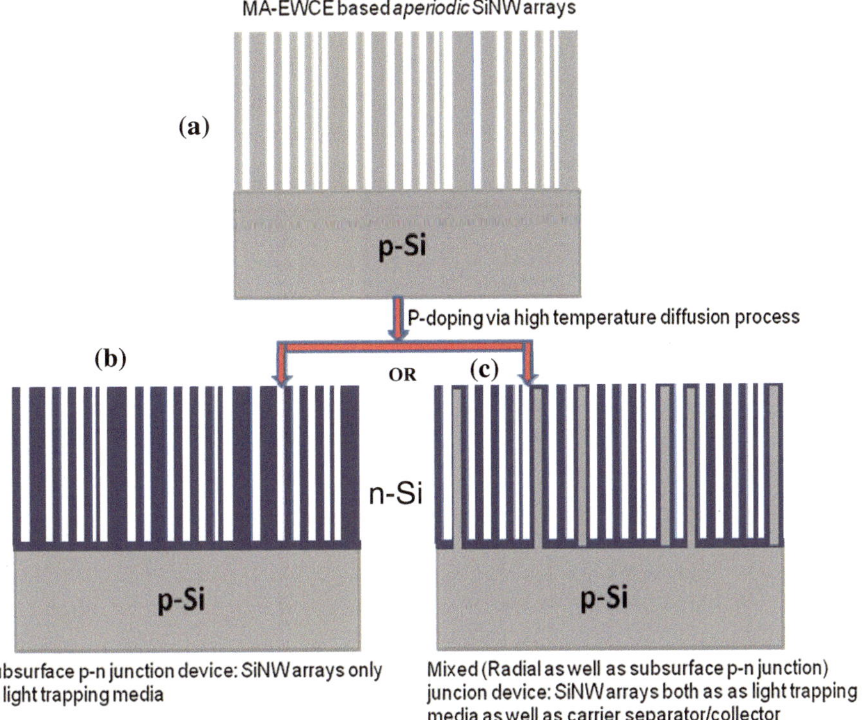

Fig. 9.13 Schematic illustration of the conventional fabrication of p-n junctions in aperiodic SiNW arrays based solar cells. **a** Fabrication of vertically aligned SiNW arrays layer by MA-EWCE Method using aqueous HF + $AgNO_3$ solution. After phosphorus dopant diffusion by using high temperature diffusion process, there are two possibilities. **b** Formation of a subsurface p-n junction in SiNW arrays wherein SiNWs are fully become n-layer and the arrays just act as an effective light trapping media. **c** It may result in combination of subsurface p-n junction as in (**b**) as well as formation of radial p-n junction in the SiNWs particularly large diameter NWs

SiNW arrays acting as an antireflection coating for enhanced optical absorption as shown in Fig. 9.13b. However, in the aperiodic SiNW arrays prepared by MA-EWCE method, there is a wide distribution in the NWs diameter and possibility of radial p-n junction in NWs (particularly, in large diameter NWs) cannot be ruled out. Therefore, there could be possibility of combination of both radial p-n junction as well as fully diffused (n-type) NWs as schematically shown in Fig. 9.13c. Though the percentage of purely radial p-n junction based NWs may be quite less. However, this aspect has not been addressed convincingly so far. In addition, dopant diffusion kinetics may not be the same in the case of SiNWs as that in conventional planar substrates. Therefore, the process requires further optimization of various parameters related with SiNW formation, optimum arrays depth, dopant diffusion, contact formation, surface recombination/passivation etc. which may result in improved cell performance.

9.10 Effective Surface Passivation of SiNW Arrays Based Solar Cells

A number attempt have been made to develop effective surface passivation schemes employing single layer or double layer stacking of dielectrics such as SiO_x, SiN_x, $SiO_x + SiN_x$, atomic layer deposition (ALD) based Al_2O_3 and so on [112–115]. For example, Lin et al. [112] have demonstrated an effective and industrially compatible technique of dielectric passivation to improve the electrical properties of SiNW based solar cells without compromising with the optical properties of the SiNW arrays. It was shown that the SiNx based passivation could reduce defect state density quite effectively and also suppressed the Auger recombination due to the well known surface passivation and hydrogen-diffusion-induced bulk passivation. They further combined the SiN_x with the SiO_2 surface passivation (SiO_2/SiN_x stack) to obtain the best Shockley–Read–Hall (SRH) and Auger recombination suppression in the SiNWs based solar cells to realize the high performance SiNW based solar cells with a conversion efficiency of 17.11 % on a large size of $125 \times 125\ mm^2$. The key factor for success lies in an efficient approach of dielectric passivation to greatly enhance the electrical properties while keeping the advantage of excellent light trapping of the SiNW structure. The schematic of the proposed passivation schemes and cell fabrication protocol are presented in Fig. 9.14. Similarly, Zhao et al. [113] investigated the performance of SiNWs based silicon solar cells with single and double stack layer of dielectrics passivation on large area ($156 \times 156\ mm^2$). It was observed that the conversion efficiency of SiNWs based silicon solar cell without passivation could be 13.8 % which was enhanced to 16.1 and 16.5 % respectively with SiO_2 and SiN_x:H passivation of SiNWs surfaces. Further, it was shown that the SiO_2/SiN_x:H stacks exhibited better efficiency (17.1 %) as compared to single dielectric layer based surface passivation of such solar cells. It was suggested that the SiO_2/SiN_x:H stacks layer decreased the Auger recombination through reducing the surface doping concentration and surface state density of the Si/SiO_2 interface, and SiN_x:H layer suppresses the SRH recombination in the SiNWs based silicon solar cell yielding the best electrical performance.

Wang et al. [114] developed efficient SiNW arrays based black silicon solar cells passivated by highly conformal Al_2O_3 layer deposited by thermal ALD. The Al_2O_3 passivation layer contributes to the suppressed surface recombination, which was explored in terms of the chemical and field-effect passivation effects. The nanostructured black silicon wafer covered with the Al_2O_3 layer exhibited a total reflectance of ~1.5 % in a broad spectrum from 400 to 800 nm. An 8 % increment of J_{sc} and 10.3 % enhancement of efficiency were achieved due to the ALD-Al_2O_3 surface passivation and forming gas annealing. They could realize a high efficiency up to 18.2 % in the ALD Al_2O_3-passivated nano-structured silicon solar cells.

Recently, Oh et al. [115] suggested slightly different approach to achieve high efficient SiNW arrays based bulk silicon solar cells. They used a wafer [300 μm thick, p-type (100)] based cell with SiNW arrays wherein a junction was formed by

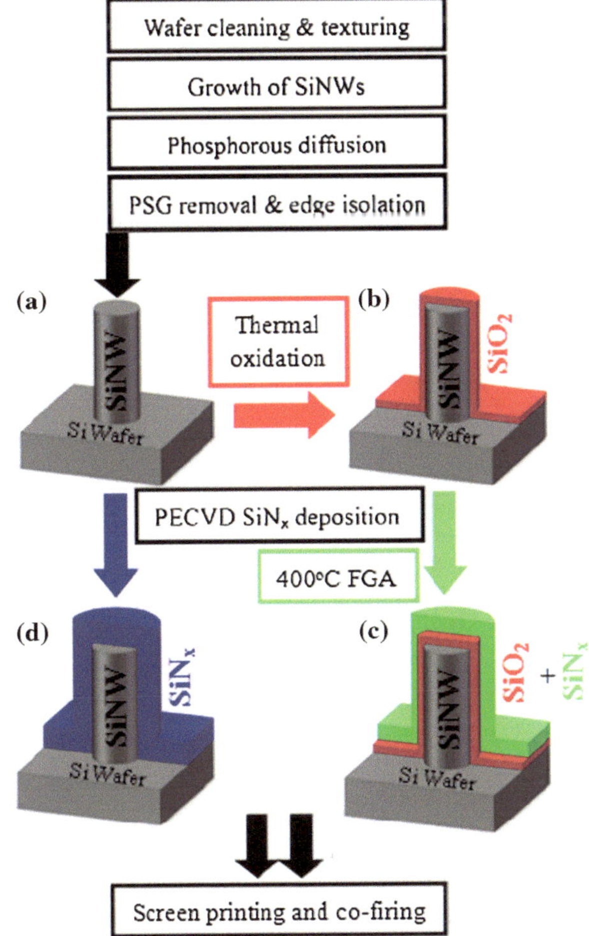

Fig. 9.14 Schematic fabrication process diagrams of the SiNW based solar cells with different passivations for series A (in *black*, bare SiNWs without dielectric shell); B (in *red*, thermal oxidation of ~10 nm SiO_2 layer); C (in *green*, thermal oxidation ~10 nm SiO_2 layer combined with 60 nm SiN_x deposition); and D (in *blue*, 70 nm SiN_x deposition only). Ref. [112]. © IOP Publishing. Reproduced by permission of IOP Publishing. All rights reserved

conventional phosphorus diffusion. It was shown that not simply the surface recombination associated with the high surface area of the NWs, but Auger recombination caused by excessive doping also limit the photogenerated charge collection and the efficiency. By suppressing Auger recombination efficiency of 18.2 % was achieved without any additional ARC. The J_{sc} was 36.45 mA/cm^2, with a V_{oc} of 628 mV and an *F.F.* of 79.6 %. Based on this work, design rule for radial p-n junction cell were also proposed suggesting that in addition to excellent surface passivation, Auger recombination needs to be suppressed for efficient PV devices based on aperiodic SiNW arrays. These results are quite encouraging towards the way to develop cost effective efficient silicon solar cells employing SiNWs.

9.11 Solar Cells Based on a Vertically Aligned Radial Hetero p-n Junction SiNW Arrays

Examples of purely radial p-n junction SiNW arrays based solar cells using the MA-EWCE based aperiodic SiNWs are very limited. However, there are few successful demonstrations of radial p-n junction on Si wafer substrates using hetero emitter such as amorphous Si layer deposited by PECVD. In the year 2008, Garnett and Yang [5] reported the first such radial hetero p-n junction solar cells (see Fig. 9.15). They fabricated SiNWs of ~18 μm length in an n-type Si wafer using MA-EWCE method followed by coating the NWs with p-type amorphous silicon yielding radial junctions. Such solar cell had efficiencies only up to 0.5 %. The limited performance was ascribed to interfacial recombination and high series resistance. Four years later, Jia et al. [116] reported transparent conducting oxide (TCO)/a-Si/SiNW heterojunction solar cells on arrays also prepared by MA-EWCE of an n-type silicon wafer. They fabricated NWs with diameter ranging up to ~300 nm and lengths of ~900 nm. This was followed by deposition of the intrinsic (i-layer) and p-type a-Si layers around the NWs by PECVD and finally TCO layer to complete the device. The mesa-structured 7 mm^2 area cells yielded a quite impressive conversion efficiency of 7.29 % with a V_{oc} of 476 mV, J_{sc} of 27 mA/cm^2 and *F.F.* of 56.2 %. Further, electron beam induced current measurements demonstrated that the NWs played an active role in the cell's PV response.

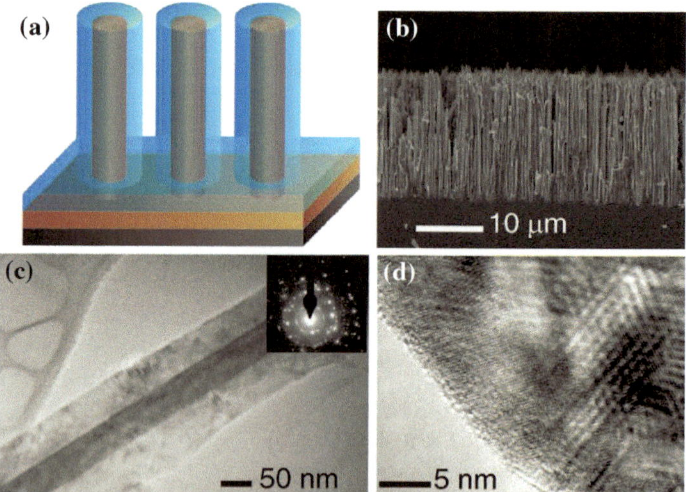

Fig. 9.15 Silicon nanowire solar cell structure. **a** Schematic cell design with the single crystalline n-Si NW core in brown, the polycrystalline p-Si shell in *blue*, and the back contact in *black*. **b** Cross-sectional SEM of a completed device demonstrating excellent vertical alignment and dense wire packing. **c** TEM image showing the single crystalline n-Si core and polycrystalline p-Si shell. The inset is the selected area electron diffraction pattern. **d** TEM image from the edge of the core-shell nanowire showing nano-crystalline domains. Reprinted with permission from Ref. [5]. Copyright (2008) American Chemical Society

9.12 Solar Cells Based on a Vertically Aligned p-n Junction SiNW Arrays on Cheap Glass Substrates

As described earlier, for a conventional planar Si solar cells, the active Si layer has to be sufficiently thick (≥200 μm) in order to harvest incident photons significantly. Experiment and theory have shown that vertically aligned SiNW arrays enable efficient broadband absorption of light, less than 1 % of the Si material (having NW arrays) would have the same absorption efficiency as in the conventional Si-wafer based devices. The reduced Si materials requirement in SiNW arrays based solar cells would decrease the production cost, since the Si material is a major cost for Si based PV cells. The work employing this concept using aperiodic SiNWs on glass substrates are limited. Sivakov et al. [17] demonstrated a solar cell from SiNWs with (non-radial) axial p^+-n-n^+ junctions on glass substrates, as shown in Fig. 9.16. They first fabricated a large-grained 2.5–3 μm thick multi-crystalline p^+-n-n^+ Si layer stack on a glass substrate by electron beam evaporation deposition followed by laser crystallization. Vertical SiNWs with axial p^+-n-n^+ junctions were then fabricated from the mc-Si layer by MA-EWCE process in HF–$AgNO_3$ solution. Under AM 1.5G illumination, the cells exhibited a maximum V_{oc} of 450 mV and a J_{sc} of 40 mA cm^{-2} resulting to overall efficiency of 4.4 %. It was suggested that severe shunting and large series resistance are responsible for the relatively low device performance. In this series, a group at Institute of Photonic Technology (IPHT) Jena, Germany [117, 118] has made significant development. Recently, they reported an efficiency of 8.8 % with a V_{oc} of 530 mV in radial p-n junction based cells on glass substrates. The SiNWs were fabricated by MA-EWCE of an n-type, 8 μm thick laser crystallized multi-crystalline Si thin film on glass [117]. PECVD was used to deposit an a-Si hetero-emitter around the NWs. In this device structure, ALD based Al_2O_3 passivation layer was also utilized. Finally, ZnO:Al was conformally deposited as a TCO in the NWs arrays to have a planar transparent conducting layer as shown in Fig. 9.17. Later, the same group reported realization of 10 % efficient prototype solar cell employing SiNWs on a polycrystalline (pc) -Si thin film with a thickness of only 8 μm formed on glass [118]. By improved nanowires engineering (cleaning of the metal contamination from the metal-assisted wet chemical etching and passivation by a-Si:H emitter as well as by an ultrathin Al_2O_3 tunnel layer on the emitter surface) and by appropriate design of the cell structure (superstrate configuration), the detrimental effects related to the NWs can be avoided while retaining the light trapping properties. Though the efficiencies of such solar cells are not so impressive at present compared to the conventional planar silicon solar cells, nevertheless, such solar cells with SiNW arrays as effective light absorber on cheap glass substrates could impact next generation PV technologies particularly the device cost.

In summary, the full potential of aperiodic SiNWs cells is yet to be realized. The cells on Si wafer substrates have better performance reflecting the superior wire

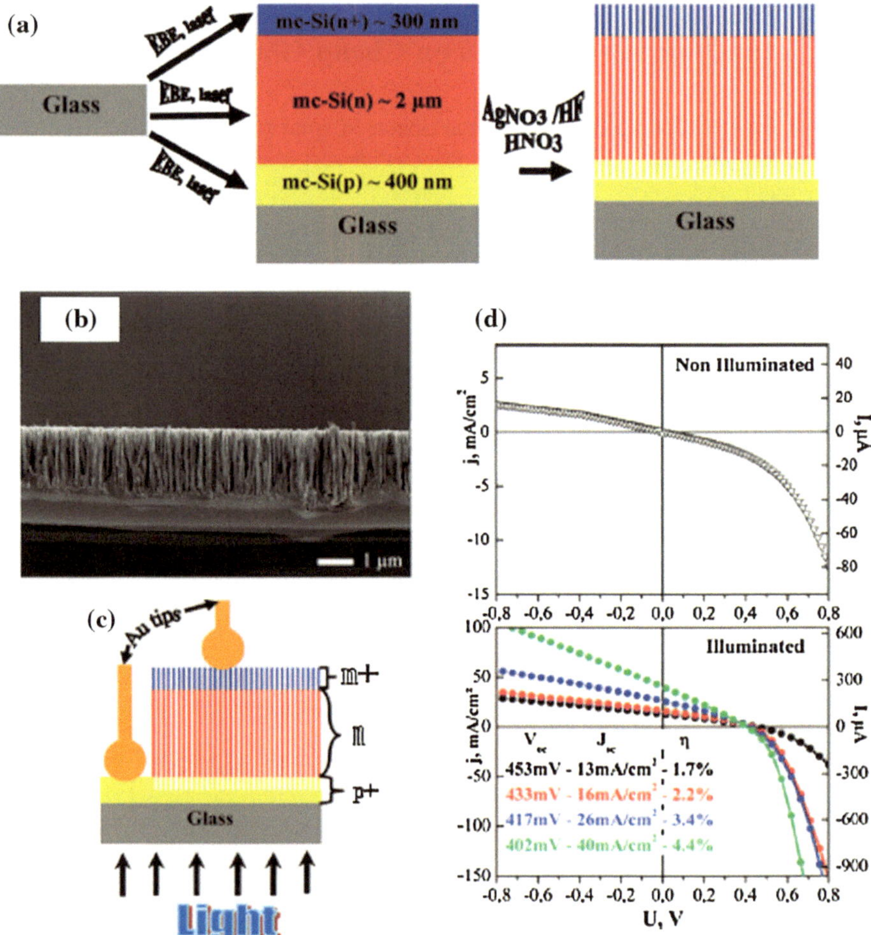

Fig. 9.16 **a** Schematic cross sectional view of the mc-Si p-n junction layer stack on a glass substrate (*left*) and the SiNWs after wet chemical etching (*right*). **b** Cross sectional SEM micrograph of the $AgNO_3$/HF etched mc-p^+nn^+-Si layer on glass. **c** Schematic representation of the I-V curve measurements of SiNW based p-n junctions. **d** Non-illuminated and illuminated (AM1.5) I-V curves of SiNWs etched into a mc-p^+nn^+-Si layer on glass. SiNW are irradiated through the glass substrate (superstate configuration) and contacted by metal tips at four different sample positions. The right-hand scale gives real measured current values and the left-hand scale current density values. Reprinted with permission from Ref. [17]. Copyright (2009) American Chemical Society

material quality. Efficiencies in the range of 17–18 % have been demonstrated in the cells but photogeneration in the Si substrate is likely to be contributing in most cases. The performance of MA-EWCE based SiNWs cells on glass has also reached a promising level of 10 % which is quite encouraging for next generation PV technologies.

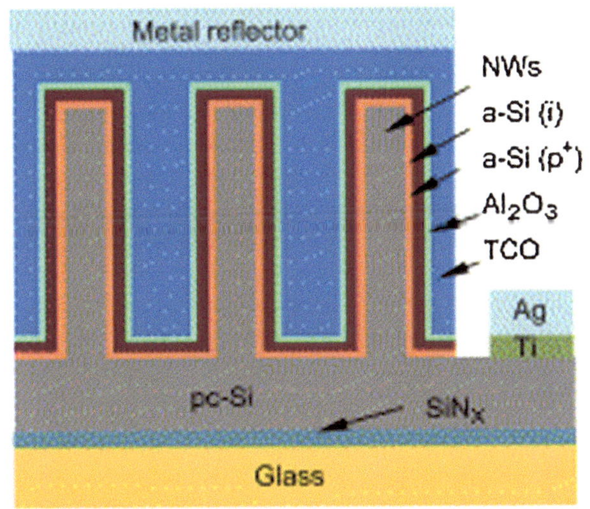

Fig. 9.17 Schematic of thin film solar cells using SiNW arrays on polycrystalline (pc) Si thin film on glass substrate using Ag reflector. Reprinted from Ref. [118]. Copyright (2014), with permission from Elsevier

9.13 Concluding Remarks and Future Perspective

SiNWs have been shown both experimentally and theoretically to be an important and a promising class of nanoscale building blocks for cost-effective and high-performance solar cells, primarily because of their superior light absorption and charge separation. The MA-EWCE is very simple and inexpensive top-down method for the fabrication of large area aperiodic SiNW arrays of desired length, optical and electronic properties. The SiNW arrays act as tunable antireflection layer that is controlled by the NWs length and surface reflection loss as low as ~ 1 % in the wide spectral range have been achieved with omni-directional optical properties. The simplicity of the process may lead to the advancements and applications of SiNWs in opto-electronics and nano-electronic devices. Owing to its simplicity, it has potential to be implemented in nano-structured based next generation silicon solar cells, both as an effective anti-reflection surface as well as an active component of the solar cells in the radial p-n junction geometry. A lot of efforts are being put world over in this new exciting area of research. Several examples of MA-EWCE fabricated SiNWs based solar cells with improved performances have been demonstrated both in conventional Si wafer based solar cells as well as in Si thin films on glass substrates. Power conversion efficiencies >17 % have been achieved depending on the processes and device structure. Most of the efforts are based on wafer based silicon solar cells wherein the exact role of SiNW arrays is yet to be identified, i.e., whether it serves only as an effective optical media for enhanced light absorption or it has also an active role in electrical characterization of solar cells. In this series, the radial p-n junction SiNW arrays may offer a viable approach to cost-effective thin solar cells with performance competitive against conventional planar Si solar cells and significant progress have been made in this novel concept achieving $\sim$10 % efficiency.

However, despite the promising potential of aperiodic SiNW arrays based solar cells, many questions are yet to be worked out. In particular, the excessive recombination at the high surface and interface areas of SiNWs must be reduced by developing effective surface passivation scheme in order to capitalize on the advantages of SiNWs in light absorption and charge collection (in case of radial junction geometry). Besides optimal fabrication of high-quality p-n junctions, better device design and quantification of NWs' role in PV performance is needed to improve device performance. Another major constraint on device performance is excessive shunting due to poor contacts, fragile nature of the NWs or impurities at the junctions. In this regard, optimal contacts for SiNW fabrication should be designed. Overall, intensive efforts to improve surface passivation, device design, contacts, light trapping etc., for improved PV device efficiency are going on around the world. In view of the exciting developments in recent years, researchers are quite optimistic to make the SiNW arrays based solar cells as a cost-effective and efficient alternative to the conventional planar Si solar cells.

Acknowledgments Authors would like to thank the Council of Scientific & Industrial Research (CSIR), India for financial support under CSIR-TAPSUN programme (project code NWP-55) and Supra Institutional Project (SIP-17). Financial supports under CSIR YSA Research project (Grant code: OLP 142732; P-81-113) from CSIR and the BOYSCAST fellowship (Award No. SR/BY/P-03/10) from the Department of Science and Technology (DST), Government of India, are also acknowledged.

References

1. M. Kelzenberg, S. Boettcher, J. Petykiewicz, D. Turner-Evans, M. Putnam, E. Warren, J. Spurgeon, R. Briggs, N. Lewis, H. Atwater, Enhanced absorption and carrier collection in Si wire arrays for photovoltaic applications. Nat. Mater. **9**, 239 (2010)
2. E. Garnett, P. Yang, Light trapping in silicon nanowire solar cells. Nano Lett. **10**, 1082 (2010)
3. L. Tsakalakos, J. Balch, J. Fronheiser, M. Shih, S. LaBoeuf, M. Pietrzykowski, P. Codella, B. Korevaar, O. Sulima, J. Rand, A. Davuluru, U. Ropol, Strong broadband absorption in silicon nanowire arrays with a large lattice constant for photovoltaic applications. J. Nanophoton. **1**, 013552 (2007)
4. B. Tian, X. Zheng, T. Kempa, Y. Fang, J. Huang, C. Lieber, Coaxial silicon nanowires as solar cells and nanoelectronic power sources. Nature **449**, 885 (2007)
5. E. Garnett, P. Yang, Silicon nanowire radial p-n junction solar cells. J. Am. Chem. Soc. **130**, 9224 (2008)
6. B. Kayes, H. Atwater, N. Lewis, Comparison of the device physics principles of planar and radial p-n junction nanorod solar cells. J. Appl. Phys. **7**, 114302 (2005)
7. M. Putnam, S. Boettcher, M. Kelzenberg, D. Turner-Evans, J. Spurgeon, E. Warren, R. Briggs, N. Lewis, H. Atwater, Si microwire-array solar cells. Energy Environ. Sci. **3**, 1037 (2010)
8. L. Hu, G. Chen, Analysis of optical absorption in silicon nanowire arrays for photovoltaic applications. Nano Lett. **7**, 3249 (2007)

9. C. Kenrick, H. Yoon, Y. Yuwen, G. Barber, H. Shen, T. Mallouk, E. Dickey, T. Mayer, J. Redwing, Radial junction silicon wire array solar cells fabricated by gold-catalyzed vapor-liquid-solid growth. Appl. Phys. Lett. **97**, 143108 (2010)
10. K. Peng, S. Lee, Silicon nanowires for photovoltaic solar energy conversion. Adv. Mater. **20**, 1 (2010)
11. O. Gunawan, K. Wang, B. Fallahazad, Y. Zhang, E. Tutuc, S. Guha, High performance wire-array silicon solar cells. Prog. Photovolt. Res. Appl. **10**, 1002 (2010)
12. J. Zhu, Z. Yu, G. Burkhard, C. Hsu, S. Connor, Y. Xu, Q. Wang, M. McGehee, S. Fan, Y. Cui, Optical absorption enhancement in amorphous silicon nanowire and nanocone arrays. Nano Lett. **9**, 279 (2009)
13. C. Lin, M. Povinelli, Optical absorption enhancement in silicon nanowire arrays with a large lattice constant for photovoltaic applications. Nano Lett. **7**, 3249 (2007)
14. B. Ozdemir, M. Kulakci, R. Turan, H.E. Unalan, Silicon nanowire–poly(3, 4ethylenedioxythiophene): poly(styrenesulfonate) heterojunction solar cells. Appl. Phys. Lett. **99**, 113510 (2011)
15. E.A. Dalchiele, F. Martín, D. Leinen, R.E. Marotti, J.R. Ramos Barrado, Single crystalline silicon nanowire array based photoelectrochemical cells. J. Electrochem. Soc. **156**, K77 (2009)
16. L. Tsakalakos, J. Balch, J. Fronheiser, B.A. Korevaar, O. Sulima, J. Rand, Silicon nanowire solar cells. Appl. Phys. Lett. **91**, 233117 (2007)
17. V. Sivakov, G. Andrä, A. Gawlik, A. Berger, J. Plentz, F. Falk, S.H. Christiansen, Silicon nanowire based solar cells on glass: synthesis, optical properties, and cell parameters. Nano Lett. **9**, 1549 (2009)
18. R.S. Wagner, W.C. Ellis, Vapor liquid solid mechanism of single crystal growth. Appl. Phys. Lett. **4**, 89 (1964)
19. J. Westwater, D.P. Gosain, S. Tomiya, S. Usui, H. Ruda, Growth of silicon nanowires via gold silane vapor liquid solid reaction. J. Vac. Sci. Technol. B **15**, 554 (1997)
20. R.L. Latu, C. Mouchet, C. Cayron, E. Rouviere, J.P. Simonato, Growth parameters and shape specific synthesis of silicon nanowires by the VLS method. J. Nanopart. Res. **10**, 1287 (2008)
21. B. Fuhrmann, H.S. Leipner, H.R. Höche, Ordered arrays of silicon nanowires produced by nanosphere lithography and molecular beam epitaxy. Nano Lett. **5**, 2524 (2005)
22. A.M. Morales, C.M. Lieber, A laser ablation method for the synthesis of crystalline semiconductor nanowires. Science **279**, 208 (1998)
23. Y.H. Yang, S.J. Wu, S.H. Chiu, P. Lin, Y.T. Chen, Catalytic growth of silicon nanowires assisted by laser ablation. J. Phy. Chem. B **108**, 846 (2004)
24. H. Pan, S. Lim, C. Poh, H. Sun, X. Wu, Y. Feng, J. Lin, Growth of Si nanowires by thermal evaporation. Nanotechnology **16**, 417 (2005)
25. S.K. Srivastava, P.K. Singh, V.N. Singh, K.N. Sood, D. Haranath, V. Kumar, Large-scale synthesis, characterization and photoluminescence properties of amorphous silica nanowires by thermal evaporation of silicon monoxide. Physica E **41**, 1545 (2009)
26. S.D. Hutagalung, K.A. Yaacob, A.F.A. Aziz, Oxide assisted growth of silicon nanowires by carbothermal evaporation. Appl. Surf. Sci. **254**, 633 (2007)
27. J.D. Holmes, K.P. Johnston, R.C. Doty, B.A. Korgel, Control of thickness and orientation of solution-grown silicon nanowires. Science **287**, 1471 (2000)
28. J. Mart, R. Garcia, Silicon nanowire circuits fabricated by AFM oxidation nanolithography. Nanotechnology **21**, 245301 (2010)
29. Y.Q. Fu, A. Colli, A. Fasoli, J.K. Luo, A.J. Flewitt, A.C. Ferrari, W.I. Milne, Deep reactive ion etching as a tool for nanostructure fabrication. J. Vac. Sci. Technol. B **27**, 1520 (2009)
30. K.Q. Peng, Y.J. Yan, S.P. Gao, J. Zhu, Synthesis of large-area silicon nanowire arrays via self-assembly nanoelectrochemistry. Adv. Matter. **14**, 1164 (2002)
31. K.Q. Peng, Y.J. Yan, S.P. Gao, J. Zhu, Dendrite assisted growth of silicon nanowires in electroless metal deposition. Adv. Funct. Mater. **13**, 127 (2003)
32. T. Qiu, X.L. Wu, G.G. Siu, P.K. Chu, Intergrowth mechanism of silicon nanowires and silver dendrites. J. Electron. Mater. **35**, 1879 (2006)

33. D. Kumar, S.K. Srivastava, P.K. Singh, K.N. Sood, V.N. Singh, N. Dilawar, M. Husain, Room temperature growth of wafer-scale silicon nanowire arrays and their Raman characteristics. J. Nanopart. Res. **12**, 2267 (2010)
34. K. Peng, M. Zhang, A. Lu, N.-B. Wong, R. Zhang, S.-T. Lee, Ordered silicon nanowire arrays via nanosphere lithography and metal induced etching. Appl. Phys. Lett. **90**, 163123 (2007)
35. Z. Huang, N. Geyer, P. Werner, J.D. Boor, U. Gösele, Metal-assisted chemical etching of silicon: a review. Adv. Mater. **23**, 285 (2011)
36. K. Peng, J. Zhu, Simultaneous gold deposition and formation of silicon nanowire arrays. J. Electroanal. Chem. **558**, 35 (2003)
37. K. Peng, J. Zhu, Morphological selection of electroless metal deposits on silicon in aqueous fluoride solution. Electrochim. Acta **49**, 2563 (2004)
38. K. Peng, Y. Wu, H. Fang, X. Zhong, Y. Xu, J. Zhu, Uniform axial orientation alignment of one dimensional single crystal silicon nanostructure arrays. Angew. Chem. Int. Ed. **44**, 2737 (2005)
39. K. Peng, J. Hu, Y. Yan, Y. Wu, H. Fang, Y. Xu, S.T. Lee, J. Zhu, Fabrication of single-crystalline silicon nanowires by scratching a silicon surface with catalytic metal particles. Adv. Funct. Mater. **16**, 387 (2006)
40. H. Fang, Y. Wu, J. Zhao, J. Zhu, Silver catalysis in the fabrication of silicon nanowire arrays. Nanotechnology **17**, 3768 (2006)
41. K. Peng, H. Fang, J. Hu, Y. Wu, J. Zhu, Y. Yan, S.T. Lee, Metal-particle-induced, highly localized site-specific etching of Si and formation of single-crystalline Si nanowires in aqueous fluoride solution. Chem. Eur. J. **12**, 7942 (2006)
42. T. Qui, X.L. Wu, Y.F. Mei, G.J. Wan, P.K. Chu, G.G. Siu, From Si nanotubes to nanowires: synthesis, characterization, and self-assembly. J. Cryst. Growth **277**, 143 (2005)
43. S.K. Srivastava, D. Kumar, S.W. Schmitt, K.N. Sood, S.H. Christiansen, P.K. Singh, Large area fabrication of vertical silicon nanowire arrays by silver-assisted single-step chemical etching and their formation kinetics. Nanotechnology **25**, 175601 (2014)
44. X. Li, P.W. Bohn, Metal-assisted chemical etching in HF/H_2O_2 produces porous silicon. Appl. Phys. Lett. **77**, 2572 (2000)
45. S.K. Gandhi, *VLSI Fabrication Principles: Silicon and Gallium Arsenide*, 2nd edn. (Willey, India, 2009), p. 195
46. W.M. Bullis, Properties of gold in silicon. Solid State Electron. **9**, 143 (1996)
47. C. del Cañizo, I. Tobías, R. Lago-Aurrekoetxea, A. Luque, Influence of depth-inhomogeneity of lifetime in silicon solar cells. J. Electrochem. Soc. **149**, G522 (2002)
48. M.-L. Zhang, K.-Q. Peng, X. Fan, J.-S. Jie, R.-Q. Zhang, S.-T. Lee, N.-B. Wong, Preparation of large-area uniform silicon nanowires arrays through metal-assisted chemical etching. J. Phys. Chem. C **112**, 4444 (2008)
49. Z.P. Huang, T. Shimizu, S. Senz, Z. Zhang, X.X. Zhang, W. Lee, N. Geyer, U. Gosele, Ordered arrays of vertically aligned [110] silicon nanowires by suppressing the crystallographically preferred <100> etching directions. Nano Lett. **9**, 2519 (2009)
50. V.A. Sivakov, G. Brönstrup, B. Pecz, A. Berger, G.Z. Radnoczi, M. Krause, S.H. Christiansen, Realization of vertical and zigzag single crystalline silicon nanowire architectures. J. Phys. Chem. C **114**, 3798 (2010)
51. W. Wang, D. Li, M. Tian, Y.-C. Lee, R. Yang, Wafer-scale fabrication of silicon nanowire arrays with controllable dimensions. Appl. Surf. Sci. **258**, 8649 (2012)
52. H.-C. Chang, K.-Y. Lai, Y.-A. Dai, H.-H. Wang, C.-A. Lin, J.-H. He, Nanowire arrays with controlled structure profiles for maximizing optical collection efficiency. Energy Environ. Sci. **4**, 2863 (2011)
53. B. Ozdemir, M. Kulakci, R. Turan, H.E. Unalan, Effect of electroless etching parameters on the growth and reflection properties of silicon nanowires. Nanotechnology **22**, 155606 (2011)
54. B.S. Kim, S. Shin, S.J. Shin, K.M. Kim, H.H. Cho, Control of superhydrophilicity/superhydrophobicity using silicon nanowires via electroless etching method and fluorine carbon coatings. Langmuir **27**, 10148 (2011)

55. S.L. Cheng, C.H. Chung, H.C. Lee, A study of the synthesis, characterization, and kinetics of vertical silicon nanowire arrays on (001) Si substrates. J. Electrochem. Soc. **155**, D711 (2008)
56. J.E. Allen, E.R. Hemesath, D.E. Perea, J.L. Lensch-Falk, Z.Y. Li, F. Yin, M.H. Gass, P. Wang, A.L. Bleloch, R.E. Palmer, L.J. Lauhon, High-resolution detection of Au catalyst atoms in Si nanowires. Nature Nanotech. **3**, 168 (2008)
57. O. Gunawan, S. Guha, Characteristicsofvapor–liquid–solid grown silicon nanowire solar cells. Sol. Energy Mater. Sol. Cells **93**, 1388 (2009)
58. C.Y. Chen, C.S. Wu, C.J. Chou, T.J. Yen, Morphological control of single crystalline silicon nanowire arrays near room temperature. Adv. Mater. **20**, 3811 (2008)
59. M.A. Green, *Silicon Solar Cells: Advanced Principles and Practice* (Bridge Printery, Sydney, 1995)
60. P.K. Singh, R. Kumar, M. Lal, S.N. Singh, B.K. Das, Effectiveness of silicon in aqueous alkaline solutions. Sol. Energy Mater. Sol. Cells **70**, 103 (2001)
61. J. Zhao, A. Wang, M.A. Green, F. Ferrazza, 19.8% efficient "honeycomb" textured multicrystalline and 24.4% monocrystalline silicon solar cells. Appl. Phys. Lett. **73**, 1991 (1998)
62. S. Wang, X.Z. Yu, H.T. Fan, Simple lithographic approach for subwavelength structure antireflection. Appl. Phys. Lett. **91**, 061105 (2007)
63. M.E. Motamedi, W.H. Southwell, W.J. Gunning, Antireflection surfaces in silicon using binary optics technology, reflection properties of nanostructure-arrayed silicon surfaces. Appl. Opt. **31**, 4371 (1992)
64. K. Hadobas, S. Kirsch, A. Carl, M. Acet, E.F. Wassermann, Reflection properties of nanostructure-arrayed silicon surfaces. Nanotechnology **11**, 161 (2000)
65. C.H. Sun, W.L. Min, N.C. Linn, P. Jiang, Templated fabrication of large area subwavelength antireflection gratings on silicon. Appl. Phys. Lett. **91**, 231105 (2007)
66. Z.N. Yu, H. Gao, W. Wu, H.X. Ge, S.Y. Chou, Fabrication of large area subwavelength antireflection structures on Si using trilayer resist nanoimprint lithography and lift-off. J. Vac. Sci. Technol. B **21**, 2874 (2003)
67. G. Zhang, J. Zhang, G.Y. Xie, Z.F. Liu, H.B. Shao, Cicada wings: a stamp from nature for nanoimprint lithography. Small **2**, 1440 (2006)
68. Y. Kanamori, E. Roy, Y. Chen, Antireflection sub-wavelength gratings fabricated by spin-coating replication. Microelectron. Eng. **78–79**, 287 (2005)
69. P. Lalanne, G.M. Morris, Antireflection behavior of silicon subwavelength periodic structures for visible light. Nanotechnology **8**, 53 (1997)
70. K.Q. Peng, Y. Xu, Y. Yu, Y.J. You, S.T. Lee, J. Zhu, Aligned single-crystalline Si nanowire arrays for photovoltaic application. Small **1**, 1062 (2005)
71. S.K. Srivastava, D. Kumar, P.K. Singh, V. Kumar, Silicon nanowire arrays based "black silicon" solar cells, in *Proceeding of 34th IEEE Photovoltaic Specialists Conference, Philadelphia, USA*, pp. 1851–1856 (2009)
72. S.K. Srivastava, D. Kumar, P.K. Singh, M. Kar, V. Kumar, M. Husain, Excellent antireflection properties of vertical silicon nanowire arrays. Sol. Energy Mater. Sol. Cells **94**, 1506 (2010)
73. Y. Kato, S. Adachi, Synthesis of Si nanowire arrays in AgO/HF solution and their optical and wettability properties. J. Electrochem. Soc. **158**, K157 (2011)
74. R.-C. Wang, C.-Y. Chao, W.-S. Su, Electrochemically controlled fabrication of lightly doped porous Si nanowire arrays with excellent antireflective and self-cleaning properties. Acta Mater. **60**, 2097 (2012)
75. C.-Y. Chen, W.-J. Li, H.-H. Chen, Tailoring broadband antireflection on a silicon surface through two-step silver-assisted chemical etching. ChemPhysChem **13**, 1415 (2012)
76. N. Nafie, M.A. Lachiheb, M. Bouaicha, Effect of etching time on morphological, optical, and electronic properties of silicon nanowires. Nanoscale Res. Lett. **7**, 393 (2012)

77. A. Yamaguchi, T. Shimizu, Y. Morosawa, K. Takase, T.-L. Chen, S.-M. Lu, H.-C. Chien, S. Shingubara, Morphology dependence of optical reflectance properties for a high-density array of silicon nanowires. Jpn. J. Appl. Phys. **53**, 06JF10 (2014)
78. M.A. Lachiheb, M.A. Zrir, N. Nafie, O. Abbes, J. Yakoubi, M. Bouaïcha, Investigation of the effectiveness of SiNWs used as an antireflective layer in solar cells. Sol. Energy **110**, 673 (2014)
79. Z. Zuo, K. Zhu, G. Cui, W. Huang, J. Qu, Y. Shi, Y. Liu, G. Ji, Improved antireflection properties and optimized structure for passivation of well-separated, vertical silicon nanowire arrays for solar cell applications. Sol. Energy Mater. Sol. Cells **125**, 248 (2014)
80. Y.-J. Hung, S.-L. Lee, K.-C. Wu, Y. Tai, Y.-T. Pan, Antireflective silicon surface with vertical aligned silicon nanowires realized by simple wet chemical etching processes. Opt. Express **19**, 15792 (2011)
81. Y.-J. Hung, K.-C. Wu, S.-L. Lee, Y.-T. Pan, Realization and characterization of aligned silicon nanowire array with thin silver film. IEEE Photon. J. **3**, 617 (2011)
82. T.-H. Pei, S. Thiyagu, Z. Pei, Ultra high-density silicon nanowires for extremely low reflection in visible regime. Appl. Phys. Lett. **99**, 153108 (2011)
83. J.-Y. Jung, Z. Guo, S.-W. Jee, H.-D. Um, K.-T. Park, J.-H. Lee, A strong antireflective solar cell prepared by tapering silicon nanowires. Opt. Express **18**, A286 (2010)
84. Y. Kanamori, M. Sasaki, K. Hane, Broadband antireflection gratings fabricated upon silicon substrates. Opt. Lett. **24**, 1422 (1999)
85. H. Sai, H. Fujii, K. Arafune, Y. Ohshita, M. Yamaguchi, Y. Kanamori, H. Yugami, Antireflective sub-wavelength structures on crystalline Si fabricated using directly formed anodic porous alumina masks. Appl. Phys. Lett. **88**, 201116 (2006)
86. C.C. Striemer, P.M. Fauchet, Dynamic etching of silicon for broadband antireflection applications. Appl. Phys. Lett. **81**, 2980 (2002)
87. S. Koynov, M.S. Brandt, M. Srutzmann, Black non-reflecting silicon surfaces for solar cells. Appl. Phys. Lett. **88**, 203107 (2007)
88. M.L. Kuo, D.J. Poxon, Y.S. Kim, F.W. Mont, J.K. Kim, E.F. Schubert, S.Y. Lin, Realization of a near-perfect antireflection coating for silicon solar energy utilization. Opt. Lett. **33**, 2527 (2008)
89. L. Tian, K.B. Ram, I. Ahmad, L. Menon, M. Holtz, Optical properties of Si nanopore arrays. J. Appl. Phys. **97**, 026101 (2005)
90. L.L. Ma, Y.C. Zhou, N. Jiang, X. Lu, J. Shao, W. Lu, J. Ge, X.M. Ding, X.Y. Hou, Wide-band "black silicon" based on porous silicon. Appl. Phys. Lett. **88**, 171907 (2006)
91. J.S. Li, H.Y. Yu, S.M. Wong, G. Zhang, X.W. Sun, P.G.Q. Lo, D.L. Kwong, Si nanopillar array optimization on Si thin films for solar energy harvesting. Appl. Phys. Lett. **95**, 033102 (2009)
92. C.X. Lin, M.L. Povinelli, Optical absorption enhancement in silicon nanowire arrays with a large lattice constant for photovoltaic applications. Opt. Express **17**, 19371 (2009)
93. R.A. Street, W.S. Wong, C. Paulson, Analytic model for diffuse reflectivity of silicon nanowire mats. Nano Lett. **9**, 3494 (2009)
94. H. Bao, X. Ruan, Optical absorption enhancement in disordered vertical silicon nanowire arrays for photovoltaic applications. Opt. Lett. **35**, 3378 (2010)
95. W.Q. Xie, J.I. Oh, W.Z. Shen, Realization of effective light trapping and omnidirectional antireflection in smooth surface silicon nanowire arrays. Nanotechnology **22**, 065704 (2011)
96. J. Nelson, *Physics of Solar Cells* (Imperial College Press, London, 2003)
97. M.A. Green, K. Emery, Y. Hishikawa, W. Warta, Solar cell efficiency tables (version 37). Prog. Photovolt. Res. Appl. **19**, 84 (2011)
98. C.A. Wolden, J. Kurtin, J.B. Baxter, I. Repins, S.E. Shaheen, J.T. Torvik, A.A. Rockett, J.M. Fthenakis, E.S. Aydil, Photovoltaic manufacturing: present status and future prospects. J. Vac. Sci. Technol. A **29**, 030801 (2011)
99. M.D. Kelzenberg, M.A. Filler, B.M. Kayes, M.C. Putnam, D.B. Turner-Evans, N.S. Lewis, H.A. Atwater, Single-nanowire Si solar cells, in *Proceeding of 33rd IEEE Photovoltaic Specialists Conference*, pp. 1–6 (2008)

100. M.D. Kelzenberg, D.B. Turner-Evans, M.C. Putnam, S.W. Boettcher, R.M. Briggs, J.Y. Baek, N.S. Lewis, H.A. Atwater, High-performance Si microwire photovoltaics. Energy Environ. Sci. **4**, 866 (2011)
101. H. Fang, L. Xudong, S. Shuang, X. Ying, Z. Jing, Fabrication of slantingly-aligned silicon nanowire arrays for solar cell applications. Nanotechnology **19**, 255703 (2008)
102. D. Kumar, S.K. Srivastava, P.K. Singh, M. Husain, V. Kumar, Fabrication of silicon nanowire arrays based solar cell with improved performance. Sol. Energy Mater. Sol. Cells **95**, 215 (2011)
103. H.-D. Um, K.-T. Park, J.-Y. Jung, X. Li, K. Zhou, S.W. Jee, J.-H. Lee, Incorporation of a self-aligned selective emitter to realize highly efficient (12.8 %) Si nanowire solar cells. Nanoscale **6**, 5193 (2014)
104. C. Chen, R. Jia, H. Li, Y. Meng, X. Liu, T. Ye, S. Kasai, H. Tamotsu, N. Wu, S. Wang, J. Chu, Electrode-contact enhancement in silicon nanowire-array-textured solar cells. Appl. Phys. Lett. **98**, 143108 (2011)
105. H.F. Li, R. Jia, C. Chen, Z. Xing, W. Ding, Y. Meng, D. Wu, X. Liu, T. Ye, Influence of nanowires length on performance of crystalline silicon solar cell. Appl. Phys. Lett. **98**, 151116 (2011)
106. J.-Y. Jung, Z. Guo, S.-W. Jee, H.-D. Um, K.-T. Park, M.S. Hyun, J.M. Yang, J.-H. Lee, A wafer-scale Si wire solar cell using radial and bulk p-n junctions. Nanotechnology **21**, 445303 (2010)
107. C. Chen, R. Jia, H. Yue, H. Li, X. Liu, D. Wu, W. Ding, T. Ye, S. Kasai, H. Tamotsu, J. Chu, S. Wang, Silicon nanowire-array-textured solar cells for photovoltaic application. J. Appl. Phys. **108**, 094318 (2010)
108. S.H. Baek, H.S. Jang, J.H. Kim, Characterization of optical absorption and photovoltaic properties of silicon wire solar cells with different aspect ratio. Current Appl. Phys. **11**, S30 (2011)
109. C. Chen, R. Jia, H.H. Yue, H.F. Li, X.Y. Liu, T.C. Ye, K. Seiya, T. Hashizume, S.L. Wang, J.H. Chu, B.S. Xu, Silicon nanostructure solar cells with excellent photon harvesting. J. Vac. Sci. Technol. B **29**, 021014 (2011)
110. B.-R. Huang, Y.-K. Yang, T.-C. Lin, W.-L. Yang, A simple and low-cost technique for silicon nanowire arrays based solar cells. Sol. Energy Mater. Sol. Cells **98**, 357 (2012)
111. M. Kulakci, F. Es, B. Ozdemir, H.E. Unalan, R. Turan, Application of Si nanowires fabricated by metal-assisted etching to crystalline Si solar cells. IEEE J. Photovoltaics **3**, 548 (2013)
112. X.X. Lin, X. Hua, Z.G. Huang, Z. Shen, Realization of high performance silicon nanowire based solar cells with large size. Nanotechnology **24**, 235402 (2013)
113. Z. Zhao, B. Zhang, P. Li, W. Guo, A. Liu, Effective passivation of large area black silicon solar cells by SiO_2/SiN_x:H stacks. Int. J. Photoenergy **2014**, 6 pp, Article ID 683654 (2014). http://dx.doi.org/10.1155/2014/683654
114. W.-C. Wang, C.-W. Lin, H.-J. Chen, C.-W. Chang, J.-J. Huang, M.-J. Yang, B. Tjahjono, J.-J. Huang, W.C. Hsu, M.-J. Chen, Surface passivation of efficient nanotextured black silicon solar cells using thermal atomic layer deposition. ACS Appl. Mater. Interfaces **5**, 9752 (2013)
115. J. Oh, H.-C. Yuan, H.M. Branz, An 18.2%-efficient black-silicon solar cell achieved through control of carrier recombination in nanostructures. Nature Nanotech. **7**, 743 (2012)
116. G. Jia, M. Steglich, I. Sil, F. Falk, Core-shell heterojunction solar cells on silicon nanowire arrays. Sol. Energy Mater. Sol. Cells **96**, 226 (2012)
117. G. Jia, A. Gwalik, J. Bergmann, B. Eisenhawer, S. Schonherr, G. Andra, F. Falk, Silicon nanowire solar cells with radial p-n heterojunction on crystalline silicon thin films: light trapping properties. IEEE J. Photovoltaics **4**, 28 (2014)
118. G. Jia, G. Andrä, A. Gawlik, S. Schönherr, J. Plentz, B. Eisenhawer, T. Pliewischkies, A. Dellith, F. Falk, Nanotechnology enhanced solar cells prepared on laser-crystallized polycrystalline thin films (<10 μm). Sol. Energy Mater. Sol. Cells **126**, 62 (2014)

Chapter 10
Recent Trends of Gelatin Nanoparticles in Biomedical Applications

Prem Prakash Sharma, Anshu Sharma and Pratima R. Solanki

Abstract Today's scenario is nanoscience and nanotechnology performing significant contributions of protein nanoparticles (PNPs) for delivery of various life savings biomolecules. Among the various PNPs, gelatin nanoparticles (GNPs) exhibit good capabilities for biomedical application because of their biodegradability, biocompatibility, non-toxicities and its cost-effectiveness. In this chapter, we have discussed different methods for GNPs preparation and their functionalization for target delivery of drugs and genes for a range of diseases like cancer, malarial and infectious. We also discussed ocular, pulmonary drugs as well as nutraceutical, proteins and peptides delivery. We also come across their application in tissue engineering.

Keywords Gelatin nanoparticles · Drug delivery · Diseases · Tissue engineering

10.1 Introduction

Proteins are a class of natural biomacromolecule of living beings that have unique advantages which makes them potential candidate for both biomedical and material sciences applications. Protein attracts researcher for nanoparticles (NPs) preparation owing to their amphiphilic nature and availability of surface functional group, to enhance their high binding affinity towards other molecules which allow attachment of drugs and targeting ligands [1]. The protein nanoparticles (PNPs) have many advantages such as biocompatibility, biodegradability, high loading capacity, simplicity of manufacture and surrounding protein environment, over synthetic polymers. PNPs have been synthesized from water-soluble proteins like bovine

P.P. Sharma · A. Sharma · P.R. Solanki (✉)
Special Centre for Nano Sciences, Jawaharlal Nehru University,
New Delhi 110067, India
e-mail: pratimarsolanki@gmail.com; partima@mail.jnu.ac.in

A. Sharma
School of Physical Sciences, Jawaharlal Nehru University, New Delhi 110067, India

M. Husain and Z.H. Khan (eds.), *Advances in Nanomaterials*,
Advanced Structured Materials 79, DOI 10.1007/978-81-322-2668-0_10

serum albumin (BSA), human serum albumin (HSA) and insoluble proteins like zein and gliadin. It has been observed that very few review articles reported on PNPs and most of them are focused on the preparation and characterization of NPs derived from albumin, gelatin, and gliadin [2]. Thus, in this chapter, we have discussed gelatin nanoparticles (GNPs) synthesis and their biomedical applications including drug, gene delivery and tissue engineering.

Gelatin is a well known natural biopolymer having long history, derived from collagen by irreversible acidic or basic hydrolysis and is being use as plasma expander as its is biocompatible [3]. The gelatin term was originated from Latin called gelatus means stiff or frozen. This biopolymer is a brittle solid when dry, without flavor, translucent and colorless biomaterial substance. In ancient years it was using as glue and now a days it still under consideration as pharmaceutical and food products including gummy candies as well as other products such as marshmallows, gelatin dessert etc. [4, 5]. US Food and Drug Administration (FDA) categorized gelatin into "Generally Recognized as Safe" (GRAS) due to natural abundance biocompatibility and biodegradability in physiological environment [6].

Recently, GNPs have received a growing attention for biomedical application because it's potential for efficient delivery and release of drugs in controlled manner [7]. They are biodegradable, nontoxic, bioactive and economical which makes GNPs compatible candidate for delivery of many therapeutic agents. The crosslinking between anionic and cationic group has significant influential role on GNPs swelling and thermal properties [8, 9]. Thus GNPs have been used in different form such as micro or nanoparticles and hydrogels [10, 11]. It has been observed that most of studies reported in literature are based on drug delivery and tissue engineering (Fig. 10.1).

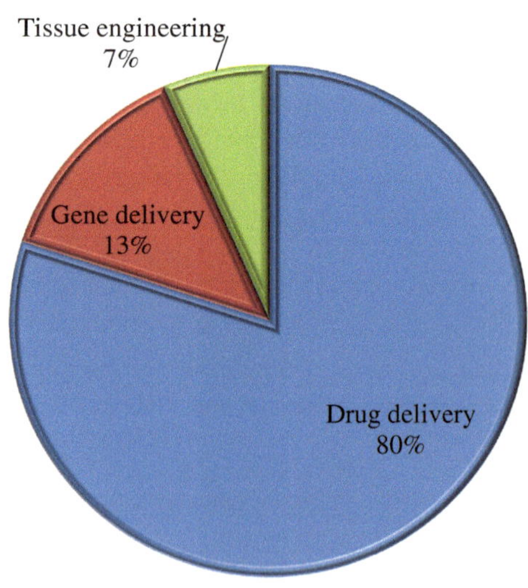

Fig. 10.1 GNPs applications in various biomedical fields

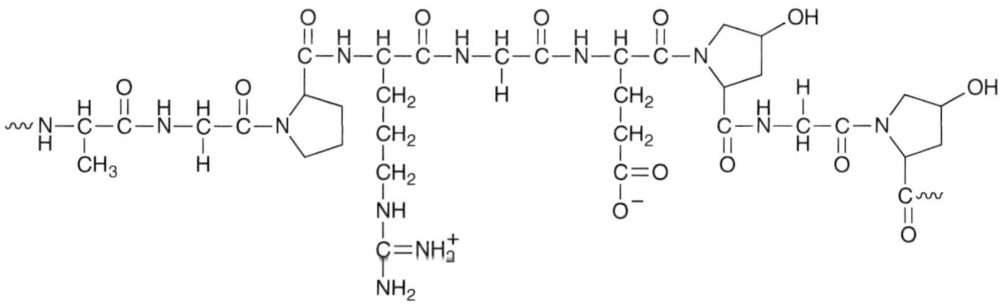

Fig. 10.2 Molecular structure of gelatin

10.2 Chemical Structure

Gelatin is a polyampholyte derived from collagen by the process of acid or alkaline hydrolysis. Gelatin is a heterogeneous fusion of single- or multistranded polypeptides containing amino acid residues range from hundred to few thousand (300–4000). It has both positive and negative charged residues as well as hydrophobic groups (approximate ratio 1:1:1) that drive this polypeptide especially for biomedical applications. The gelatin molecule contains ~25 % charged amino acids (both positive and negative) residues. Positive charged amino acids includes lysine (K) and arginine (R) while negative charged contains glutamate (E) and aspartate (D). Nearby 11 % of gelatin chain contains hydrophobic amino acids residues such as leucine (L), isoleucine (I), methionine (M) and valine (V). The remaining gelatin chain composed of glycine (~33 %) and proline (P) or hydroxyproline (O) (~33 %). Commercially, the cationic gelatin type A is prepared from pig skin type 1 collagen by an acid hydrolysis. On other side anionic gelatin type B prepared from bovine collagen by alkaline hydrolysis. Both commercial proteins are synthesized without any additional functionalization [8, 9, 12]. Figure 10.2 shows the molecular structure and amino acid composition of gelatin.

10.3 Preparation of GNPs

Several methods have been used to synthesize GNPs and its derivatives as NPs including desolvation, coacervation, solvent evaporation method, spontaneous emulsification/solvent diffusion method, nanoprecipitation method, salting out method etc.

10.3.1 Desolvation Method

In desolvation method, NPs are synthesized by thermodynamically driven self-assembly process. The size of NPs influenced by various factors such as protein content, ionic strength, pH, agitation speed, concentration of crosslinking agent and desolvating agents etc. The aqueous disolvation of protein and polysaccharide are driven by the 3 factors such as change in pH, temperature and counter ions concentration. Even these factors may also influence cross linking to the desolvation step. Protein desolvation contains three steps; dissolution, aggregation and disaggregation.

For GNPs, gelatin is initially solubilized in distilled water in the presence of invariable heating temperature range. Then acetone (DMK) or ethanol (EtOH) was added as desolvating agents to turn gelatin solution into high molecular weight (HMW) gelatin precipitate. After discarding supernatant, the HMW gelatin was further dissolved into distilled water at uniform heating. The gelatin solution pH should be maintained in the range of from 2.5 to 12. Then a drop-wise desolvating agent was added for NPs formation. Glutaraldehyde was also added in the last as cross-linking agents which affects the particle size [13].

10.3.2 Coacervation

Coacervation is well known for a liquid-liquid phase separation method where homogenous charged macromolecules solution undergoes phase separation where bottom region is polymer rich and a transparent supernatant above. In this method, both two incompatible and immiscible liquid phases are in equilibrium. In simple polyelectrolyte coacervation, adding up salt or alcohol normally promotes coacervation and NPs present in supernatant in dilute liquid phase [14].

10.3.3 Solvent Evaporation Method

In this method, initially polymer is solubilised in organic solution [dichloromethane (DCM), chloroform (CHF) or ethyl acetate (EA)] then drug is dissolved or dispersed within it followed by polymer solution emulsification. During emulsification surfactant/emulsifying agents such as gelatin and polyvinyl alcohol are used to form oil/water (o/w) emulsion. Once the stable emulsion formed it undergoes solvent evaporation process under applied temperature and pressure Fig. 10.3 [15]. The stirring speed, stabilizers and polymers concentration are the influencing factors for NPs size determination.

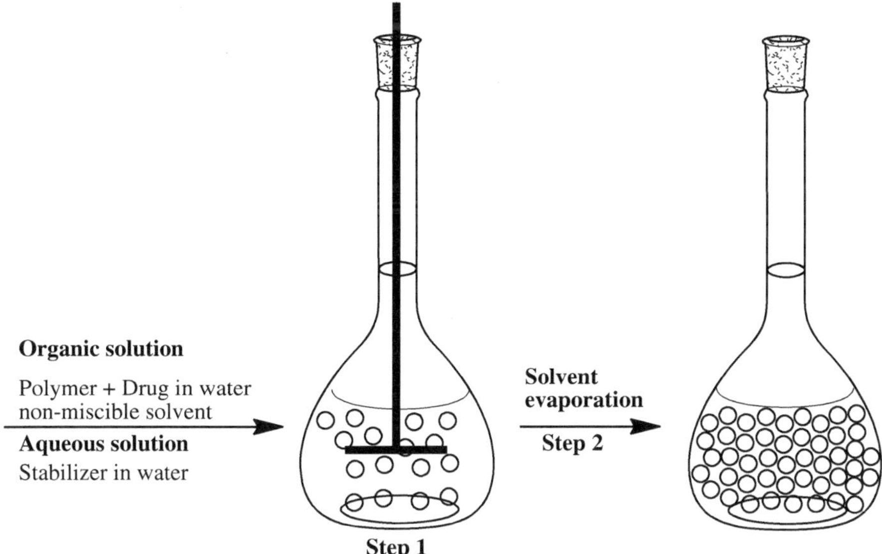

Fig. 10.3 Schematic representation of the emulsification evaporation technique

10.3.4 Spontaneous Emulsification/Solvent Diffusion Method

This is also known as modified solvent diffusion method where both water-miscible solvents [DMK or methanol (MeOH)] and water insoluble organic solvent (DCM or CHF) are considered as oil phase. During the process an interfacial turbulence is developed between two phases due to spontaneous diffusion of solvents that leads to the small particle formation. Smaller particle can be prepared by increasing the water-soluble solvent concentration [9].

10.3.5 Nanoprecipitation Method

Nanoprecipitation method can be used for both hydrophobic and hydrophilic drug entrapment. Polar, water-miscible solvent [DMK, acetonitrile (ACN), ethanol (EtOH), or MeOH] are used for dissolving polymers and drugs then the solution poured into an aqueous solution with surfactant in dropwise manner. Finally NPs are formed immediately by rapid solvent diffusion along with solvent removal under reduced pressure [9].

10.3.6 Salting Out Method

In this method, the polymer is solubilised in the water-miscible organic solvent, like acetone or tetrahydrofuran. Then solution is emulsified in an aqueous phase, under applied strong mechanical shear stress. The organic phase insoluble emulsifier and salts (magnesium chloride hexahydrate ($MgCl_2.6H_2O$) and magnesium acetate tetrahydrate ($(CH_3COO)_2Mg.4H_2O$ are presents in aqueous phase. This method is somewhat different from emulsion diffusion method because of no solvent diffusion due to salts occurrence. The ionic strength of o/w emulsion reduces upon addition of pure water that induces nanosphere formation by migration of the water-soluble organic solvent to the aqueous phase. Finally NPs are purified by cross flow filtration or centrifugation to remove the salting out agent [9].

10.4 Crosslinking with GNPs

Crosslinking is requiring for modification of GNPs which provide stability, shape with improved circulation time on in vivo administration over bared (uncrosslinked nanoparticles). Because bared NPs can easily undergo aggregate with disturbing its 3D pattern upon ageing. Thus different types of crosslinker have been utilized for functionalization of GNPs: (i) **Glutaraldehyde** (GA) is an effective crosslinker which keep GNPs stable for more than 10 months at 2–8 °C. GA induces crosslinks during polymer formation by joining lysine or hydroxylysine residues via their free amino groups. GA has no side effects because of it completely removed from GNPs crosslinking. **D,L-Glyceraldehyde** is a non-toxic cross linking agent as compared to GA and well known material for pharmaceutical applications. (ii) **Genipin** is also used as a crosslinker which is natural extracted from gardenia fruit. Genipin is approximately 10,000 times less cytotoxic than GA but requires longer crosslinking time. Both amino groups of lysine residue bind with one molecule of genipin. Genipin-crosslinked recombinant human gelatin (rHG) nanoparticles are significantly non-toxic for the host cell once efficiently internalized. (iii) **Carbodiimide and N-hydroxysuccinimide (CDI/NHS)** is a mixture carbodiimide and N-hydroxysuccinimide (CDI/NHS) which is water soluble and used as a non-toxic crosslinker for GNPs. CDI/NHS cross-linking has more advantage over GA in terms of smoother and homogenous NPs with small size and narrow size distribution. It also provides higher drug entrapment and loading efficiencies than GA crosslinked GNPs. (iv) **Microbial transglutaminase (MTG)** is an acyltransferase which forms intra- and intermolecular isopeptide bonds by crosslinking the ε-amino groups of lysine to the side chain amide group of glutamine that results in releasing of one molecule of NH_3 per crosslink. Crosslinking reactions should be at room temperature, neutral pH, ion-free solvent and at last terminated after 12 h. Particles below 250 nm size and narrow size distribution stable for a short duration. Double-crosslinked gelatin–chitosan NPs are also reported where both polymers are ionically-crosslinked with sodium sulfate [16]. PEGylated GNPs conjugate were prepared in the presence of glyoxal cross-linker. These conjugate was studied for biodistribution in subcutaneous Lewis lung carcinoma (LLC)-bearing female C57BL/6J mice. It was found that PEGylated GNPs conjugate was occurred in the blood pool for long lasting due to the steric repulsion effect of PEG chains as compared to the GNPs alone [17].

10.5 Characterization

Characterization of bare and drug loaded GNPs a drug molecule becomes extremely important which can be characterized using different techniques. Scanning electron microscopy (SEM), atomic force microscopy and high resolution transmission electron microscope (HRTEM) can be used to monitor shape and size of loaded and unloaded GNPs. Figure 10.4 exhibits spherical shape GNPs. The particle size analyzer and zeta potential can be utilized to study the particle size and surface charge, respectively. The conformation of functionalization of bare GNPs can be studied by using Fourier transform infrared spectrophotometer and UV-Vis spectrophotometer.

10.6 Applications of GNPs

GNPs have been used for various applications such as drug deliver, gene delivery, protein and vaccine, gene, ocular drug delivery etc.

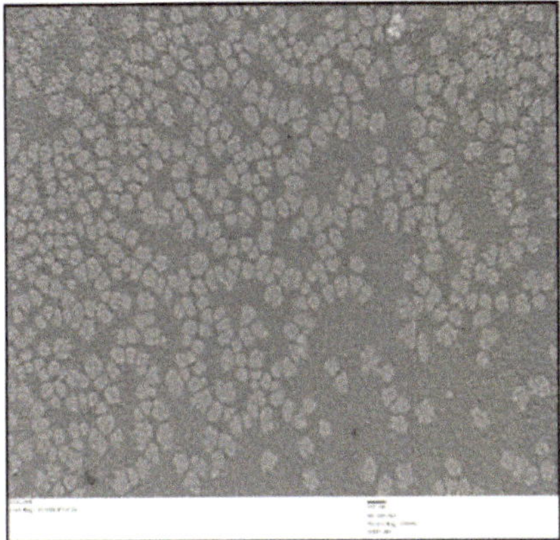

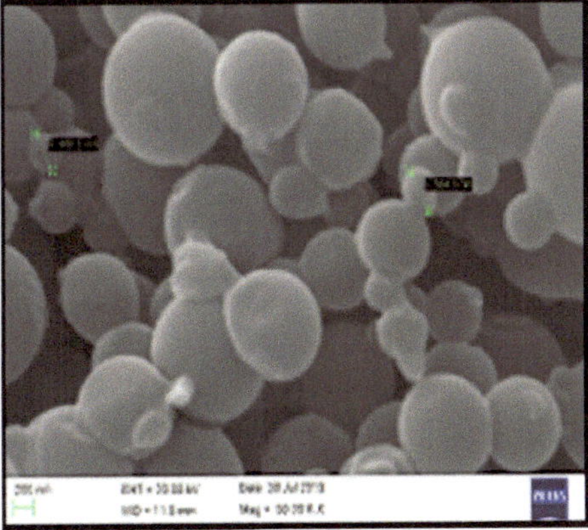

Fig. 10.4 TEM image (*left side*) and SEM image (*right side*) the GNPs synthesized by reverse method

10.6.1 GNPs for Drug Delivery for Cancer and Infectious Diseases

GNPs are promising vector to deliver drug and gene due to their potential possibility for chemical modification and cross-linking, which can increase anti-tumor efficacy. Even controlled and target delivery of drugs to the tumor site reduces unwanted toxicity. The drug release from GNPs mostly depends on the three important phenomenon including desorption, biodegradation and diffusion. GNPs are simple, reproducible, cost effective, very low cytotoxic and can be produced in large scale. It is also reported that the GNPs passive target delivery of drug through EPR effects, where nanoparticles stay at the tumor region for enough longer that leads to the accumulation of drug at the target site with high concentration even with low doses and low frequency. Lu et al. loaded paclitaxel into GNPs and administered intravesically into dogs. They found that that paclitaxel-GNPs formulation has more than 2.5 times concentration into bladder tumor compared to commercial cremophor/EtOH formulation. The gelatin-drug nanoparticles remarkably lower cytotoxicity as well as impeded tumor growth and defeated pulmonary metastasis as compared with free drug and the superior efficacy of drug loaded in both in vitro and in vivo system [18]. The drug loaded GNPs shows higher cytotoxicity against different cancer cell lines compared to free drug which might be explained by higher endocytotic uptake of GNPs in cancer cells. Even noscapine-loaded GNPs has lower IC50 i.e., 26.3 μM on human breast cancer cell line (MCF7) than free noscapine (40.5 μM) [19].

GNPs and its hybrid with other NPs have been extensively used for the delivery of various anti-cancer drugs both hydrophilic and hydrophobic including doxorubicin (DOX) [20], methotrexate [21], cytarabine [22], camptothecin [23], 17-AAG [24], curcumin [25], cycloheximide [26], resveratrol [27], paclitaxel [18], cisplatin [28] and noscapine [19]. Harsha entrapped amoxicillin antibiotic which known for β-lactum semi synthetic penicillin into GNPs. They found that this nanoarchitecture is stabled for 12 months at 25 ± 2 °C and 60 ± 5 % relative humidity and amoxicillin was released in sustained manner for up to 12 h. The original drug was released by 90 % in first 30 min while only 15.9 % was released for this new formulation [29]. Kuntworbe and Al-Kassas synthesized cryptolepine hydrochloride-loaded GNPs a novel approach for malaria treatment. Their formulation was less haemolytic that pure compound. They found sustained released drug from all formulation types for more than 192 h period which is important in terms of increased contact time between drug and infected RBCs. Nanosuspension and freeze dried samples were found to be stable at 4 and 25 °C, respectively. Over one year period, the former was less stable at room temperature [30]. The matrix metalloproteinase like MMP-2 and MMP-9 have GNPs hydrolysis properties or even some bacteria also secretes gelatinase. Li et al., worked in the development of gelatinase responsive antibiotic delivery system (Van⊂SGNPs@RBC) which could deliver drug to the bacterial infection site i.e. "on-demand" antibiotic delivery. The core, crosslinked supramolecular GNPss (SGNPs) where its surface was decorated by RBC membranes (SGNPs@RBC) along with vancomycin (Van) was encapsulated within it. The shell coated with uniform red blood cell membranes represents whole structure as part of host body that reduces the clearance of the antibiotic delivery by the immune system simultaneously absorbs the exotoxin secreted during bacterial infection [31].

10.6.2 GNPs for Protein and Vaccine Delivery

Since the late 20th century numerous therapeutically important proteins and peptides have emerged in the pharmaceutical market. The biocompatibility as well as biological activity of many proteins and peptides entities makes them competent therapeutic agents. Enzymes are the best group of protein drugs delivery for example lysosomal enzymes are require for a patient who suffer from deficiency of some lysosomal enzymes (so-called storage diseases), that can be treated with exogenous enzymes therapeutics. Moreover, some other enzymes required for antitumor activity by disrupting tumor requiring amino acids which can reduce the further growth of tumor cells.

BSA also used as a model protein for delivery and other proteins including insulin, alkaline phosphatase (ALP), tissue-type plasminogen activator (t-PA), angiogenic basic fibroblast growth factor (bFGF) and bone morphogenetic protein-2 (BMP-2), reported for effectively encapsulated into GNPs with the biological activity and used for delivery of protein drug in vivo system. Azimi et al. [32] reported BSA loaded in GNPs with 87.4 % efficiency and found that its released was followed biphasic release modulation where initially released by rapid then slower. Kaintura et al. [7] reported hydrophilic GNPs for delivery of BSA and HSA and obtained encapsulation efficiency as 90 % (w/w) of BSA and 80 % (w/w) of HAS. It was found that murine bone marrow dendritic cells uptake antigen loaded GNPs provide specific immunoadjuvant reported by Coester et al. [33]. Toxoid loaded GNPs subcutaneously injected in BALB/c mice where the formulation shows significant immune response over IgG [34]. Chen et al., developed gelatin nanocomposites (DG-NPs) based on glycidyl methacrylate derivatized dextrans (DEX-GMA) with particle size 20–100 nm. DG-NPs possessed good swelling and degradation properties against dextranase environment. They also entrapped bone morphogenetic protein (BMP) and noticed its releasing from DG-NPs in sustained manner where 90 % BMP was released for more than 12 days without any positive cytotoxic results [35].

10.6.3 GNPs for Gene Delivery

Gene therapy is a novel approach to treat disease at genetic level. In gene therapy, target gene expression is developed in administering or injecting foreign genes. GNPs are considered as a non-viral gene delivery vector. These GNPs makes conjugate with other compounds that stimulate receptor-mediated endocytosis. GNPs have multiple plasmids encapsulation efficiency as well as DNA bioactivity could be improved by preventing digestion from nucleases and by using long-circulating PEGylated NPs. Kaul et al., studied tumor-targeted gene delivery, capability of PEGylated GNPs. They encapsulated plasmid DNA encoding for β-galactosidase (pCMV-β) in PEGylated GNPs and transfected Lewis lung carcinoma (LLC) cells with pCMV-β. The pCMV-β expression was measured quantitatively in LCC cells using an enzymatic assay for the conversion of o-nitrophenyl-β-D-galactopyranoside to o-nitrophenol. Additionally they also administered plasmid DNA-encapsulated gelatin and nanoparticle intravenously and intratumorally to LLC-bearing female C57BL/6J mice and noticed that this systems were significantly superior in transfecting tumor mass but intratumor administration had better transfection than intravenous administration [36].

Kaul and Amiji [37] initially developed type B GNPs as noncondensing gene delivery systems. DNA condensing positive charged lipids and polymers are opposed by negative charged type B gelatin (encapsulated reporter and therapeutic nucleic acid) at neutral pH. These plasmid DNA (pDNA) encapsulated matrix protects DNA in the systemic circulation and upon cellular transport. The released pDNA should have super coiled structure at the nuclear membrane surface for its efficient uptake and transfection by host cell [38]. The surface of GNPs can be modified with quaternary amine (e.g. cholamine) to increase ionic interaction for negative charged nucleic acids [39].

Amiji co-workers also studied non-condensing type B GNPs potential for systemic and oral gene therapy. GNPs loaded tetramethylrhodamine-labeled dextran (TMR–dextran) as a model hydrophilic drug used to studies cell trafficking in BT-20 cells where the formulation is taken up by endocytosis which ultimately reach to the perinuclear area in the cytoplasm [37, 40]. Kommareddy and Amiji synthesized PEG modified thiolated GNPs (PEG-SHGel) to deliver plasmid DNA driven by intracellular glutathione concentration (Fig. 10.1). The glutathione reduces disulphide crosslinks and drive DNA release. They observed that PEG-SHGel (40–45 %) released greater fraction of DNA than unmodified gelatin particle (25–30 %). The encapsulated DNA stability was intact confirmed by agarose gel electrophoresis. Finally, both PEG-Gel and PEG-SHGel nanoparticles shown highest transfection efficiency of plasmid *in* murine fibroblast cells (NIH3T3) (in vitro) [41].

Magadala and Amiji synthesized gelatin-based engineered nanocarrier systems (GENS) where epidermal growth factor receptor—targeting peptides are placed on nanocarrier surface by using a heterobifunctional poly(ethylene glycol) spacer. They encapsulated plasmid DNA code for enhanced green fluorescent protein into surface-modified GENS. It also noted that when the pDNA was loaded with EGFR cells enhance the both quantitative and qualitative transgene expression [38].

10.6.4 GNPs for Ocular Drug Delivery

The ocular drugs delivery remains a great challenge due to several barriers and hurdles faced by this kind of administration. The GNPs were used successful as ocular delivery as it derived from collagen which is present in the stroma of eye cornea. The cationic colloidal nanoparticles are expected to penetrate ocular tissues more efficiently than anionic carriers because of negative surface charge of cornea and conjunctiva. GNPs used for encapsulate hydrophilic (pilocarpine HCl) and hydrophobic (hydrocortisone) drugs for topical ophthalmic use [42]. They have reported that the transport of NPs form the cornea cell layers (first) depends on their nanometric size and reported that sustained release of both drugs from GNPs was observed that about zero order release kinetics. The dry eye causes multifactorial decencies related to ocular surface and tear film. The tear film responds to very delicate neuroregulatory phenomena and hormonal stimulation. The dry eye syndrome is associated with many factors such as lacrimal gland secretion blockage, decreased tear production and hyperosmolar tears. Among all of the factors related to tear film stability under normal circumstances, a type of mucin called MUC5AC have special interest because its low level are associated with conditions like dry eye syndrome. Recently, the elevation of MUC5AC level is achieved for the first time using a new plasmid and cationized gelatin NPs as vehicles. These GNPs and a new plasmid successfully transfected the ocular epithelial cells in vitro and in vivo encoding a modified human MUC5AC mucin protein [43]. Corneal cell lines showed detectable MUC5AC mRNA expression in cells exposed to the NPs. It also observed that the cationic GNPs useful for transfecting the epithelial layer of ocular cells and protect pDNA which increase transfection efficacy and in vitro studies experiment reported that spermine-cationized GNPs is efficient to transfect human corneal epithelial (HCE) cells [44].

10.6.5 GNPs for Pulmonary Drug Delivery

Pulmonary drug delivery is a tremendous scientific and biomedical interest in the health well fare care research area. The lung is capable of absorbing drugs either by local deposition or by systemic delivery. It has been reported that the pulmonary route is a potential non-invasive pathway for systemic and local delivery of therapeutic agents. The high permeability, great absorptive surface area of lungs, ($\sim$70–140 m^2 in adult humans) and good blood supply make lung potential for non-invasive administration of drugs. Pulmonary route has further advantages such as rapid absorption of drug, low enzymatic activity and the capacity for overcoming first-pass metabolism over per oral adsorption [45].

GNPs are under consideration for its promising vector capabilities to transport drugs efficiently to the lung via inhalation. As gelatin itself biocompatible, so that inhalation of GNPs does not cause any lung inflammation. They delivered cisplatin via GNPs as nanocarriers to treat lung cancer reported by Tseng et al. [46]. On one side nanocarriers make target drug delivery more efficient and on other side its nano-sizes becomes obstacle in its pulmonary application. Due to mass median aerodynamic diameter (MMAD) of nanosized, large fraction of carriers will be exhaled and little fraction will reach the site of action. Sham et al. [47] took forward step to deal with these issues and prepared lactose containing spray-dried micrometer-sized carrier NPs which facilitate the immediate delivery of GNPs to the lung via a dry powder inhaler. An aerosol formulation of cationized GNPs also played significant role to enhance towards immunotherapy efficacy for treatment of equine recurrent airway obstruction over hypersensitivity [48, 49] in horses.

10.6.6 GNPs for Nutraceutical Delivery

Nutraceutical are food or products derived from foods that have health benefits. These do not come under FDA guidelines so vast number of foods or their derived products is come under nutraceuticals. In addition to drug and gene delivery, GNPs showing great potential for efficient delivery of nutraceuticals with better their bioavailability, stability and bioactivity in addition to providing controlled release. Chen et al. [50] reported the antioxidant activity of tea catechins in encapsulated GNPs can be retained after three weeks of storage. Both catechins and gelatin are prevented from enzymatic digestion and oxidation, thus tea catechin-GNPs might be a functional antioxidant carrier. Natural polyphenols, tannic acid, epigallocatechin gallate (EGCG), the aflavin and curcumin are also encapsulated into layer-by-layer (LbL) coated GNPs also reported to improve their bioavailability and half-life prolongation [51]. The gelatin–dextran micelles have high loading capacity for tea polyphenol (TPP) and sustained release TPP (in vitro). The gelatin–dextran micelles have significantly higher cytotoxicity against breast cancer cells over free TPP [52]. Nanoencapsulation of cocoa procyanidins (CPs) into GNPs has significantly superior CPs strength with good apoptotic activity in human acute monocytic leukemia cells over free CPs [53].

10.6.7 GNPs for Enzyme Immobilization

Enzyme immobilization is a process in which an enzyme is attach to several carrier or support nanomaterial or material which can improve the enzyme activity and stability as compared to free enzyme in solution. Various methods have been reported for enzyme immobilization such as physical adsorption or ionic binding, covalent binding and crosslinking. Physical adsorption or ionic binding is the simplest and easy method based on the nonspecific adsorption via non-covalent interaction such as hydrogen bonding, hydrophobic interactions *and van der Waals forces* for enzyme immobilized/cross linking on the carrier matrix. The enzyme immobilized by this method preserves the catalytic activity however the adsorption based on the physiochemical conditions for example pH, temperature, ionic strength etc. But this method suffers from the problem of enzyme leakage. So researcher is being used cross linking or covalent binding using different linkers which discuses in previous section (crosslinking of GNPs). The enzyme immobilization has been widely used in various biomedical and food industries.

Gan et al. [54] reported the GA-crosslinked GNPs for glucoamylase immobilization and optimized enzyme activity with temperature. They have studied the repeated heating and cooling effect of the reversible swelling and contracting of the crosslinked GNPs with several cycles of heating and cooling and found that this system could be used for enzyme immobilization and release. They have optimized the temperature for immobilized enzyme release and found that above 40° performed high activity similar to free enzyme, while no enzyme release below 40 °C. The loading efficiencies obtained as 59.9 and 24.7 % for immobilized glucoamylase by entrapment and adsorption methods, respectively and efficiency of temperature-triggered release reported as 99.3 % for adsorption method. The authors described that GA react with glucoamylase where gelatin and glucoamylase linked together by covalent bond which results in elevated immobilization efficiency for entrapment method. However, the covalently linked enzyme was hardly to be released. In case of adsorption method, the glucoamylase get adsorb on the gelatin matrix which facilitated its release [54].

10.7 Tissue Engineering

Tissue engineering is emerging as a significant hope for patients having tissue and organ failure conditions. Tissue engineering addressed these conditions by implanting natural, synthetic or semi synthetic tissue/organ mimics. These mimics are fully efficient from the start even develop into the required functionality [55]. Day by day, number of research papers is increasing on GNPs application in tissue engineering which shows good potential in this field.

Hydroxyapatite (HAp) is a calcium-containing biomaterial composed of chemical elements similar natural bone mineral components. The HAp possesses excellent properties that are required for bone grafts such as great stiffness, cell affinity, biocompatibility and osteoconductivity. Chang et al., prepared and characterized gelatin-HAp nanocomposite. The chemical bonding between calcium ions and carboxyl ions of HAp confirmed by FT-IR analysis and found red-shift of the 1339 cm^{-1} band of gelatin. The gelatin solution exhibits significant influential role in size of Hap nanocrystals development i.e., as the concentration of gelatin decreases the size of Hap nanocrystals increases [56]. Chiu et al., reported that this gel-Hap nanocomposites mimics natural bone microstructure. They have developed HAp-gelatin modified siloxane nanocomposite. The successfully coated HAp-GNPs with amino silane to provide strength that preserves them from damage. They observed that only titania additives revealed good in vivo bone formation in rat with calvarium defects [57]. Binding of gelatin onto titanium was significantly enhanced by dopamine conjugation [58]. Day by day titanium application in biomedical fields is continuously increasing so in tissue engineering is necessary that tissue should attach to the titanium. Yang et al., synthesized recombinant human gelatin-dopamine conjugate (D-rhG) using carbodiimide as a surface modifier. The affinity of human recombinant gelatin for titanium surface is significantly enhanced by conjugation with dopamine. They used human umbilical endothelial cells to assay the cell attachment and growth on this surface modified gelatin-dopamine conjugate where gelatin directly enhanced cell attachment. The D-rhG-coated surface is suitable for the capture and sequestering of a specific growth factor that supports cell proliferation [58].

Taguchi and Endo designed novel biodegradable amphiphilic polymer to assemble liposomes and cells. They synthesized cholesteryl group-modified tilapia gelatin (Chol-T-Gltns) using standard nucleophilic substitution reaction. The above formulation has gelatin-based backbone. They formed liposome gel where the Chol groups in Chol-T-Gltns anchor into the lipid membranes of dimethyldioctadecyl ammonium bromide liposomes to form liposome gel. The Chol-T-Gltn contains gelatin as backbone where gelatin has the cell adhesion sequence RGD (Arg-Gly-Asp). Once Chol groups attach to the phospholipid cell membrane then integrin in HepG2 cells recognizes the RGD sequence of Chol-T-Gltn, resulting in cell assembly [59]. Yang et al., prepared succinylated gelatin (s-GL) NPs synthesized by crosslinked aldehyde heparin using Schiff's base linkage. These NPs exhibit the possible nanotherapeutic application for injectable hydrogel or use as potential multifunctional devices like nano [60].

10.8 Conclusions

GNPs are receiving growing attention as a result of their increased used in biomedical applications. Due to its biocompatibility, biodegradability, non-toxic and cost-effectiveness making more robust transporter for target and controlled delivery of drugs. GNPs are in its initial phase of its research so there is a need to explore the opportunities for GNPs application in other biomedical areas.

References

1. N.M. Molino, S.W. Wang, Caged protein nanoparticles for drug delivery. Curr. Opin. Biotechnol. **28**, 75–82 (2014)
2. W. Lohcharoenkal, L. Wang, Y.C. Chen, Y. Rojanasakul, Protein nanoparticles as drug delivery carriers for cancer therapy. Biomed. Res. Int. **2014**, 12 (2014)
3. K.D. Schwenke, The science and technology of gelatin. Herausgegeben von A. G. Ward u. A. Courts, XVI und 564 Seiten mit zahlreichen Abb. u. Tab. (Academic Press London, New York, San Francisco 1977). Preis: 18,00 £; 39,50 $. Food/Nahrung **22**(4), 444–445 (1978)
4. H. Shah, F. Yusof, Gelatin as an ingredient in food and pharmaceutical products: an Islamic perspective. Adv. Environ. Biol. **8**(3), 7 (2014)
5. C. Pena, K. de la Caba, A. Eceiza, R. Ruseckaite, I. Mondragon, Enhancing water repellence and mechanical properties of gelatin films by tannin addition. Bioresour. Technol. **101**(17), 6836–6842 (2010)
6. R. Schrieber, H. Gareis, *Gelatine Handbook: Theory and Industrial Practice* (New York, Wiley, 2007)
7. R. Kaintura, P. Sharma, S. Singh, K. Rawat, P.R. Solanki, Gelatin nanoparticles as a delivery system for proteins. J. Nanomed. Res. **2**(1), 00018 (2015)
8. K. Rawat, P.R. Solanki, K. Arora, H.B. Bohidar, Response of gelatin modified electrode towards sensing of different metabolites. Appl. Biochem. Biotechnol. **174**(3), 1032–1042 (2014)
9. A. Mahapatro, D. Singh, Biodegradable nanoparticles are excellent vehicle for site directed in-vivo delivery of drugs and vaccines. J. Nanobiotechnol. **9**(1), 55 (2011)
10. K. Singh, A. Mishra, Gelatin nanoparticle: preparation, characterization and application in drug delivery. Int. J. Pharm. Sci. Res. **5**(6), 2149–2157 (2014)
11. L. Jeong, W.H. Park, Preparation and characterization of gelatin nanofibers containing silver nanoparticles. Int. J. Mol. Sci. **15**(4), 6857–6879 (2014)
12. S. Heydarkhan-Hagvall, K. Schenke-Layland, A.P. Dhanasopon, F. Rofail, H. Smith, B.M. Wu, R. Shemin, R.E. Beygui, W.R. MacLellan, Three-dimensional electrospun ECM-based hybrid scaffolds for cardiovascular tissue engineering. Biomaterials **29**(19), 2907–2914 (2008)
13. S. Azarmi, Y. Huang, H. Chen, S. McQuarrie, D. Abrams, W. Roa, W.H. Finlay, G.G. Miller, R. Lobenberg, Optimization of a two-step desolvation method for preparing gelatin nanoparticles and cell uptake studies in 143B osteosarcoma cancer cells. J. Pharm. Pharm. Sci. Publ. Can. Soc. Pharm. Sci. Soc. Can. Des Sci. Pharm. **9**(1), 124–132 (2006)
14. B. Mohanty, V.K. Aswal, J. Kohlbrecher, H.B. Bohidar, Synthesis of gelatin nanoparticles via simple coacervation. J. Surface Sci. Technol. **21**(3–4), 12 (2005)
15. C. Pinto Reis, R.J. Neufeld, A.J. Ribeiro, F. Veiga, Nanoencapsulation I. Methods for preparation of drug-loaded polymeric nanoparticles. Nanomed. Nanotechnol. Biol. Med. 2(1), 8–21 (2006)
16. A.O. Elzoghby, Gelatin-based nanoparticles as drug and gene delivery systems: reviewing three decades of research. J. Controlled Release **172**(3), 1075–1091 (2013)
17. G. Kaul, M. Amiji, Biodistribution and targeting potential of poly (ethylene glycol)-modified gelatin nanoparticles in subcutaneous murine tumor model. J. Drug Target. **12**(9–10), 585–591 (2004)
18. Z. Lu, T.K. Yeh, M. Tsai, J.L. Au, M.G. Wientjes, Paclitaxel-loaded gelatin nanoparticles for intravesical bladder cancer therapy. Clin. Cancer Res. Official J. Am. Assoc. Cancer Res. **10**(22), 7677–7684 (2004)
19. J. Madan, N. Dhiman, S. Sardana, R. Aneja, R. Chandra, A. Katyal, Long-circulating poly (ethylene glycol)-grafted gelatin nanoparticles customized for intracellular delivery of noscapine: preparation, in-vitro characterization, structure elucidation, pharmacokinetics, and cytotoxicity analyses. Anti-cancer Drugs **22**(6), 543–555 (2011)
20. E. Leo, M. Angela Vandelli, R. Cameroni, F. Forni, Doxorubicin-loaded gelatin nanoparticles stabilized by glutaraldehyde: involvement of the drug in the cross-linking process. Int. J. Pharm. **155**(1), 75–82 (1997)

21. M.G. Cascone, L. Lazzeri, C. Carmignani, Z. Zhu, Gelatin nanoparticles produced by a simple W/O emulsion as delivery system for methotrexate. J. Mater. Sci. Mater. Med. **13**(5), 523–526 (2002)
22. A.K. Bajpai, J. Choubey, In vitro release dynamics of an anticancer drug from swellable gelatin nanoparticles. J. Appl. Polym. Sci. **101**(4), 2320–2332 (2006)
23. W.-M. Li, D.-M. Liu, S.-Y. Chen, Amphiphilically-modified gelatin nanoparticles: self-assembly behavior, controlled biodegradability, and rapid cellular uptake for intracellular drug delivery. J. Mater. Chem. **21**(33), 12381–12388 (2011)
24. Y.-W. Won, S.-M. Yoon, C.H. Sonn, K.-M. Lee, Y.-H. Kim, Nano self-assembly of recombinant human gelatin conjugated with α-tocopheryl succinate for Hsp90 inhibitor, 17-AAG, delivery. ACS Nano **5**(5), 3839–3848 (2011)
25. H. Ai, S. Jones, Y. Lvov, Biomedical applications of electrostatic layer-by-layer nano-assembly of polymers, enzymes, and nanoparticles. Cell Biochem. Biophys. **39**(1), 23–43 (2003)
26. A. Saxena, K. Sachin, H.B. Bohidar, A.K. Verma, Effect of molecular weight heterogeneity on drug encapsulation efficiency of gelatin nano-particles. Colloids Surf., B **45**(1), 42–48 (2005)
27. S. Karthikeyan, N. Rajendra Prasad, A. Ganamani, E. Balamurugan, Anticancer activity of resveratrol-loaded gelatin nanoparticles on NCI-H460 non-small cell lung cancer cells. Biomed. Prev. Nutr. **3**(1), 64–73 (2013)
28. D. Ding, Z. Zhu, Q. Liu, J. Wang, Y. Hu, X. Jiang, B. Liu, Cisplatin-loaded gelatin-poly (acrylic acid) nanoparticles: synthesis, antitumor efficiency in vivo and penetration in tumors. Eur. J. Pharm. Biopharm. **79**(1), 142–149 (2011)
29. S. Harsha, Pharmaceutical suspension containing both immediate/sustained-release amoxicillin-loaded gelatin nanoparticles: preparation and in vitro characterization. Drug Des. Dev. Ther. **7**, 1027–1033 (2013)
30. N. Kuntworbe, R. Al-Kassas, Design and in vitro haemolytic evaluation of cryptolepine hydrochloride-loaded gelatine nanoparticles as a novel approach for the treatment of malaria. AAPS PharmSciTech **13**(2), 568–581 (2012)
31. L.-L. Li, J.-H. Xu, G.-B. Qi, X. Zhao, F. Yu, H. Wang, Core-shell supramolecular gelatin nanoparticles for adaptive and "on-demand" antibiotic delivery. ACS Nano **8**(5), 4975–4983 (2014)
32. B. Azimi, P. Nourpanah, M. Rabiee, S. Arbab, Producing gelatin nanoparticles as delivery system for bovine serum albumin. Iran. Biomed. J. **18**(1), 34 (2014)
33. C. Coester, P. Nayyar, J. Samuel, In vitro uptake of gelatin nanoparticles by murine dendritic cells and their intracellular localisation. Eur. J. Pharm. Biopharm. **62**(3), 306–314 (2006)
34. M.S. Sudheesh, S.P. Vyas, D.V. Kohli, Nanoparticle-based immunopotentiation via tetanus toxoid-loaded gelatin and aminated gelatin nanoparticles. Drug Delivery **18**(5), 320–330 (2011)
35. F.M. Chen, Z.W. Ma, G.Y. Dong, Z.F. Wu, Composite glycidyl methacrylated dextran (Dex-GMA)/gelatin nanoparticles for localized protein delivery. Acta Pharmacol. Sin. **30**(4), 485–493 (2009)
36. G. Kaul, M. Amiji, Tumor-targeted gene delivery using poly (ethylene glycol)-modified gelatin nanoparticles: in vitro and in vivo studies. Pharm. Res. **22**(6), 951–961 (2005)
37. G. Kaul, M. Amiji, Long-circulating poly (ethylene glycol)-modified gelatin nanoparticles for intracellular delivery. Pharm. Res. **19**(7), 1061–1067 (2002)
38. P. Magadala, M. Amiji, Epidermal growth factor receptor-targeted gelatin-based engineered nanocarriers for DNA delivery and transfection in human pancreatic cancer cells. AAPS J. **10**(4), 565–576 (2008)
39. K. Zwiorek, C. Bourquin, J. Battiany, G. Winter, S. Endres, G. Hartmann, C. Coester, Delivery by cationic gelatin nanoparticles strongly increases the immunostimulatory effects of CpG oligonucleotides. Pharm. Res. **25**(3), 551–562 (2008)
40. G. Kaul, C. Lee-Parsons, M. Amiji, Poly (ethylene glycol)-modified gelatin nanoparticles for intracellular delivery. Pharm. Eng. **23**(5), 108–117 (2003)

41. S. Kommareddy, M. Amiji, Poly (ethylene glycol)-modified thiolated gelatin nanoparticles for glutathione-responsive intracellular DNA delivery. Nanomed. Nanotechnol. Biol. Med. **3**(1), 32–42 (2007)
42. J. Vandervoort, A. Ludwig, Preparation and evaluation of drug-loaded gelatin nanoparticles for topical ophthalmic use. Eur. J. Pharm. Biopharm. **57**(2), 251–261 (2004)
43. G. Konat Zorzi, L. Contreras-Ruiz, J.E. Párraga, A. López-García, R. Romero Bello, Y. Diebold, B. Seijo, A. Sánchez, Expression of MUC5AC in ocular surface epithelial cells using cationized gelatin nanoparticles. Mol. Pharm. **8**(5), 1783–1788 (2011)
44. G.K. Zorzi, J.E. Párraga, B. Seijo, A. Sánchez, Hybrid nanoparticle design based on cationized gelatin and the polyanions dextran sulfate and chondroitin sulfate for ocular gene therapy. Macromol. Biosci. **11**(7), 905–913 (2011)
45. J.S. Patil, S. Sarasija, Pulmonary drug delivery strategies: a concise, systematic review. Lung India Official Organ Indian Chest Soc. **29**(1), 44–49 (2012)
46. C.-L. Tseng, W.-Y. Su, K.-C. Yen, K.-C. Yang, F.-H. Lin, The use of biotinylated-EGF-modified gelatin nanoparticle carrier to enhance cisplatin accumulation in cancerous lungs via inhalation. Biomaterials **30**(20), 3476–3485 (2009)
47. J.O.H. Sham, Y. Zhang, W.H. Finlay, W.H. Roa, R. Löbenberg, Formulation and characterization of spray-dried powders containing nanoparticles for aerosol delivery to the lung. Int. J. Pharm. **269**(2), 457–467 (2004)
48. S. Fuchs, J. Klier, A. May, G. Winter, C. Coester, H. Gehlen, Towards an inhalative in vivo application of immunomodulating gelatin nanoparticles in horse-related preformulation studies. J. Microencapsul. **29**(7), 615–625 (2012)
49. J. Klier, S. Fuchs, A. May, U. Schillinger, C. Plank, G. Winter, H. Gehlen, C. Coester, A nebulized gelatin nanoparticle-based CpG formulation is effective in immunotherapy of allergic horses. Pharm. Res. **29**(6), 1650–1657 (2012)
50. Y.-C. Chen, S.-H. Yu, G.-J. Tsai, D.-W. Tang, F.-L. Mi, Y.-P. Peng, Novel technology for the preparation of self-assembled catechin/gelatin nanoparticles and their characterization. J. Agric. Food Chem. **58**(11), 6728–6734 (2010)
51. T.G. Shutava, S.S. Balkundi, P. Vangala, J.J. Steffan, R.L. Bigelow, J.A. Cardelli, D. P. O'Neal, Y.M. Lvov, Layer-by-layer-coated gelatin nanoparticles as a vehicle for delivery of natural polyphenols. ACS Nano **3**(7), 1877–1885 (2009)
52. H. Zhou, X. Sun, L. Zhang, P. Zhang, J. Li, Y.-N. Liu, Fabrication of biopolymeric complex coacervation core micelles for efficient tea polyphenol delivery via a green process. Langmuir **28**(41), 14553–14561 (2012)
53. T. Zou, S.S. Percival, Q. Cheng, Z. Li, C.A. Rowe, L. Gu, Preparation, characterization, and induction of cell apoptosis of cocoa procyanidins–gelatin–chitosan nanoparticles. Eur. J. Pharm. Biopharm. **82**(1), 36–42 (2012)
54. Z. Gan, T. Zhang, Y. Liu, D. Wu, Temperature-triggered enzyme immobilization and release based on cross-linked gelatin nanoparticles. PLoS ONE **7**(10), e47154 (2012)
55. S. Palanisamy, S. Cheemalapati, S.-M. Chen, Enzymatic glucose biosensor based on multiwalled carbon nanotubes-zinc oxide composite. Int. J. Electrochem. Sci. **7**, 8394 (2012)
56. M.C. Chang, C.-C. Ko, W.H. Douglas, Preparation of hydroxyapatite-gelatin nanocomposite. Biomaterials **24**(17), 2853–2862 (2003)
57. C.-K. Chiu, J. Ferreira, T.-J. Luo, H. Geng, F.-C. Lin, C.-C. Ko, Direct scaffolding of biomimetic hydroxyapatite-gelatin nanocomposites using aminosilane cross-linker for bone regeneration. J. Mater. Sci. Mater. Med. **23**(9), 2115–2126 (2012)
58. X. Yang, L. Zhu, S. Tada, D. Zhou, T. Kitajima, T. Isoshima, Y. Yoshida, M. Nakamura, W. Yan, Y. Ito, Mussel-inspired human gelatin nanocoating for creating biologically adhesive surfaces. Int. J. Nanomed. **9**, 2753–2765 (2014)
59. T. Taguchi, Y. Endo, Crosslinking liposomes/cells using cholesteryl group-modified tilapia gelatin. Int. J. Mol. Sci. **15**(7), 13123–13134 (2014)
60. Y. Yang, H. Tang, A. Köwitsch, K. Mäder, G. Hause, J. Ulrich, T. Groth, Novel mineralized heparin–gelatin nanoparticles for potential application in tissue engineering of bone. J. Mater. Sci. Mater. Med. **25**(3), 669–680 (2014)

Chapter 11
Deployment of New Carbon Nanostructure: Graphene for Drug Delivery and Biomedical Applications

Mahe Talat and O.N. Srivastava

Abstract Graphene, the well awarded and popularly known as 2D carbon allotrope, is a versatile material. Its unique physico-chemical properties finds its application in a wide range of areas ranging from quantum physics, nanoelectronics, energy research, catalysis and engineering of nanocomposites, biomaterials and drug delivery. In principle, it is possible to produce graphene from high purity graphite sheet. Since graphite is stacked layers of many graphene sheets, bonded together by week van der Waals force and if these forces are disrupted individual graphene sheets can be separated out. In the present chapter we will discuss different methods of the synthesis of graphene like LPCVD, CVD (Chemical Vapor Deposition), thermal exfoliation, arc discharge, electrochemical exfoliation and chemical reduction. Since graphene having an innate property of hydrophobicity have the tendency to agglomeration and insolubility. To overcome the aforementioned insolubility some functionality needs to be attached to the graphene sheet. Therefore, different methods of functionalization like covalent and non-covalent approaches are also discussed. The last part of the chapter will deal with the application of graphene particularly in the drug delivery and biomedical field.

Keywords Graphene · Exfoliation · Functionalization · Application · Drug delivery

11.1 Introduction

Graphene, the superstar of the 2010 bagged the Nobel Prize in physics and has been a constant source of attraction in the material science in the recent past. Graphene, a single layer 2-D structure nanomaterial with unique physicochemical properties has

M. Talat (✉) · O.N. Srivastava
Nanoscience and Nanotechnology Unit, Physics and Bioscience Division, Banaras Hindu University, Varanasi 221005, India
e-mail: mahetalat04@gmail.com

M. Husain and Z.H. Khan (eds.), *Advances in Nanomaterials*,
Advanced Structured Materials 79, DOI 10.1007/978-81-322-2668-0_11

found vast applications in a wide range of areas. Having high surface area, excellent electrical mobility, strong mechanical strength, unparalleled thermal conductivity [1] together with remarkable biocompatibility and ease of functionalization, has received increasing attention in physical, chemical and biomedical fields.

Graphene, a single layer of sp^2-hybridized single layer of carbon atoms arranged in a honeycomb lattice, has evoked enormous inquisitiveness throughout the scientific community since its debut in 2004 [1].

The distinctive properties of graphene enabled it to be considered as an ideal candidate for broad range applications right from quantum physics to nanoelectronics, energy research, nanocomposites, nanomedicine and bionanomaterials [2–5]. The discovery of graphene has also gained an increased research attention and it is now being explored as a new material for drug delivery applications. In the field of nanomedicine, graphene and its other carbon variant have emerged as new vector for drug delivery. This provides new avenues for the development of extensive applications including fabrication of biosensors, nano-carriers for drug delivery, and probes for cell and biological imaging [6–9]. Graphene and other graphene derivatives including graphene oxide (GO) have been widely explored in the last five years for drug delivery applications by many research groups world wide. Although graphene-based delivery of nanomedicine, is still in its infancy, however initial reports appear to be motivating and may appear as novel opportunities for future disease diagnosis in medical science.

If Graphene is compared with the Carbon Nano Tubes (CNTs), it appears to be superior to CNTs in terms of its properties which are desirable in many applications. Therefore, the dominance of Graphene is visible now because of its significant properties like having large surface area which is a prerequisite for high drug loading capacity, lower toxicity, higher biocompatibility and ease of functionalization. Some workers have used functionalized graphene sheet as a vehicle for in vitro intracellular drug delivery of anticancer chemotherapy. It was demonstrated that polyethylene glycol (PEG)-functionalized graphene oxide (NGO) can be used as a novel drug carrier to transport anticancer drugs via non-covalent interaction and have in vitro cellular uptake capacity [10, 11].

In this chapter, we will discuss the latest progress and developments of using graphene for various biomedical applications, including drug delivery, and biosensing. This chapter will also summarize some of the work of our group on the applications of Graphene focusing on drug delivery and biomedical aspects. But before discussing the application of graphene sheet, it is important to understand the characteristics and methods of synthesis of various graphene materials from different routes and its methods of functionalization. First segment of the chapter will give a brief introduction of method of synthesis and functionalization of graphene and the last part of the chapter will deal with the applications of graphene in medicine and biomedical done in our laboratory.

11.2 Synthesis of Graphene

In principle, it is possible to synthesize graphene sheet from high purity graphite. Since graphite is stacked layers of many graphene sheets, held together by week Van der Waals forces and with the help of exfoliation these bonds can be broken and individual graphene sheets can be separated out. Different methods of synthesis of graphene used in our lab are LPCVD, CVD (Chemical Vapor Deposition), thermal exfoliation, arc discharge, electrochemical exfoliation and chemical reduction method. Table 11.1 summarizes the methods, precursors, experimental set-up and product obtained through different method of graphene synthesis. An optical image of the unit employed in our lab is also given in Fig. 11.1.

11.3 Functionalization of Graphene Sheets

Biomolecules play indispensable roles in all life processes including disease development, so the accurate detection of biomolecules is critical to disease diagnosis, drug delivery and therapy. Graphene having an innate property of hydrophobicity have the tendency to agglomeration and insolubility. To overcome the insolubility and hydrophobicity, covalent and non-covalent functionalization approaches have been developed. This kind of approach not only improves the solubility and dispersion but also facilitates the attachment of ligand/moiety of interest which is further used for biomedical or drug delivery applications etc. Broadly, the functionalization methods could be divide into two categories: covalent and non-covalent. Covalent functionalization is based on the formation of a covalent bond between functional entities and the graphene sheet while non-covalent functionalization is mainly through non covalent forces such as Van der Waals force, hydrogen bonding, electrostatic force and pi-pi interactions [12].

Choice of functionalization depends upon the requirement of the application. In covalent functionalization, aggressive chemical such as the use of neat acids at high temperatures is employed, which might introduce structural defects resulting in the change in unique properties of Graphene. Non-covalent treatment is particularly attractive because it is based on interactions of the hydrophobic part of the adsorbed molecules with Vander-waals, pi-pi, CH-pi, and other interactions, and aqueous solubility is provided by the hydrophilic part of the molecules. Also treatment with harsh chemicals is overcome thereby preserving the innate property of graphene and leaving the electronic structure intact [13]. We have functionalized graphene using both the methods of i.e., covalent and non-covalent. However, in order to avoid the effect of chemicals, non-covalent functionalization method is preferred in drug delivery and biomedical applications.

Table 11.1 The methods, precursors, experimental set-up and product obtained through different method of graphene synthesis

S. No.	Methods	Precursor	Set-up temperature/atmosphere	Product
1	LPCVD	At an elevated temperatures gaseous hydrocarbons like methane, ethylene etc. used as the carbon source	Low pressure chemical vapor deposition (LPCVD) technique employed for the synthesis of graphene by catalytic decomposition of hydrocarbon gases like methane and ethylene onto the Ferritin/Fe-SiO_2 substrates	A highly programmed LPCVD Unit is used to grow Single and few-layer graphene films
2	Thermal/chemical exfoliation	In this method oxidation of graphite in the presence of neat acids and oxidants is done	The graphite oxide is placed in a quartz tube (diameter $\sim$25 mm and length $\sim$1.3 m). The sample is flushed with inert gas for e.g.: Ar for 10–15 min and the quartz tube is inserted into a furnace preheated to 1050 °C and held in the furnace for 30 s Single/few-layer graphene oxide via stirring or mild sonication in water is obtained. Severe oxidation treatment converts graphite to hydrophilic graphite oxide	Functionalized graphene oxide is obtained having functionalities of hydroxyl, epoxy and carboxylic groups. These groups provide, reactive sites for chemical modifications and ligand attachment
3	Arc discharge method	In the inner wall region of the arc chamber Graphite rod is used as an electrodes	A current of 100 A passes through the electrodes, creating plasma between them. The temperature of this plasma typically reaches the temperature of $\sim$4000 K under relatively high pressure of hydrogen in the absence of catalyst	Because of the high temperature the carbon on the anode is vaporized. This vaporized carbon later gets deposited onto the cathode

(continued)

Table 11.1 (continued)

S. No.	Methods	Precursor	Set-up temperature/atmosphere	Product
4	Electrochemical exfoliation	Involve the application of cathodic or anodic potentials or currents in either aqueous (acidic or other media) or non-aqueous electrolytes	A set up of anode and cathode is prepared where graphite is used as anode and platinum is used as a cathode and voltage is applied. Within few minutes Graphite foil is exfoliated into thin graphene sheets. These exfoliated graphene sheet floating on electrolyte can be collected	Single- or multi-layered GN flakes can easily be produced in less time A high quality graphene with improved electrical properties superior to reduced graphene oxide is obtained

Setup of low pressure chemical vapor deposition (LPCVD) (Atomate ,USA)

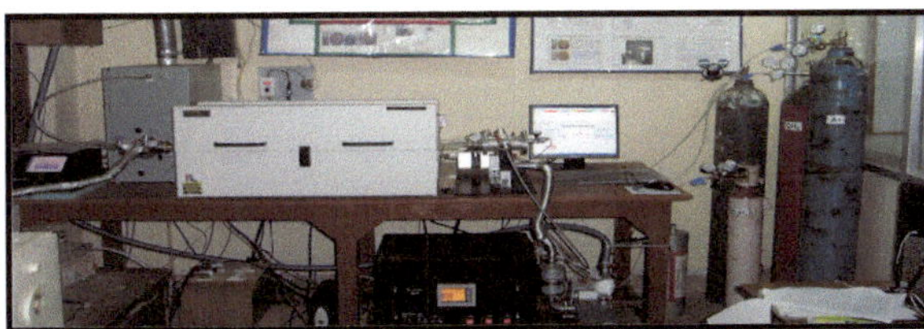

Optical photograph of electric arc discharge method

Optical photograph of tubular furnace for thermal exfoliation of Graphite Oxide

Fig. 11.1 An optical image of the units employed in our lab

In this method Graphene sheets were functionalized with 0.1 M cysteine using gentle steps like ultrasonication, filtration, magnetic stirring and centrifugation etc. simple scheme below depicts the functionalization of graphene sheet Fig. 11.2.

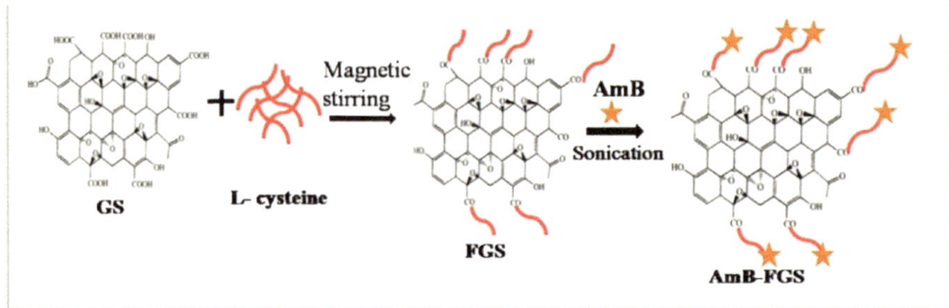

Fig. 11.2 Animated scheme of functionalization of graphene sheet with the amino acid L-cysteine non-covalently

11.4 Applications of Graphene Sheet

Applications of graphene are currently being investigated by many research groups in the scientific community and some of them are producing a lot of promising results although mostly it is preliminary results. The exploration of graphene in drug delivery and biomedicine was first initiated by the Hongjie Dai group at Stanford University (CA, USA) in 2008 [14, 15]. After getting encouraging results on CNTs for drug delivery our group extended this work for Graphene based drug delivery. Since graphene has approximately double the surface area of (600 m^2/g) CNT i.e., (2630 m^2/g), which could prove an efficient carrier for drug loading. Similarly in biotechnological applications enzyme will get more free sites for immobilization on Graphene sheet owing to its high surface area. Some of the applications of Graphene done in our lab are given below.

11.4.1 Graphene in Drug Delivery: Amine-Functionalized Graphene for the Delivery of Amphotericin B for the Treatment of Visceral Leishmaniasis: In Vivo and in Vitro Studies

Visceral leishmaniasis (VL) is one of the most neglected tropical diseases caused by intracellular protozoan parasites of the *Leishmania donovani* complex. Amphotericin B (AmB) is used as a first-line drug for VL in India and constitutes an alternate treatment for patients resistant to antimonials elsewhere. However, it has major drawbacks in terms of protracted hospitalization and significant nephrotoxicity [16, 17]. Previously, we described that nanoparticles of AmB have significantly greater efficacy than conventional AmB, with an improved safety profile. Subsequently, we also reported functionalized carbon nanotubes (f-CNTs) as an effective nanovector for AmB for oral administration in the treatment of VL

[18, 19]. Functionalized graphene oxide has been shown to be a promising nanovector for the efficient delivery of drugs into cells. In spite of having great application potential, it is important to mention that graphene itself possesses zero band gap as well as inertness to reaction. The π-π interaction between graphene sheets (GS) results in the stacking of graphene sheets which eventually results in the formation of multilayered graphite. In this work we explored the efficacy and toxicity of a novel Amp-B formulation as a conjugate (f-Gr AmB) with amine-modified graphene (f-Gr) in the treatment of VL. The results of this work are published in Drug Design, Development and Therapy 2014:8. These results were so well received since it went online that within 4 weeks of uploading there was 420 views.

This novel formulation of AmB conjugated to amine-modified graphene (f-Gr) is preferred for safety and efficacy over conventional AmB. The f-Gr was prepared in a gentle one-step process of noncovalent (amine) functionalization with the help of amino acid L-cysteine (Fig. 11.3). This f-Gr was further conjugated to AmB by peptide bond. It was found that f-Gr-AmB exhibited lesser cytotoxicity toward J774A.1 cells than AmB, and did not induce any hepatic or renal toxicity in Swiss albino mice. In vitro antileishmanial assay in J774A.1 cells showed significantly enhanced efficacy of f-Gr-AmB over AmB (Fig. 11.4). Furthermore, percentage inhibition of amastigote replication in a hamster model of VL was significantly higher in the f-Gr-AmB treated group (87.8 %) compared to the AmB group (70.4 %) [20].

11.5 Biomedical Application of Graphene

11.5.1 Immobilization of Beta-Galactosidase onto Functionalized Graphene Nano-sheets for Analytical Applications

ß-Galactosidase is commercially important enzyme and has two main commercial applications in food industry: reduction of lactose in dairy products for safe consumption by the patients who are lactose intolerant and secondly producing galacto-oligosaccharides (GOS) for maintaining gastrointestinal flora via trans-galactosylation reaction. Lactose is an integral component of breast milk which causes flatulence and uneasiness in such patients and this condition aggravates with the aging due to reduced secretion of the gastric beta-galactosidase. Galacto-oligosaccharides (GOS) are containing two to five molecules of galactose held together through glycosidic bonds. Also they fall under the category of non-digestible sugars. Thus enzyme ß-Galactosidase can be used for the above

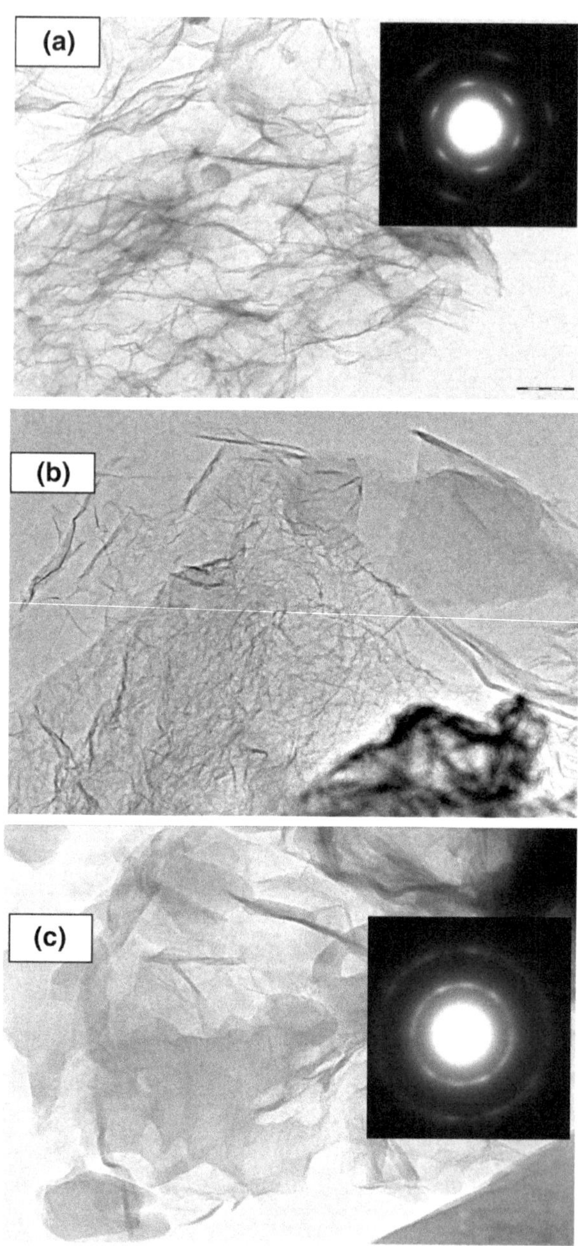

Fig. 11.3 Micrograph (TEM) of GS (**a**), f-Gr (**b**), and f-Gr-AmB (**c**) Ref. [20]

purpose but enzymes are too expensive to be discarded after single use, which makes them commercially non-viable. However, this obstacle can be tackled by coupling of enzyme with a suitable support material. This kind of immobilization

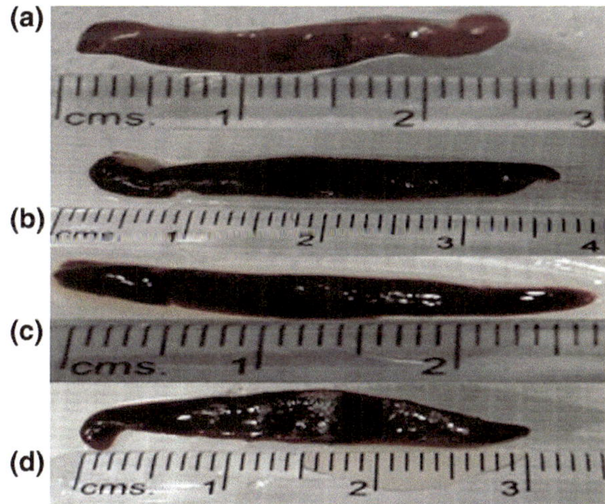

Fig. 11.4 Spleen sizes of hamsters of different experimental groups. **a** Healthy control (2.4 cm); **b** vehicle control (3.4 cm); **c** f-Gr-AmB treated (2.6 cm); **d** AmB treated (3 cm) Ref. [20]

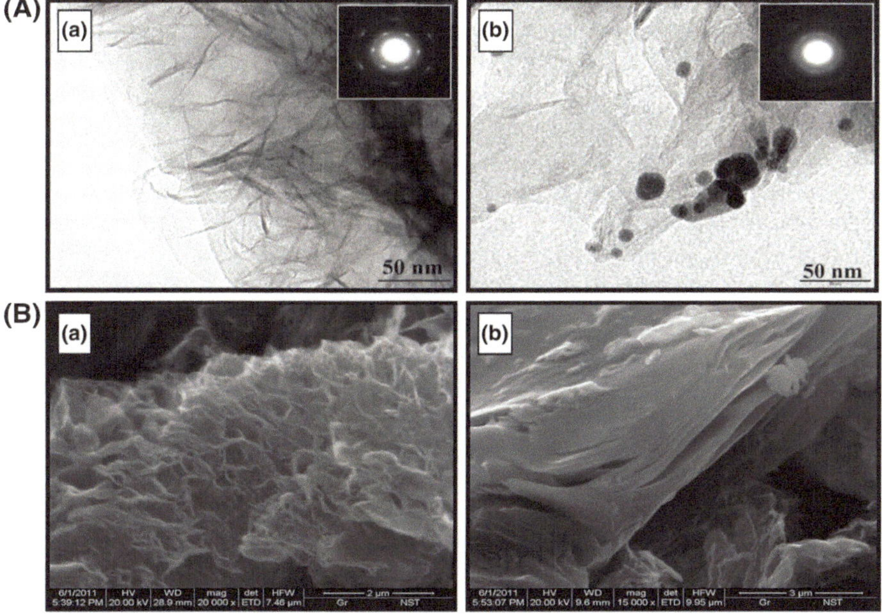

Fig. 11.5 Micrographs of functionalized and immobilized graphene nano sheets (*A*) TEM images of functionalized (**a**) coupled (**b**) graphene showing fine transparent sheets with inset showing the Selected Area electron Diffraction pattern (SAD). Amine functionalized graphene sheets appear transparent whereas islands of immobilized enzyme can be seen in coupled graphene (*B*) SEM images of functionalized (**a**) and coupled (**b**) graphene Ref. [21]

provides many advantages such as high enzyme reusability and shelf life, high yield, improvement of thermal stability, continuous operation, controlled product formation, etc. Out of several support matrices, nano-materials, particularly carbon variant graphene is preferred over bulk materials. Graphene, because of their miniature size, large surface area with high enzyme loading capacity and aqueous suspendibility prove to be most appropriate support system for enzyme immobilization. Graphene, unlike other regular nanoparticles, provide single carbon atoms thick graphite sheets with enormous surface area; perfect for uniform attachment of enzymes. Enzyme beta-galactosidase from plant source (*Cicer arietinum*) has been isolated by our group and further this isolated and purified enzyme is used for covalent attachment to functionalized graphene (Fig. 11.5). The enzyme kinetics and other parameters has also been studied to obtain best catalytic performance. The immobilization efficiency was significantly very high and was recorded 84.2 % and enhanced reusability with greater storage ability was also observed [21].

11.5.2 Functionalized Graphene Sheets for the Immobilization of Pharmaceutically Important Enzyme, β-Amylase

B-Amylase enzyme has an important role in mashing and brewing process. It is essential enzyme for the generation of maltose from cereal grain starch. *β*-amylase also plays an important role in the production of diphtheria pertussis tetanus vaccine, in being the exclusive source of carbon. In this work graphene sheets were functionalized and customized as a matrix which is further used for the covalent immobilization of Fenugreek *β*-amylase using glutaraldehyde as a cross-linker. Statistical software was used to optimize the factors affecting the process using Response Surface Methodology based Box-Behnken design of experiment which resulted in 84 % immobilization efficiency. The results are published in PLOS-1 (2014) [22]. Below given is an FTIR spectra of different stages of immobilization, which is an important technique to confirm that the immobilization /attachment of enzyme has occurred (Fig. 11.6). A unique signature of the functional groups in the spectra is recorded and which was analyzed in the results. Detailed analysis of the results are given in Ref. [22].

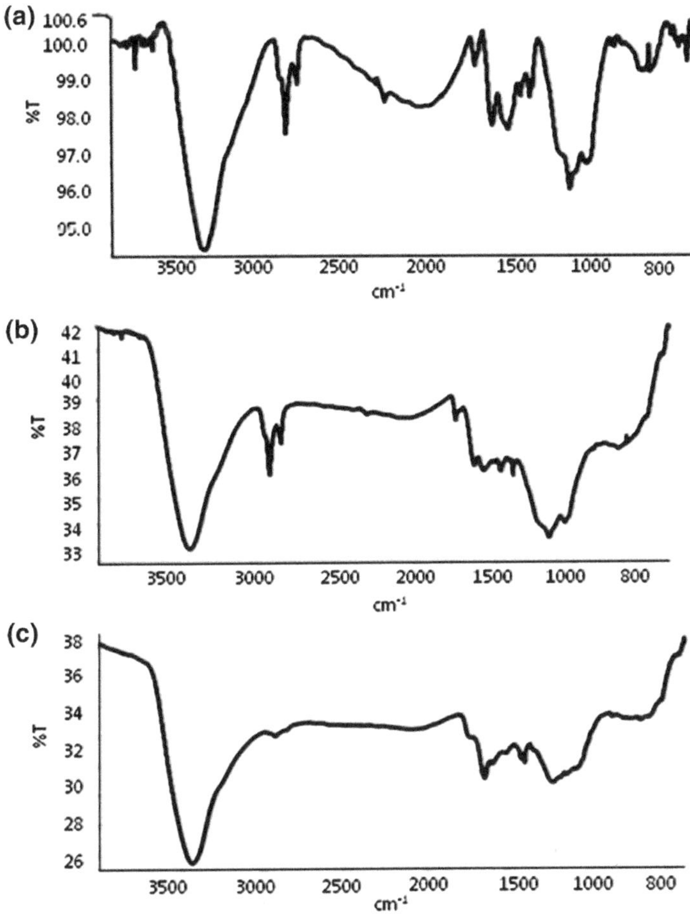

Fig. 11.6 FTIR spectra of graphene sheets during different stages of immobilization Ref [22]

11.6 Conclusions

Today the development of improved drug delivery systems with efficient therapeutic profile and efficacy is one of key issues faced by modern medical science, which may be satisfied with this new wonder material called Graphene. Knowing the fact that there are still many challenges in graphene-assisted drug delivery, the distinctive physico-chemical properties of many graphene variants are still very attractive platform for various innovative applications in biological sensing, drug delivery and other related biomedical field. The research on graphene and its derivatives emerge as a novel nanomaterials platform for application and is relatively a new direction that deserves special attention, yet many challenges need to be addressed and solved by effective collaborations crossing multiple disciplines including chemistry, biology, medicine etc. Certainly some concrete research work is needed and will hopefully appear in the next few years as the potential carbon material to offer very powerful new tools for the treatment and diagnosis of diseases and other applications.

Acknowledgments Authors are grateful to Nano Science and Technology Initiative, Department of Science and Technology (DST), India for financial support. Authors also thank Prof. C.N.R. Rao, Prof. Shyam Sunder, Prof. A.M. Kayastha and all the lab members who are engaged and contributed directly and indirectly in the nanoscience research. Authors gratefully acknowledge the financial support from CSIR, and MNRE, New Delhi. One of the co-author M.T thanks WOS-A fellowship for the support.

References

1. A.K. Geim, K.S. Novoselov, The rise of graphene. Nat. Mater. **6**, 183–191 (2007)
2. K.S. Novoselov, A.K. Geim, S.V. Morozov, D. Jiang, Y. Zhang, S.V. Dubonos et al., Electric field effect in atomically thin carbon films. Science **306**, 666–669 (2004)
3. K.S. Kim, Y. Zhao, H. Jang, S.Y. Lee, J.M. Kim, K.S. Kim et al., Large-scale pattern growth of graphene films for stretchable transparent electrodes. Nature **457**, 706–710 (2009)
4. C. Berger, Z. Song, X. Li, X. Wu, N. Brown, C. Naud et al., Electronic confinement and coherencein patterned epitaxial graphene. Science **312**, 1191–1196 (2006)
5. S. Stankovich, D.A. Dikin, R.D. Piner, K.A. Kohlhaas, A. Kleinhammes, Y. Jia et al., Synthesis of graphene-based nanosheets via chemical reduction of exfoliated graphite oxide. Carbon **45**, 1558–1565 (2007)
6. L. Feng, L. Wu, X. Qu, New horizons for diagnostics and therapeutic applications of graphene and graphene oxide. Adv. Mater. **25**, 168–186 (2013)
7. G. Gonçalves, M. Vila, M.T. Portolés, M. Vallet-Regi, J. Gracio, P.A.A. Marques, Nano-graphene oxide: a potential multifunctional platform for cancer therapy. Adv. Healthcare Mater. (2013). doi:10.1002/adhm.201300023
8. K. Yang, L. Feng, X. Shi, Z. Liu, Nano-graphene in biomedicine: theranostic applications. Chem. Soc. Rev. **42**, 530–547 (2013)
9. H. Shen, L. Zhang, M. Liu, Z. Zhang, Biomedical applications of graphene. Theranostics **2**, 283–294 (2012)
10. Z. Liu, J.T. Robinson, X.M. Sun, H.J. Dai, PEGylated nanographene oxide for delivery of water-insoluble cancer. drugs. J. Am. Chem. Soc. **130**, 10876–10877 (2008)
11. X.M. Sun, Z. Liu, K. Welsher et al., Nano-graphene oxide for cellular imaging and drug delivery. Nano Res. **1**, 203–212 (2008)
12. V. Georgakilas, M. Otyepka, A.B. Bourlinos, V. Chandra, N. Kim, K.C. Kemp, P. Hobza, R. Zboril, K.S. Kim, Functionalization of graphene: covalent and non-covalent approaches, derivatives and applications. Chem. Rev. **112**(11), 6156–6214 (2012). doi:10.1021/cr3000412
13. B. Long, M. Manning, M. Burke, B.N. Szafranek, G. Visimberga, D. Thompson, J.C. Greer, I. M. Povey, J. MacHale, G. Lejosne, D. Neumaier, J.Q. Aidan, Non-covalent functionalization of graphene using self-assembly of alkane-amines. Adv. Funct. Mater. **22**, 717–725 (2012)
14. Z. Liu, J.T. Robinson, X.M. Sun, H.J. Dai, PEGylated nanographene oxide for delivery of water-insoluble cancer drugs. J. Am. Chem. Soc. **130**, 10876–10877 (2008)
15. X.M. Sun, Z. Liu, K. Welsher et al., Nano-graphene oxide for cellular imaging and drug delivery. Nano Res. **1**, 203–212 (2008)

16. C.P. Thakur, A.K. Pandey, G.P. Sinha, S. Roy, K. Behbehani, P. Olliaro, Comparison of three treatment regimens with liposomal amphotericin B (AmBisome) for visceral leishmaniasis in India: a randomized dose-finding study. Trans. R. Soc. Trop. Med. Hyg. **90**(3), 319–322 (1996)
17. S. Sundar, M. Chatterjee, Visceral leishmaniasis—current therapeutic modalities. Indian J. Med. Res. **123**(3), 345–352 (2006)
18. V.K. Prajapati, K. Awasthi, S. Gautam et al., Targeted killing of Leishmania donovani in vivo and in vitro with amphotericin B attached to functional-ized carbon nanotubes. J. Antimicrob. Chemother. **66**(4), 874–879 (2011)
19. V.K. Prajapati, K. Awasthi, T.P. Yadav, M. Rai, O.N. Srivastava, S. Sundar, An oral formulation of amphotericin B attached to functionalized carbon nanotubes is an effective treatment for experimental visceral leishmaniasis. J. Infect. Dis. **205**(2), 333–336 (2012)
20. S.L. Mudavath, M. Talat, M. Rai, O.N. Srivastava, S. Sundar, Characterization and evaluation of amine-modified graphene amphotericin B for the treatment of visceral leishmaniasis: in vivo and in vitro studies. Drug Des. Dev. Ther. **8**, 1235–1247 (2014)
21. D. Kishore, M. Talat, O.N. Srivastava, A.M. Kayastha, Immobilization of b-Galactosidase onto functionalized graphene nano-sheets using response surface methodology and its analytical applications. PLoS ONE **7**(7), e40708 (2012). doi:10.1371/journal.pone.0040708
22. G. Srivastava, K. Singh, M. Talat, O.N. Srivastava, Kayastha AM functionalized graphene sheets as immobilization matrix for fenugreek β-amylase: enzyme kinetics and stability studies. PLoS ONE **9**(11), e113408 (2014). doi:10.1371/journal.pone.0113408

Chapter 12
Optical Coherence Tomography as Glucose Sensor in Blood

Hafeez Ullah, Ejaz Ahmad and Fayyaz Hussain

Abstract Optical coherence tomography is a modern imaging modality that can visualize the biological tissues on micron levels. This chapter describes the use of OCT technique for measuring glucose in liquid phantoms, whole blood (in vitro *and* in vivo) based on temporal dynamics of light scattering. Whole blood smears imaged with microscope reveal the effect of red blood cells deformation and aggregation with white light microscope for animal and human blood. We found the changes in the shape of individual cells from biconcave discs to spherical shapes and eventually the lysis of the cells at optimum concentration of glucose. The increase of glucose in blood causes the changes in diffusion coefficients and shapes of the erythrocytes of glucose in stagnant and flowing fluids. The relative contributions of these competing effects have been studied by examining the motion dynamics of deformable asymmetrical RBCs and non deformable symmetrical PMS as flowing scattering particles. These systematic studies are aimed at eventual in vivo tissue imaging scenarios with speckle-variance OCT to visualize normal and malignant blood microvasculature in three and two dimensions and to monitor the glucose levels in blood by analyzing the Brownian motion of the red blood cells.

Keywords Optical coherence tomography · Glucometry · Blood rheology · 2D and 3D imaging · Blood vessels

H. Ullah (✉) · E. Ahmad
Laser and Optronics Laboratory, Department of Physics, Bahauddin Zakariya University, Multan 60800, Pakistan
e-mail: hafeezullah79@gmail.com

E. Ahmad
e-mail: drejazch@hotmail.com

F. Hussain
Material Simulation Research Laboratory (MSRL), Department of Physics, Bahauddin Zakariya University, Multan 60800, Pakistan
e-mail: fayyazhussain248@yahoo.com

M. Husain and Z.H. Khan (eds.), *Advances in Nanomaterials*,
Advanced Structured Materials 79, DOI 10.1007/978-81-322-2668-0_12

12.1 Introduction

The OCT 1-15 μm axial resolution (in situ and in real time) is a high resolution imaging modality of biological tissues, transparent tissue that uses low-coherence interferometry to perform high-resolution imaging [1]. A low coherent light source of coherence length ($\sim$10–15 μm) for measuring an interferometeric signal is used in the system of OCT with moving mirror in the reference arm, the sample arm, and a photodetector [2]. Reference mirror and the light signals from the sample are combined in the beam splitter and the axial resolution specifying system interferometeric signal coherence length is acquired within the photodetector. Depth scan can be obtained up to $\sim$2 mm limit by scanning the reference mirror. Tissue surface can be scanned, if the sample is introduced in laterally movable mirror scanning; therefore, by making the two-dimensional image with a high resolution [3, 4]. Notable applications of OCT systems include cardiovascular, joints, gastrointestinal tract, the bladder, the female reproductive tract, respiratory tract, surgical and dental clinical diagnosis guidance [5–12].

The leading cause of death in developed countries is low-density lipoprotein (LDL). The lipid bearing materials, calcium and other substances cause blood to accumulate in atherosclerotic disease of the arterial wall. Blood by finger prick for blood glucose control is the prerequisite that makes it unpleasant and painful. Therefore, it is necessary to introduce a non-invasive method of monitoring blood glucose from the skin can provide a wonderful relief to patients with diabetes [13, 14]. These methods include for example, optical methods, ultrasonic, impedance, heat capacitance, etc. [15, 16]. However, in these methods, the measurement can be different according to the skin because the skin topography and has a non-uniform structure of the blood vessel biophysical characterization of the complex, thereby relying on an upper layer of skin (the epidermis and upper dermis) only [13].

Therefore, the development of the technology for monitoring and quantification of vascular molecular diffusion can promote new therapeutic agents, drug delivery technologies, as well as a novel method for clinical diagnosis [17]. Many techniques have been used for this purpose in the past, such as spectrophotometry, fluorescence spectrophotometry, fluorescence microscopy, microdialysis, optical coherence tomography [18], Raman spectroscopy, near infrared absorption and scattering and photoacoustics [19, 20]. These techniques, for example low sensitivity, specificity, and accuracy in the measurement of physiological glucose result in limitations. OCT has been considered in functional imaging, monitoring, and quantification of epithelial and connective tissue in the diffusion process to be applied due to the importance of the diagnosis and treatment of various diseases for in vivo applications [17]. In OCT, the scattering signal's main reason is the refractive index of scattering particles such as collagen and elastic fibers and extracellular fluid [21]. Thus the denser medium (i.e. extra cellular fluid causes the decrease in mismatch and hence scattering coefficient decreases as whole for the tissue [22]. The OCT's capability to image the structure ~0.2–2 mm in depth and 1500 μm^2 make it potentially applicable to visualize the epithelium, subepithelium, and part of the dermal structure that could measure the capillary regions [20].

12.2 OCT's Basic Principle: Mechanism of Time Domain OCT

In OCT, back reflection of light intensity is not easy to measure due to the high speed and spread of light. Measurements for interfering signals are made by Michelson interferometer in general. Figure 12.1 shows a schematic view for the time domain (TD-OCT) with a simple Michelson interferometer. Reference arm strength is to be measured indirectly assess interference back reflection intensity. In TD-OCT, light from a light source is split at beam splitter, half towards the sample and half towards the movable mirror [23]. Light is reflected back from the mirror and the sample, recombine the two signals in the beam splitter and is usually detected by a photodiode or a CCD camera. Interference signal depends on the difference in path length and the coherence length of the system. If the path length is less than the coherence length of the light source, the intensity of the interference is generated as interferogram [24, 25].

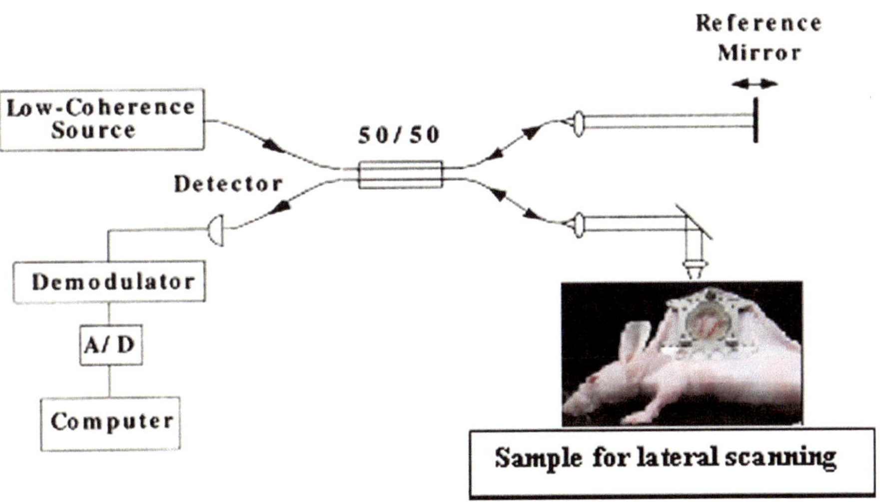

Fig. 12.1 A conceptual diagram of Michelson interferometer in OCT system. The resultant interferogram is sent to detector

12.3 Glucose Levels in Blood

Hypoglycemia and hyperglycemia in diabetes are life-threatening condition. For Patients at critical stage, the glucose quantification in the blood of the rapid and non-invasive measurement is important. In recent years, many methods have been used to monitor blood glucose, comprising of test strips and reflectometer. Although, these methods are reliable and cost effective glycemic, but piercing the skin can cause skin irritation and injury. Different blood glucose test strips and quality control comparison is other problems. OCT explores the blood glucose as noninvasive monitoring methods [24, 25].

12.4 Application of OCT for Glucose Monitoring

If, $\mu_s \gg \mu_a$ then only the scattering (ballistic) photon is enough to explore the scattering centers (such as cell membranes, cellular components, and protein aggregates) caused by changes in the optical characteristics, n_s, and interstitial fluid (ISF), n_{ISF}. The 10 mg/dl increase in glucose increase by a factor of 1.52×10^{-5} and consequently, ns is decreased by decreasing the index mismatch $\Delta n = \frac{n_s}{(n_{ISF} + 1.52 \times 10^{-5}/10\ \text{mg/dl})}$ [2, 24, 25].

OCT glucose monitoring for average structured OCT image (Fig. 12.2a) for a 1D optical temporal distribution (Fig. 12.2b), we can get information about the changes in optical parameters, for example, by optical attenuation contour analysis of scattering from porcine aorta [26, 27]. OCT has the ability to monitor the diffusion of glucose from in vitro porcine aortic health epithelial tissue in the eye [17, 28, 29].

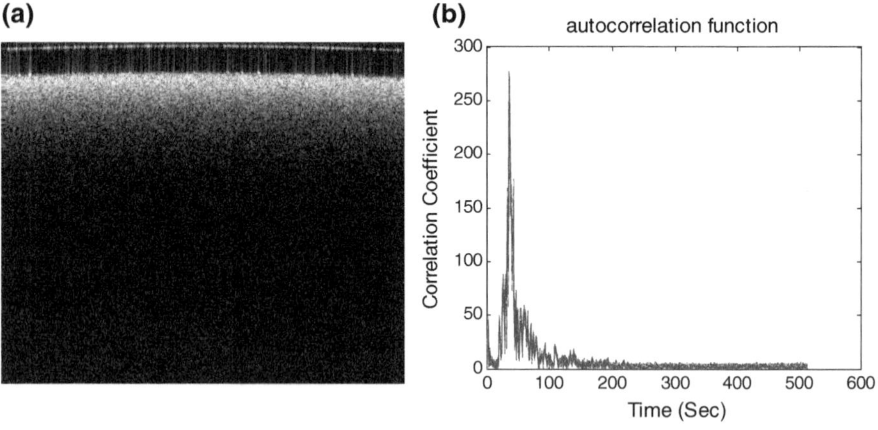

Fig. 12.2 OCT data from healthy pig aorta (**a**) structured OCT image (**b**) and corresponding OCT signal

12.5 SS-OCT in Glucose Sensing

Dynamic light scattering (DLS) or ballistic photon (single) scattering underlying Brownian motion can be analyzed using SS-OCT in M-mode scanning [30] to quantify the glucose levels in fluids [30]. Translation relax and rotation relaxation: two types of motions are relate to the Brownian motion of particles in turbid media. Translational relaxation or light contrast to the relevant time provides liquid and solid tissues. This quantitative data is base line for translating the in vitro study into in vivo environment where blood viscosity can be measured to monitor glucose levels [24, 25].

12.6 Glucose Quantification in Stagnant Blood

OCT techniques have been used in the determination of glucose, based on a dynamic light scattering from liquid phantom and whole blood (in vitro) in time domain. Adding glucose affects scatterers's mirage and erythrocyte's Brownian motion in the blood, polystyrene microspheres (PMS) carrying fluids [24, 25, 31].

12.6.1 Samples Preparation

Three types of scattering samples with varying glucose concentrations have been investigated:

1. The first set was water phantoms with 1.4 μm diameter PMS and six different concentrations of glucose (0, 100, 200, 300, 400, and 500 mM). The microsphere concentrations were increased from 0.69 % (weight/volume) for glucose-free suspension to 0.76 % for the 0.5 mM phantom, in order to keep the scattering coefficient μ_s constant in the presence of the refractive index matching effect. Assuming the refractive index of water ~1.33, μ_s was calculated from Mie theory at 1310 nm to be 100 cm^{-1} [24, 25, 31, 32].
2. The second experimental series was a symmetric scatterer set [PMS as in (1)] suspended in blood plasma instead of water. Figure 12.3 shows the photo of a centrifuged blood plasma heparin tube in which ~55 % of plasma can be observed while remaining 45 % consists of red blood cells (erythrocytes). In these plasmas we added a fixed concentration 0.0073 % of 1.4 μm diameter PMS and varying glucose amounts (0, 80, 160, 240, 320, and 400 mM). Here, our Mie scattering calculations assumed a refractive index value ~1.34 for rat plasma, extrapolated from previously-reported *n* ~ 1.35 at 630 nm for human blood plasma [24, 25, 31, 33, 34].
3. Finally, asymmetric RBC scatterers in the natural blood environment were investigated. Whole blood drawn from five months old Lewis rats was admixed

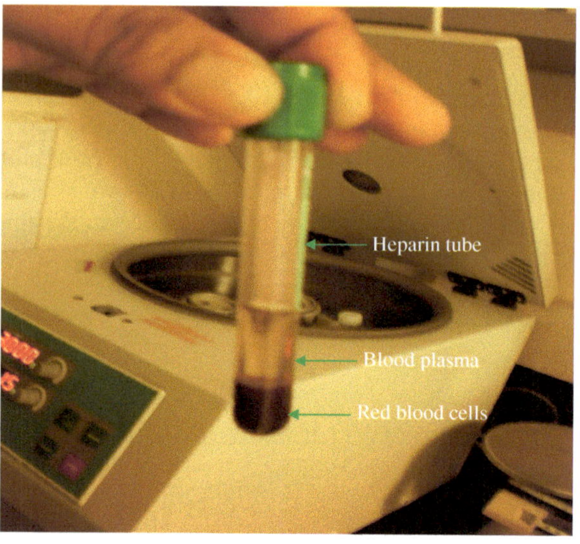

Fig. 12.3 Centrifuged plasma from drawn rat blood in heparin tube. We can see that ~55 % of blood consists of plasma and other 45 % is retained by red blood cells [24, 25]

with additional glucose to 0, 20, 40, 60 and 80 mM levels. These add glucose levels were lower than in sets (1) and (2), and indeed closer to the level above human physiology; this was aimed is to better gauge the applicability of this approach to finally in vivo situation [24, 25, 31].

12.6.2 Measurement System

A Custom built 36 kHz of SS-OCT system was used for M-mode scanning. For M-mode measurements a snapshot (Fig. 12.4) of the experimental apparatus is shown for a typical blood sample. Briefly, the SS-OCT system consists of a frequency-domain mode-locked (FDML) fiber ring laser sources, including polygon-based tunable filter. The FDML cavity length configuration includes 3.3–4.5 km. A fiber Bragg grating is used for A-scan (depth scan) triggering. Coherence length and spectral scanning range were respectively 6 and 112 nm, the center wavelength of 1310 nm. Average output power and axial resolution were 48 mW and 8 microns respectively [24, 25, 31].

Figure 12.5 shows OCT images for a typical glucose concentration of 80 mM in whole blood. Figure 12.5a depicts a raw 2-D OCT image (640 × 512 pixels and 5 mm × 2 mm, width × depth) and Fig. 12.5b shows an M-mode 2-D OCT image with (64 × 512 pixels and 0 mm × 3 mm, width × depth). Figure 12.5 ensures that ballistic scattering region near top surface (above ~70 μm) below glass cover is more suitable for dynamic light scattering. To obtain M-mode data, it was

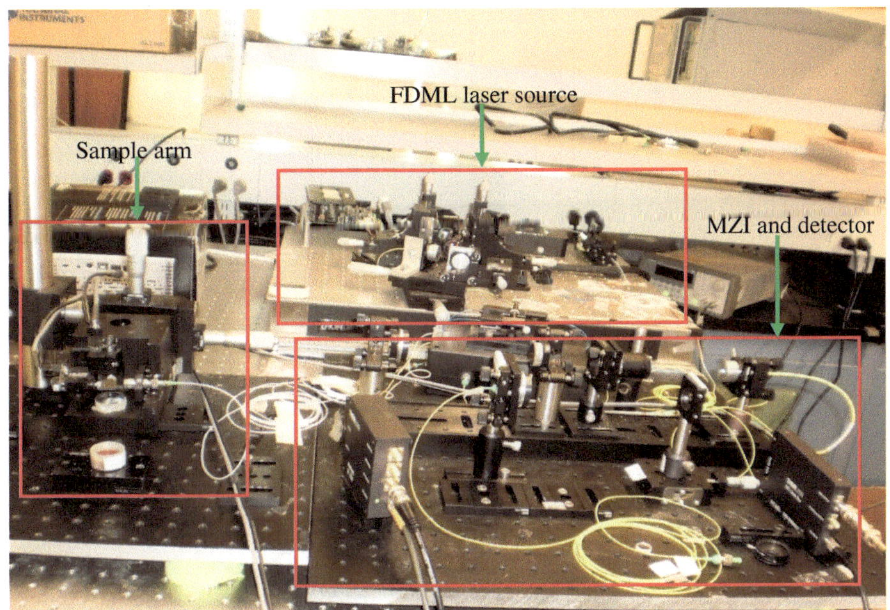

Fig. 12.4 Photograph of experimental setup during M-mode measurements of a typical blood sample set (3) [24, 25]

repeatedly acquired full depth (~2 mm) A-scans for ~5 s at the same central sample lateral location and averaged 64 consecutive A-scans to obtain average signal intensity at our selected 70 μm depth. The experimental temperature was 21 °C [24, 25, 31].

12.6.3 Signal Processing and Results

The signal recorded at the detector of the SS-OCT system is given by [24, 25, 35];

$$I_D = \frac{1}{2}\int_0^{\infty} I_{so}(k)(1+\cos kx)dk = \frac{1}{2}I_{so} + \frac{1}{4}\int_{-\infty}^{\infty} I_{so}(k)\mathrm{e}^{ikx}dk, \qquad (12.1)$$

where x is the path length difference, $I_{SO}(k)$ is the source intensity and the integral $\int_0^{\infty} I_{so}(k)\mathrm{e}^{ikx}dk$ is called autocorrelation function (*ACF*). The power spectrum *P*(*ω*) and *ACF* are related by Weiner-Khinchin theorem as a Fourier pair [24, 25]:

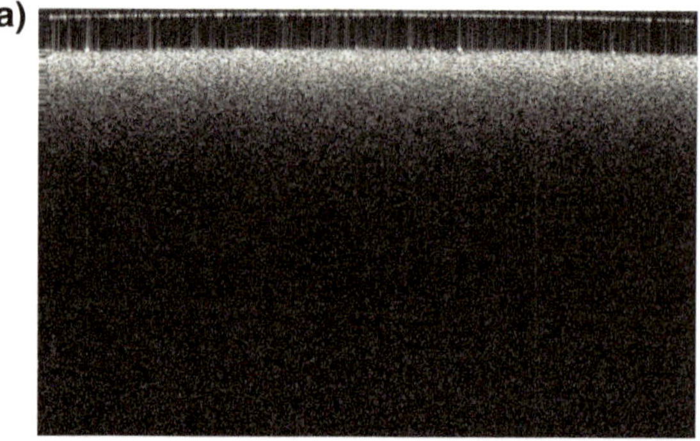

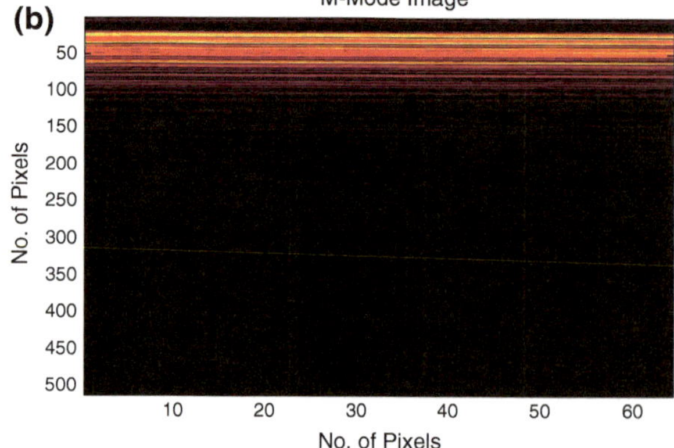

Fig. 12.5 **a** Srtructural 2-D OCT image with pixels (640 × 512) and dimensions (5 mm × 2 mm) along x and z-axis for blood with 80 mM glucose in it. **b** An M-mode 2-D OCT image with (64 × 512) pixels and dimensions (0 mm × 3 mm) along x and z-axis [24, 25]

$$ACF(x) = \int P(\omega)\mathrm{e}^{-i\omega x/c} d\omega. \tag{12.2}$$

Raw OCT intensity M-mode data at a depth of 70 μm is shown in Fig. 12.6a for six different glucose concentrations in the blood plasma + PMS phantom set (2) up to ~2.0 s for the purpose of illustration. Equation (12.2) yielded The *ACF*s as processed using Matlab and is shown in Fig. 12.6b. As can be seen, increased blood glucose levels can lead to longer relaxation OCT signal attenuation, from the slow Brownian motion of the scattering particles as expected. Differences can be observed in the decay very careful, because the decay time interval, $\Delta\tau$ is within the range of milliseconds. Figure 12.7a, b display the OCT intensity and *ACF* for whole blood phantom set (3) similar to plasma as stated above for all five concentrations of glucose. Figure 12.7 shows the characteristic behavior of the medium (whole blood + glucose) like Fig. 12.6 of medium (plasma + PMS + glucose). Figure 12.8a–c shows ACFs as a function of time for ~5 s extracted from OCT (power spectrum) data for samples sets (1,2, and 3) in glucose free conditions [24, 25, 31].

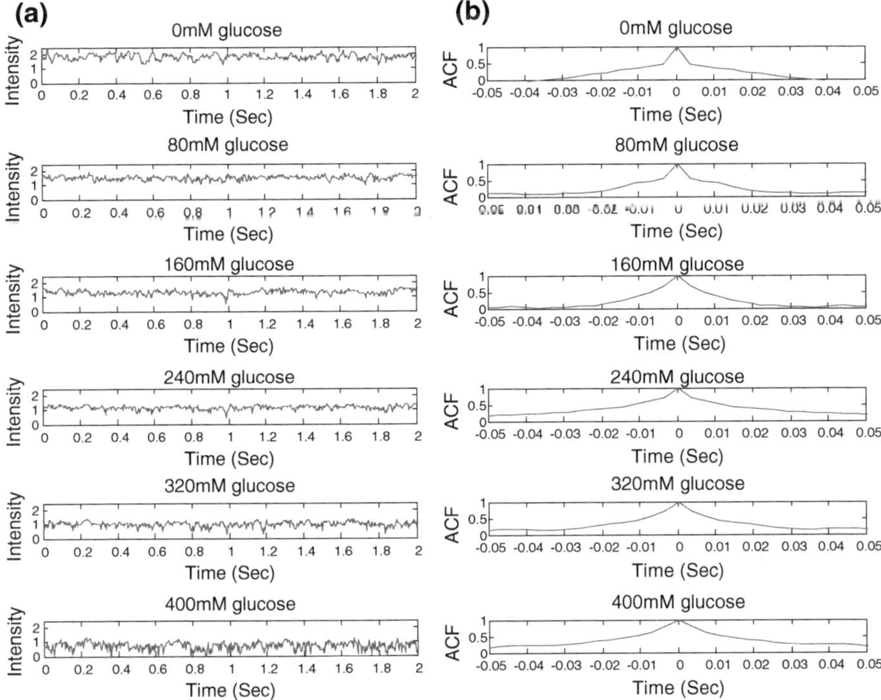

Fig. 12.6 **a** OCT M-mode intensity signals at a depth of 70 μm for six different glucose concentrations in the blood plasma + PMS phantom set (2) at 21 °C. The decrease in OCT signal can be exclusively observed for higher glucose concentrations specially at 400 mM [24, 25]. **b** Signal autocorrelation functions (*ACF*) obtained from the OCT M-mode data. *ACF* curves exhibit exponential decay, which is characteristic of Brownian motion of the scatterers. Slower relaxation is seen in samples with higher glucose concentration [24, 25]

Two distinct signal decorrelation mechanisms i.e. translation and rotation for scattering particles undergoing Brownian motion have been investigated for symmetrical and asymmetrical particles carrying samples respectively [24, 25, 31].

A single exponential fit function $f_1 = A*exp(-t/\tau_T)$ was applied to each *ACF* to extract the translational decorrelation time τ_T to sample water and plasma samples with the correlation coefficient r^2, typically ~ 0.98. A typical exponential to the plasma + PMS sample is shown in Fig. 12.9 prior to glucose addition. Translational decorrelation time τ_T derived obtained from single exponential fit is related to the translational diffusion coefficient D_T, which in turn depends on the medium viscosity. For spherical scatterers of radius *R* [24, 25, 31, 36];

$$\tau_T = \frac{1}{2k^2 D_T} \quad \text{and} \quad D_T = \frac{k_B T}{6\pi\eta R} \tag{12.3}$$

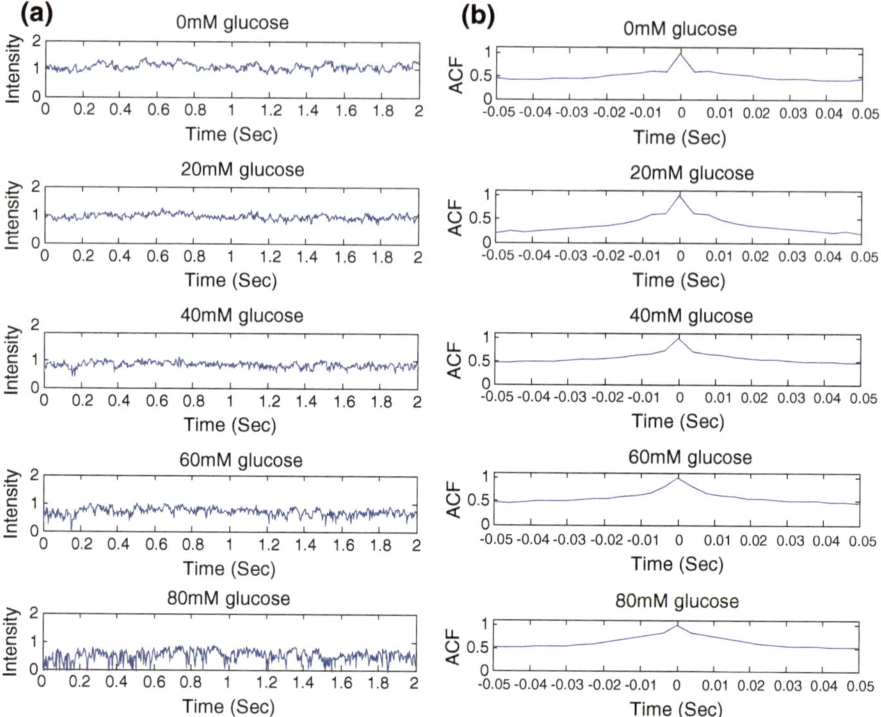

Fig. 12.7 **a** OCT M-mode intensity signals at a depth of 70 μm for six different glucose concentrations in the whole blood phantom set (3) at 21 °C. The OCT signal can be clearly observed in decreasing trend for glucose concentrations. The effect is more pronounced at 80 mM, where RBCs are more likely to burst up after deformation [24, 25]. **b** Signal autocorrelation functions (*ACF*) obtained from the OCT M-mode data. *ACF* curves exhibit exponential decay, which is characteristic of Brownian motion of the scatterers. Slower relaxation is seen in samples with higher glucose concentration [24, 25]

where $k = 4\pi n/\lambda$ is wave number, n is the refractive index, k_B is Boltzmann constant, T is absolute temperature, and η is the viscosity of the medium [24, 25, 31].

For whole blood, results having asymmetric RBCs, a double exponential fit form $f_2 = B^* \exp(-t/\tau'_T) + C^* \exp(-t/\tau_R)$ (where τ'_T and τ_R are the translational and rotational decorrelation times of erythrocytes) to account for both translational and rotational relaxations was applied to the OCT's ACFs. By considering the erythrcytes as a flattened ellipsoid of radius R we can write ½ of the average thickness a, in analogy with Eq. (12.3) [24, 25, 31, 36];

$$\tau'_T = \frac{1}{2k^2 D'_T} \quad \text{and} \quad D'_T = \frac{k_B T}{6\pi\eta a} G(\rho) \tag{12.4}$$

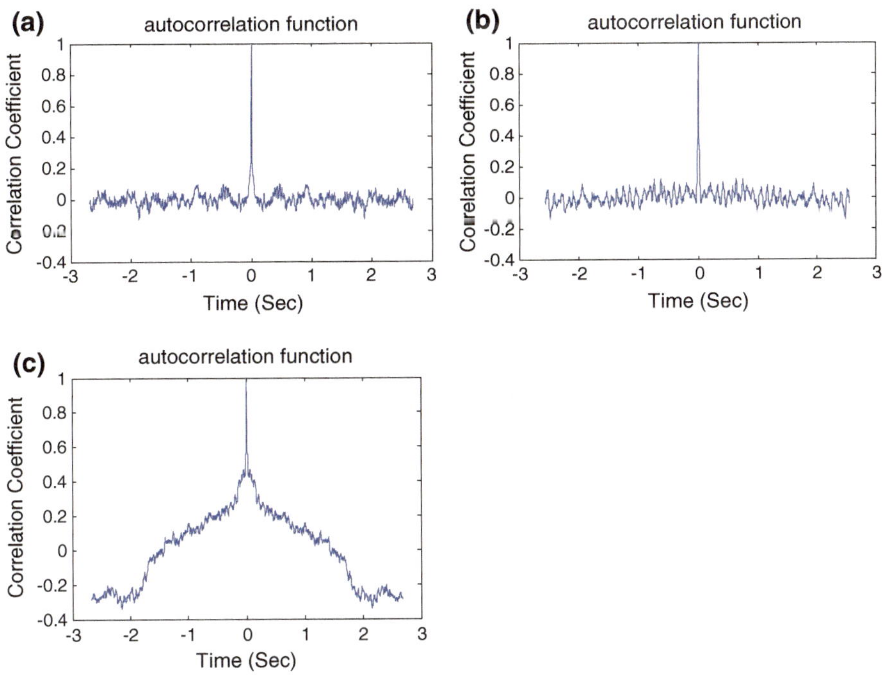

Fig. 12.8 Autocorrelation functions obtained from OCT average intensity data for sample sets **a** for 1, **b** for 2, and **c** for (3) prior to glucose addition [24, 25]

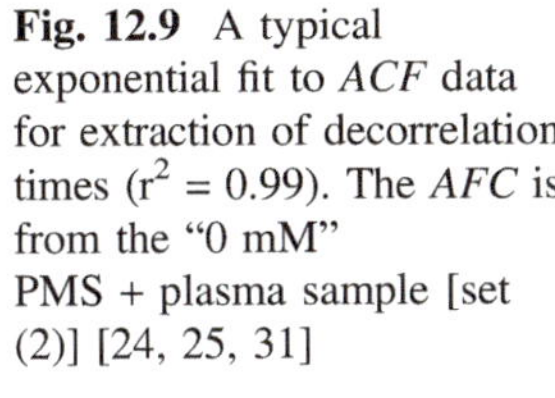

Fig. 12.9 A typical exponential fit to *ACF* data for extraction of decorrelation times ($r^2 = 0.99$). The *AFC* is from the "0 mM" PMS + plasma sample [set (2)] [24, 25, 31]

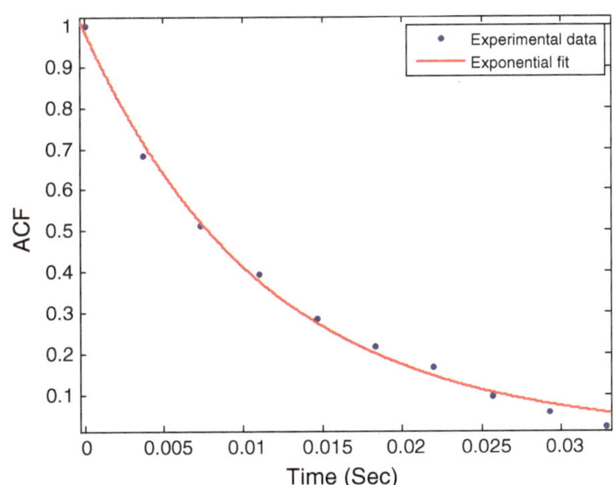

where G is a geometrical factor given by $G(\rho) = (\rho^2 - 1)^{1/2} * \rho * \text{arc}\tan(\rho^2 - 1)$ and $\rho = \dfrac{R}{a} > 1$. For a red blood cell, R $\sim$ 3.5 μm and $a \sim$ 1.0 μm [31]. Looking at the ratio of B/C to estimate the relative importance of translational against rotational relaxation for the five blood samples experimental group (3), this ratio was ~2.7, indicates that translational decorrelation is dominant over the rotational relaxation [24, 25, 31].

12.6.4 Discussions

Table 12.1 summarizes the quantitative study findings for the samples investigated in this work [31]. Prior to glucose in water + PMS sample, the obtained viscosity η_{water} = 0.88 MPa s, agreeing with the literature i.e. 0.98 MPa s at 20 °C [37]. The 10 % difference may be due to due to the imprecise experimental temperature because the viscosity is highly temperature dependent. Comparing with literature, our derived viscosity value of 1.44 MPa s for the 500 mM sample (9 % w/w) is in reasonable agreement with the previously-reported η = 1.24 MPa s for 10 % w/w glucose solution [24, 25, 38].

Table 12.1 Summary of the OCT experimental results and analysis for the three phantom sets, demonstrating the quantification of glucose-induced viscosity changes [24, 25, 31]

Sample glucose concentration (mM)	Translational decorrelation time (ms)	Viscosity	r^2-value
Phantom set (1)—single exponential fit			
0	8.77 ± 0.04	0.88	0.97
100	9.25 ± 0.06	0.93	0.98
200	10.21 ± 0.06	1.03	0.98
300	11.20 ± 0.10	1.13	0.99
400	12.20 ± 0.06	1.24	0.97
500	14.10 ± 0.05	1.44	0.95
Phantom set (2)—single exponential fit			
0	11.46 ± 0.14	1.16	0.99
80	12.27 ± 0.16	1.25	0.99
160	15.70 ± 0.12	1.60	0.98
240	21.12 ± 0.21	2.17	0.97
320	26.4 ± 0.16	2.72	0.94
400	44.1 ± 0.29	4.61	0.99
Phantom set (3)—double exponential fit[a]			
0	8.20 ± 0.04	9.03	0.93
20	9.52 ± 0.08 (8.41 ± 0.04)	10.26	0.99
40	12.00 ± 0.05 (10.40 ± 0.06)	10.81	0.91
60	26.30 ± 0.13 (12.2 ± 0.06)	24.61	0.93
80	63.00 ± 0.40 (25.00 ± 0.14)	51.05	0.99

[a]For the blood samples, the bracketed decorrelation times represent measurement baseline drift due to blood exposure to air, in the absence of glucose aliquots (see text for details). It is corrected for this prior to using Eq. (12.4) to derive the displayed viscosity values in column (3), via τ'_T (corrected) = τ'_T (0 mM) + [difference between the τ'_T values in column (2)]

The plasma + PMS phantom set results exhibit a similar trend to the water + PMS phantoms. Its baseline viscosity is of 1.16 MPa s ~ 28 % higher than that of water, and agrees reasonably well with literature values of 1.29 MPa s [39]. Comparing the η_{plasma} glucose dependence to water-based results as displayed in Fig. 12.10, a slightly higher increase in viscosity has been observed.

The derived viscosity value for the blood sample without added glucose is ~8 times that of water. This static η_{blood} value also agrees well with the literature, where a value of ~10 MPa s is reported [40, 41]. It shows that the addition of glucose in whole blood has a more drastic change compared to the average viscosity of water or plasma being the non-Newtonian fluid. The blood is difficult to explain because of supra- exponential behavior. This, perhaps, is not surprising, given the complex nature of the environment—the viscosity changes, irregular shape of the RBC and its changes, exposure to air complications, may RBC aggregations, etc. [24, 25, 31].

Further study includes the potential complications caused by the red blood cell shape change and aggregation after increases in glucose concentrations in a whole blood smear are shown in Fig. 12.11 in microscopy results. Normal rat blood's erythrocytes behavior in Fig. 12.11a is expected, and is seen as ~7 μm diameter and a double concave disc-like shape. With addition of glucose (20 mM in Fig. 12.11b, 40 mM in 12.11c), deformation of individual erythrocytes into spheroidal shapes, and their multi-particle aggregation behavior are becoming evident. The former is a well known osmotic effect, which may be an indication of the formation of rouleaux [24, 25, 31, 42–46].

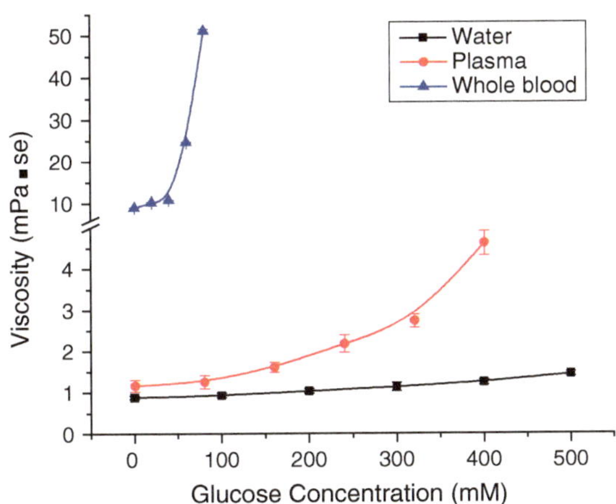

Fig. 12.10 Viscosities of water and plasma phantom sets [with PS microsphere scatterers, analyzed via Eq. (4.3)], and of whole blood [with RBC scatterers, analyzed via Eq. (12.4)]. **a** Linear scale **b** Semi-logarithmic scale, showing exponential dependence of viscosity on glucose levels in water and plasma systems, and a more complicated (supra-exponential) behavior in whole blood. Symbols are the results of experimental measurements (with errors bars = standard deviations) [24, 25, 31]

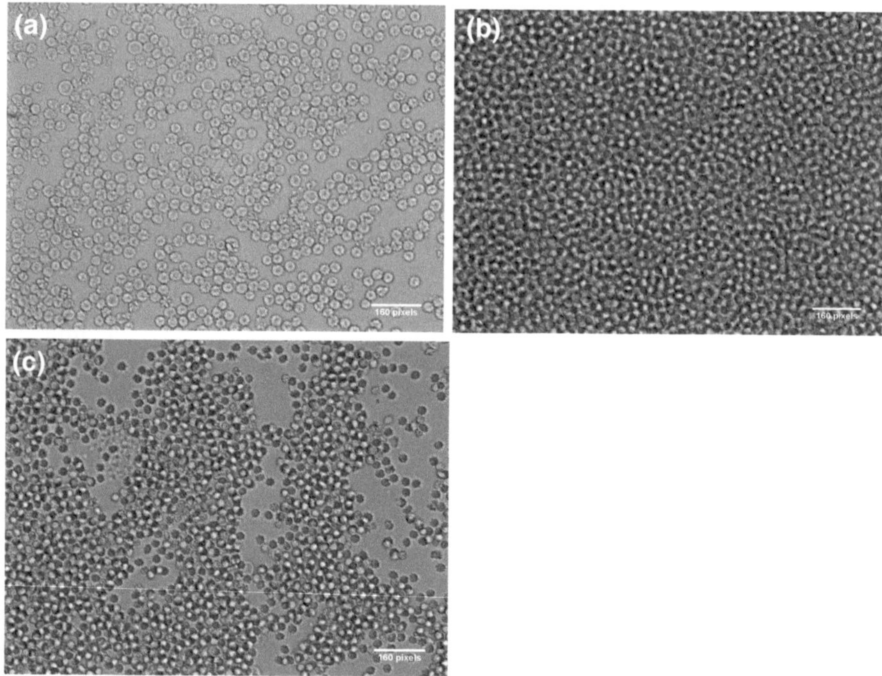

Fig. 12.11 Microscopy results demonstrating deformation and rouleaux formation of RBCs in rat whole blood samples with the addition of glucose. **a** No glucose added, **b** 20 mM, and **c** 40 mM. The shape of individual cells changes from biconcave discs in **a** to more spheroidal shapes in **b** and **c**. In addition, the collective aggregation in **c** is suggestive of rouleaux formation. (Image size: width × height = 1392 × 1040 pixels, 1 pixel = 6.45 μm). Field of view of microscope objective was 8.98 × 6.71 mm [24, 25, 31]

12.7 Microscopic Images for Qualitative Monitoring

White light microscope in transmission mode has been used to examine the behavior of rupturing of RBCs after admixing of three types of glucose concentration in blood (in vitro) [47].

12.7.1 Materials and Methods

Glucose ($C_6H_{12}O_6$) was dissolved in the whole blood sample. For prevention of clotting of 29 years human blood, 9 heparin tubes were used and anticoagulant (citrate phosphate dextrose adenine (CPDA-1) was added in each tube. The sample consisted of [whole blood + nine different concentrations of glucose (0, 50, 100, 150, 200, 250, 300, 350, 400 mM)] that were admixed in each of heparin tube and shake well gently to mix-up the glucose completely. After getting mixed the glucose, a blood smear was prepared on a glass slide for all aforementioned concentrations and was kept under a transmission mode microscope with 40X magnification [47].

12.7.2 Results and Discussions

Figure 12.12 shows a clear effect of cell's deformations from its natural shape and aggregations as the concentrations of glucose is increased from 0 to 400 mM. The complete burst-up starting from biconcave disc like shape to sphereoidal shape and ultimately damaging of the whole structure has been observed. Figure 12.12a shows the native structure of erythrocytes i.e. a complete pan cake like shape [44] is visible. Figure 12.12b, c show the clumping of RBCs tending towards the rouleaux formation. After, the addition of 150–200 mM of glucose, the RBCs are deformed and there shape start transition from biconcave discs to more spheroidal shape, getting swellings perhaps due to hyperglycaemic shock and consequently, drastically the burst up of RBCs have been observed as shown in Fig. 12.12d, e. In Fig. 12.12f–h, for higher concentration i.e. 250–350 mM a very pronounced change of destruction/lyses of the RBCs can be observed. It gets totally deformed in size and shape in drastic shrivelling as shown in Fig. 12.12i creating the irreversible situation where RBCs become unable to re-function and die out ultimately [47].

The, normal glucose levels in human blood lie in the range of 70–140 mg/dl before meals (fasting level) and fluctuates up to 200 mg/dL after two hours of the meal. According to American diabetes association (ADA) [48] report, the hyperglycaemic condition for the diabetes is greater than 240 mg/dl. The patients at this level need the proper medication, dieting and exercise to bring this level down [48]. We have investigated the optimum level of glucose in the human blood after those the erythrocytes become non-functional [47].

In order to examine potential complications due to RBC shape changes and aggregation in the presence of higher glucose concentrations, the images with the help of microscope from blood smears are shown in Fig. 12.12. Normal human blood in Fig. 12.12a exhibits expected biconcave (disc like) shapes of RBCs $\sim$ 7 μm diameter with a central void in this microscopic image [44]. Figure 12.12b, c show the just start-up of the aggregation/clumping of RBCs after addition of 50 and 100 mM glucose concentration respectively. Evidently, the aggregation of multi-particle (indicative of the rouleaux formation) and deformation of individual erythrocytes from disc like shape to spheroidal shapes have been initiated. This transition is due to a well known osmotic pressure effect [43, 45, 46]. Figure 12.1b, c with glucose concentrations of 50–100 mM lies in the physiologically levels [49] (that forms the surrounding of RBC's cytoplasm is compatible with the blood plasma) tries to keep the erythrocytes in its original size and shape. After, the addition of 150–200 mM of glucose, the RBCs have been deformed and there shape has started to change from biconcave discs to more spherulated (spheroidal shape),

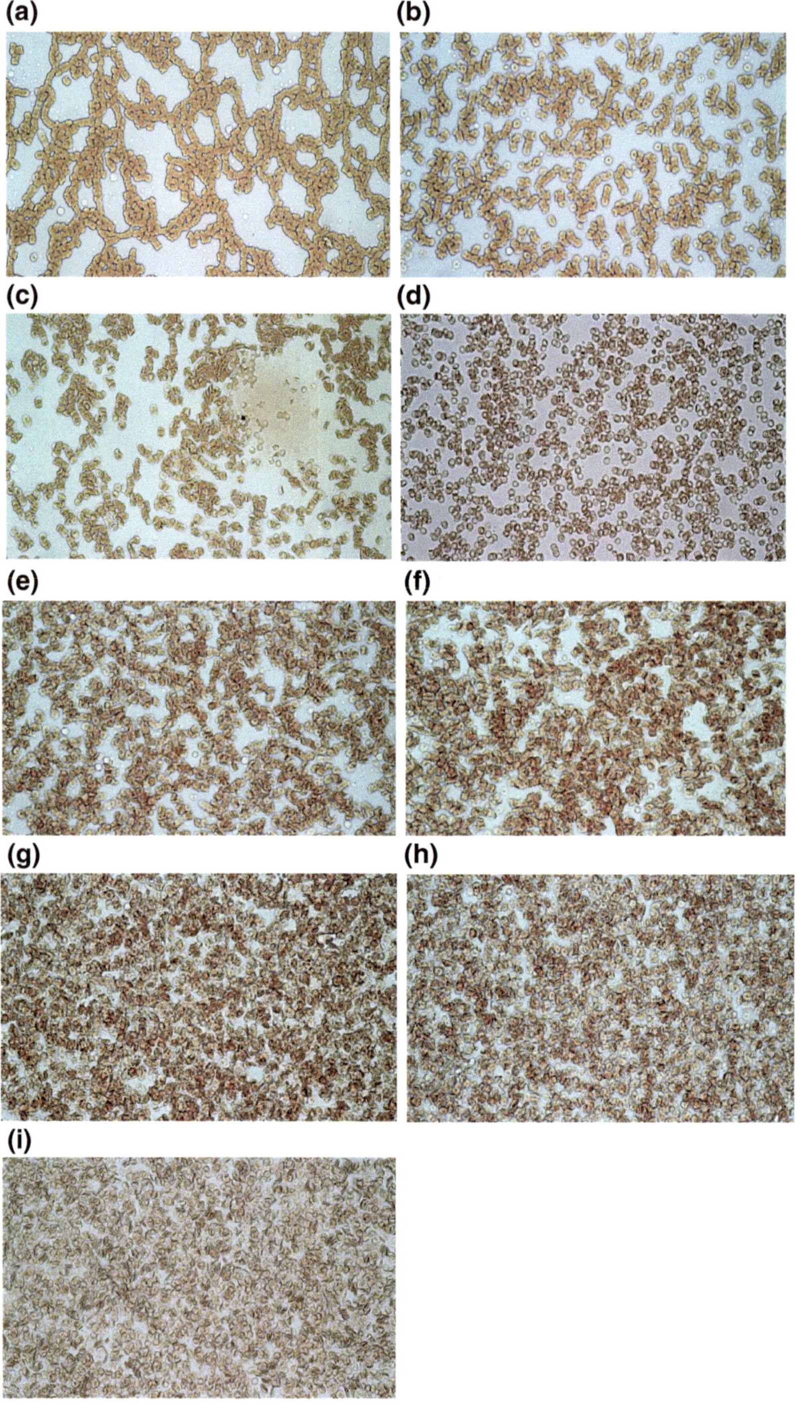
(a)
(b)
(c)
(d)
(e)
(f)
(g)
(h)
(i)

◀ **Fig. 12.12** 2D images of erythrocytes in whole blood of 19 years old human obtained with light microscope in transmission mode for glucose concentrations of **a** 0 mM, **b** 50 mM, **c** 100 mM, **d** 150 mM, **e** 200 mM, **f** 250 mM, **g** 300 mM, **h** 350 mM, and **i** 400 mM [47]

perhaps due to hyperglycaemic shock and consequently a drastic change tending towards the burst up of RBCs is shown in Fig. 12.12d, e. These results have good agreement with literature [44] where 100 and 200 mM of glucose concentration made a significant change in the size and shape of RBCs in which the spheroidal shape has been reported. For higher concentrations of glucose i.e. more than 250 mM, the well known effect of hyperglycemic shock of erythrocyte is observed. This is an evidence of morphological changes of the whole blood underlying the physiological and hyperglycemic conditions. In Fig. 12.12f–h, for 250–350 mM concentration, a very pronounced change of destruction (drastic shrivelling) can be observed i.e. getting totally deformed in shape and causing the reduction in number of normocytes as well. Figure 12.12i shows the situation for 400 mM glucose creating the irreversible situation where RBCs become unable to re-function and die out ultimately [44, 50]. This is actually due to the reduction in the diameter of RBCs, while their outer membrane rings got thicker and ultimately gets burst [44]. These changes in shapes support the characteristic of transformation from an essentially flattened shape to a spherical shape as well as rod like shape. This might be supported by reduction in the optical attenuation due to refractive indices mismatching effect [49, 51]. The effect is dominant in scattering properties with negligible absorption of light in whole blood. The effective optical scattering will gradually decrease due to increase in mismatch of refractive indices between plasma serum and RBCs [52]. Thus, by losing the immunity, erythrocytes couldn't face the severe hyperglycemic conditions and after swelling the outer membrane were permanently damaged. In diabetics, these changes can be measured for in vivo cases in accurate glucose monitoring as a significant *non-invasive* application [47, 53].

12.8 Flow Phantom Measurements

This work is an extension of stagnant blood quantification results where glucose was monitored via its effect on medium viscosity modulation [31]. Brownian motion plays a crucial role in explanation of the flowing fluid diffusion theory [54–56]. In this work, the diffusion coefficient of the PMS from the relaxation time in the water and plasma's OCT M-mode measurements, and whole blood's erythrocytes have been carried out. The resulting relaxation/decorrelated time's changes were caused by glucose addition by modulating flow rate. These results are enabling steps towards further elucidating the source of glucose-induced contrast in various OCT imaging modes (M-mode, SV-OCT). As we have demonstrated an initial SV-OCT imaging of polymer capillary tube mimic to blood vessels in which blood was flown under the action of gravity [24, 25, 57, 58].

Like stagnant experiments, three types of phantoms (water, blood plasma and whole blood) have been investigated flowing under the action of gravity trough a capillary polymer tube of 600 μm diameter. This study includes the measurements of translational diffusion coefficient D_T for spherical and disc shaped particles with different glucose concentrations under the effect of dynamics light scattering [24, 25, 58].

12.8.1 Samples Preparation

(1) The first set was water phantoms with 1.4 μm diameter PMS and four different concentrations of glucose (0, 100, 200, and 300 mM). The PMS concentration was increased from 0.69 % (weight/volume) for glucose-free suspension to 0.74 % for the 0.3 M phantom, in order to keep the scattering coefficient μ_s constant in the presence of the glucose-caused refractive index matching effect [59]. Assuming the refractive index of water ~1.33, μ_s was calculated from Mie theory at 1310 nm to be 100 cm^{-1} [32]. (2) The second sample was blood plasma with PMS suspensions. A fixed concentration 0.0073 % of 1.4 μm microspheres was added to plasma along with varying glucose amounts (0, 100, 200, and 300 mM). Its refractive index was assumed to be ~1.34, extrapolated from previously-reported n ~ 1.35 at 630 nm for human [34] blood plasma. (3) Finally, whole blood containing asymmetric RBCs scatterers was admixed with additional glucose of 0, 20, 40, 60 and 80 mM levels [24, 25, 58].

12.8.2 Measurements and Signal Processing

Phantom fluid was allowed to flow under gravity through the inner diameter 600 μm polymer capillaries with different flow rates and velocity due to changes in glucose concentration. Figure 12.13 shows a snapshot of the experiment for the blood flow measurement under the SS-OCT scanning head [24, 25].

For a typical water phantom set (1) with a 0 mM glucose, as shown in Fig. 12.14, the flow rate, Q was determined from time data versus the mass flow. The average flow velocity v was calculated from measured mass flow, density, from the tube cross-sectional area with the equation [24, 25];

$$v = Q\,(\text{g/s}) \times \frac{1}{\rho\,(\text{g/ml})} \times \frac{1}{Area\,(\text{mm}^2)}. \tag{12.5}$$

where Q is flow rate and ρ is the density of the fluid. Shear rate $\dot{\gamma}$ is an important parameter in specific flow dynamics in hemodynamic as it is correlated with the shape deformation behavior of red blood cells [60]. Shear rate of each sample set

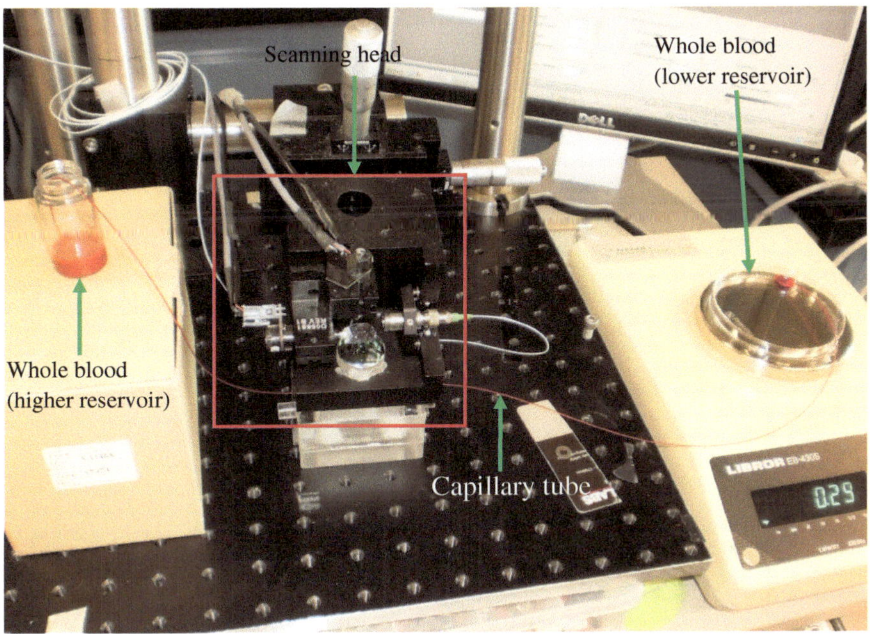

Fig. 12.13 Experimental setup snapshot consisting on higher reservoir, lower reservoir, balance and SS-OCT sample arm for a typical sample set (3) [24, 25]

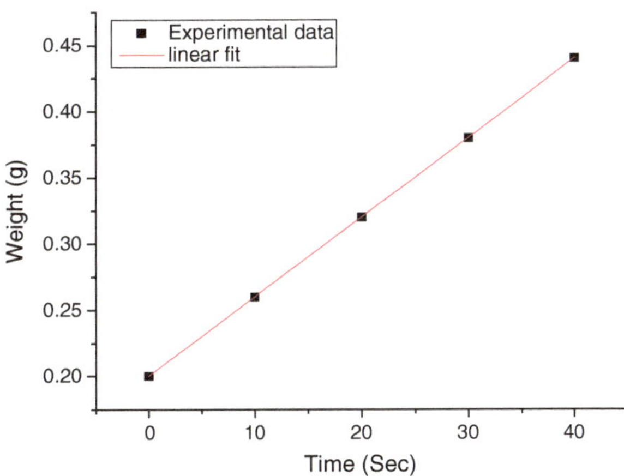

Fig. 12.14 Flow rate evaluation: The slope gives the mass flow rate for a typical sample of water phantom with 0 mM glucose in it which is 6.0 mg/s [24, 25]

for different glucose concentrations was calculated by $\overset{\bullet}{\gamma} = \frac{8v}{d}$ [61], where v is the velocity of the fluid and d is the diameter of the capillary tube.

A Typical SV-OCT image of polymer tube with flown whole blood prior to glucose addition is shown in Fig. 12.15a (6400 × 1520 × 512 pixels). This is mainly (display perhaps typically observed parabolic velocity profile) that is intended to

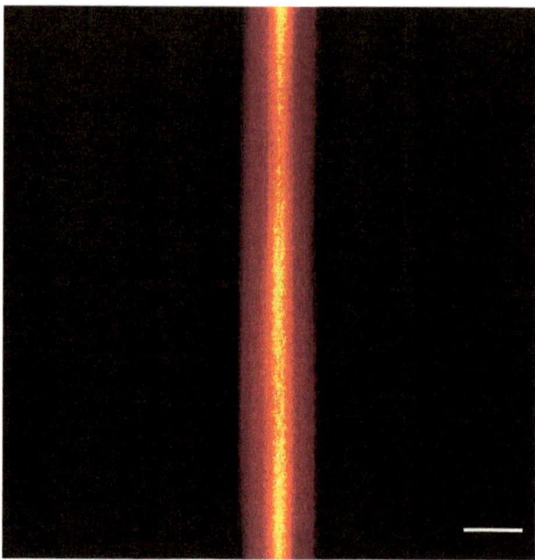

Fig. 12.15 Speckle variance OCT image of blood mimic capillary tube (600 μm) carrying whole blood (with 80 mM glucose concentration) flowing through it depicts that velocity of fluid is higher in the center of tube. Dimensions: (6400 × 1520 × 512, pixels) and (6 mm × 6 mm × 3.2 mm). *Scale bar* 750 μm [25, 58]

indicate a velocity gradient in the tube under a higher rate of medium flow conditions [24, 25, 60]. The M-mode analyses were performed at a depth of 40 μm below the liquid surface (lower interface of the upper wall of the tube). The scanning protocol and signal analysis was same as described in previous section for stagnant phantoms [25, 31, 58].

For sample set (1) and (2) a single exponential fit function $f_1 = A * \exp\left(-{}^{t}/_{\tau_T}\right)$ was applied to each *ACF* for extraction of the translational decorrelation time τ_T. The mono-exponential decay with correlation coefficient r^2 (for goodness of fit) was ~0.98 for samples in sets (1) and (2) yielding τ_T to derive the translational diffusion coefficient D_T. For spherical scatterers of radius R, D_T is evaluated using [24, 25, 31, 58],

$$\tau_T = \frac{1}{2k^2 D_T}. \tag{12.6}$$

where $k = 4\pi n/\lambda$ is wave number, n is the refractive index, and λ is wavelength.

D_T for phantom set (1 and 2) was evaluated with Eq. (12.6) for different glucose concentrations.

Whole blood samples (3) were analyzed with double exponential fit function $f_1 = B * \exp\left(-{}^{t}/_{\tau'_T}\right) + C * \exp\left(-{}^{t}/_{\tau_R}\right)$, where τ'_T and τ_R are the translational and rotational decorrelation times of erythrocytes [31, 58]. There is a possibility that has been pursued for another estimate of as well the corresponding rotational relaxation analysis τ_R. However, we have analyzed the translation by examining the B/C ratio (relative importance to the rotational relaxation). This ratio was suggesting the importance of the translational over the rotation relaxation to 1.4. Blood D'_T was evaluated using the same Eq. (12.6) at different glucose concentrations [24, 25, 58].

12.8.3 Results and Discussion

Table 12.2 summarizes the quantitative results including density, flow rate, velocity of three phantom sets and shear rates with different glucose additions in flowing phantom set (1, 2, and 3). Table 12.3 contains the quantitative parameters including derived translational decorrelation times for (1), (2), and (3), diffusion coefficients for (1), (2), and (3) and the fitting correlation coefficients r^2. Table 12.4 compares these flowing-fluid findings with our previous results for non-flowing phantoms [24, 25, 58].

Considering the water + PSM phantom set (1) first, our determined $D_T = 2.14 \times 10^{-12}$ m^2/s for the 0 mM glucose sample can be compared with a reported literature value with a value 4.0×10^{-13} m^2/s [62], although the experiment was performed at significantly lower shear rates (280/s for our result versus 21/s in [62]). Speed of flow, perhaps indeed we have observed as, for faster decorrelation, will be yield large diffusion coefficient. The possible reasons include the difference in microsphere size and concentration (1.4 μm, 0.71 % vs. 1.1 μm, 0.11 %) for the

Table 12.2 Summary of the flow phantom OCT experimental results and analysis for the three phantom sets, demonstrating the density, flow rate and velocity [24, 25, 58]

Sample glucose concentration (mM)	Density (g/ml)	Flow rate (mg/s)	Avg. speed (mm/s)	Shear rate (s^{-1})
Phantom set (1) [water]				
0	1.009	6.0	21.0	280.1
100	1.012	5.6	19.5	260.7
200	1.023	5.3	18.3	244.1
300	1.025	5.1	17.2	229.8
Phantom set (2) [blood plasma]				
0	1.025	3.2	11.0	147.1
100	1.028	2.7	9.3	123.7
200	1.030	2.4	8.2	109.8
300	1.035	1.9	6.6	87.4
Phantom set (3) [whole blood]				
0	1.060	4.8	1.6	21.3
20	1.069	4.3	1.4	18.9
40	1.077	3.9	1.3	17.1
60	1.086	3.3	1.1	14.3
80	1.099	2.0	0.6	8.6

Table 12.3 Summary of the OCT experimental results and analysis for the three phantom sets, demonstrating the quantification of glucose-induced diffusion coefficient changes [24, 25, 58]

Sample glucose concentration (mM)	Translational decorrelation time (ms)	Translational diffusion coefficient (m^2/s)	r^2-value
Phantom set (1) [water]—single exponential fit			
0	1.44 ± 0.002	2.14×10^{-12}	0.99
100	2.30 ± 0.003	1.33×10^{-12}	0.99
200	2.95 ± 0.005	1.03×10^{-12}	0.99
300	3.37 ± 0.007	9.12×10^{-13}	0.99
Phantom set (2) [Plasma]—single exponential fit			
0	1.90 ± 0.006	1.59×10^{-12}	0.99
100	2.74 ± 0.006	1.10×10^{-12}	0.97
200	3.34 ± 0.001	8.99×10^{-13}	0.96
300	5.80 ± 0.006	5.16×10^{-13}	0.86
Phantom set (3) [Blood]—double exponential fit			
0	34.01 ± 0.69	8.65×10^{-14}	0.99
20	46.30 ± 0.40	6.39×10^{-14}	0.99
40	58.12 ± 0.55	5.06×10^{-14}	0.99
60	68.43 ± 0.57	4.31×10^{-14}	0.99
80	129.02 ± 0.94	2.27×10^{-14}	0.99

Table 12.4 Comparison of flow phantom results with previous results of static fluids [24, 25, 31, 58]

Sample glucose concentration (mM)	Translational diffusion coefficient-static fluids (m^2/s)	Translational diffusion coefficient-flowing fluids (m^2/s)	Flowing/static ratio
Phantom set (1) [water]—single exponential fit			
0	3.51×10^{-13}	2.14×10^{-12}	6.11
100	3.31×10^{-13}	1.33×10^{-12}	4.02
200	2.99×10^{-13}	1.03×10^{-12}	3.45
300	2.71×10^{-13}	9.12×10^{-13}	3.34
Phantom set (2) [plasma]—single exponential fit			
	2.64×10^{-13}@0 mM glucose	1.59×10^{-12} @0 mM glucose	6.02
	2.46×10^{-13}@80 mM glucose	1.10×10^{-12} @100 mM glucose	4.47
	1.92×10^{-13}@160 mM glucose	8.99×10^{-13} @200 mM glucose	4.68
	1.13×10^{-13}@320 mM glucose	5.16×10^{-13} @300 mM glucose	4.56
Phantom set (3) [blood]—double exponential fit			
0	3.59×10^{-13}	8.65×10^{-14}	0.24
20	3.09×10^{-13}	6.39×10^{-14}	0.21
40	2.45×10^{-13}	5.06×10^{-14}	0.21
60	1.12×10^{-13}	4.31×10^{-14}	0.39
80	4.65×10^{-14}	2.27×10^{-14}	0.49

~5X difference in the measured diffusion coefficient values for the water + PMS glucose-free medium. A similar trend of glucose-caused decrease in D_T is observed in plasma + PMS phantom set (2). Note that the corresponding flow speeds are lower in plasma than in water phantoms (1), likely owing to greater viscosity of the plasma compared to water. For example, water phantom without glucose flows with 21 mm/s than plasma phantom which has speed of 11 mm/s and corresponding diffusion coefficients are 2.14×10^{-12} m^2/s and 1.59×10^{-12} m^2/s. These values are ~6X greater than our previous stagnant-fluids findings [24, 25, 58].

It's reassuring to note that the value of D'_T for whole blood (without added glucose) agrees well with the literature, where a translational diffusion coefficient of ~6.5×10^{-14} m^2/s is reported [24, 25, 56, 58]. Thus, many possibilities by forced convection caused modulations in the shear rate of the phantom flow and the glucose concentration is significantly indicates a change of D_T. Therefore, OCT has the potential to quantify the glucose levels in the blood stream, based on the dynamic light scattering underlying the Brownian motion [24, 25, 58, 63].

A significant application of these methods lies in the patients of hyperglycemia where oxygen carrying capability of blood is enhanced by lowering blood viscosity after fibrinogen depletion. the occurrence of atherothrombotic vascular disease is monitored with the increased viscosity of blood or plasma an indicator. This works has reported the reference value of glucose diffusion by measuring diffusion coefficient of healthy animals that can be implemented clinically and lab experiments as a non invasive method. The OCT's capability for monitoring glucose in blood vessels of diameters in micron ranges has a worth of non invasive light scattering methods as we imaged animal's blood vessels without injection of glucose. It provides the route to enhance the application for further glucose concentrations quantifications in the presence of RBCs [24, 25, 58].

12.9 In Vivo Blood Glucose Quantification

SV-OCT and structured OCT have been utilized to map out the normal and malignant blood microvasculature in three and two dimensions and to monitor the glucose levels in blood by analyzing the Brownian motion of erythrocytes underlying dynamic light scattering. The technique has been applied on nude live mouse's skin for which the microscopic images have also been obtained for both types of blood vessels i.e. normal and tumor [24, 25, 64].

12.9.1 The Dorsal Skinfold Mouse Window Chamber Model

Dorsal skinfold mouse windows chamber (DS-MWCM) has been widely used in microscopy for the investigation of striated skin muscle of nude mouse in immunoincompetent [65] and for quantitative analysis of the microhemodynamic parameters such as microvessel diameter, RBC's velocity in arterioles and the autochthonous microvasculature [66]. The DS-MWCM has been implemented to image normal micro blood vessels, tumor angiogenesis with SV-OCT [67]. The crucial parameter required for SV-OCT includes the minimization of tissue motion artifacts; therefore, DS-MWCM provides a stationary platform (free from bulk motion) and is shown in Fig. 12.16a implanted on a nude mouse. In Fig. 12.16b the white light microscopy image of window chamber model have been depicted for illustration purpose that delineates the blood vessels for SV-OCT mapping [24, 25, 64].

(a)

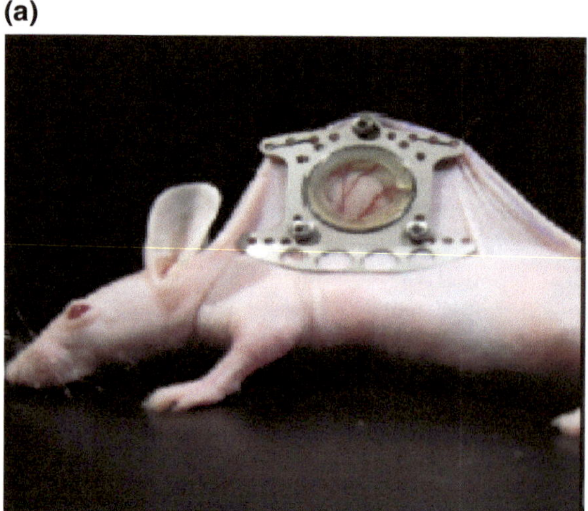

(b)

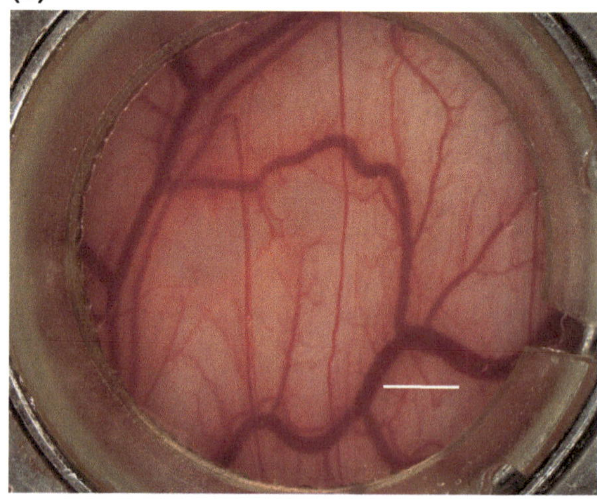

Fig. 12.16 **a** The dorsal skin-fold window chamber model implanted in a female NcrNu.nude mouse, and **b** white light microscopy image of Window chamber model that delineates the blood vessels for SV-OCT mapping. *Scale bar* 2 mm [24, 25]

12.9.2 Animal Model

The SV algorithm has been used to image microvasculature net of normal and tumor induced mice (in vivo). A DS-MWCM was applied on anesthetized mouse under. The mouse was anesthetized with a ketamine–xylazine (90–10 mg/kg) mixture. A 10 mm diameter region of skin is placed between the titanium plates of DS-MWCM. A 12 mm diameter and 250 μm thick cover slip is used to protect the exposed fascia and vasculature. OCT imaging was performed by fixing the DS-MWCM into the removable light weight aluminum plate to keep the animal at 37 °C after recovery of the animal from surgery. The tumor ME180 human cervical carcinoma cells were transfected with the DsRed2 fluorescent protein into the mice. We waited for one week so that tumor would grow before imaging after injection the tumor into the fascia. The use of fluorescent protein enhances the visibility of tumor through microscope for subsequent coregistration of fluorescent and SV-OCT images [24, 25, 64].

12.9.3 OCT Imaging System

OCT data were obtained from 36 kHz SS-OCT system as described earlier [24, 64]. OCT system, using a broadband light source is based on dispersion of a number of the k-values signal. In SS-OCT, the laser source does not directly sweep linearly into k-space but rather time space. Thus, firstly re-calibration is required from the data of the linear-in-time to data in a linear -k. The general approach to achieve this goal is to fix the optical path length for interferometeric signal. Therefore, it can be written the SS-OCT signal in k-space like [24, 25, 64];

$$I(k) \approx S(k)\cos(2knz_0), \tag{12.7}$$

where, $S(k)$ is the incident intensity for each k-value used (the swept source spectrum) and z_0 = offset distance between reference plane and object surface. A photograph of scanning head with animal under measurements is shown in Fig. 12.17a [24, 25, 64].

12.9.4 Speckle Variance OCT: Blood Vessel Images

The interframe OCT fluid contrast algorithm termed as SV-OCT is based on the idea that the image speckle/texture of relatively solid regions will persist between consecutive images, whereas speckle will show greater inter-image speckle washout in regions of greater fluidity [68]. The algorithm for generation of speckle variance

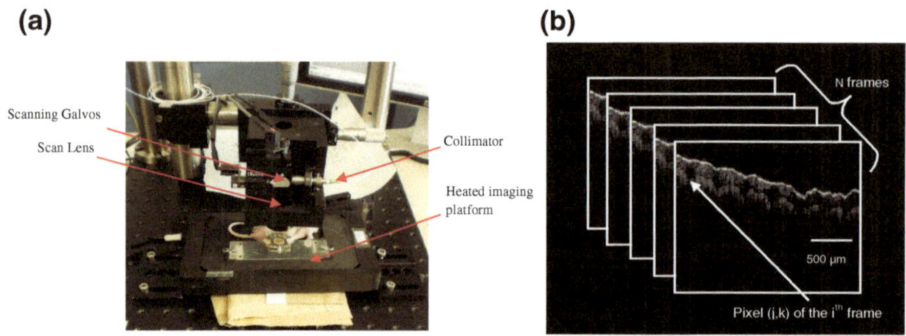

Fig. 12.17 **a** A nude mouse with WCM implanted to image blood vessels with SV-OCT under the scanning head of OCT system, **b** A conceptual diagram of an acquired speckle variance data set of *N* frames and corresponding indices used to label the frame (*i*), transverse pixel location (*j*) and the axial pixel location (*k*) [24, 64]

images of OCT data needs to calculate the variance of pixels from a set of *N*, B-mode images (N is gate length), acquired from the same spatial location [24, 25, 64];

$$SV_{ijk} = \frac{1}{N}\sum_{i=1}^{N}\left(I_{ijk} - \frac{1}{N}\sum_{i=1}^{N} I_{ijk}\right)^2. \tag{12.8}$$

where *i, j* and *k* are indices for the frame (up to N), transverse, and axial pixels, and *I* is the corresponding pixel intensity value at the imaging speeds of 20 frames per second. A more clearly schematic representation of the data set and pixel indices for three dimensional frames stacking according to Eq. (12.8) is shown in Fig. 12.17b.

12.9.5 Results and Discussions

Figure 12.18a shows a blood vessels image of normal nude mouse obtained with SV-OCT. this 3D processed image of blood microvasculature shows the distinction between the individual blood vessels with dimensions of 6 mm × 6 mm × 2.2 mm. Figure 12.18b shows a light microscopic image displaying the micro vessels with high lateral resolution with dimension of 6.2 × 6.2 mm in DS-MWCM. Figure 12.19a shows blood vessels having tumor induced in nude mouse obtained with SV-OCT with dimensions of 6 mm × 6 mm × 2.2 mm. Figure 12.19b shows corresponding maximum-intensity projection image of a fluorescence confocal z-stack image obtained using 500 kD Fluorocein labeled dextran with dimensions of 2.2 × 2.2 mm [24, 25, 64].

Different layers under laying the fascia can be observed in an image (Fig. 12.20a) that depicts the original structured OCT image but the image is not clearer for diagnostic point of view. Anyhow, sufficient information is available to

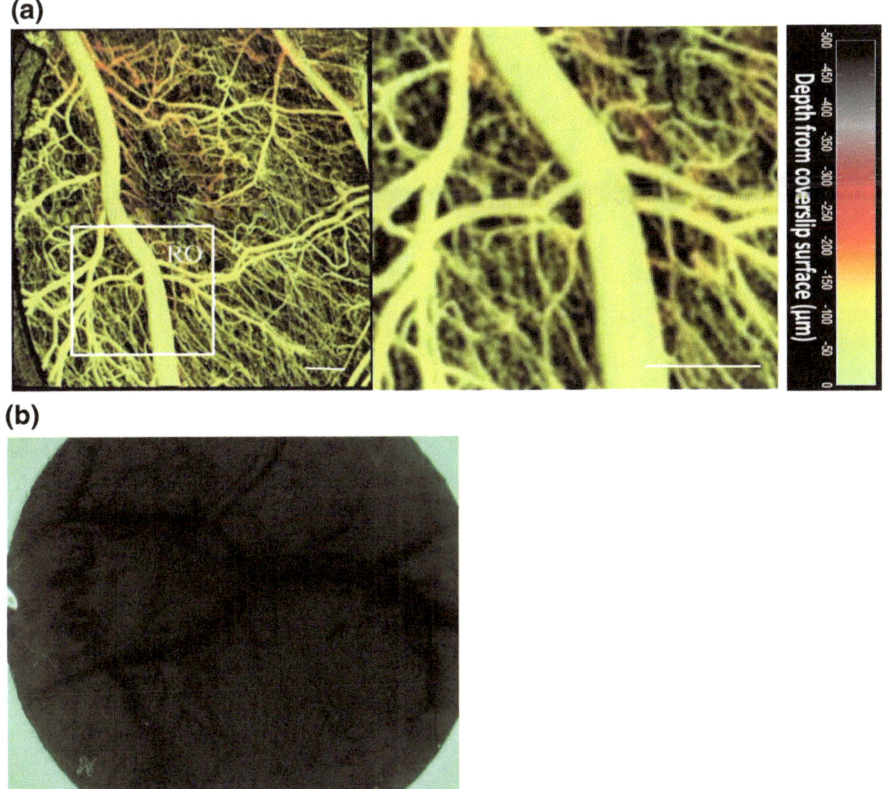

Fig. 12.18 **a** SV-OCT microvasculature image of a normal mouse in WCM, a low bulk tissue motion situation with gate length N = 8. This shows a 3D processed image of blood microvasculature to distinguish the individual blood vessels. Dimensions: (6400 × 1520 × 512, pixels) and/or (6 mm × 6 mm × 2.2 mm). The ROI is represented with magnification and depth encoded color bar used to specify the depth of the vessel. *Scale bar* 250 μm, **b** a microscopic z-stack image displaying the micro vessels with high lateral resolution but minimal axial information. *Dimension* (6.2 × 6.2 mm) and *Scale bar* 1 mm [24, 64]

the limit for layers differentiations. The corresponding OCT signal decaying exponentially for nude female mouse skin in vivo is shown in Fig. 12.20b. The first peak corresponds to the backscattering of cover glass and second peck represents the surface of the tissue [24, 25, 64].

The SV-OCT image was used to select the ROI for in vivo study of glucose monitoring in five animals. With the help of M-mode imaging the average value of D_T was yielded ~5.85 × 10^{-14} m^2/s (Table 12.5). The only 10 % difference was observed from theoretical value of diffusion coefficient (5.85 × 10^{-14} m^2/s vs. 6.50 × 10^{-14} m^2/s [56]). The reason for this difference may include the forced convection, shear effect and technique employed for measurements. A clearer 3D image of glucose free mouse's microvasculature bed helps to visualize the suitable

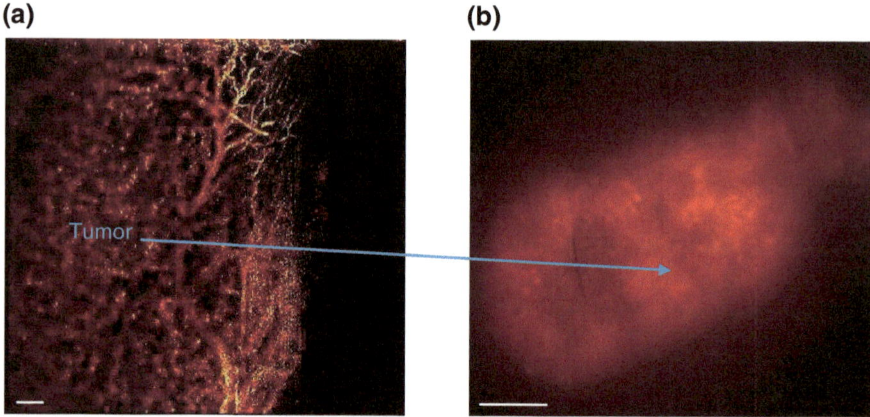

Fig. 12.19 **a** Blood vessels of nude tumor induced mouse obtained with SV-OCT. This shows a 3D processed image of blood microvasculature to distinguish the individual blood vessels. Dimensions: (6400 × 1520 × 512, pixels) and/or (6 mm × 6 mm × 3.2 mm) and Scale bar: 250 μm and **b** corresponding maximum-intensity projection image of a fluorescence confocal z-stack obtained using 500 kD Fluorocein labeled dextran. *Dimensions* (2.2 × 2.2 mm) and *Scale bar* 500 μm [24, 64]

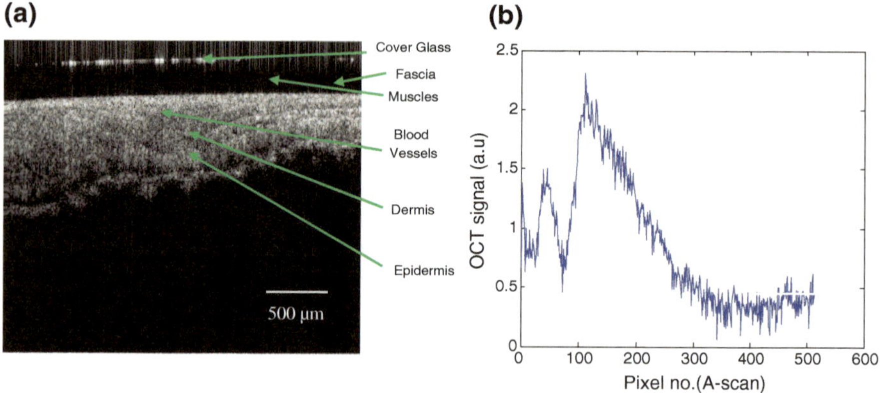

Fig. 12.20 **a** original gray scale structured OCT image of mouse skin. Dimensions (640 × 512 pixels) pixels, where 1 pixel = 3.9 μm. **b** The OCT signal decaying exponentially for nude female mouse skin in vivo. The first peak is due to the backscattering of cover glass and second peck represents the surface of the tissue. The depth contains a total of 512 pixels in ~3 mm depth, again 1 pixel ~3.9 μm [24, 64]

blood microvessells provides easy selection of the ROI to obtain M-mode data (Fig. 12.18a). In this way, we have achieved the threshold measurement to quantify glucose levels for in vivo case scenario of injected higher glucose concentrations intravenously [24, 25, 64].

Table 12.5 Summary of Brownian motion analysis results in dynamic light scattering regime including translational decorrelation time, translational diffusion coefficient and r^2 values for all five animals [24, 25, 64]

Animals	Translational decorrelation time (ms)	Translational diffusion coefficient (m^2/s)	r^2-value
Animal 1	38.20	1.02×10^{-14}	0.93
Animal 2	44.00	6.68×10^{-14}	0.95
Animal 3	48.80	6.03×10^{-14}	0.98
Animal 4	46.70	6.30×10^{-14}	0.99
Animal 5	52.20	9.20×10^{-14}	0.96
Average	45.98	5.85×10^{-14}	0.96

OCT is increasing the effectiveness of dramatically preclinical studies to monitor the disease in real-time. Especially in SV-OCT, speckle contrast levels provide a better 3D understanding than original structured unprocessed images for in vivo skin. The OCT technique, by monitoring the vascular system to a depth of a few millimeters in the case of superficial tumors, can be applied during the photodynamic therapy (PDT) procedures [69]. But during PDT and post PDT, there are multiple tissue reactions. Thus, the data response will interfere with the signal quality and resolution in ensuring the SV-OCT imaging after PDT session [24, 25, 64].

Acknowledgments Our own contributions in this chapter were supported by Higher Education Commission Pakistan, Islamabad, Pakistan and Canadian Institutes of Health Research, Ottawa, Canada. We would like to acknowledge all those authors whose results are included/cited in this work. We specially pay our thanks to Dr. Prof. Alex Vitkin, Department of Medical Biophysics, University of Toronto, Canada, who allowed me to conduct the experiments and discussed the results about the quantification of glucose levels in blood in his OCT laboratory.

References

1. C.A. Puliafito, M.R. Hee, J.S. Schuman, J.G. Fujimoto, *Optical Coherence Tomography of Ocular Diseases*, 2nd, illustrated ed. (SLACK Inc., New Jersey, 2004), p. 714
2. M.M.K.V. Larin, M.S. Eledrisi, R.O. Esenaliev, Noninvasive blood glucose monitoring with optical coherence tomography, a pilot study in human subjects. Diabetes Care **25**, 2263–2267 (2002)
3. P.A.M.W. Lindner, F. Kiesewetter, G. Häusler, *Hand book of Optical Coherence Tomography*, ed. by B. Bouma, E. Tearney (Marcel Dekker Inc., New York, 2002)
4. M. Atif, H. Ullah, M.Y. Hamza, M. Ikram, Catheters for optical coherence tomography. Laser Phys. Lett. **8**(9), 629–646 (2011)
5. M.E. Brezinski, G.J. Tearney, N.J. Weissman, S.A. Boppart, B.E. Bouma, M.R. Hee, A.E. Weyman, E.A. Swanson, J.F. Southern, J.G. Fujimoto, Assessing atherosclerotic plaque morphology: comparison of optical coherence tomography and high frequency intravascular ultrasound. Heart **77**(5), 397 (1997)

6. G.J. Tearney, M.E. Brezinski, J.F. Southern, B.E. Bouma, S.A. Boppart, J.G. Fujimoto, Optical biopsy in human gastrointestinal tissue using optical coherence tomography. Am. J. Gastroenterol. **92**(10), 1800–4 (1997)
7. C. Pitris, M.E. Brezinski, B.E. Bouma, G.J. Tearney, J.F. Southern, J.G. Fujimoton, High resolution imaging of the upper respiratory tract with optical coherence tomography. A feasibility study. Am. J. Respir. Crit. Care Med. **157**(5), 1640 (1998)
8. G.J. Tearney, M.E. Brezinski, J.F. Southern, B.E. Bouma, S.A. Boppart, J.G. Fujimoto, Optical biopsy in human urologic tissue using optical coherence tomography. J. Urol. **157**(5), 1915 (1997)
9. C.A. Jesser, S.A. Boppart, C. Pitris, D.L. Stamper, G.P. Nielsen, M.E. Brezinski, J.G. Fujimoto, High resolution imaging of transitional cell carcinoma with optical coherence tomography: Feasibility for the evaluation of bladder pathology. Br. J. Radiol. **72**(864), 1170 (1999)
10. C. Pitris, A.K. Goodman, S.A. Boppart, J.J. Libus, J.G. Fujimoto, M.E. Brezinski, High resolution imaging of cervical and uterine malignancies using optical coherence tomography. Obstect. Gyn. **93**, 135 (1999)
11. J.B.W. Colston, M.J. Everett, L.B. Da Silva, L.L. Otis, P. Stroeve, H. Nathel, Imaging of hard- and soft-tissue structure in the oral cavity by optical coherence tomography. Appl. Opt. **37** (16), 3582 (1998)
12. X.-J. Wang, T.E. Milner, J.F. de Boer, Y. Zhang, D.H. Pashley, J.S. Nelson, *Characterization of Dentin and Enamel by use of Optical Coherence Tomography*. Appl. Opt. **38**(10), 2092 (1999)
13. P. Zakharov, M.S. Talary, I. Kolm, A. Caduff, Full-field optical coherence tomography for the rapid estimation of epidermal thickness: study of patients with diabetes mellitus type 1. Physiol. Meas. **31**(2), 193 (2010)
14. O.S. Khalil, Non-invasive glucose measurement technologies: an update from 1999 to the dawn of the new millennium. Diabetes Technology & Therapeutics **6**(5), 660–697 (2004)
15. C. Ok Kyung, Y.O. Kim, H. Mitsumaki, K. Kuwa, Noninvasive measurement of glucose by metabolic heat conformation method. Clin. Chem. **50**, 1894–1898 (2004)
16. E.-H. Yoo, S.-Y. Lee, Glucose biosensors: an overview of use in clinical practice. Sensors **10**, 4558–4576 (2010)
17. M.G. Ghosn, V.V. Tuchin, K.V. Larin, Depth-resolved monitoring of glucose diffusion in tissues by using optical coherence tomography. Opt. Lett. **31**(15), 2314–2316 (2006)
18. G.L. Cote, M.D. Fox, R.B. Northrop, Noninvasive optical polarimetric glucose sensing using a true phase measurement technique. IEEE Trans.Biomed. Eng. **39**(7), 752–756 (1992)
19. M.R. Prausnitz, J.S. Noonan, Permeability of cornea, sclera, and conjunctiva: a literature analysis for drug delivery to the eye. J. Pharm. Sci. **87**(12), 1479–1488 (1998)
20. R.A. Gabbay, S. Sivarajah, Optical coherence tomography-based continuous noninvasive glucose monitoring in patients with diabetes. Diabet. Technol. Ther. **10**(3), 188–193 (2008)
21. H. Xiong, Z. Guo, C. Zeng, L. Wang, Y. He, S. Liu, Application of hyperosmotic agent to determine gastric cancer with optical coherence tomography ex vivo in mice. J. Biomed. Opt. **14**(2), 024029 (2009)
22. M. Kohl, M. Cope, M. Essenpreis, D. Böcker, Influence of glucose concentration on light scattering in tissue-simulating phantoms. Opt. Lett. **19**(24), 2170–2172 (1994)
23. M. Brezinski, *Optical Coherence Tomography: Principles and Applications*. (Academic Press, Cambridge, 2009)
24. U. Hafeez, *Imaging of Biological Tissues using Diffuse Reflectance and Optical Coherence Tomography*. (Department of Physics, Pakistan Institute of Engineering and Applied Sciences, Islamabad, 2012), p. 152
25. H. Ullah, M. Ikram, *Optical Coherence Tomography for Glucose Monitoring in Blood*. (LAP Lambert Academic Publishing, Saarbrücken, 2012)
26. K.V. Larin, M.G. Ghosn, S.N. Ivers, A. Tellez, J.F. Granada, Quantification of glucose diffusion in arterial tissues by using optical coherence tomography. Laser Phys. Lett. **4**(4), 312–317 (2007)

27. H. Ullah, M. Atif, S. Firdous, M.S. Mehmood, M. Ikram, C. Kurachi, C. Grecco, G. Nicolodelli, V.S. Bagnato, Femtosecond light distribution at skin and liver of rats: analysis for use in optical diagnostics. Laser Phys. Lett. **7**(12), 889–898 (2010)
28. K.V. Larin, M.G. Ghosn, S.N. Ivers, A. Tellez, J.F. Granada, Quantification of glucose diffusion in arterial tissues by using optical coherence tomography. Laser Phys. Lett. **4**(4), 312 (2007)
29. K.V. Larin, V.V. Tuchin, Functional imaging and assessment of the glucose diffusion rate in epithelial tissues in optical coherence tomography. Quantum Electron. **38**(6), 551 (2008)
30. H. Ullah, G. Gilanie, M. Attique, M. Hamza, M. Ikram, M-mode swept source optical coherence tomography for quantification of salt concentration in blood: an in vitro study. Laser Phys. **22**(5), 1002–1010 (2012)
31. H. Ullah, A. Mariampillai, M. Ikram, I. Vitkin, Can temporal analysis of optical coherence tomography statistics report on dextrorotatory-glucose levels in blood? Laser Phys. **21**(11), 1962–1971 (2011)
32. S. Prahl. *Mie Scattering Calculator* (2011), (cited 11 Apr 2011), Available from: http://omlc.ogi.edu/calc/mie_calc.html
33. X. Guo, Z.Y. Guo, H.J. Wei, H.Q. Yang, Y.H. He, S.S. Xie, G.Y. Wu, H.Q. Zhong, L.Q. Li, Q.L. Zhao, In vivo quantification of propylene glycol, glucose and glycerol diffusion in human dkin with optical coherence tomography. Laser Phys. **20**, 1849–1855 (2010)
34. Y.L. Jin, J.Y. Chen, L. Xu, P.N. Wang, Refractive index measurement for biomaterial samples by total internal reflection. Phys. Med. Biol. **51**(20), N371 (2006)
35. M. Brezinski, *Optical coherence tomography principles and applications* (Elsevier, San Diego, USA, 2006)
36. B.J. Berne, R. Pecora, *dynamic light scattering with applications to chemistry, biology, and physics* (Dover Publications, Inc., Mineola, New York, 2000)
37. website. *Physical characteristics of water (at the atmospheric pressure)*. (2011) (cited 22 Feb 2011), Available from: http://www.thermexcel.com/english/tables/eau_atm.htm
38. Telisa, J. Telis-Romeroa, H.B. Mazzottia, A.L. Gabasb, Viscosity of aqueous carbohydrate solutions at different temperatures and concentrations. Int. J. Food Prop. **10**(1), 185–195 (2007)
39. S. Kim, S. Yang, D. Lim, Effect of dextran on rheological properties of rat blood. J. Mech. Sci. Technol. **23**(3), 868–873 (2009)
40. N. Dobrovol'skii, Y. Lopukhin, A. Parfenov, A. Peshkov, A blood viscosity analyzer. Biomed. Eng. **31**(3), 140–143 (1997)
41. (2011) (cited 2011 28th January), Available from: http://www.epakmachinery.com/products/viscosity-chart
42. O.S. Zhernovaya, V.V. Tuchin, I.V. Meglinski, Monitoring of blood proteins glycation by refractive index and spectral measurements. Laser Phys. Lett. **5**(6), 460–464 (2008)
43. G. Barshtein, I. Tamir, S. Yedgar, Red blood cell rouleaux formation in dextran solution: dependence on polymer conformation. Eur. Biophys. J. **27**(2), 177–181 (1998)
44. A.A. Bednov, E.V. Savateeva, A.A. Oraevsky. Opto-acoustic monitoring of blood optical properties as a function of glucose concentration (2003)
45. T.W. Secomb, B. Styp-Rekowska, A.R. Pries, Two-dimensional simulation of red blood cell deformation and lateral migration in microvessels. Ann. Biomed. Eng. **35**, 755–765 (2007)
46. R. Skalak, P.R. Zarda, K.M. Jan, S. Chien, Mechanics of Rouleau formation. Biophys. J. **35**(3), 771–781 (1981)
47. H. Ullah, F. Hussain, M.A. Abdul, J. Malik, M.A. Sial, E. Ahmed, Durr-e-Sabeeh, *Qualitative monitoring of glucose, salt and distilled water in whole blood: an in vitro study*. Unpublished data (2015)
48. A.I. Joseph, S. Yazdanfar, V. Westphal, S. Radhakrishan, A.M. Rollins, Real-time and functional optical coherence tomography. In *IEEE*, p. 110 (2002)
49. H. Ullah, A. Mariampillai, M. Ikram, I.A. Vitkin, Can temporal analysis of optical coherence tomography statistics report on dextrorotatory-glucose levels in blood? Laser Phys. **21**(11), 1962–1971 (2011)

50. A.A. Bednov, A.A. Karabutov, E.V. Savateeva, W.F. March, A.A. Oraevsky. Monitoring glucose in vivo by measuring laser-induced acoustic profiles (2000)
51. H. Ullah, B. Davoudi, A. Mariampillai, G. Hussain, M. Ikram, I.A. Vitkin, Quantification of glucose levels in flowing blood using M-mode swept source optical coherence tomography. Laser Phys. **22**(4), 797–804 (2012)
52. V.V. Tuchin, *Laser Fiber Optics in Biomedical Research.* (Saratov State Univ. Publ., Russia, 1998), 383p
53. J.A. Jacquez, Red blood cell as glucose carrier: significance for placental and cerebral glucose transfer. Am. J. Physiol. Regul. Integr. Comparative Physiol. **246**(3), R289–R298 (1984)
54. J.D. Ramshaw, Brownian motion in flowing fluids. Phys. Fluids **22**, 1595–1601 (1979)
55. D.B. Kunimasa Miyazaki, Brownian motion in a fluid in simple shear flow. Phys. A **217**, 53–74 (1995)
56. M. Ninck, M. Untenberger, T. Gisler, Diffusing-wave spectroscopy with dynamic contrast variation: disentangling the effects of blood flow and extravascular tissue shearing on signals from deep tissue. Biomed. Opt. Express **1**(5), 1502–1513 (2010)
57. B.D.H. Ullah, A. Mariampillai, G. Hussain, M. Ikram, I.A. Vitkin, Quantification of glucose levels in flowing blood using M-mode swept source optical coherence tomography. Laser Phy. (2011, article in press)
58. H. Ullah, B. Davoudi, A. Mariampillai, G. Hussain, M. Ikram, I. Vitkin, Quantification of glucose levels in flowing blood using M-mode swept source optical coherence tomography. Laser Phys. **22**(4), 797–804 (2012)
59. M. Kinnunen, R. Myllyla, S. Vainio, Detecting glucose-induced changes in in vitro and in vivo experiments with optical coherence tomography. J. Biomed. Opt. **13**(2), 021111–021117 (2008)
60. J. Moger, S.J. Matcher, C.P. Winlove, A. Shore, Measuring red blood cell flow dynamics in a glass capillary using Doppler optical coherence tomography and Doppler amplitude optical coherence tomography. J. Biomed. Opt. **9**(5), 982–994 (2004)
61. R. Darby, *Chemical Engineering Fluid Mechanics.* (Marcel Dekker, Inc., New York, NY 2001), p. 10016
62. D. Rusu, D. Genoe, P. van Puyvelde, E. Peuvrel-Disdier, P. Navard, G.G. Fuller, Dynamic light scattering during shear: measurements of diffusion coefficients. Polymer **40**(6), 1353–1357 (1999)
63. Z. Li, H. Li, J. Li, X. Lin, Feasibility of glucose monitoring based on Brownian dynamics in time-domain optical coherence tomography. Laser Phys. **21**(11), 1995–1998 (2011)
64. H. Ullah, E. Ahmed, M. Ikram, Human cervical carcinoma detection and glucose monitoring in blood micro vasculatures with swept source OCT. JETP Lett. **97**(12), 690–696 (2013)
65. M.L. Hans-Anton Lehr, Michael D. Menger, Dirk Nolte, Konrad Messmer, Dorsal Skinfold Chamber Technique for Intravital Microscopy in Nude Mice. Am. J. Pathol. **143**(4), 1055–1062 (1993)
66. S.J. Md Menger, P. Walter, F. Hammersen, K. Messmer, A novel technique for studies on the microvasculature of transplanted islets of Langerhans in vivo. Int. J. Microcirc. Clin. Exp. **9**, 109–117 (1990)
67. A. Mariampillai, B.A. Standish, E.H. Moriyama, M. Khurana, N.R. Munce, M.K.K. Leung, J. Jiang, A. Cable, B.C. Wilson, I.A. Vitkin, V.X.D. Yang, Speckle variance detection of microvasculature using swept-source optical coherence tomography. Opt. Lett. **33**(13), 1530–1532 (2008)
68. N. Sudheendran, S.H. Syed, M.E. Dickinson, I.V. Larina, K.V. Larin, Speckle variance OCT imaging of the vasculature in live mammalian embryos. Laser Phys. Lett. **8**(3), 247–252 (2011)
69. G. Hüttmann, Optical coherence tomography (OCT) for early diagnosis of tumors and online-control of photodynamic therapy (PDT). Photodiagn. Photodyn. Ther. **8**(2), 152 (2011)

Author Index

M. Husain and Z.H. Khan (eds.), *Advances in Nanomaterials*,
Advanced Structured Materials 79, DOI 10.1007/978-81-322-2668-0